Volker Beckmann and Chris Shrader

Active Galactic Nuclei

Related Titles

von Berlepsch, R. (ed.)

Reviews in Modern Astronomy Vol. 22
Deciphering the Universe through Spectroscopy

2011
ISBN: 978-3-527-41055-2

Röser, S. (ed.)

Formation and Evolution of Cosmic Structures
Reviews in Modern Astronomy Vol. 21

2009
ISBN: 978-3-527-40910-5

Salaris, M., Cassisi, S.

Evolution of Stars and Stellar Populations

2005
ISBN: 978-0-470-09219-4

Phillipps, S.

The Structure and Evolution of Galaxies

2005
ISBN: 978-0-470-85506-5

Rüdiger, G., Hollerbach, R.

The Magnetic Universe
Geophysical and Astrophysical Dynamo Theory

2004
ISBN: 978-3-527-40409-4

Volker Beckmann and Chris Shrader

Active Galactic Nuclei

WILEY-VCH Verlag GmbH & Co. KGaA

The Authors

Dr. Volker Beckmann
APC Laboratory, CNRS/IN2P3
Université Paris Diderot
10 rue Alice Domon et Léonie Duquet
75013 Paris
France

Dr. Chris Shrader
NASA Goddard Space Flight Center and
Universities Space Research Association
Mail Code 661
Greenbelt, MD 20771
USA

Library of Congress Card No.:
applied for

British Library Cataloguing-in-Publication Data:
A catalogue record for this book is available from the British Library.

Bibliographic information published by the Deutsche Nationalbibliothek
The Deutsche Nationalbibliothek lists this publication in the Deutsche Nationalbibliografie; detailed bibliographic data are available on the Internet at http://dnb.d-nb.de.

Cover Design Grafik-Design Schulz, Fußgönheim
Typesetting le-tex publishing services GmbH, Leipzig
Printing and Binding Markono Print Media Pte Ltd, Singapore

Printed in Singapore
Printed on acid-free paper

Hardcover ISBN 978-3-527-41091-0
Softcover ISBN 978-3-527-41078-1

Contents

Preface

Active galactic nuclei (AGN) are the most energetic persistent objects in the Universe. Our understanding of them is roughly summarized in Chapter 1. Such a summary will be insufficient, but the goal is to provide an overview of the topic for the newcomer in the field. In Chapter 2 we take a quick tour through the radiative processes which are common in AGN and give the relevant formulas in order to make the arguments in the book understandable. Chapter 3 then discusses our understanding of what mechanisms drive the AGN emission, and what the main elements are, such as the black hole itself, the accretion disk, the broad and narrow line regions, outflowing jets and absorbing material. The different types of AGN are discussed in Chapter 4, including an attempt to explain all different types by the most simple model possible. In Chapter 5 we take a look at AGN in different energy bands, from the radio to the gamma-ray domain, and examine the overall energy output for beamed and nonbeamed sources. In Chapter 6 we will discuss what one can learn from variability studies of AGN. Up to that point we deal with the central engine and its closest surroundings. In Chapter 7 we will then have a look at in what types of galaxies and galaxy clusters supermassive black holes reside and how the AGN is influenced by the host galaxy and how in turn the central engine might affect the star formation in the surrounding medium. In order to understand the role of AGN in the Universe, we briefly discuss the current cosmological model in Chapter 8 and show how AGN can be used as tools for cosmological studies. We then turn in Chapter 9 to the ultimate question of where quasars come from, how they might be formed, and how they evolve. This also includes the aspects of AGN density evolution in time and how this might depend on the type of AGN or the energy range we study. Chapter 10 summarizes the open issues remaining in AGN research, that is, the big questions which still lack a satisfying answer. We hope that the reader will find that stimulating – and that she or he through their own research will contribute to further progress in this thriving field of astrophysics.

The literature about AGN is full of acronyms and abbreviations. We list the most important ones and those which are used in this book starting after the Preface. Finally, the physical and astronomical constants applied in this book are listed. Throughout the book we use cgs units rather than SI, as it is still common practice in astrophysics.

As for any such undertaking as writing a text book covering a broad scientific topic, we heavily relied on the experience and publications of a large number of scientists. Many of those publications are listed in the bibliography. Some of the textbooks we had at hand when compiling this book, were Osterbrock (1989), Peterson (1997), Kembhavi and Narlikar (1999), Krolik (1999), De Young (2002), and of course Rybicki and Lightman (1986). Throughout the book we point the reader to review articles on certain topics, and give some bibliographic references for further reading. Here we have put an emphasis on recent publications over more established ones. The idea is that the reader will usually find references to earlier work in the most recent publications in a field. We would like to apologize to all the colleagues whose work we have not mentioned or have not given the proper weight in this book.

This book would not have been possible to write without the help, advice of and discussions with many colleagues. Here we would like to thank in particular Chiara Caprini (CEA/Saclay) and Olaf Wucknitz (AIfA Bonn) for their advice on the cosmology chapter, Markos Georganopoulos (UMBC) on jets and radiative processes, Knud Jahnke (MPIA Heidelberg) for many comments on the environment of AGN, Demos Kazanas (NASA/GSFC) for interesting discussions and advice, Ralf Keil for digging out the NLS1 spectrum, Dirk Lorenzen, who knows how to get a book ready for publication, Piotr Lubiński (CAMK Torun) for reading the entire manuscript, Fabio Mattana (APC Paris) for the discussion about synchrotron self-absorption, Marie-Luise Menzel for the artwork on the unified model, Pierre-Olivier Petrucci (LAOG Grenoble) who provided advice on the typology of AGN, Michael Punch (APC Paris) for improving the VHE discussion, Tapio Pursimo (IAC Canary Islands), discussing the AGN phenomenology and appearance, Fabrizio Tavecchio (OAB Merate) for valuable advice on blazars and jets, Marc Türler (ISDC Geneva) who provided input on several topics, Jane Turner (UMBC) for sharing her abundant expertise on the X-ray properties of AGN, Lisa Winter (University of Colorado) for her advice on ultraluminous X-ray sources, and Christian Wolf (Oxford University) for explaining the multiband surveys. Volker thanks his mother for pointing out that NGC 4889 hosts the most massive black hole known to mankind. We also would like to mention the support of our institutions and in particular Pierre Binétruy (APC Paris), Neil Gehrels (NASA/GSFC) and Mike Corcoran (USRA) for their support and encouragement in enabling us to write this book. A very special thank you goes to Simona Soldi (CEA/Saclay), for encouraging the project from the beginning, discussing the context on a daily basis, and correcting the whole manuscript several times.

Of the team at Wiley we would like to thank Oliver Dreissigacker, for contacting us in the first place and thus initiating the AGN book project, Ulrike Fuchs, the commissioning editor, Anja Tschörtner, for helping in the early phase of the project, and Nina Stadthaus, who was a very efficient and friendly editor helping with the technical realization of the project. Finally, we would like to thank Petra Möws and the team at le-tex for their support in the proof-reading process.

Paris and Greenbelt, May 2012 *Volker Beckmann, Chris R. Shrader*

Abbreviations and Acronyms

Here we give the list of the most common acronyms and abbreviations used in this book and in AGN science in general. We caution that the use of abbreviations is not consistent throughout the literature.

ACF Autocorrelation function
ADAF Advection-dominated accretion flow
AGN Active galactic nucleus
ALMA Atacama Large Millimeter/submillimeter Array
ASCA Advanced Satellite for Cosmology and Astrophysics
ATHENA Advanced Telescope for High ENergy Astrophysics
BAL Broad absorption line quasar
BAO Baryon acoustic oscillations
BHXRB Galactic black hole X-ray binary
BLR Broad-line region
BLRG Broad-line radio galaxy
BSC ROSAT All-Sky Survey Bright Source Catalogue
CADIS Calar Alto Deep Imaging Survey
CCD Charge-coupled device
CCF Cross-correlation function
CDF Chandra Deep Field (X-rays)
CGRO (Arthur Holley) Compton Gamma-Ray Observatory
CMB Cosmic microwave background
COMBO-17 Classifying Objects by Medium-Band Observations in 17 Filters
COSMOS Cosmic Evolution Survey
CTA Cherenkov Telescope Array
CXB Cosmic X-ray background
DCF Discrete cross-correlation function
EBL Extragalactic background light
EC External Compton scattering model
EGRET Energetic Gamma-Ray Experiment Telescope (CGRO)
EMSS Einstein Observatory Extended Medium Sensitivity Survey
eROSITA extended ROentgen Survey with an Imaging Telescope Array
EUVE Extreme Ultraviolet Explorer

EW	Equivalent width
FIRST	Faint Images of the Radio Sky at twenty-centimeters
FR	Fanaroff and Riley
FSRQ	Flat-spectrum radio quasar
FUSE	Far Ultraviolet Spectroscopic Explorer
FWHM	Full-width at half-maximum
GALEX	Galaxy Evolution Explorer
GBH	Galactic black hole
GPS	Gigahertz-peaked spectrum source
HBL	High-frequency cutoff BL Lac object/high frequency peaked BL Lac object
HEAO-2	EINSTEIN satellite
HES	Hamburg/ESO Survey
HPQ	Highly polarized quasar
HQS	Hamburg Quasar Survey
HR	Hardness ratio
HRI	ROSAT High-Resolution Imager
HST	Hubble Space Telescope
HzRG	High-redshift radio galaxy
IACT	Imaging Atmospheric Cherenkov Telescope
IC	Inverse Compton scattering
ICM	Intracluster medium
IBL	Intermediate BL Lac object
IDV	Intraday variability
IGM	Intergalactic medium
IMF	Initial mass function
INTEGRAL	International Gamma-Ray Astrophysics Laboratory
IPC	EINSTEIN Imaging Proportional Counter
IRAS	Infrared Astronomical Satellite
ISO	Infrared Space Observatory
IUE	International Ultraviolet Explorer
LAT	Large Area Telescope (Fermi)
LBL	Low-frequency cutoff BL Lac object/low-frequency peaked BL Lac object
LDDE	Luminosity-dependent density evolution
LECS	Low Energy Concentrator Spectrometer (BeppoSAX)
LF	Luminosity function
LINER	Low-ionization nuclear emission-line region
LISA	Laser Interferometer Space Antenna
LMC	Large Magellanic Cloud
LOFAR	Low Frequency Array (radio)
LSST	Large Synoptic Survey Telescope (optical)
MACHO	Massive compact halo object
mas	Milliarcseconds
MECS	Medium Energy Concentrator Spectrometer

MHD	Magnetohydrodynamics
NED	NASA/IPAC Extragalactic Database
NELG	Narrow-emission-line galaxy
NLR	Narrow-line region
NLRG	Narrow-line radio galaxy
NLS1	Narrow-line Seyfert 1 galaxy
NRAO	National Radio Astronomy Observatory (USA-VA)
NVSS	NRAO VLA Sky Survey
OSSE	Oriented Scintillation Spectrometer Experiment (CGRO)
OVV	Optical violent variable blazar
PAH	Polycyclic aromatic hydrocarbon
PDS	Power Density Spectrum or Phoswich Detector System (BeppoSAX)
PLE	Pure-luminosity evolution
PSF	Point spread function
PSPC	Position Sensitive Proportional Counter (ROSAT)
QPO	Quasiperiodic oscillation
RASS	ROSAT All-Sky Survey
RBL	Radio-selected BL Lac object
ROSAT	Röntgensatellit
RXTE	Rossi X-ray Timing Explorer
SDSS	Sloan Digitized Sky Survey
SED	Spectral energy distribution
SF	Structure function
SIMBAD	Set of Identifications, Measurements, and Bibliography for Astronomical Data
SIS	Solid-State Imaging Spectrometer (ASCA)
SKA	Square Kilometer Array (radio telescope)
SL	Superluminal
SMBH	Supermassive black hole
SMC	Small Magellanic Cloud
SNR	Supernova remnant
SOFIA	Stratospheric Observatory for Infrared Astronomy
SRG	Spectrum-Roentgen-Gamma
SRSQ	Steep radio spectrum quasar
SSC	Synchrotron self-Compton scattering
UHBL	Ultrahigh-frequency peaked BL Lac object
UHECR	Ultrahigh-energy cosmic rays
ULIRG	Ultraluminous infrared galaxy
ULX	Ultraluminous X-ray source
VHE	Very high energy
VLA	Very Large Array (radio)
VLBA	Very Long Baseline Array (radio)
VLBI	Very Long Baseline Interferometry (radio)
VSOP	VLBI Space Observatory Programme

WFI	Wide Field Imager or Wide Field Instrument
WHIM	Warm-hot intergalactic medium
WIMP	Weakly interacting massive particles
XBL	X-ray-selected BL Lac object
XBONG	X-ray bright optically inactive galaxy
XMM	X-ray Multimirror Mission
2MASS	Two-Micron All-Sky Survey

Astronomical and Physical Constants

We list here some astronomical and physical constants which are used throughout the book.

Physical Constants[a]

Speed of light	c	$2.997\,924\,58 \times 10^{10}\ \mathrm{cm\,s^{-1}}$
Planck constant	h	$6.626\,07 \times 10^{-27}\ \mathrm{g\,cm^2\,s^{-1}}$
Boltzmann's constant	k	$1.380\,65 \times 10^{-16}\ \mathrm{erg\,K^{-1}}$
Stefan–Boltzmann constant	σ	$5.6704 \times 10^{-5}\ \mathrm{erg\,cm^{-2}\,s^{-1}\,K^{-4}}$
Gravitational constant	G	$6.6738 \times 10^{-8}\ \mathrm{cm^3\,g^{-1}\,s^{-2}}$
Electron mass	m_e	$9.109\,383 \times 10^{-28}\ \mathrm{g}$
Proton mass	m_p	$1.672\,621\,8 \times 10^{-24}\ \mathrm{g}$
Neutron mass	m_n	$1.674\,927\,4 \times 10^{-24}\ \mathrm{g}$
Atomic mass unit	1 u	$1.660\,538 \times 10^{-24}\ \mathrm{g}$
Elementary charge	e	$1.602\,176\,6 \times 10^{-19}\ \mathrm{C}$
Electron volt	1 eV	$1.602\,176\,6 \times 10^{-12}\ \mathrm{erg}$
Fine-structure constant	α	$7.297\,352\,57 \times 10^{-3}$
Wien's displacement constant	b	$0.289\,777\ \mathrm{cm\,K}$

[a] Values taken from the on-line version of Mohr *et al.* (2008).

Astronomical Constants

Solar mass	$1\ M_\odot$	$1.989 \times 10^{33}\ \mathrm{g}$
Solar luminosity	$1\ L_\odot$	$3.839 \times 10^{33}\ \mathrm{erg\,s^{-1}}$
Astronomical unit	1 AU	$1.4960 \times 10^{13}\ \mathrm{cm}$
Siderial year	1 yr	$3.155\,815 \times 10^{7}\ \mathrm{s}$
Light year	1 ly	$9.4605 \times 10^{17}\ \mathrm{cm}$
Parsec	1 pc	$3.0857 \times 10^{18}\ \mathrm{cm}$
		3.2616 ly
Hubble constant	H_0	$70.8\ \mathrm{km\,s^{-1}\,Mpc^{-1}}$

Color Plates

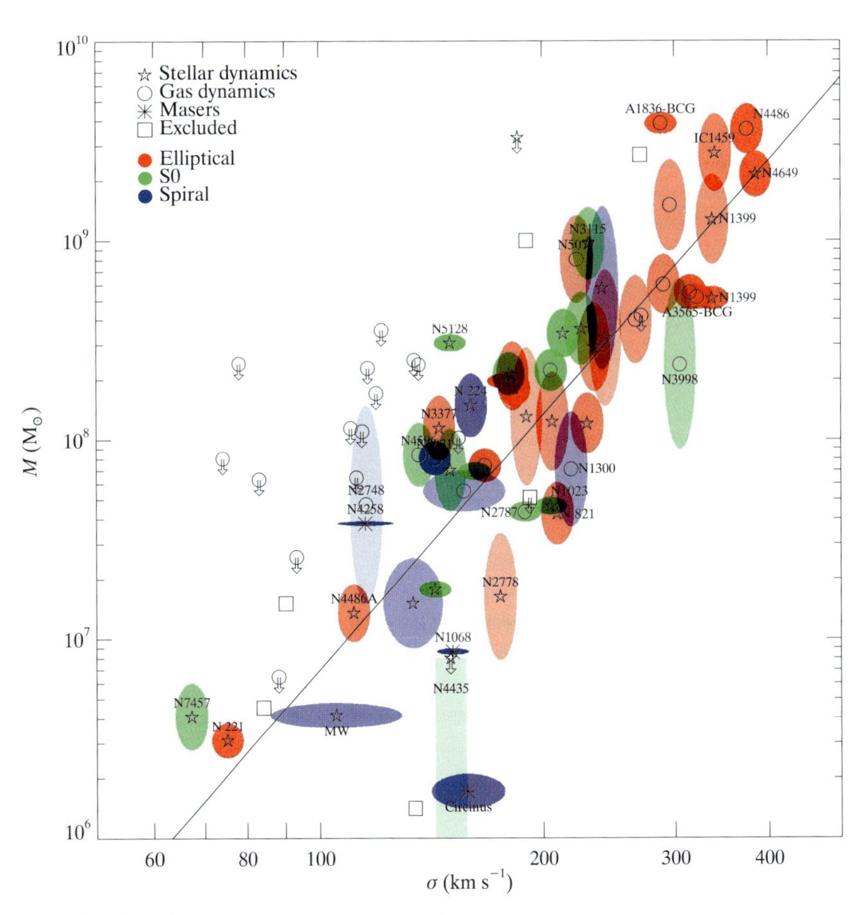

Figure 3.3 This figure illustrates a recent calibration of the $M-\sigma$ relation from Gültekin *et al.* (2009). The colors indicate different galaxy types and the symbols (star, circle and asterisk), indicate the type of black-hole measurement.

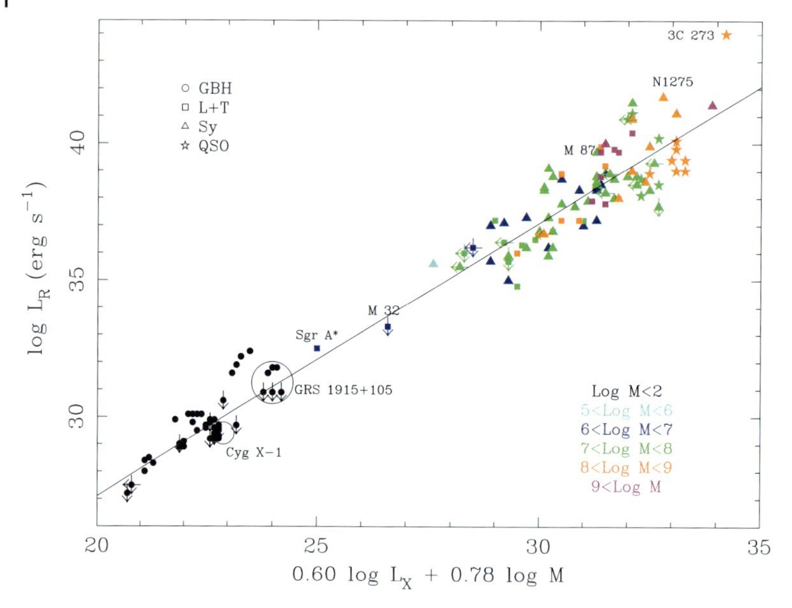

Figure 3.4 The so called Fundamental Plane for accreting black holes with which jet outflows are also associated (Merloni *et al.*, 2003). The observable quantities, L_R and L_X are proxies for the jet and disk power respectively. Objects with mass determinations from other methods were used in constructing this diagram. The tight correlation over many decades of central black hole masses suggests a commonality in the physics underlying the disk-accretion and jet launching mechanism across object sub-classes.

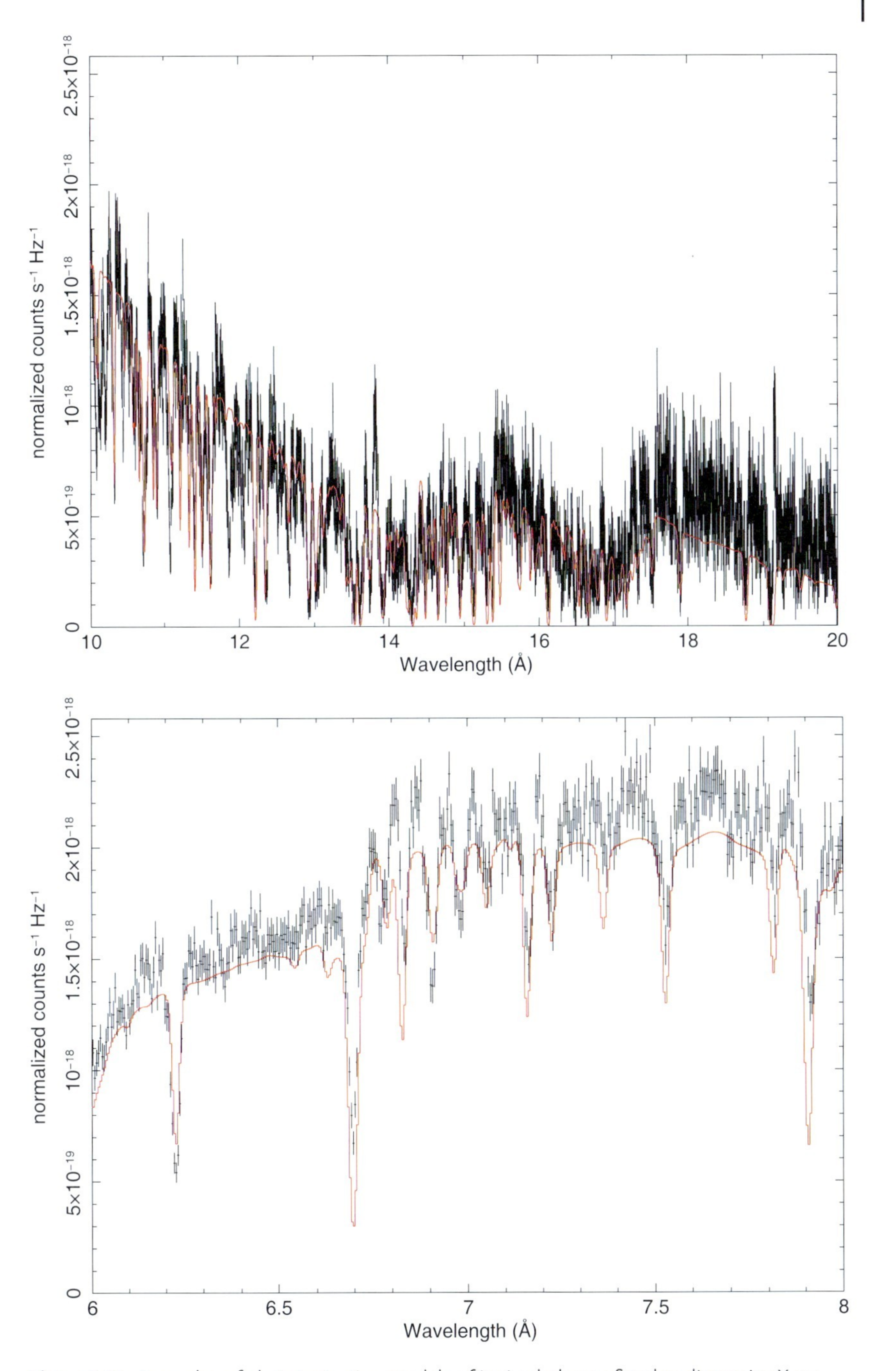

Figure 3.16 Examples of photoionization models of ionized plasma fitted to dispersive X-ray spectra obtained with the Chandra X-ray observatory (Kallman, 2010). The fits led to refinements in iron recombination rates and revealed multiple gas velocity components.

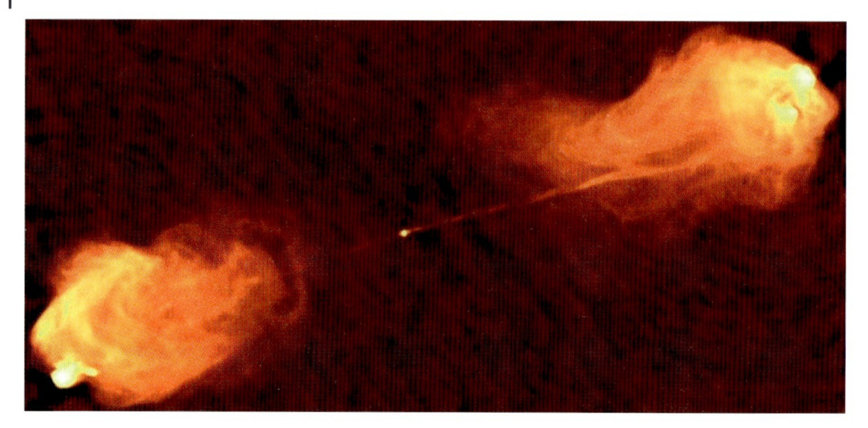

Figure 4.9 The FR-II type radio galaxy Cygnus A is the brightest extragalactic radio source. The picture shows the 5 GHz image taken with the VLA telescope array with 0.4″ resolution (Carilli and Barthel, 1996). The AGN core is located at the bright spot at the center, the radio lobes extend out to about 50 kpc from the core, far beyond the host galaxy which is not visible in the radio domain.

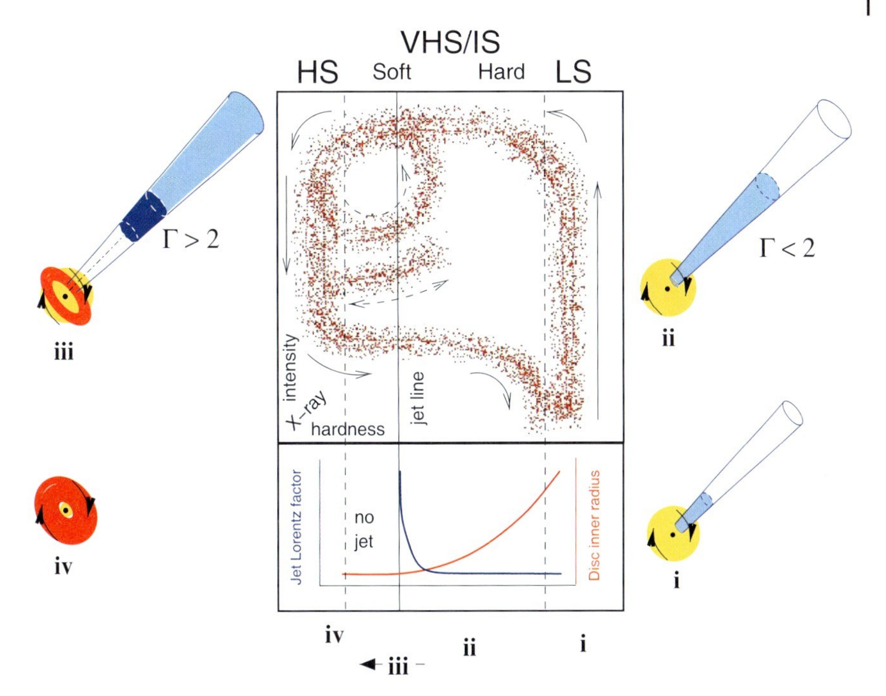

Figure 4.18 Schematic representation of the connection of jet emission, inner disk radius, and spectral state in galactic black hole binaries. The top panel shows the hardness-intensity diagram, in which black holes seems to follow the paths indicated by the arrows. A flat-spectrum radio flux appears and increases with X-ray intensity in the hard state – the rith-hand vertical track of the "q" diagram. The radio emission becomes optically thin and the jet appears as the emission transitions leftwards along the upper horizontal track. The jet disappears after the source moves into the high-soft state. In the bottom panel we see the dependence of jet speed and the inner disk radius on the hardness of the spectrum. X-ray states are indicated with HS (high/soft state), VHS/IS (very high and intermediate state), and LS (low/hard state). The sketches around the outside illustrate the concept of the relative contributions of jet, corona (light grey) and accretion disc (dark grey) at these different stages. Graphic from Fender *et al.* (2004).

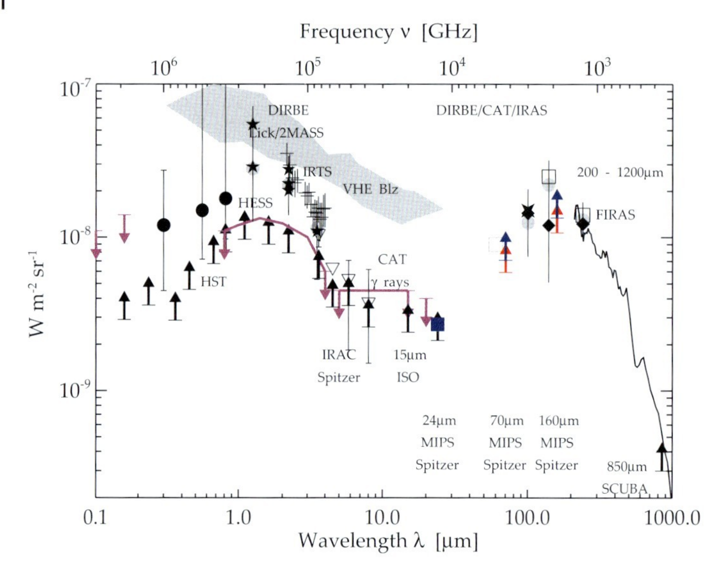

Figure 5.16 The Extragalactic Background Light (EBL) in the optical to infrared range, based on photometric measurements and on indirect techniques (Dole *et al.*, 2006; reproduced with permission © ESO). The HESS TeV data of blazars put strong constraints on the EBL in the range $\lambda = 0.8–4\ \mu m$ (indicated by a line).

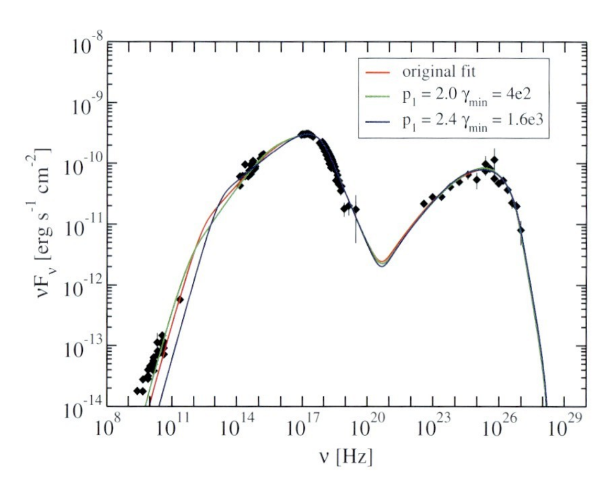

Figure 5.19 SSC model fit of the spectral energy distribution of Mrk 421 based on simultaneous data during a quiescent state of the source. Graphic from Abdo *et al.* (2011b), reproduced by permission of the AAS.

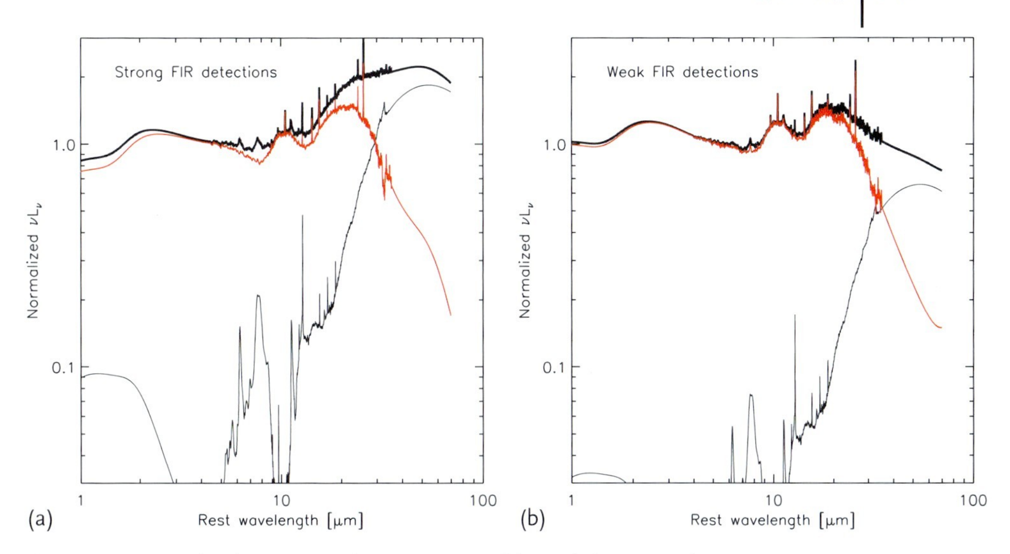

Figure 5.22 Normalized mean SEDs for strong far-infrared (FIR) emitting quasars (a, top curve) and weak FIR quasars (b, top curve). The adjacent red SED curves show "intrinsic" AGN SEDs obtained by the subtraction of the scaled mean starburst (ULIRG) spectrum (shown as the lowest curve in black) from the mean SEDs. From Netzer *et al.* (2007), reproduced by permission of the AAS.

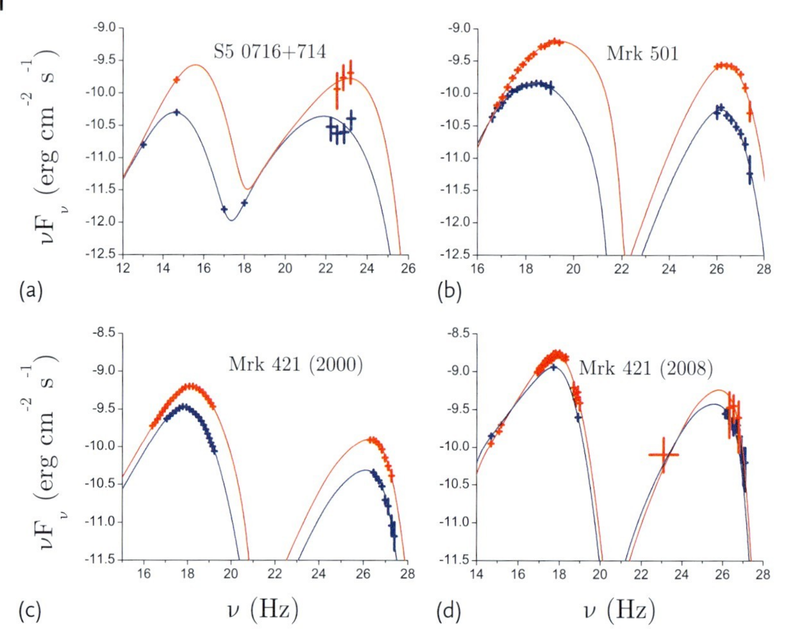

Figure 6.7 Examples for variability of the spectral energy distribution in blazars in low and high emission states. Blazars tend to shift the peak of the synchrotron and inverse Compton branch to higher frequencies, while when comparing different blazars, usually the ones with higher peak frequencies are the less luminous ones as discussed in Section 5.8.1. Graphic from Paggi *et al.* (2009), reproduced with permission © ESO.

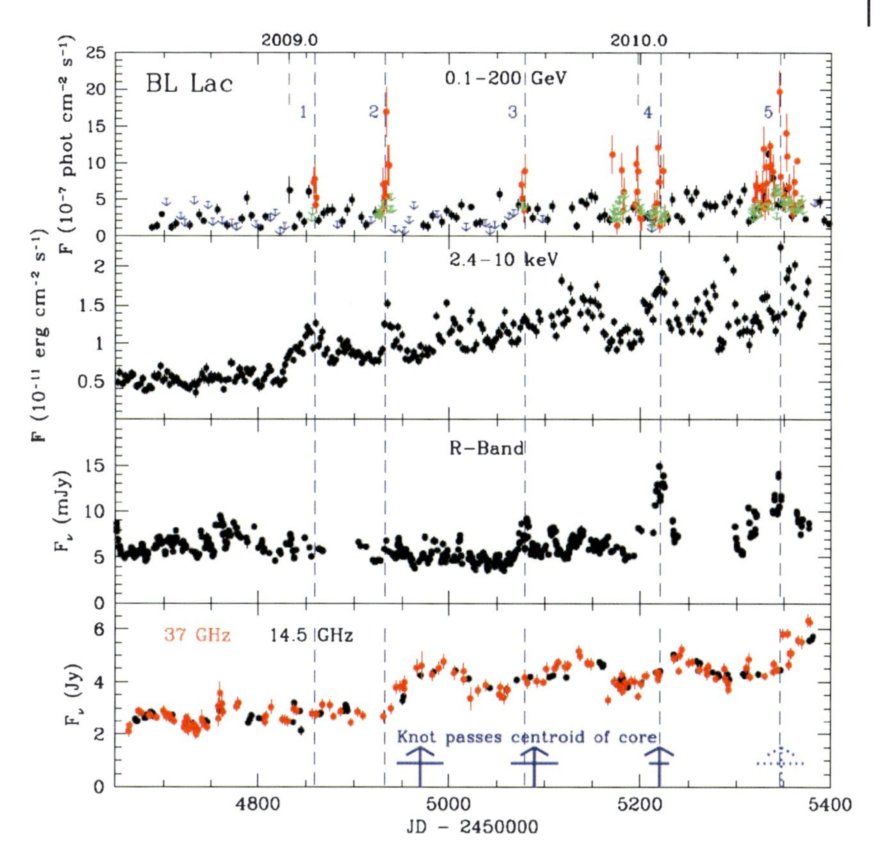

Figure 6.9 Multi-wavelength lightcurve of BL Lac (Marscher *et al.*, 2011). As described in the text, patterns are beginning to emerge between the radio intensity, morphological evolution and flaring at X- and gamma-ray bands. Additional information is potentially forthcoming from comparing polarization measurements of individual jet components to optical polarization of flaring features; specifically, discrete increases in polarization are expected to coincide with gamma-ray flares. The gamma-ray peaks are sharp, and coincide with the rising phase of the radio flux. The radio peaks lag behind the gamma-ray peaks by days to weeks. The VLBA images, not shown here, indicate that the jet knots usually pass through the position corresponding to the radio core concurrently with the gamma-ray flares (vertical arrows).

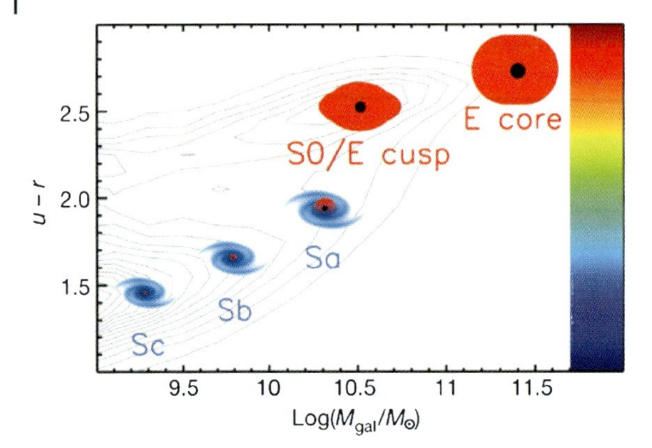

Figure 7.4 Color $u-r$ versus galaxy mass. The larger the $u-r$ value, the redder the galaxy. The central bulge of the spirals are similar to small elliptical galaxies. Ellipticals and spiral bulges have similar stellar mass to black hole mass ratio of the order of 0.1%. Ellipticals are located above a critical mass of $M_{crit} \simeq 10^{12}\, M_{\odot}$. Graphic from Cattaneo *et al.* (2009).

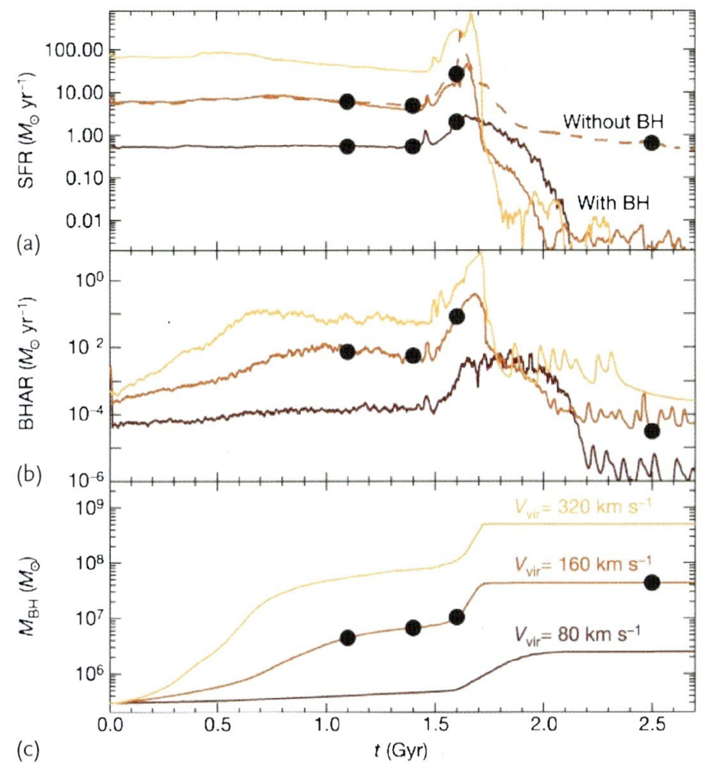

Figure 7.6 Star formation rate (SFR), black hole mass accretion rate (BHAR) and black hole mass for a simulated merging event between two galaxies containing each a super massive black hole about the size of the one in our Galaxy. For comparison, in the upper most panel also the evolution of SFR for a merger without central black holes is shown. Cases of different virial velocity are given. The dots indicate specific times in the merging process. From left to right: first passage of the two galaxies, tidal interaction just before merging, coalescence, and after the merging process has finished. In this graphic from Di Matteo *et al.* (2005) a feedback on the SFR is assumed in the sense that the AGN shuts off the creation of new stars, a scenario disputed for example by Debuhr *et al.* (2010) who argue that the AGN has an impact only on the innermost part of the bulge.

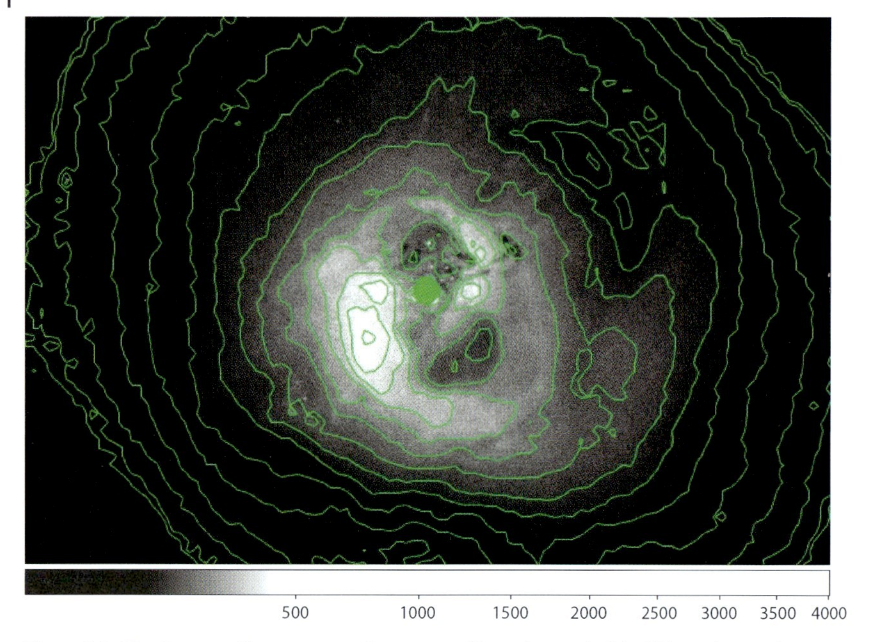

Figure 7.7 The Perseus Cluster as seen by the Chandra X-ray telescope. The center of the cluster is dominated by the emission of the Seyfert and radio galaxy NGC 1275. North and south of the AGN, cavities are visible as well as a larger dark bubble in the northwest. Chandra data obtained from the High Energy Astrophysics Science Archive Research Center (HEASARC), provided by NASA's Goddard Space Flight Center.

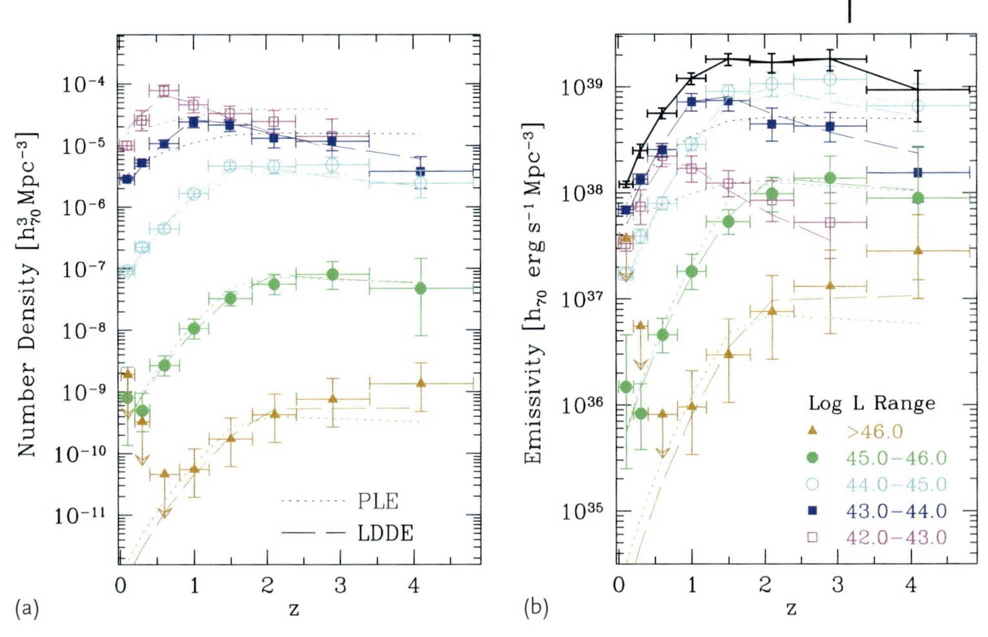

Figure 9.5 X-ray luminosity function as presented by Hasinger *et al.* (2005); reproduced with permission © ESO. Left panel: the space density of AGN as a function of redshift for several X-ray luminosity bins. The lines shown are for a pure-luminosity evolution (PLE) and for a luminosity dependent density evolution (LDDE). Right panel: emissivity of AGN as a function of redshift. The uppermost curve shows the summed emissivity of all luminosity classes considered here.

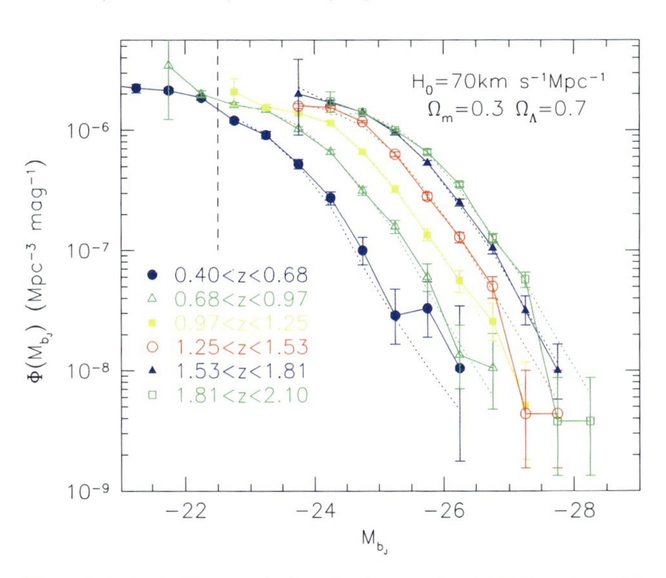

Figure 9.6 Optical luminosity function in several redshift bins derived from SDSS data. The dotted lines denote the predictions of the best-fitting double power law exponential evolution mode. From Croom *et al.* (2004).

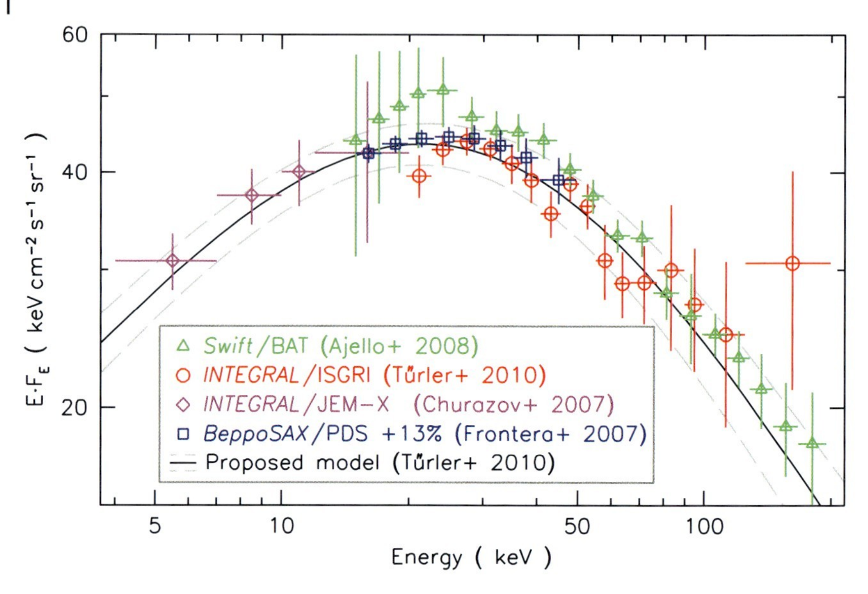

Figure 9.8 Recent measurements of the cosmic X-ray background by BeppoSAX, Swift, and INTEGRAL. The continuous line shows the best fitting model derived by Türler *et al.* (2010). Figure courtesy of Marc Türler.

1
The Observational Picture of AGN

1.1
From *Welteninseln* to AGN

The dawn of extragalactic astronomy can be attributed to the year 1750, in which Thomas Wright speculated that some of the nebulae observed in the sky were not actually part of the Milky Way, but rather independent Milky Ways themselves (Wright, 1750). A few years later, Immanuel Kant introduced the term "Welteninseln" for these distant nebulae ("island universes"; Kant, 1755). It was François Arago in 1842 who first called the attention of astronomers to Kant,[1] whom he calls "the Astronomer of Königsberg," and declared that his name in that connection did not deserve the oblivion into which it had fallen (Arago and Barral, 1854). Thus the extragalactic hypothesis spread rapidly in the scientific community, although it was still not completely accepted as true. One main difficulty was the fact that some of the nebulae were actually of galactic origin, such as planetary nebulae and globular clusters. A significant step forward was the compilation of a large catalog of some 5000 nebulae assembled by William Herschel in the late eighteenth and early nineteenth century. Another advance was made by Lord Rosse, who constructed in 1845 a new 72″ telescope in Ireland, managing to distinguish individual point sources in some of the nebulae, and therefore giving further support to Kant's and Wright's hypothesis. Spectroscopic observations by Vesto Slipher of nebulae in the early twentieth century revealed that some of these show redshifted lines indicating they are moving relative to the Milky Way at velocities exceeding the escape velocity of our Galaxy (Slipher, 1913).

The issue, whether some of the observed nebulae are actually extragalactic, was finally settled in the 1920s. In 1920 Heber Curtis summarized a number of arguments why the Andromeda "nebula" M31 is a galaxy of its own, similar to the Milky Way (Curtis, 1920). For example, he noticed a Doppler shift in M31 due to its rotation and absorption by dust similar to what was observed in our Galaxy. Finally, a distance estimation of M31 was given with $d = 450\,\mathrm{kpc}$ (Öpik, 1922), about a factor of 2 lower than its actual value, but placing the Andromeda nebula clearly as an ex-

1) Arago wrote: "Kant condensait ses idés dans le moindre nombre de mots possible, quand il appelait la Voie lactée le Monde des Mondes."

Active Galactic Nuclei, First Edition. Volker Beckmann and Chris Shrader.

tragalactic object. Using the 100″ Mt. Wilson telescope, Edwin Hubble was able to observe Cepheids in M31 and M33. Cepheids are variable stars with characteristic light curves which allow to determine their absolute brightness. Using the distance modulus of several Cepheids in these nearby galaxies, Hubble confirmed the large distance of these objects, although again underestimating their distance by a factor of ~ 3 to be about 285 kpc. Based on his observations, he also established a system of classifying galaxies, the so-called Hubble sequence (Hubble, 1926), and laid the starting point for cosmology assuming an expanding Universe (Hubble, 1929).

The first evidence that some galaxies were hosting some additional strongly emitting component in their center was found by Carl Seyfert in the 1940s. He obtained spectra of six galaxies, showing high-excitation nuclear emission lines superposed on a normal star-like spectrum (Seyfert, 1943). He also noticed that some galaxies showed broad emission lines, while others exhibited only narrow ones. The nature of the strong emission from the center of some galaxies remained a mystery. A common hypothesis was the assumption that a large number of stars would produce the observed features. Woltjer (1959) pointed out though that the observed concentration of the emission within the central 100 pc of the galaxies would require a mass of a few $10^8\ M_\odot$. A step closer to current understanding was the idea that in the center of these galaxies resides a stellar type object of very large mass, which then would emit mainly by accretion processes of a surrounding disk of gas (Hoyle and Fowler, 1963). It was not until a year later that the idea was put forward to assume that in the center of an AGN there could lie a black hole as opposed to a hypermassive star (Salpeter, 1964; Zel'Dovich and Novikov, 1964).

The hypothesis that there might exist objects in the Universe whose gravity would be sufficient to trap even light was discussed first by John Mitchell[2] in the late eighteenth century (Mitchell, 1784). Independently Pierre-Simon Laplace developed the concept of "dark stars," speculating that the most massive stars would be invisible due to their strong gravity (Laplace, 1796). The concept of the black hole was ignored though in later years, as light was considered to be made of massless particles with no interaction with a gravitational field. When Albert Einstein (1916) formulated the general relativity theory the possible existence of black holes was shown to be a solution for the gravitational field of a point mass and of a spherical mass by Karl Schwarzschild (1916). Nevertheless, this solution to Einstein's theory was thought to be merely hypothetical. Only when solutions had to be found to explain phenomena like AGN, and the fact that massive stars had to collapse into a black hole (Oppenheimer and Volkoff, 1939), was the existence of black holes accepted by a continuously growing fraction of the scientific community.

The idea of a supermassive black hole in the center of active galactic nuclei (Salpeter, 1964; Zel'Dovich and Novikov, 1964; Lynden-Bell, 1969) and also in the center of our own galaxy (Lynden-Bell and Rees, 1971) was a powerful model. It

2) Mitchell wrote: "If the semidiameter of a sphere of the same density as the Sun were to exceed that of the Sun in the proportion of 500 to 1, a body falling from an infinite height towards it would have acquired at its surface greater velocity than that of light, and consequently supposing light to be attracted by the same force in proportion to its *vis inertiae*, with other bodies, all light emitted from such a body would be made to return towards it by its own proper gravity."

explained not only the large energy output based on the release of gravitational energy through accretion phenomena, but also the small size of the emitting regions and connected to it the short variability time scales of AGN. The field was now open to study the physics involved in the accretion phenomenon, to observe and explain AGN emission throughout the electromagnetic spectrum, and to study the distribution in space, the origin, the evolution and fate of these elusive objects.

1.2 Broad Lines, Narrow Lines, and the Big Blue Bump

The first notably distinct observational characteristic of AGN was the presence of emission lines with widths upwards of 1000 km s^{-1} and far in excess of any known class of objects. Furthermore, the centers of these broad emission lines did not correspond to the laboratory wavelengths of any known atomic species and certainly not to the well known hydrogen Balmer series or other common lines known to be of astrophysical origin. This dilemma was resolved in the 1960s leading to the basic AGN paradigm described in the previous section of a distant and highly luminous object powered by a massive, accreting black hole. The deep gravitational potential of the black hole was responsible for the dynamical broadening of the observed lines and for radiatively efficient accretion leading to the extreme luminosities. The line identification dilemma was solved with the realization that the distances involved were of such magnitude that the cosmological expansion of the Universe redshifted atomic emission lines to the observed values including some high-ionization UV lines were.

Fast forwarding ahead several decades, it became evident that these broad emission line spectra could be exploited as a diagnostic of the physical conditions in the environment ambient to the central black hole. As we discuss in some detail later in the text, correlated variability of line and continuum emission components have been applied to "reverberation mapping" analyses leading to constraints on the broad-line emission region size and on the mass of the central black hole, for example Peterson and Horne (2004); Bentz *et al.* (2009a). This led to a dramatic revision of our basic understanding of AGN. Additionally, this knowledge has been used to cross-calibrate alternative black hole mass estimation methods and to better constrain physical models of the broad-line emission media as virialized gas clouds in photoionization equilibrium with the central engine radiation field.

Another distinguishing observational feature of some AGN is the presence of narrow, nonvariable forbidden emission lines. The similarity to nebular line emission in our galaxy was noted and some of the atomic physics and computational formalism developed and to study those objects was employed (Osterbrock, 1989). There were some significant differences between the galactic nebulae and the AGN as well. In particular, the AGN narrow lines required a broad-band ionizing continuum extending far bluewards of the stellar radiation fields responsible for photoionizing the galactic nebulae.

At present, with an improved understanding of the narrow-line region morphologies and thermodynamics, insight into the gas and dust distributions of the central regions of AGN can be gained. The dynamics of the inner AGN region can also be probed in cases using the narrow lines. For example, approximately 1% of low-redshift ($z \simeq 0.3$) optically selected type 2 AGNs exhibit a double-peaked [O III] narrow-line profile in spatially resolved spectra Shen *et al.* (2011). These types features have been interpreted in the context of kinematics, such as biconical outflows (Kraemer *et al.*, 2008) or rotation of the narrow-line region about the central black hole, or to the relative motion of two distinct NLRs in an ongoing AGN merger event.

Another observational aspect of AGN that was evident in early observations was that the continuum spectral distribution was very distinct from an integrated stellar continuum characteristic of normal galaxies (Oke and Sargent, 1968). Observationally, AGN were comparatively very blue. In fact, radio-quiet AGN would later be identified and cataloged primarily by performing multicolor imaging of sky regions, sorting the results in color-color plots, and performing spectroscopic follow-ups on the blue excess subpopulation thus identified (Green *et al.*, 1986, it is now known that this approach omits redder objects that are picked up in X-ray surveys).

The blue colors were due to both the fact that the continuum emission extended into the UV and beyond and that structure was often seen in the blue continua – the so-called "big blue bump" (Richstone and Schmidt, 1980). The big blue bump spectral component was a positive flux excess relative to an underlying power-law continuum. It exhibited curvature that suggested a thermal origin. This was interpreted as the first observational evidence for the presence of an accretion disk (Malkan and Sargent, 1982a), thus lending support to the basic paradigm of Lynden-Bell (1969). The quest to further corroborate this basic paradigm and to gain a deeper fundamental understanding of putative AGN accretion disks was the driver behind many subsequent observational campaigns and theoretical efforts during the decades which followed.

1.3 Jets and Other Outflows

The sixteenth century French astronomer Charles Messier published a catalog comprising 103 spatially extended or *nebular* objects. Among the most prominent, roughly spherical, examples in his catalog was object number 87, thus its designation as M87. In 1918 the American astronomer Heber Curtis noted that the presence of a "curious straight ray" that protruded from the nebula and apparently traced back to its nucleus (Curtis, 1918). As noted, there was at that time still disagreement as to whether or not the nebulae were external to or contained within our galaxy, but as that issue was soon resolved and it became clear that M87 was a giant elliptical galaxy, we consider Heber's observation to be the first documented example of an AGN *jet*.

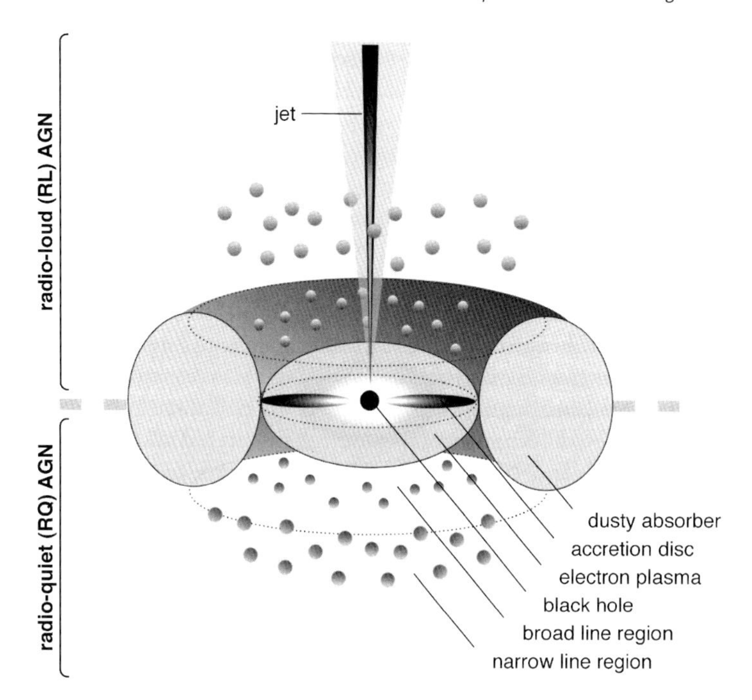

Figure 1.1 Schematic representation of our understanding of the AGN phenomenon and its main components. Note that this is a simplified view and not to scale. Graphic courtesy of Marie-Luise Menzel.

There were relatively few of examples of these highly collimated, bipolar outflows or jets from the centers of AGN until the observational techniques of radio interferometry matured in the decades subsequent to the Second World War. There are now AGN subclasses, which are believed to be related to each other within the context of the unification scenarios described in subsequent chapters. The physics underlying the launching and propagation of these AGN jets – their remarkable energetics, enormous size and plethora of associated phenomenology (hot spots, knots, bends) – remains enigmatic in many regards and its pursuit may be considered a "holy grail" of modern astrophysics.

Our general understanding of the main components of an AGN is shown in Figure 1.1. The schematic representation distinguishes between sources which display a jet and are therefore bright in the radio band, and those which do not show strong radio dominance and in which case one assumes no or weak jet emission.

1.4
X-ray Observations: Probing the Innermost Regions

During the same decade of the 1960s when the basic AGN paradigm was being developed, the first cosmic X-ray source, known as Scorpios X-1, was discovered using a rocket-borne detector (Giacconi *et al.*, 1962). It quickly became evident that X-ray emission was characteristic of compact, accretion-powered sources associated with galactic binaries. With the realization that the deep gravitational wells of massive black holes were likely the source for the extreme energetics exhibited by quasars, the generation of X-rays seemed natural. However, the rocket-borne experiments had limited capabilities and the first significant breakthrough came with the launch of the Uhuru satellite (also known as SAS-A) in 1970. A catalog of sources detected with Uhuru ultimately included nearly 340 objects, mostly galactic binaries, but also including about a dozen AGN (Forman *et al.*, 1978).

The field progressed rapidly during the 1970s, with detectors having larger collecting area, increased spectral coverage and improved spectral resolution, for example the Ariel-5 satellite launched in 1975 and the OSO-7 (1974), OSO-8 (1975) and HEAO-1 (1977) (see, e.g., Tucker and Giacconi, 1985). This led to the characterization of AGN as a class of X-ray sources and to the first detection of the iron K_α line emission from an extragalactic source. The biggest breakthrough however came later in that decade with the launch of the Einstein Observatory (originally called HEAO-2). This was the first true orbiting X-ray *telescope*, in that it utilized a concentric array of grazing incidence mirrors to focus $\sim$ keV photons onto its focal plane detectors. The resulting images provided, in addition to vastly improved spatial localization of sources on the sky, a large leap in sensitivity since the source and celestial background could be effectively separated.

It had become clear that X-ray emission was a common property of different subclasses of active galaxies with the X-ray flux comprising a significant fraction (about 5–40%) of the bolometric emission from such objects (Ward *et al.*, 1987). Rapid variability was also found to be a prominent feature of the X-ray emission with kilosecond timescale X-ray flux variations seen in local Seyfert galaxies. This imposed new and increasingly stringent constraints on the size of the X-ray emission region and strongly supported the idea that it occurs very close to the active nucleus (e.g., Pounds *et al.*, 1986). The origin of X-rays from close to the central black hole means that X-ray data offer a chance to study the immediate environs of supermassive black holes and the poorly understood accretion process that fuels them. Although the angular scale of the X-ray emission region is too small to image with current instrumentation, timing analysis and spectroscopy offered methods to probe these regions indirectly.

Specific spectral signatures were attributed to characteristics of the gas inflow and outflow near the central most regions in AGN. The X-ray observations also provided signatures of reprocessing of radiation in material within approximative distance of hundreds of gravitational radii and thus the potential for discerning signatures of the accretion disk at even smaller radii. Features such as the weak, broad emission lines due to low-ionization states of iron as well as other structured deviations from simple power laws had been identified in the spectra of AGN. In 1991, George and Fabian (1991) offered an interpretation of these features in terms of X-irradiation of relatively cold, dense gas in the vicinity of the central black hole. The emergent spectrum then consists of direct radiation from the central source plus a scattered or "reflected" spectrum that includes imprinted photoabsorption, fluorescent emission and Compton scattering from matter within the surrounding accretion flow. This basic idea has withstood the scrutiny of improved observational data and has become a tenant of the AGN paradigm.

X-ray observations of AGN are also being applied to address issues of fundamental black hole physics. The shapes of line profiles have also been applied to models which in principle allow one to infer an intrinsic property of the central black hole, namely its intrinsic angular momentum or *spin* (e.g., Brenneman and Reynolds, 2009). The basic idea is that the asymmetry of a line profile produced in the inner AGN accretion disk depends in a predictable manner on the shape of the gravitational potential which in turn depends on the black hole *spin*.

In the chapters that follow, we discuss these issues in further detail highlighting a number of results from the current astrophysics literature. We also speculate on the possibilities offered by future orbiting X-ray observatories, which are currently under discussion.

1.5 Up, Up and Away: from Gamma-Rays toward the TeV Range

The impact of gamma-ray astronomy on AGN research did not emerge as rapidly as did X-ray astronomy, although the fields were initiated more or less concurrently with 1960s rocket flights followed by satellite-borne experiments in the 1970s. The reasons for this are several-fold. There are fewer gamma-ray photons than lower energy photons emitted even though the overall energy budget for some AGN may be dominated by the gamma rays. There are substantial instrumental and celestial backgrounds at gamma-ray energies that need to be understood and modeled or subtracted. Gamma-ray detectors tend to be more massive for a given effective collection area than X-ray detectors and gamma rays cannot be focused. Additionally, it became apparent that only the radio-loud AGN, which comprise $\sim$ 5% of the overall population are prolific emitters of gamma radiation. We should note here that the term "gamma rays" encompasses a huge swath of the electromagnetic spectrum. Here we will designate photons with energies above $\sim$ 100 keV as gamma rays. AGN have been detected at $\sim$ TeV, thus we are considering over 7 decades in our discussion of gamma-ray studies. The energy range above about 100 MeV has, somewhat surprisingly, provided the richest bounty of results as we will further discuss in later chapters.

In the 1970s the ESA mission COS-B, along with NASA's SAS-2, provided the first detailed views of the Universe in gamma rays. COS-B, launched in August 1975, was originally projected to last two years, but it operated successfully for nearly seven. It made the first gamma-ray measurement of an AGN, that being 3C 273 (Swanenburg *et al.*, 1978). However, it was not until nearly 20 yr later with the launch of the Compton Gamma-Ray Observatory (CGRO) that additional gamma-ray detections were made, starting with the discovery in 1991 of bright gamma-ray emission from 3C 279 (Hartman *et al.*, 1992). New results came quickly after that leading ultimately to the identification of some 70 high-latitude CGRO gamma-ray sources with radio-loud AGN. Specifically, BL Lac objects and flat-spectrum radio quasars (FSRQs), known collectively as blazars, comprised the entire gamma-ray sample. It was also clear that the radiative output of the blazars was typically dominated by the gamma rays. The gamma-ray emission was also found to be variable on time scales less than a day.

These observations had several immediate implications for physical models. The emission had to emanate from a compact region. For example, a factor of 2 flux variation limits, approximately, the size r of a stationary, isotropic emitter to $r \lesssim c\delta t_{\text{var}}/(1+z)$ where δt_{var} is the variation time scale. The implications from the early CGRO results, which by this line of reasoning necessitated a very compact emission region, were problematic in any scenario in which the gamma-ray production involves such a stationary isotropic source. The problem involved the transparency of a compact region such as inferred here. If X-rays are produced cospatially with the gamma rays, attenuation of the gamma rays due to the process $\gamma\gamma \rightarrow e^+e^-$ for which the cross-section for attenuation of $\sim$ 100 MeV gamma rays is in the X-ray range $\sim$ keV X-ray range. The inferred gamma-ray opacity from the CGRO observations would exceed unity in many instances. Either the radiating particles were strongly beamed or the emitting plasma was undergoing bulk relativistic motion. Thus beaming was very strongly implied.

Models that had been previously favored to explain the radio-to-optical continua in these objects, for example Blandford and Konigl (1979), implied that we are viewing nearly along an axis of a relativistic plasma jet ejected from near the central black hole, involving nonthermal synchrotron emission as we will discuss in later chapters. An extension of this scenario invoking a distinct second spectral component was now clearly required. The basic idea was that gamma rays emitted by blazars are produced by the same population of electrons that produced the synchrotron emission via Compton scattering of ambient low-energy photons. The ambient photon field could be the synchrotron photons themselves (e.g., Maraschi *et al.*, 1992) or from an external source such as the accretion disk or broad-line clouds (e.g., Dermer *et al.*, 1992).

Shortly after the CGRO results began to emerge, another major discovery followed from ground-based Cherenkov gamma-ray telescopes, which measured gamma rays in the $\sim$ TeV range. Blazars such as Markarian 421 (Punch *et al.*, 1992) and Markarian 501 (Quinn *et al.*, 1996) were detected during a high-amplitude variability episode. These discoveries established this subclass of AGN as emitters over $\sim$ 20 decades of the electromagnetic spectrum. As such, they were a striking example of the value, indeed the necessity, inherent in the multiwavelength approach to studying AGN. The high-energy gamma-ray observations also fit in naturally with the synchrotron plus Comptonization model scenarios. They also had other potentially significant implications, not only on the blazar AGN themselves, but on the gamma-ray transparency of the universe and thus in turn the background radiation fields to the cosmic star-formation history.

In the two decades since these discoveries, gamma-ray studies of AGN have expanded enormously. The Fermi Gamma-Ray Space Telescope, launched in 2008, has cataloged approximately 900 gamma-ray AGN. Advances in ground-based

Cherenkov telescope facilities, as well as in detection and analysis methodologies, has produced a similar order-of-magnitude increase in the TeV gamma-ray sample. Multiwavelength campaigns have begun to reveal how jet formation and propagation may be correlated with the gamma-ray flux variations. Clearly, gamma-ray astronomy will continue to be a vital component of our quest to better understand the AGN phenomenology for the foreseeable future.

2
Radiative Processes

A basic understanding of the interaction between photons and particles and of particles with other particles or fields is essential for the understanding of emission and absorption processes. In AGN, we observe a large variety of processes. Synchrotron emission plays a key role in AGN jets, their accretion disk seems to emit thermal radiation, and the highest photon energies are reached through inverse Compton processes, for example in the jet or in a plasma close to the accretion disk. In this chapter we will give a brief overview of the physics underlying these processes. This shall be by no means a comprehensive review, and we refer the reader for example to Rybicki and Lightman (1986) for a more detailed discussion.

2.1
Scattering of Photons

Electromagnetic radiation produced in any kind of physical process may not travel undisturbed to the observer. In many cases, photons will be scattered on particles, losing or gaining energy on their way. This has important effects on our interpretation of an observed photon spectrum.

2.1.1
Thomson Scattering

Thomson scattering describes the nonrelativistic case of an interaction between an electromagnetic wave and a free charged particle. The effect was first described by Sir Joseph John Thomson, who discovered the electron when studying cathode rays in the late nineteenth century. The process can be understood as elastic or coherent scattering, as the photon and the particle will have the same energy after the interaction as before. For this process the energy E of the photon has to be much smaller than the rest energy of the particle:

$$E = h\nu \ll mc^2 \tag{2.1}$$

with ν being the frequency of the photon and m the mass of the particle. Another requirement for Thomson scattering is that the particle must be moving at nonrelativistic speed ($v \ll c$). In the classical view of this process, the incoming photon is absorbed by the particle with charge q, which is set into motion and then re-emits a photon of the same energy.

Using the classical electron radius $r_0 = q^2/(mc^2)$ the differential cross-section of this elastic scattering process can be written as

$$\frac{d\sigma}{d\Omega} = \frac{1}{2}(1 + \cos^2\theta)r_0^2 \tag{2.2}$$

This is symmetric with respect to the angle θ, thus the amount of radiation scattered in the forward and backward direction is equal. The total cross-section is then given by

$$\sigma_T = 2\pi \int_0^\pi \frac{d\sigma}{d\Omega} \sin\theta \, d\theta = \frac{8\pi}{3} r_0^2 = \frac{8\pi}{3} \left(\frac{q^2}{mc^2}\right)^2 \tag{2.3}$$

In the case of an electron, this gives a Thomson cross-section of $\sigma_T \simeq 6.652 \times 10^{-25}\,\text{cm}^2$. The cross-section for a photon scattering on a proton is a factor of $(m_p/m_e)^2 \simeq 3.4 \times 10^6$ smaller.

Since in the classical view of this process, the electron has no preferred orientation, the cross-section is independent of the polarization of the incoming electromagnetic wave. The polarization of the scattered radiation depends, however, on the polarization of the incoming photon wave. Unpolarized radiation becomes linearly polarized in the Thomson scattering process with the degree of polarization being

$$\Pi = \frac{1 - \cos\theta^2}{1 + \cos\theta^2} \tag{2.4}$$

Therefore, polarization of the observed emission can be a sign that the emergent radiation has been scattered.

Thomson scattering is important in many astrophysical sources. Any photon which will be produced inside a plasma can be Thomson scattered before escaping in the direction of the observer. The chance for the single photon to be Thomson scattered and how many of the photons will be scattered out of or into the line of sight is quantified in terms of the optical depth τ of the plasma:

$$\tau = \int \sigma_T n_e dx \tag{2.5}$$

Here n_e is the electron density, and dx is the differential line element. The mean free path λ_T of the photon, that is, the mean distance traveled between scatterings will thus be $\lambda_T = (\sigma_T n_e)^{-1}$.

2.1.2 Compton Scattering

Thomson scattering describes the interaction between the photon and the charged particle as an elastic scattering process. In this treatment, the incoming and outgoing photon have the same energy. In a quantum-mechanical treatment of the problem, the photon carries momentum as well as energy, and there will be an energy transfer from the photon to the electron, thus leading to a recoil of the electron. The larger the change of direction of the photon, the larger the energy transfer toward the electron will be.

Another condition for Thomson scattering is the low energy of the photon in comparison with the charged particle. If this requirement is dropped and the photon can have an energy comparable to the rest mass of the charged particle, it is necessary to consider quantum mechanical effects. This will reduce the effective cross-section relative to the Thomson case.

Including both effects, the energy transfer to the charged particle and the possibility of high photon energies, leads to the case of Compton scattering. The energy loss of the photon is $\Delta E = h\Delta\nu = hc/\Delta\lambda$ with

$$\Delta\lambda = \frac{h}{mc}(1 - \cos\theta) \tag{2.6}$$

The resulting energy E_2 of the photon after the Compton scattering is

$$E_2 = \frac{E_1}{1 + \frac{E_1}{mc^2}(1 - \cos\theta)} \tag{2.7}$$

with E_1 being the energy of the photon before the scattering. Thus, in the limit of $E_1 \ll mc^2$ this results in the case described by Thomson scattering, where no energy is transferred to the electron.

The cross-section for the high-energy limit of Thomson scattering is described by the Klein–Nishina formula derived by applying quantum electrodynamics. Here, the differential cross-section is described by

$$\frac{d\sigma}{d\Omega} = \frac{1}{2} r_0^2 f(\epsilon, \theta)^2 (f(\epsilon, \theta) + f(\epsilon, \theta)^{-1} - \sin^2\theta) \tag{2.8}$$

with

$$f(\epsilon, \theta) = \frac{1}{1 + \epsilon(1 + \cos\theta)} \tag{2.9}$$

In this notation, the energy of the incoming photon E_1 has been expressed in units of the rest mass energy of the charged particle:

$$\epsilon = \frac{E_1}{mc^2} = \frac{h\nu_1}{mc^2} \tag{2.10}$$

The total cross-section is then

$$\sigma_{\mathrm{KN}} = \frac{3\sigma_{\mathrm{T}}}{4} \left\{ \frac{1+\epsilon}{\epsilon^3} \left[\frac{2\epsilon(1+\epsilon)}{1+2\epsilon} - \ln(1+2\epsilon) \right] + \frac{\ln(1+2\epsilon)}{2\epsilon} - \frac{1+3\epsilon}{(1+2\epsilon)^2} \right\} \tag{2.11}$$

A useful approximation for photon energies lower than $E_1 \simeq 200\,\text{keV}$ is

$$\sigma_{\text{KN}} \simeq \sigma_{\text{T}} \frac{1 + 2\epsilon + 1.2\epsilon^2}{(1 + 2\epsilon)^2} \tag{2.12}$$

Again, when $\epsilon \ll 1$ and thus the energy of the incoming photon is small compared to the rest mass energy of the particle, this reduces to the Thomson cross-section. For energies $E \gg mc^2$ the cross-section is approximately

$$\sigma_{\text{KN}} \simeq \frac{3}{8\epsilon} \sigma_{\text{T}} \left(\ln 2\epsilon + \frac{1}{2} \right) \tag{2.13}$$

The Klein–Nishina effect results in a reduction of the Compton scattering cross-section at high photon energies. In other words, at the highest frequencies Compton scattering becomes less efficient. To illustrate this effect we show in Figure 2.1 the cross-section in units of the Thomson value as a function of the photon energy. Only above a few keV ($\nu \simeq 10^{18}$ Hz) does the Klein–Nishina effect become

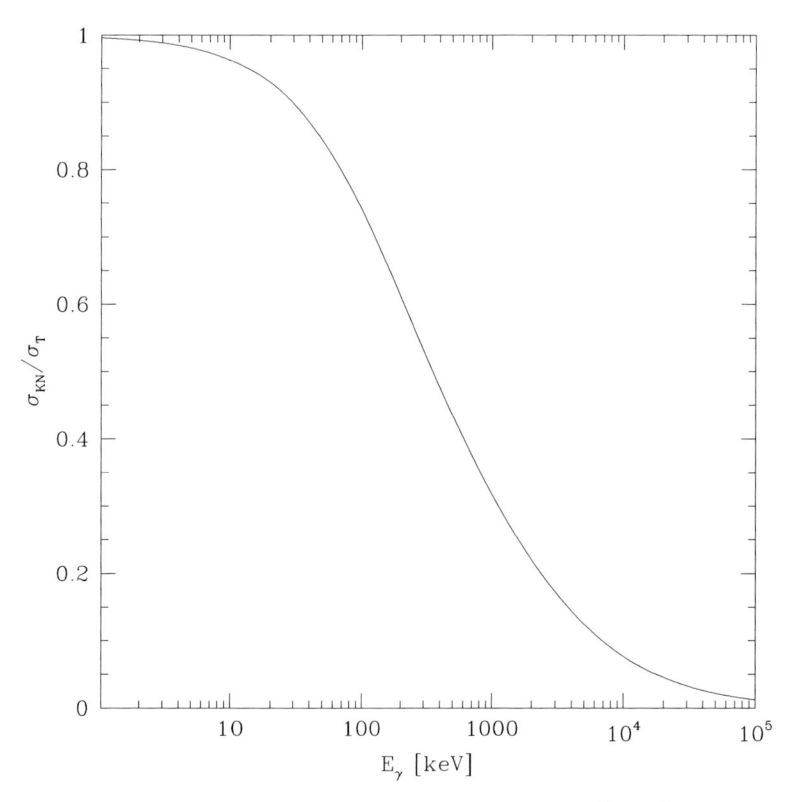

Figure 2.1 The cross-section for Compton scattering decreases with increasing photon energy due to the Klein–Nishina effect. The plot shows the cross-section in units of the Thomson cross-section $\sigma_{\text{T}} = 6.65 \times 10^{-25}\,\text{cm}^{-2}$ as a function of the photon energy, in keV, for Compton scattering on an electron. Up to the X-ray band, Klein–Nishina effects are negligible.

important. At 350 keV the cross-section is reduced to $1/2\sigma_T$ and it falls below 10% of the Thomson cross-section at ~ 7 MeV. This effect has important consequences regarding the interpretation of AGN spectral energy distributions as we will see in subsequent chapters.

2.1.3 Inverse Compton Scattering

In the case of Compton scattering the photon loses energy which is transferred to the electron. In cases where the electron is moving at relativistic speeds, the Compton process can lead to the opposite effect: the low-frequency photon can gain in energy through the so-called inverse Compton effect. Because in this case the condition $h\nu \ll m_e c^2/\gamma$ is fulfilled, one can apply the Thomson scattering cross-section.

Assume that the laboratory frame in which we observe the scattering event is L, and the frame of the relativistic electron with Lorentz factor γ is L' in which it is at rest before the event. The Lorentz factor is defined as

$$\gamma \equiv \frac{c}{\sqrt{c^2 - v^2}} \tag{2.14}$$

In the electron's frame L' the energy of the photon is much smaller than the rest energy of the electron: $h\nu' \ll mc^2$. Thus, the center of momentum is close to that of the relativistic electron. Following the relativistic Doppler shift formula the energy of the photon in L' is given by

$$h\nu' = \gamma h\nu \left(1 + \frac{v_e}{c}\cos\theta\right) \tag{2.15}$$

Note that the angle θ is the angle between the direction of the photon and of the incoming electron, traveling at speed v_e, in the laboratory frame L. In the electron's frame, this angle appears smaller:

$$\sin\theta' = \frac{\sin\theta}{\gamma\left(1 + \frac{v_e}{c}\cos\theta\right)} \tag{2.16}$$

Because the photon energy is small compared to the electron rest energy, we can treat the event as Thomson scattering. As this is an elastic scattering process, the energy of the photon in the electron's frame does not change, that is, $E_2' \simeq E_1'$. If we transform this expression back into the laboratory frame, we get

$$E_2 \simeq \gamma^2 E_1 \tag{2.17}$$

thus the photon gains energy in proportion to the square of the Lorentz factor γ of the electron. For relativistic electrons this means that a photon in the radio domain can be up-scattered to the optical or even X-ray range in the inverse Compton process. The maximum energy gain of the photon is determined by the energy conservation of the process as seen in the laboratory frame:

$$E_2 \le E_1 + \gamma m_e c^2 \tag{2.18}$$

Thus, the maximum change in photon frequency is

$$\Delta\nu \leq \gamma m_e c^2 h^{-1} \tag{2.19}$$

The total power (in other words, the luminosity) of the inverse Compton process will naturally depend on the density of photons n_{ph} available for the scattering. Thus, the luminosity L_{IC} of the inverse Compton component will be

$$L_{IC} \propto n_{ph}\gamma^2 E_1 \tag{2.20}$$

In other words, the luminosity will be proportional to the energy density of the photon field:

$$L_{IC} \propto \gamma^2 U_{ph} \tag{2.21}$$

We will not present the exact derivation of the luminosity of the single Compton scattering, which can be studied in Rybicki and Lightman (1986), but instead point out some general considerations.

The resulting L_{IC} will depend on the cross-section for the inverse Compton process. We recall that this scattering can be treated as elastic, and thus the Thomson cross-section σ_T applies. The single electron will "see" per unit time an incoming energy from the photons, which is the product of the speed of the incoming photons c, the cross-section of the process σ_T, and the energy density of the photon field U_{ph}:

$$\frac{dE}{dt} = c\sigma_T U_{ph} \tag{2.22}$$

In addition we have to consider the relativistic Doppler shift, and thus the resulting luminosity is

$$L_{IC} = \frac{4}{3}\frac{v_e^2}{c^2} c\sigma_T \gamma^2 U_{ph} = \frac{4}{3}\frac{v_e^2}{c}\sigma_T \gamma^2 U_{ph} \tag{2.23}$$

In cases where the condition $h\nu' \ll mc^2$ is not fulfilled in the electron's frame, the assumption of Thomson scattering is no longer valid. Thus, for high photon energies Klein–Nishina effects have to be considered, which involves a dependence on photon energy and scattering angle. Following for example Blumenthal and Gould (1970), the resulting luminosity for the inverse Compton process is

$$L_{IC,KN} = \frac{4}{3}\frac{v_e^2}{c}\sigma_T \gamma^2 U_{ph}\left(1 - \frac{63}{10}\frac{\gamma\langle E_1^2\rangle}{mc^2\langle E_1\rangle} + \ldots\right) \tag{2.24}$$

The first term in the brackets gives the luminosity considering Thomson scattering. Since in the Klein–Nishina realm the cross-section is reduced, the second term in the brackets reduces L_{IC} depending on the average photon energy $\langle E_1\rangle$ in the photon field. The larger $\langle E_1\rangle$, the more the luminosity of the inverse Compton process will decrease. For very high photon energies, the luminosity can be negative and the photon will lose energy to the electron.

In astrophysical sources, one has to deal with not only a single electron but with electron distributions. Thus, in order to determine the luminosity of the inverse Compton process L_{IC} one has to integrate over the single scattering described by Eq. (2.23), applying an energy distribution for the electrons. An assumption that is often valid for a distribution of electron energies is that of a simple power law:

$$N(E)\mathrm{d}E = k_E E^{-p}\mathrm{d}E = k_\gamma \gamma^{-p}\mathrm{d}\gamma \tag{2.25}$$

with $N(E)$ being the number of electrons at a given energy E, p the slope of the power law, and k_E the normalization. If we assume that the electrons are highly relativistic, that is, that $v_e/c \simeq 1$, then we can integrate Eq. (2.23) and derive the luminosity of the inverse Compton radiation for the electron distribution. Assuming that the electrons have a minimum Lorentz factor γ_{min} and a maximum of γ_{max}, this results in:

$$L_{IC} = \frac{4}{3} c \sigma_T U_{ph} k_\gamma \frac{\gamma_{max}^{3-p} - \gamma_{min}^{3-p}}{3-p} \tag{2.26}$$

2.1.4 Thermal Bremsstrahlung

Another important mechanism in astrophysics for electrons losing energy, is bremsstrahlung (from German "breaking radiation"). Here, the electron interacts with another free charged particle, for example an ion. As both particles are not bound before and after the scattering process, this is also referred to as *free–free emission*. The interaction leads to a change in the momentum of the particles involved, and through this acceleration of charges to radiation. This type of scattering takes place for example in hot, but nonrelativistic plasmas. The most prominent source of extragalactic bremsstrahlung is the intracluster gas which is bound in the gravitational wells of galaxy clusters with temperatures of $T \simeq 10^7 - 10^8$ K (Section 7.4). Interaction of two particles with the same mass, like electron–electron or positron–positron scattering does not produce dipole radiation, and the quadrupole term is comparably low. Only in the relativistic case can this produce a significant amount of radiation.

Thus, in extragalactic sources the most important bremsstrahlung is produced in interactions of electrons with ions of charge Ze. As the ion is much more massive than the electron, we can consider the scattering event with the ion being at rest, and the electron being the only particle which undergoes acceleration. One can further assume a scattering event which does not result in a large change in the electron's direction, and consider only small-angle scattering. In a classical approach to the problem, the acceleration will depend on how close the electron is passing by the ion. If d is the instantaneous distance between the electron and the ion, then the acceleration experienced by the electron is

$$\frac{F}{m} = \frac{Ze^2}{md^2} \simeq \frac{Ze^2}{mb^2} \tag{2.27}$$

where b is the minimum value of d, often referred to as the impact parameter. The luminosity as a function of the distance is then (see, e.g., De Young, 2002):

$$L(b) = \frac{2e^2 a^2}{3c^3} = \frac{2Z^2 e^6}{3c^3 m^2 b^4} \tag{2.28}$$

with a being the acceleration of the electron. As the radiated power depends strongly on the distance $L \propto d^{-4}$, only the time when the two particles near the point of closest approach will be relevant. We can thus assume that the important acceleration takes place when the distance is $d \leq \sqrt{2}b$. The longer the electron is close to the ion, the larger the effect will be, thus, the acceleration will be stronger for slower electrons. In the nonrelativistic case the time during which the electron is at distance $d \leq \sqrt{2}b$ is simply

$$\Delta t = \frac{2b}{v} \tag{2.29}$$

with v being the velocity of the electron relative to the ion. The energy radiated away through bremsstrahlung per electron will then be:

$$E(b, v) = \frac{4Z^2 e^6}{3c^3 m^2 b^3 v} \tag{2.30}$$

One can derive the energy per unit frequency from Eq. (2.30). The maximum of the radiation will be achieved for a frequency which is related to the time of encounter Δt between the two particles. The maximum frequency is at $\omega_{\text{max}} = \pi/\Delta t = \pi v/2b$. The energy per unit frequency is then

$$E_\nu(b, v) = \frac{E(b, v)}{\omega_{\text{max}}} = \frac{8Z^2 e^6}{3\pi c^3 m^2 b^2 v^2} \tag{2.31}$$

In order to derive the total energy output per unit volume and unit frequency, it is necessary to integrate Eq. (2.31) over the number of bremsstrahlung events which occur per second. The number of scatterings per electron per second is $N = n_{\text{i}}\sigma v$, with n_{i} being the ion density and σ the cross-section $\sigma = \int 2\pi b\,db$. Considering that the encounters will take place at different ion–electron distances d, ranging from b_{min} to b_{max}, and assuming an electron density n_{e}, the resulting emissivity $\epsilon(v)$ (i.e., luminosity per unit volume and unit frequency) as a function of electron velocity is

$$\epsilon(v) = \int_{b_{\text{min}}}^{b_{\text{max}}} 2\pi n_{\text{e}} n_{\text{i}} E(b, v) v b\,db = \frac{16Z^2 e^6 n_{\text{e}} n_{\text{i}}}{3c^3 m^2 v} \ln \frac{b_{\text{max}}}{b_{\text{min}}} \tag{2.32}$$

If the distance between electron and ion is too large, the effect will be negligible, which can be assumed to occur when $b \gtrsim v/\omega$. On the other hand we assumed that the change in direction for the electron should be small, thus the change in velocity should be $\Delta v \lesssim v$. Equation (2.32) is a function of the electron speed v.

If we assume a gas of temperature T which is in thermal equilibrium, the electron velocities will follow a Maxwell–Boltzmann distribution. This gives the probability density function for the speed of

$$f(v) = \sqrt{\frac{2}{\pi}\left(\frac{m}{kT}\right)^3} v^2 \exp\left(\frac{-mv^2}{2kT}\right) \tag{2.33}$$

Integrating Eq. (2.32) over the electron speeds, considering a minimum speed $v_{\rm min}$ and a maximum speed $v_{\rm max}$ gives an emissivity of

$$\begin{aligned} \epsilon &= \int\limits_{v_{\rm min}}^{v_{\rm max}} \epsilon(v) f(v) dv \\ &= \frac{16 Z^2 e^6 n_e n_i}{3c^3 m^2} \bar{g}_{\rm ff} \frac{4\pi^2}{\sqrt{3}} (2\pi)^{-3/2} \sqrt{\frac{m}{kT}} \exp\left(\frac{-m v_{\rm min}^2}{2kT}\right) \end{aligned} \tag{2.34}$$

Here, the quantum mechanical correct Gaunt factors $\bar{g}_{\rm ff}$ averaged over the velocities have replaced the expression $\ln(b_{\rm max}/b_{\rm min})$ in Eq. (2.32). In order to perform the integration, the minimum and maximum velocities have been chosen. While $v_{\rm max} = \infty$, the minimum velocity is given by the limitation that the kinetic energy of the electron has to be large enough to produce the bremsstrahlung's photon: $\frac{1}{2} m v^2 \geq h\nu$. Finally the luminosity of bremsstrahlung radiation per unit volume turns out to be

$$L = \frac{32\pi Z^2 e^6}{3c^3 m^2} \bar{g}_\nu \sqrt{\frac{2\pi m k T}{3h^2}} n_e n_i = 1.42 \times 10^{-27} \sqrt{T} \bar{g}_\nu n_e^2 \,\text{erg cm}^{-3}\,\text{s}^{-1} \tag{2.35}$$

Here $\bar{g}_\nu$ is the frequency-averaged Gaunt factor, which lies in the range of 1.1–1.5. The emissivity and therefore the luminosity of the thermal bremsstrahlung plasma thus increases with the square of the electron density and with the square root of the plasma temperature. Note that as stated before the emission from the ions is negligible compared to that of the electrons, as their acceleration is low because of their large mass. The spectrum cuts off at frequencies $\nu = kTh^{-1}$.

A characteristic time scale for a radiative plasma, referred to the astrophysics literature as the "cooling time" is given by its thermal energy per unit volume divided by its specific radiative luminosity. For a thermal bremsstrahlung plasma, the thermal energy per unit volume is $\sim n_e kT$, which gives for the cooling time $\tau_{\rm Brems}$:

$$\tau_{\rm Brems} \sim \frac{n_e kT}{\epsilon} \sim \frac{1}{n_e \alpha \sigma_T c} \sqrt{\frac{kT}{mc^2}} \tag{2.36}$$

Here $\sigma_T = 6.65 \times 10^{-25}\,\text{cm}^{-2}$ is the Thomson cross-section as described in Eq. (2.3) and $\alpha = 1/137.04$ is the fine structure constant. For plasma trapped in the gravitational potential at the center of a massive galaxy cluster, the electron density is of the order of $10^{-2}\,\text{cm}^{-3}$ and the temperature around $T \simeq 10^7$ K, giving a cooling time of approximately 10^9 years. Galaxy clusters seem to remain at high temperature on longer time scales, however, which leads to a *cooling time problem*: what is keeping the center of galaxy clusters hot? We will come back to this question in Section 7.4.

2.1.5
Pair Production

When the energy of a photon exceeds the sum of rest mass energy of a particle and its anti-particle, pair production can take place. Pair production has to obey conversation laws, that is, the energy, momentum, angular momentum, electric charge and lepton number must be conserved. The charge and lepton number of the product particles thus have to sum to zero, as the photon has zero values for these properties. The lowest mass particle pair which can be produced in this manner is the electron–positron pair. Other possible products are muon and anti-muon, or a tau and an anti-tau. The e^-/e^+ pair production requires a high-energy photon with $E > 2m_e c^2 = 1022\,\text{keV}$. It must occur in the vicinity of a massive particle, for example an ion, electron, or positron in order to conserve momentum. Photon energy in excess of the rest mass energy of the two particles will be carried away as kinetic energy. As the photon carried a momentum, this momentum will have to be conserved and so some momentum will be imparted to the nearby particle. The cross-section for this process depends on the particle's mass. With an interaction of the type $\gamma + e^{\pm} \rightarrow e^{\pm} + e^- + e^+$, the cross-section for a photon with energy E is (e.g., Krolik, 1999)

$$\sigma_{e^{\pm}\gamma}(E) = \frac{3}{8\pi}\alpha\sigma_T \left\{ \begin{array}{ll} \frac{\sqrt{\pi}}{324}\left(\frac{E}{m_e c^2} - 4\right)^2 & \text{for } \frac{E}{m_e c^2} - 4 \ll 1 \\ \frac{28}{9}\ln\frac{2E}{m_e c^2} - \frac{218}{27} & \text{for } \frac{E}{m_e c^2} \gg 4 \end{array} \right\} \tag{2.37}$$

If a nucleus is involved, that is, if we look at a pair production process following $\gamma + Z \rightarrow Z + e^- + e^+$, the cross-section turns out to be

$$\sigma_{Z\gamma}(E) = \frac{3}{8\pi}\alpha\sigma_T Z^2 \left\{ \begin{array}{ll} \left(\frac{2\pi}{3}\right)\left(\frac{E}{m_e c^2} - 2\right)^2 & \text{for } \frac{E}{m_e c^2} - 2 \ll 1 \\ \left(\frac{28}{9}\right)\ln\left(\frac{2E}{m_e c^2}\right) - \frac{218}{27} & \text{for } \frac{E}{m_e c^2} \gg 2 \end{array} \right\} \tag{2.38}$$

We see that for large photon energies ($E \gg 4m_e c^2$) the cross-section is the same and scales only with the square of the charge of the particles Z^2. Pair production involving two charged particles can also take place. But the cross-sections are lower compared to Eq. (2.38) by another factor of $\sim \alpha$, and thus do not contribute significantly here.

Another possibility for producing an e^-/e^+ pair involves the interaction of two photons. This purely quantum mechanical process is the inverse of the pair annihilation process which has been well studied in laboratory experiments and is observed in some astrophysical environments. The photon–photon interaction turns out to be important in the study of a subclass of AGN which emit high-energy gamma rays. It introduces a mechanism for high-energy photons in the TeV range emitted by these AGN to be effectively "absorbed" by low-energy photons, such as those characteristic of the extragalactic background light, which peaks at infrared wavelengths.

In order for the photon–photon interaction to be able to produce an e^-/e^+ pair, the energy has to be large enough and the relative directions of the motion of the photons has to be considered. Assuming that two photons travel along direction vectors $\boldsymbol{u}_1$ and $\boldsymbol{u}_2$, respectively, pair production is possible when the product of the photon energies $E_1 \cdot E_2$ fulfills the following requirement:

$$E_1 \cdot E_2 \geq \frac{2\left(m_e c^2\right)^2}{1 - \boldsymbol{u}_1 \cdot \boldsymbol{u}_2} \tag{2.39}$$

It is important to note here that the product of the photon energies has to fulfill this condition for pair production to be possible, and not the sum. Thus, pair production is impossible if the two photons travel in the same direction, and the efficiency is largest for a head-on collision. Assuming a relative angle θ, the term $\boldsymbol{u}_1 \cdot \boldsymbol{u}_2 = \cos\theta$. Assuming an isotropic distribution of photon directions, we can assume an average value of $\bar{\theta} = 90°$. In this case, the frequencies ν_1 and ν_2 of the two photons have to fulfill the condition that $\nu_2 \geq 3 \times 10^{40}\,\text{Hz}^2\,\nu_1^{-1}$. In addition to the photon energy threshold necessary to allow pair production, the cross-section for photon–photon interaction is critically energy-dependent. The cross-section starts as $\sigma = 0$ at the minimum energy described in Eq. (2.39). It reaches its maximum at twice the minimum energy value, that is, at a frequency $\nu_2 = 6 \times 10^{40}\,\text{Hz}^2\,\nu_1^{-1}$, where the cross-section is $\sigma \simeq 0.2\sigma_\text{T}$ with σ_T being the Thomson cross-section (Eq. (2.3)). When expressed in terms of photon energies, the maximum cross-section appears when $E_1 \cdot E_2 \simeq (1\,\text{MeV})^2$. Thus, for photon–photon pair production most likely "partners" are two photons at MeV energies, or a GeV photon scattering with a keV photon, a TeV photon on a eV (i.e., optical) photon, and so forth.

2.2 Synchrotron Emission

Radio emission in astrophysical sources is caused by the synchrotron process which occurs when charged particles are accelerated in a magnetic field. The electron changes direction because the magnetic field exerts a force perpendicular to its original direction of the motion. The energy of the emitted photons is a function of the electron energy, of the magnetic field strength B and of the angle between the electron's path and the magnetic field lines. Synchrotron emission was first detected in 1946 in a General Electric synchrotron accelerator (Elder *et al.*, 1947). Alfvén and Herlofson (1950) predicted synchrotron radio emission from stars but the first astrophysical synchrotron source detected was the jet of the FR-I galaxy M87 (Burbidge, 1956).

Let us start with the more general assumption of a particle with charge $q = Ze$, rest mass m and Lorentz factor

$$\gamma \equiv \frac{1}{\sqrt{1 - v^2/c^2}} \tag{2.40}$$

traveling through a magnetic field which is uniform and static. This will lead to a force

$$\frac{\mathrm{d}}{\mathrm{d}t}(\gamma m \boldsymbol{v}) = \frac{Ze}{c}(\boldsymbol{v} \times \boldsymbol{B}) \tag{2.41}$$

The acceleration is perpendicular to the direction in which the charged particle travels, thus its velocity v and Lorentz factor γ are constant. In addition, the force is perpendicular to the direction of the magnetic field, and thus also the velocity of the particle in direction of the B field $v_{\parallel}$ will be constant. Thus, as the total velocity v, and $v_{\parallel}$ are constant, the velocity perpendicular to the magnetic field will also be constant: $v_{\perp} = \sqrt{v^2 - v_{\parallel}^2}$. Therefore, the particle will travel with constant velocity in the direction of the magnetic field and will perform a circular motion with constant radius r_{g} and constant pitch angle β with respect to the B field. The radius r_{g} is called the gyroradius:

$$r_{\mathrm{g}} = \frac{v_{\perp} \gamma m c}{Z e B} \tag{2.42}$$

This also gives us the frequency of the circular motion, the so-called gyrofrequency or Larmor frequency:

$$\nu_{\mathrm{g}} = \frac{Z e B}{2\pi \gamma m c} \tag{2.43}$$

The angular gyrofrequency is then $\omega_{\mathrm{g}} = 2\pi \nu_{\mathrm{g}}$. Thus, for an electron traveling at nonrelativistic speed ($\gamma \simeq 1$) the gyrofrequency is only a function of the magnetic field strength: $\nu_{\mathrm{g}}(\gamma = 1) = B \cdot 2.8 \times 10^6\ \mathrm{Hz\,G^{-1}}$.

The total luminosity (power) of the acceleration process is, (e.g., Rybicki and Lightman, 1986)

$$L = \frac{2Z^2 e^2}{3c^3} \gamma^4 \left[\left(\frac{\mathrm{d}v_{\perp}}{\mathrm{d}t} \right)^2 + \gamma^2 \left(\frac{\mathrm{d}v_{\parallel}}{\mathrm{d}t} \right)^2 \right] \tag{2.44}$$

The acceleration is perpendicular to the velocity, thus $\mathrm{d}v_{\parallel}/\mathrm{d}t = 0$. Because the charged particle follows a circular motion, the perpendicular acceleration is $\mathrm{d}v_{\perp}/\mathrm{d}t = \omega_{\mathrm{g}} v_{\perp}$. The perpendicular velocity can be expressed as a function of the constant particle velocity v and the pitch angle β between the direction of the particle (i.e., the direction of $\boldsymbol{v}$) and the magnetic field: $v_{\perp} = v \sin\beta$. Thus, the luminosity is

$$L = \frac{2Z^4 e^4 B^2 \gamma^2}{3c^5 m^2} v_{\perp}^2 = \frac{2Z^4 e^4 B^2 \gamma^2 v^2 \sin^2\beta}{3c^5 m^2} \tag{2.45}$$

If we assume an isotropic distribution of particle velocities, we can average the power over all possible pitch angles β resulting in different values for the perpendicular velocity:

$$\langle v_\perp^2 \rangle = \frac{v^2}{4\pi} \int \sin^2 \beta \mathrm{d}\Omega = \frac{2v^2}{3} \tag{2.46}$$

This gives a synchrotron luminosity of

$$L = \frac{4Z^4 e^4 B^2 \gamma^2 v^2}{9c^5 m^2} = -\frac{\mathrm{d}E}{\mathrm{d}t} = \frac{4Z^4 e^4 B^2 E^2}{9m^4 c^7} \tag{2.47}$$

Here E is the energy of the particle. Because $L \propto m^{-2}$, this is most efficient for electrons and positrons. The proton-to-electron mass ratio is $m_\mathrm{p}/m_\mathrm{e} \simeq 1836.2$, and therefore the radiative power of a proton traveling at speed v through the magnetic field will be lower by a factor of 3×10^{-7} compared to an electron under the same conditions. If we consider an electron and a proton with the same energy E, the emitted power of the proton is even lower by a factor of $(m_{\mathrm{e}^-}/m_\mathrm{p})^4 \simeq 9 \times 10^{-14}$. This is an important consideration in the discussion of particles as to which particles dominate the synchrotron emission in relativistic jet plasmas. For a given synchrotron luminosity, a proton jet would have to be significantly faster and/or more massive than an electron or positron jet.

For an electron or positron we can write this formula in a more compact form, using the Thomson cross-section σ_T (Eq. (2.3)) and the magnetic energy density $U_B = B^2/8\pi$:

$$L_{\mathrm{e}^-} = \frac{4}{3} \sigma_\mathrm{T} \frac{v^2}{c} \gamma^2 U_B \tag{2.48}$$

In the case of highly relativistic electrons, one can approximate that $v/c \simeq 1$ and the synchrotron luminosity becomes a function only of the Lorentz factor and the energy density in the magnetic field:

$$L_{\mathrm{e}^-,\mathrm{rel}} = \frac{4}{3} \sigma_\mathrm{T} c \gamma^2 U_B \simeq 2.7 \times 10^{-14}\,\mathrm{cm}^3\,\mathrm{s}^{-1} \gamma^2 U_B \tag{2.49}$$

The synchrotron radiation will not be emitted isotropically, but because of the beaming effects it will be concentrated in a narrow cone symmetric about the particle's velocity vector, as shown in Figure 2.2. The opening angle of the radiation cone will be $\phi \simeq \gamma^{-1}$. As the particle describes a helical motion in a uniform magnetic field, the observer will see a lighthouse effect and detect radiation only when the narrow cone is pointing toward the line of sight. The observer will receive a pulsed emission with the width of the pulses being much smaller than the gyration period ω_g. The pulse duration is of the order $\Delta t \simeq 0.5\gamma^{-2}\omega_\mathrm{g}^{-1}$ thus about a factor of γ^{-2} shorter than the gyration period ω_g which applies in the nonrelativistic case. The observed frequency is then:

$$\nu \sim \Delta t^{-1} \sim \gamma^2 \nu_\mathrm{g} = \gamma^3 \nu_r = \frac{\gamma^3 v}{2\pi r_0} \tag{2.50}$$

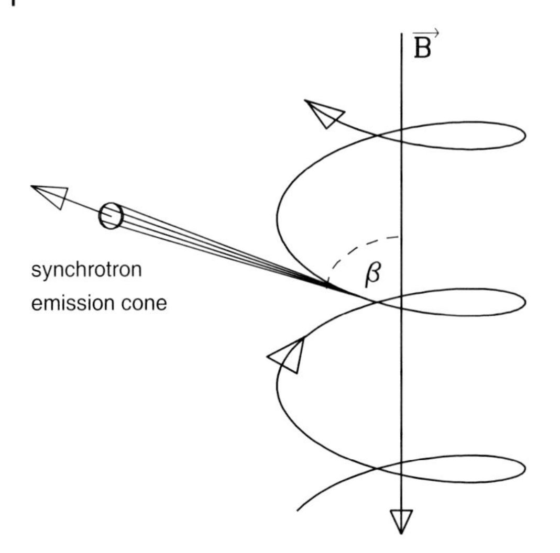

Figure 2.2 The synchrotron emission of an electron with pitch angle β between magnetic field and velocity will form a cone with opening angle $\phi \simeq \gamma^{-1}$. The higher the Lorentz factor γ, the more narrow the cone will be. The synchrotron radiation is emitted in the direction of the electron's motion. Averaged over the rotation, this gives a solid angle in the forward direction (i.e., toward the top in this illustration).

with ν_r being the non-relativistic gyrofrequency and r_0 the radius of curvature of the particle's orbit (gyroradius) in the non-relativistic case.

Deriving the spectrum of the single particle emitting synchrotron radiation involves a complex calculation, so we proceed directly to the result here. Details on the derivation can be found in Rybicki and Lightman (1986). The spectrum emitted by a single charged particle averaged over the orbit has a maximum at $\nu = 0.29\nu_c$, where ν_c is the *critical frequency*:

$$\nu_c = \frac{3Ze\gamma^2 B \sin\beta}{4\pi mc} = \frac{3ZeB\sin\beta}{4\pi mc}\left(\frac{E}{mc^2}\right)^2 = \frac{3}{2}\gamma^3 \nu_g \sin\beta \tag{2.51}$$

Because the gyrofrequency is inversely proportional to the Lorentz factor ($\nu_g \propto \gamma^{-1}$; Eq. (2.43)), this leads to a dependency of the maximum synchrotron frequency of $\nu_c \propto E^2 \propto \gamma^2$.

The derivation of the power per unit frequency of the synchrotron emission has the form (e.g., Ginzburg, 1989):

$$L(E,\nu) = \frac{\sqrt{3}Z^3 e^3 B \sin\beta}{mc^2} F(x) \tag{2.52}$$

with $x = \nu/\nu_c$ and

$$F(x) = \frac{\nu}{\nu_c} \int_{\nu/\nu_c}^{\infty} K_{5/3}(\zeta)\mathrm{d}\zeta \tag{2.53}$$

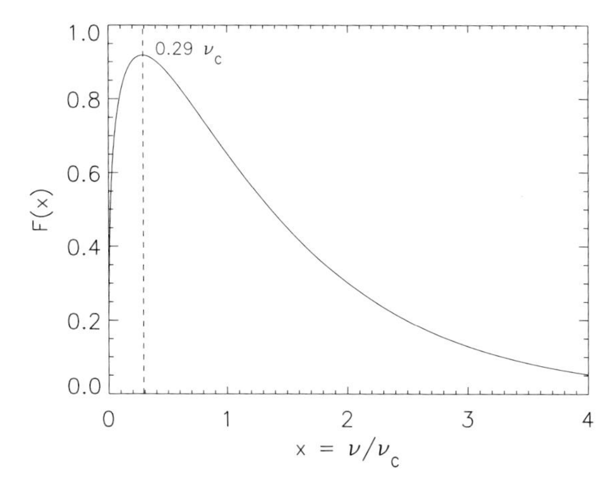

Figure 2.3 The total power F of the synchrotron emission of a single electron as a function of the frequency ν in units of the critical frequency $\nu_c = 3/2\gamma^3\nu_g \sin\beta$. The peak of the emission appears at $\nu = 0.29\nu_c$. Figure courtesy of Simona Soldi.

Here $K_{5/3}$ is the modified Bessel function of order 5/3. The resulting spectrum is shown in Figure 2.3 where F is presented as a function of frequency in units of the critical frequency.

2.2.1
Synchrotron Emission of a Particle Plasma

In astrophysical sources we observe emission from large numbers of charged particles. If we consider for example an electron plasma emitting synchrotron radiation, we can simply sum up the single particle contributions. This emission can then be absorbed again before leaving the plasma. We will consider this later in this section and start by assuming that the plasma is optically thin so that the radiation reaches the observer unaltered. We note that we base our discussion on synchrotron emission from electrons and positrons, and thus assume the charge $q = Ze = e$, noting that the formulas presented are valid for protons as well.

In the last section we saw that the spectrum of a single electron emitting synchrotron radiation is rather sharply peaked below the critical frequency ν_c. When considering an ensemble of electrons, one can thus proceed with the approximation that the single particle emits all its synchrotron power at the critical frequency. Its frequency is then

$$\nu \simeq \nu_c \simeq \gamma^2\nu_g = \left(\frac{E}{mc^2}\right)^2 \nu_g \tag{2.54}$$

Thus, the energy distribution of the synchrotron emission will depend directly on the energy distribution of the electrons. If we assume that the electrons have energies in the interval $E_1 - E_2$, the total synchrotron emissivity, that is, the power per unit frequency and per unit volume, will be

$$\epsilon(\nu) = \int_{E_1}^{E_2} L(E,\nu)n(E)\mathrm{d}E \tag{2.55}$$

Here $L(E,\nu)$ is the power per unit frequency as described in Eq. (2.52) and $n(E)\mathrm{d}E$ is the number density of electrons in the energy interval between E and $E+\mathrm{d}E$. The energy distribution of charged particles in astrophysical sources often has a power-law form. In this case we can write the particle density as a function of energy as

$$n(E)\mathrm{d}E = kE^{-p}\mathrm{d}E \tag{2.56}$$

where k is the normalization of the power law and p is the power-law index, and both, k and p are constant. The number of electrons falls toward higher energies with a power law of index $-p$.

Performing the integration of Eq. (2.55) assuming a power-law distribution of the electron energies, we derive as total synchrotron emissivity

$$\epsilon(\nu) = \frac{\sqrt{3}e^3}{2mc^2}\left(\frac{3e}{4\pi m^3c^5}\right)^{\frac{p-1}{2}} k(B\sin\beta)^{\frac{p+1}{2}}\nu^{\frac{1-p}{2}} G\left(\frac{\nu}{\nu_{c1}}, \frac{\nu}{\nu_{c2}}, p\right) \tag{2.57}$$

with ν_{c1} and ν_{c2} being the critical frequencies as defined in Eq. (2.51) at energies E_1 and E_2, respectively. Setting $x = \nu/\nu_c$, the function G is given by

$$G(x_1, x_2, p) = \int_{x_1}^{x_2} x^{\frac{p-3}{2}} F(x)\mathrm{d}x \tag{2.58}$$

with $F(x)$ as defined in Eq. (2.53). For frequencies well within the interval defined by E_1 and E_2 with $\nu_1 \ll \nu \ll \nu_2$, we have $x_1 = \nu/\nu_{c1} = 0$ and $x_2 = \nu/\nu_{c2} = \infty$ and G reduces to

$$G(0,\infty,p) = \frac{2^{\frac{p-3}{2}}}{3}\left(\frac{3p+7}{p+1}\right)\Gamma\left(\frac{3p-1}{12}\right)\Gamma\left(\frac{3p+7}{12}\right) \tag{2.59}$$

with Γ being the gamma function. Thus, $G(\nu/\nu_{c1}, \nu/\nu_{c2}, p)$ will no longer depend on ν anymore and the emissivity will have the form

$$\epsilon(\nu) \propto \nu^{-\alpha_R}, \quad \text{with spectral index in the radio band} \quad \alpha_R = \frac{p-1}{2} \tag{2.60}$$

Measuring the gradient of a synchrotron spectrum in the radio domain in the form of a power law with energy index α_R, we can derive directly the slope of the electron energy distribution: $p = 2\alpha_R + 1$. In the case of flat spectrum radio quasars (FSRQ) where the radio continuum has a slope of $\alpha_R \simeq 0.5$, this means that the spectrum is of the form $f(\nu) \propto \nu^{-0.5}$ and the electron energy distribution satisfies $n(e) \propto E^{-2}$.

The high-energy cutoff of the synchrotron spectrum is then defined by the maximum critical frequency $\nu_{c,\max}$ of the most energetic electrons.

2.2.2 Polarization

Synchrotron emission of a single charged particle is polarized. This means, the electromagnetic wave's electric field has a specific orientation. We recall that there are two types of polarization. The electric field can be oriented in a single direction (linear polarization), or the polarization vector can rotate (circular or elliptical polarization). In the latter case the electric field vector describes an ellipse in a plane perpendicular to the propagation direction of the wave. The synchrotron radiation of the single charged particle is thus elliptically polarized. The sense of polarization will depend on whether the line of sight lies inside the circle described by the cone, or outside of it. If one's line of sight lies within the cone of the synchrotron radiation, one will observe linearly polarized emission. In most astrophysical scenarios we can assume that we observe a plasma of particles with different pitch angles. Thus, the elliptical polarization contributions of the single particles will cancel and an ensemble of electrons as described in the previous section, will not produce elliptically polarized light. Instead it will result in partially linear polarized radiation. The degree of polarization then depends on the ratio between the power per unit frequency emitted in parallel and perpendicular to the projection of the magnetic field on the plane of the sky:

$$\Pi = \frac{L_{\perp}(\nu) - L_{\parallel}(\nu)}{L_{\perp}(\nu) + L_{\parallel}(\nu)} \tag{2.61}$$

If we assume an energy distribution of the particles in power-law form with index p as in the previous section, the degree of linear polarization will be (e.g., Ginzburg, 1989):

$$\Pi = \frac{p + 1}{p + 7/3} \tag{2.62}$$

This gives us a second method for determining the power-law index p of the electron energy distribution. Equation (2.62) also shows that the polarization is independent of the frequency at which we measure, under the assumption that the magnetic field is perfectly ordered. In a real plasma, this is probably not the case, but the magnetic field will be ordered on small length scales. In this case, one expects a frequency dependence of the measured polarization. At high frequencies, the polarization will be higher because we probe only a small volume of the emitting plasma. The reason is that at higher frequencies the cooling is more efficient and particles will not travel a long way. Thus, it is more likely to find an ordered magnetic field. As one moves to lower frequencies with less efficient cooling and probes a larger volume, the polarization will decrease.

2.2.3 Faraday Rotation

Michael Faraday discovered in 1845 that the angle of polarization of an electromagnetic wave changes when the wave is sent through a medium with a magnetic field. The so-called Faraday rotation can also affect the synchrotron emission. Faraday rotation can be understood as the different effect the magnetized plasma has on the left and right circularly polarized light. Depending on the orientation with respect to the magnetic field, the components will "see" a different refractive index. Thus, the phase velocity of the two components will be affected slightly differently and lead to a shift of their relative phases. This causes the plane of polarization to rotate, depending on how strong the magnetic field is and what distance the wave has to travel through the plasma. A similar effect is also observed with linearly polarized light. This can be understood if we think of the linear polarization being made up out of two circular components, a left-handed and a right-handed one. Once the linearly polarized synchrotron light is emitted and traveling towards the observer, it can pass through magnetized matter causing Faraday rotation. This can be the emitting plasma itself, or any magnetized gas along the line of sight. In astrophysical applications, we can simplify the problem by considering only free electrons in magnetic fields.

The amount of rotation in the polarization angle depends then on the magnetic field strength and density of the electrons along the line of sight, but also on the frequency of the electromagnetic wave we are observing:

$$\Delta\theta = \lambda^2 \mathrm{RM} \tag{2.63}$$

Here λ is the wavelength of the polarized radiation, and RM is the *rotation measure* which is a function of the electron density n_e and of the component of the magnetic field $B_{||}$ parallel to the line of sight:

$$\Delta\theta = \lambda^2 \frac{e^3}{2\pi m^2 c^4} \int n_e(s)\, B_{||}(s) \mathrm{d}s \tag{2.64}$$

Thus, the rotation is larger for lower frequencies. This can be qualitatively understood recalling that the frequency of the wave is much larger than the gyrofrequency of the electron (Eq. (2.43)). The closer the frequency of the polarized light to the gyrofrequency, the closer the light and the electron are to a resonant state, and thus the larger the energy transfer from the wave to the electron. The light from extragalactic sources will not only have to cross the intergalactic medium, but the interstellar medium of our Galaxy as well on its path to the observer. The magnetic field along the line of sight will not be constant, and importantly, it will not be of the same orientation throughout the path of light. To determine the net effect of Faraday rotation along the line of sight, it is necessary to measure polarization at closely spaced frequency intervals over many frequencies. This has been done in the radio domain. Recently, Hovatta *et al.* (2012) showed that one can measure the transverse rotation gradient in the jet of 3C 273. This indicates a helical magnetic field enclosing and collimating the jet. Because the rotation affects the high frequencies the least, the best way to get an estimate of the intrinsic polarization of a synchrotron source is to measure at high frequencies. This makes X-ray polarization measurements an interesting prospect. As we will see in Section 5.5, AGN are prominent X-ray emitters, and in the cases where synchrotron emission stretches into the X-ray band, future polarization measurements could provide a valuable tool to probe the intrinsic properties of the synchrotron emitting plasma.

A first attempt to launch an X-ray polarimeter into space is the Gravity and Extreme Magnetism SMEX (GEMS) mission (Swank *et al.*, 2010). GEMS is a NASA satellite project currently planned to launch in 2014. It should be able to measure polarization of bright X-ray sources in the 2–10 keV energy band, among them a modest number of bright Seyfert galaxies and quasars.

2.2.4 Synchrotron Self-Absorption

When deriving the spectrum of synchrotron emission in Section 2.2.1 we assumed an optically thin plasma and that the emitted radiation will travel unabsorbed all the way to the observer. Under this condition, we saw that the observed spectrum will have the form $\epsilon(\nu) = \nu^{-\alpha}$ with α being a function of the power-law slope of the electron energy distribution only. Based on this correlation, one would expect a continuously rising spectrum as one moves to lower and lower frequencies. In reality this is not observed and the spectrum breaks at a certain frequency.

In AGN the main reason for this effect is absorption within the plasma of the compact emitting source. The source itself will be partially optically thick. A derivation of the formalism involved can be found for example in Rybicki and Lightman (1986), but here we will discuss only general arguments which explain the effect we see in the synchrotron spectra of astrophysical sources.

We assumed a power-law distribution in energy of the emitting electrons. If we allow for self-absorption, the synchrotron light produced in the plasma can be then absorbed by the same population of electrons. This will be most efficient when the frequency of the synchrotron radiation is close to the critical frequency ν_c of the absorbing electron, as described in Eq. (2.51). The frequency ν_c is proportional to the square of the electron's energy $\nu_c \propto E^2$. Toward higher energies, the density of electrons falls off with a power law with slope $-p$, thus, the chance for a photon to escape is larger the larger its energy, and the larger the emissivity $\epsilon(\nu)$ of the synchrotron radiation. For low frequencies on the other hand, the emitting plasma will be optically thick. This effect of self-absorption being higher for lower frequencies is competing with the intrinsic increase of the emissivity toward lower frequencies as shown in Figure 2.3. Deriving the spectrum which accounts for the effect of synchrotron self-absorption requires a knowledge of the absorption coefficient κ_ν as a function of frequency. We also assume that both, the magnetic field and the population of electrons are isotropic. Then, the absorption coefficient turns out to be:

$$\kappa_\nu = \frac{\sqrt{3}e^3}{8\pi m}\left(\frac{3e}{2\pi m^3 c^5}\right)^{\frac{p}{2}} \Gamma\left(\frac{3p+2}{12}\right)\Gamma\left(\frac{3p+22}{12}\right) k_\mathrm{e} B_\perp^{\frac{p+2}{2}} \nu^{-\frac{p+4}{2}} \tag{2.65}$$

where Γ is the gamma function and k_e is the normalization of the electron distribution. The optical depth τ is defined as

$$\tau = \int \kappa_\nu \mathrm{d}s \tag{2.66}$$

The path length in order to reach the optically thick case ($\tau \geq 1$) is roughly inversely proportional to the absorption coefficient κ_ν. The higher the frequency, the further the synchrotron emission can travel unaltered. The effective emissivity, also called the source function, in the optically thick case with $\tau = 1$ is then

$$\epsilon_{\tau=1} \simeq \frac{\epsilon(\nu)}{\kappa_\nu} \propto \frac{\nu^{\frac{1-p}{2}}}{\nu^{-\frac{p+4}{2}}} = \nu^{5/2} \tag{2.67}$$

Thus, for low frequencies the optically thick case will apply and the spectrum will rise proportionally to $\nu^{5/2}$, while at high frequencies in the optically thin domain, the spectrum will fall with $\nu^{(1-p)/2}$. Note that the slope of the optically thick part of the spectrum does not depend on the energy distribution of the electrons, while the optically thin branch is a function of the power-law index p of the electron energy distribution. The optically thick part of the spectrum will dominate until the frequency of the synchrotron radiation reaches a critical value ν_sa at which $\tau = 1$. Above this frequency photons travel further than the size of the emitting region before being absorbed. The transition frequency ν_sa where the emitting region changes from being optically thick to optically thin is (e.g., Marscher, 1987):

$$\nu_\mathrm{sa} \propto B^{1/5} f^{2/5}(\nu_\mathrm{sa}) \Phi^{-4/5} \tag{2.68}$$

The spectrum reaches its maximum at ν_sa, which is a function of the magnetic field strength, the observed flux at the transition frequency $f(\nu_\mathrm{sa})$, and the observed angular size Φ of the emitting region. The resulting flux as a function of frequency then follows (e.g., Stevens *et al.*, 1995)

$$f_\nu = C\left(\frac{\nu}{\nu_\mathrm{sa}}\right)^{5/2}\left\{1 - \exp\left[-\left(\frac{\nu}{\nu_\mathrm{sa}}\right)^{-\frac{p+4}{2}}\right]\right\} \tag{2.69}$$

where C is a constant and corresponds to the synchrotron flux in the optically thin case:

$$C = f_\nu^\mathrm{thin}(\nu_\mathrm{sa}) \tag{2.70}$$

An example of a resulting synchrotron spectrum accounting for the effect of synchrotron self-absorption is shown in Figure 2.4.

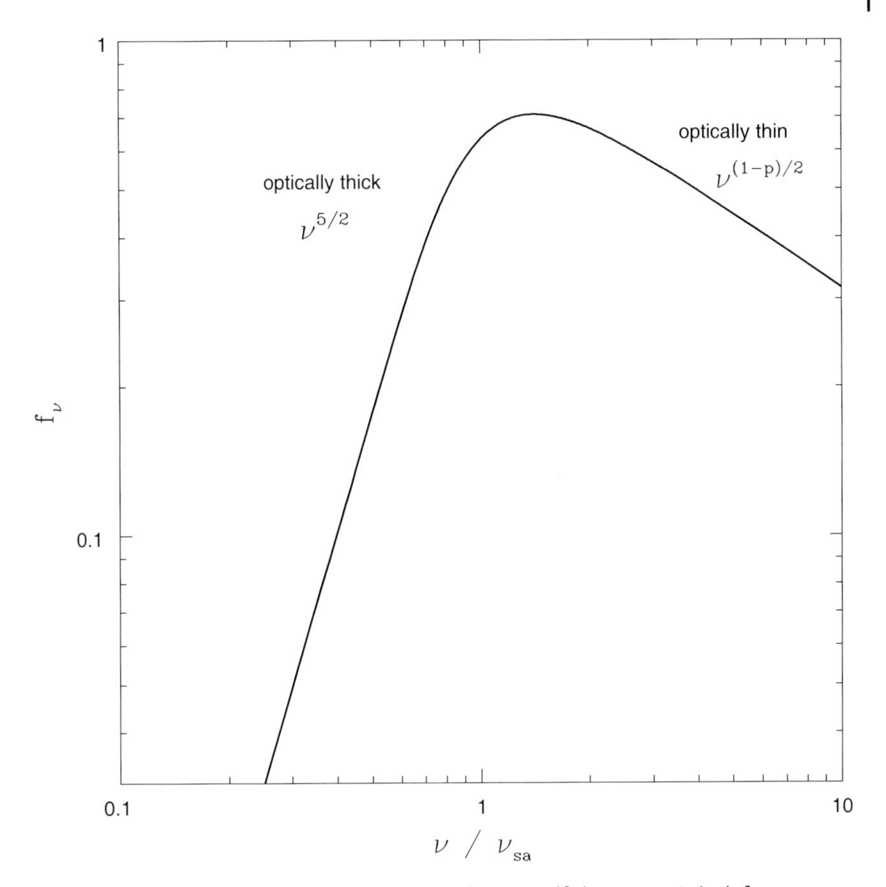

Figure 2.4 The synchrotron spectrum for an electron plasma within a uniform magnetic field following Eq. (2.69) as a function of frequency in units of the synchrotron self-absorption frequency ν_{sa}. At low frequencies the plasma is optically thick, causing synchrotron self-absorption. At high frequencies the spectrum is a function of the energy distribution of the electrons and falls with $f(\nu) \propto \nu^{(1-p)/2}$ where p is the power-law index of the electron energy distribution, which we have set to $p = 2$ in this example.

The electrons have a nonthermal spectrum. Nevertheless, it can be instructive to assign a characteristic effective temperature, because the synchrotron radiation is emitted preferentially by electrons at a certain frequency (see Figure 2.4). The brightness temperature is defined as the equivalent black body peak temperature for the measured emission. For an optically thick source the brightness temperature T_b is defined by:

$$I_\nu = \frac{f_\nu}{\pi \Phi^2} = B_\nu(T) = \frac{2k T_b \nu^2}{c^2} \tag{2.71}$$

where f_ν is the flux observed at frequency ν and Φ is the angular radius of the source, k is Boltzmann's constant, and I_ν is the intensity, i.e., the emitted energy per unit time, unit area, solid angle, and frequency. Because we treat the emission as if it originates from a perfect black body, $I_\nu = B_\nu(T)$, where $B_\nu(T)$ is the Planck function, derived by Max Planck in 1900. The kinetic temperature of the electrons is

$$T_{\text{kin}} = \frac{\gamma m_e c^2}{3k} \tag{2.72}$$

From thermodynamics we know that the energy assigned to the radiation cannot exceed the kinetic energy of the electrons, thus $k T_b \leq k T_{\text{kin}}$. It can be shown that this condition is equivalent to

$$T_b \lesssim \left(\frac{B}{1\,\mu\text{G}}\right)^{-1/2} \sqrt{\frac{\nu}{1\,\text{GHz}}}\, 10^{12}\,\text{K} \tag{2.73}$$

As we have seen already in Eq. (2.68), the peak frequency ν_{sa} of the synchrotron spectrum is not a function of the energy index p of the electron energy distribution. Thus, also the synchrotron brightness temperature T_b of a source is independent of the slope of the electron energy spectrum.

2.2.5 Synchrotron Self-Compton

In Section 2.1.3 we saw that a photon of energy E_1 can be up-scattered by relativistic charged particles of Lorentz factor γ to a higher energy $E_2 \simeq \gamma^2 E_1$. This inverse Compton scattering can push radio and infrared photons to X-ray or gamma-ray energies, depending on how relativistic the particles are. The synchrotron self-Compton (SSC) process describes the case in which the seed photons of the inverse Compton scattering are provided by the synchrotron emission itself. These photons are then up-scattered by the same electron population responsible for the synchrotron emission. In order for the photons to interact with the electron population, the plasma has to be optically thick. While the luminosity of inverse Compton scattering is proportional to the energy density of the photon field (Eq. (2.23)), the luminosity of the synchrotron radiation (i.e., the photon field for the inverse Compton process) is proportional to the magnetic field density (Eq. (2.49)):

$$\frac{L_{\text{SSC}}}{L_{\text{sync}}} = \frac{U_{\text{rad}}}{U_B} \tag{2.74}$$

The ratio between the inverse Compton versus synchrotron radiation flux for a spherical source with radius R, following Eqs. (2.67) and (2.23) as shown in Ghisellini *et al.* (1993), turns out to be

$$\frac{f_\nu^{\text{IC}}}{f_\nu^{\text{sync}}} \simeq k(\alpha) R n_0 \ln\left(\frac{\nu_{\max}}{\nu_{\text{sa}}}\right)\left(\frac{\nu_{\text{sync}}}{\nu_{\text{IC}}}\right)^\alpha \tag{2.75}$$

Here α is the spectral index in the radio band, $k(\alpha) \simeq 0.08\alpha + 0.14$ (Ghisellini *et al.*, 1993), $\nu_{\max}$ is the maximum frequency reached by the synchrotron radiation as defined by the maximum Lorentz factor $\gamma_{\max}$, ν_{sa} is the frequency where the synchrotron emission reaches its maximum (Eq. (2.68)), ν_{sync} is the frequency where we measure the synchrotron (radio) flux, and ν_{IC} is the frequency at which we measure the inverse Compton component in the X-ray or gamma-ray band. The normalization n_0 refers to the electron energy distribution of the form:

$$n = \int_{\gamma_{min}}^{\gamma_{max}} n_0 \gamma^{-p} d\gamma \tag{2.76}$$

where γ is the Lorentz factor of the electrons.

Also the up-scattered photons of the inverse Compton process can contribute again to the photon field's energy density U_{rad}, leading to multiple synchrotron self-Compton scattering. This effect is very sensitive to the source brightness, and above a certain brightness temperature the inverse Compton losses become so significant that they cool the electrons efficiently.

In the previous section we saw that the brightness temperature of the synchrotron emission does not significantly exceed $T_b \sim 10^{12}$ K, because above that temperature synchrotron self-absorption dominates. Kellermann and Pauliny-Toth (1969) derived for the ratio between inverse Compton branch and synchrotron emission the expression:

$$\frac{L_{IC}}{L_{sync}} = \left(\frac{T_b}{T_{thresh}}\right)^5 \left[1 + \left(\frac{T_b}{T_{thresh}}\right)^5\right] \tag{2.77}$$

With $T_{thresh} \simeq 10^{12}$ K, we see that as soon as $T_b > T_{thresh}$, the inverse Compton branch should dominate very rapidly, with $L_{IC}/L_{sync} \propto T_b^{10}$. Slightly exceeding the threshold temperature should result in an emergent inverse Compton branch with little or no synchrotron emission. This effect is often called the *inverse Compton catastrophe*.

Apparently some mechanism is limiting the range of possible brightness temperatures in the electron plasma in AGN to a value $T_b \lesssim T_{thresh} \sim 10^{12}$ K. Sources might still appear "hotter" when they are Doppler boosted towards the observer as we will discuss in Section 5.8.1.1. For example, the core of the blazar 3C 279 appears to have a brightness temperature of $T_b = 10^{13}$ K at 22 GHz (Wehrle *et al.*, 2001). Several explanations for the brightness temperature limit have been brought forward, see for example Readhead (1994) or Tsang and Kirk (2007).

3
The Central Engine

Very soon after the interpretation of 3C 273 as a cosmologically redshifted object astrophysicists faced the dilemma of explaining its remarkable energetic output. The observed variability time scales limited its size to dimensions of order light days. Thus, given that its inferred luminosity $L \sim 10^{47}$ erg s^{-1} is approximately one million times that of our own galaxy, it was immediately clear that some mechanism other than a dense conglomeration of stars had to be involved. The idea of a central massive object and accretion-powered radiation quickly emerged, primarily due to the lack of any other tangible explanation. Hoyle and Fowler (1963) put forth the general concept considering scenarios where a "hyperstar" accreted from its ambient medium. More detailed models soon followed: Salpeter (1964), Zel'Dovich (1964), Lynden-Bell (1969).

By the early years of the next decade the accreting black hole paradigm was widely accepted. In subsequent years it has been widely substantiated by a large body of observational evidence as we discuss in Section 3.1.2. Estimates of AGN black hole masses from several independent methods are now available in the published literature and we will discuss those results and their implications as well.

3.1
The Black Hole

It has long been assumed, primarily because no other plausible explanation has emerged, that AGN must be powered by accretion onto massive black holes at the dynamical centers of their host galaxies. These central black holes are believed to have masses upwards of a million times that of the sun. The AGN are both compact, extremely luminous and often most variable at the shortest observed wavelengths. Thus, a mechanism such as accretion that can provide highly efficient conversion of potential and kinetic energy to radiation is needed as is a large mass to accommodate high persistent Eddington luminosities (Eq. (3.9)). Indeed, central supermassive black holes are now believed to reside at the center of most galaxies and AGN-like behavior is expected whenever a supply of material comes within a critical distance of the central black hole.

Active Galactic Nuclei, First Edition. Volker Beckmann and Chris Shrader.

3.1.1
Approaching a Black Hole

The most basic characteristic defining a black hole is the presence of an *event horizon* – a boundary in through which matter and light can fall inward towards the black hole, but can never re-emerge. Mathematically stated, the gravitational escape velocity equals the speed of light at the event horizon surface which occurs at a radius

$$R_S = \frac{2G M_{BH}}{c^2} \tag{3.1}$$

Any information resulting from an "event" occurring within the boundary defined by that radius cannot be communicated to an outside observer, thus the origin of the terminology.

A detailed treatment of the environment, or the "space time" ambient to a black hole requires general relativity. However, many of the concepts relevant to an AGN study can be formulated from Newtonian approximations or discussed qualitatively without a complete general relativistic treatment. For interested readers wishing to learn more about general relativity, Misner *et al.* (1973) remains one of the best and most complete references on that subject.

Several discrepant effects would become evident to an observer approaching a black hole who remained in communication with a second, distant observer far away from the black hole. In the region surrounding a black hole, space is distorted in such a way that the paths followed by particles tend to be deflected towards the black hole. This would be evident to the distant observer. At the event horizon, this deflection becomes overwhelming in that there are no paths leading away from the black hole.

The rate at which time passes would also be perceived differently by the two observers. The distant observer would conclude that clocks near a black hole appear to tick more slowly than those in his own vicinity. This effect, known as gravitational time dilation, causes an object falling into a black hole to appear to slow down as it approaches the event horizon. In fact it would take an infinite time to reach it. The apparent rates of all processes occurring on this object would slow down causing, for example, light that it emits to appear redder since its characteristic frequencies would decrease. It would also appear dimmer since the number of photons emitted per second would be reduced. This effect is known as gravitational redshift. Specifically, the ratio of emitted frequency to that observed far from the black hole is

$$\frac{\nu_0}{\nu} = 1 - \frac{G M_{BH}}{c^2 r} \tag{3.2}$$

where r is the distance from the black hole. Just before it reaches the event horizon, the falling object becomes so dim that it can no longer be seen.

The perception of the observer falling into a black hole is quite different. He does not notice any of these effects (although he would be physically stretched by the tidal force effects of the extreme gravitational field!). According to his clock, he would cross the event horizon within a finite time, but he would be unable to determine exactly when he crosses it.

3.1.2
Evidence for Black Holes in AGN

As previously discussed, the idea that black hole accretion is responsible for the energy budgets of AGN emerged soon after they were established by observations as very luminous, cosmologically distant objects. This idea quickly gained acceptance by a majority of astrophysicists primarily due to the lack of any plausible alternative explanation. However, actual observational evidence for black holes in AGN took much longer to emerge. It is now not only widely accepted that AGN are powered by central black holes but that most galaxies, including our own, have massive central black holes.

The problem is that since a black hole does not emit any radiation it cannot be directly observed. Instead, its presence must be inferred through its interaction with matter in its surrounding environment. For example, by tracking the movement of a star or group of stars in orbit around a putative black hole and observing that their trajectories have been modified, its presence can be inferred and its mass can be estimated. In a small number of cases, there are compact radio sources associated with OH masers in close proximity to the galactic dynamical center allowing the superior resolving power of radio telescopes to be applied to the problem leading to even stronger constraints. Another possibility is to detect radiation from in-falling matter as we will discuss in Section 3.2. The gas spirals inward, becomes heated to extreme temperatures thus emitting thermal radiation that can be detected using ground-based and/or space-based telescopes. In particular, short time scale variations in this emission impose constraints on the physical scales involved. However, the information forthcoming is less complete than what can be obtained from using a stellar dynamics approach.

Two developments in the 1980s and 1990s are notable in establishing physical scales in the central regions of AGN and the probing centrally concentrated mass. The first of these is known as "reverberation mapping."

Reverberation mapping, described in more detail in Section 3.6 involves sampling the variability in the continuum and emission line fluxes on time scales expected to be smaller than the light travel time between the central black hole and the broad-line clouds. Cross-correlation analysis, combined with a continuum-to-line response function can then be used to infer the size of the broad-line region. That size, along with dynamical information inferred from the measure lines widths can be used to estimate the black hole mass by applying the virial theorem.

For a small subset of AGN, radio observations during the last several decades have revealed new and compelling evidence for the existence and characterization of AGN central black holes. This involves the discovery of water masers embedded in the orbiting material close to the central compact central objects. Intense maser action in cosmic molecular clouds was discovered in the 1960s (e.g., Weaver *et al.*,

1965) from observations of molecular lines of OH, H_2O, SiO, and CH_3OH. In the case of water vapor, the commonly observed masers emit in the centimeter (or tens of GHz) range, notably the prominent feature at 1.35 cm. Masers are generally believed to be associated with star-forming regions but they can also be associated with evolved stars (e.g., Elitzur, 1992; Reid and Moran, 1988). With the aid of very long baseline interferometry (VLBI), which provides submilliarcsecond angular resolution at centimeter wavelengths as well as sufficient spectral resolution, typically of the order of a $\mathrm{km\,s^{-1}}$ or less, the structure and dynamics of the accreting material can be studied in unprecedented detail (Goulding *et al.*, 2010; Ishihara *et al.*, 2001). For certain objects – for example NGC 4258, NGC 4945, NGC 1068, IC 2560 and the Circinus Galaxy – this has led to direct observational confirmation of the existence of disks as well as to constraints on the total mass and mass within the region enclosed by the maser orbit. Water vapor masers in the nuclei of these AGN can provide precise black hole masses with uncertainties $\lesssim$ 20%. The artists' conception and schematic is shown in Figure 3.1, and actual data for a particular galaxy are shown in Figure 3.2, illustrating the application of the method and its potential power.

A major discovery in the 1990s revealed that the mass of a galaxy's central black hole is closely related to the surrounding spheroidal distribution of stars, that is, to the so-called galactic bulge (Magorrian *et al.*, 1998). Methods for estimating M_{BH} quickly emerged and are now common in the literature (e.g., Ferrarese and Merritt, 2000; Gebhardt *et al.*, 2000; Greene and Ho, 2006). The central black hole mass is believed to be proportional to the bulge mass, which in turn scales directly with an observable quantity: the stellar velocity dispersion. The latter is often denoted as σ, thus the so-called M–σ relation is now a common part of astrophysics lexicon. The relation can be expressed as

$$\frac{M_{\mathrm{BH}}}{10^8\,M_\odot} = \alpha\left(\frac{\sigma}{200\,\mathrm{km\,s^{-1}}}\right)^{\beta} \tag{3.3}$$

The values of α and β found in the literature vary by small amounts depending on the surveys from which they are derived. For example Gebhardt *et al.* (2000) find $\alpha = 1.2 \pm 0.2$ and $\beta = 3.75 \pm 0.30$. More recently, Gültekin *et al.* (2009) find $\alpha = 1.312 \pm 0.08$ and $\beta = 4.24 \pm 0.41$ based on a sample of about 50 measurements. The correlation is shown in Figure 3.3.

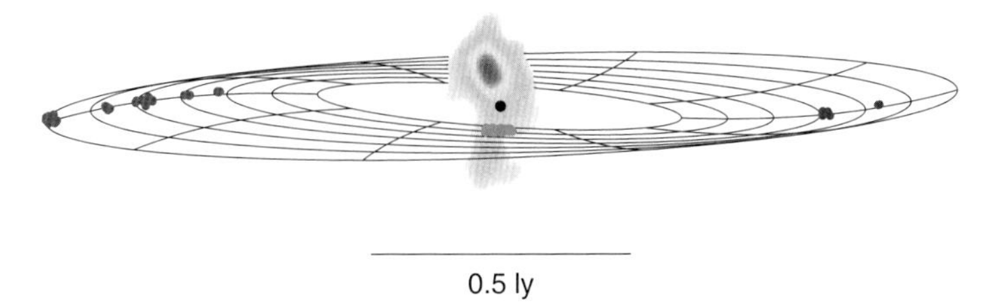

Figure 3.1 The warped disk modeled to measure maser positions. The location of the supermassive black hole is indicated by the black dot in the center and the extended radio emission is shown in contours (Moran *et al.*, 1999).

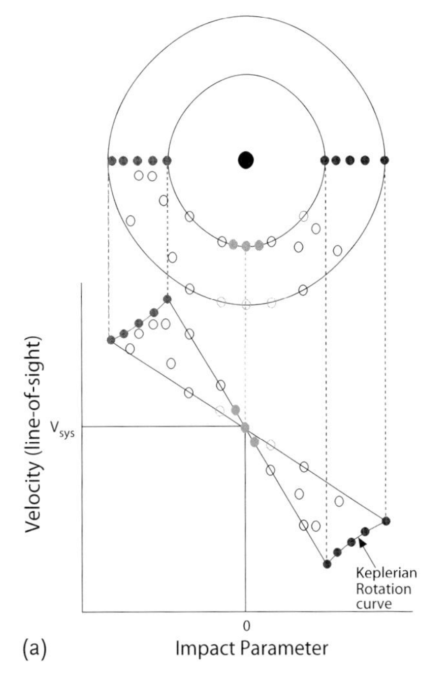

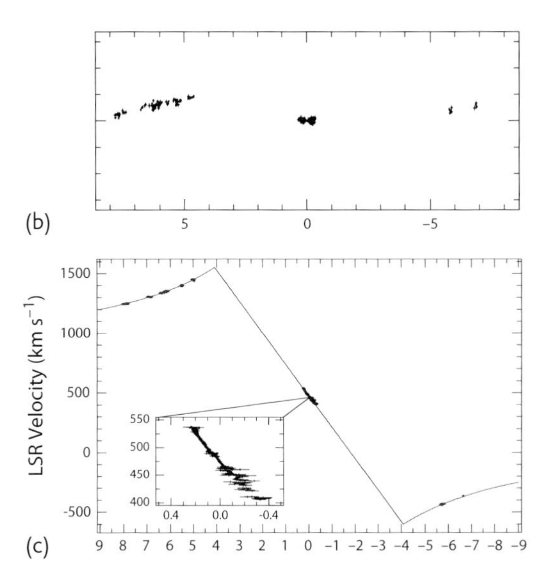

Figure 3.2 How the maser embedded disk sources lead to a rotation curve (a), which can be interpreted in terms of simple Keplerian dynamics. An image of the maser sources relative to the dynamical center of the galaxy (b). The tick marks represent 1 mas. The actual rotation curve for the Seyfert 2 radio galaxy NGC 4258 (c) (Moran *et al.*, 1999).

It should be noted that further refinements are likely in the future. For example, recent work by Gebhardt *et al.* (2011) using adaptive optics techniques has led to a very precise black hole mass determination for M87 (NGC 4486). Those authors applied axis-symmetric orbital models to the stellar kinematic measurements to derive a mass of $(6.6 \pm 0.4) \times 10^9\ M_\odot$. This value lies well above the M–σ prediction, thus suggesting a need for calibration refinements of the high-mass end of the relation.

The existence of the M–σ relation and the relatively low amount of scatter seen poses an intriguing question: how do the black hole and the stars populating the bulge "communicate" with each other? It has been suggested that some kind of "feedback" mechanism acts to regulate the black hole mass-stellar velocity dispersion relationship. Assuming approximate spherical symmetry, an individual star in the outer bulge experiences the gravitational mass its orbit encloses independent of the distribution of that mass. Furthermore, one might expect processes like accretion onto the central black hole or galaxy–galaxy interactions to increase the M–σ diagram's scatter over time. One possible feedback mechanism was suggested by Silk and Rees (1998) and has subsequently been expanded upon by others. They proposed that a supermassive black hole first forms through the collapse of giant gas clouds on rapid enough time scales so that it precedes significant star forma-

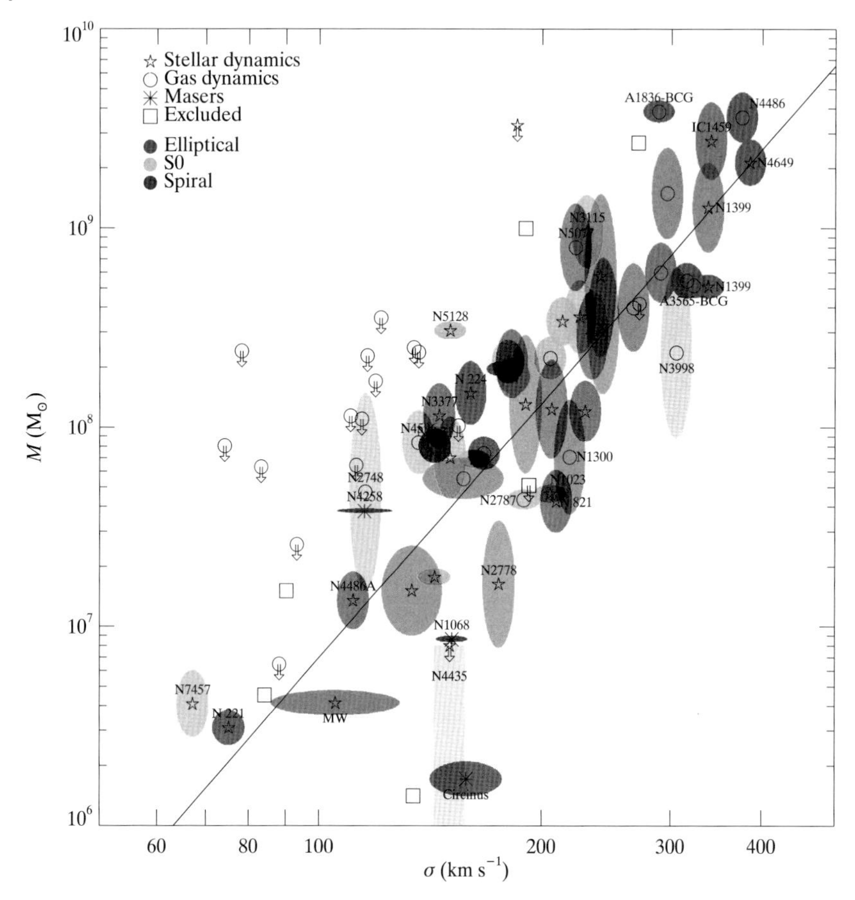

Figure 3.3 Pictured is a recent calibration of the $M-\sigma$ relation from Gültekin *et al.* (2009). The colors indicate different galaxy types and the symbols (star, circle and asterisk), indicate the type of black hole measurement. For a color version of this figure, please see the color plates at the beginning of the book.

tion in the surrounding bulge. The black holes could then accrete matter from its ambient environment. The in-falling matter outside the Schwarzschild radius could then emit radiation which could in turn drive a wind. This wind could then in effect regulate the accretion flow. The flow would cease if the rate of deposition of mechanical energy into the in-falling gas becomes large enough to disrupt the protogalaxy in one crossing time. Their model predicts a slope for the M–σ relation of $\alpha = 5$, which is somewhat larger than what is observed but not unreasonably so. They also predict the approximately correct normalization, suggesting that at least qualitatively this type of scenario is a plausible explanation. Black hole masses in thousands of galaxies have been estimated using the M–σ method. The M–σ relation in turn can be used to calibrate secondary and tertiary mass estimation techniques. In particular, methods which relate the black hole mass to the strength

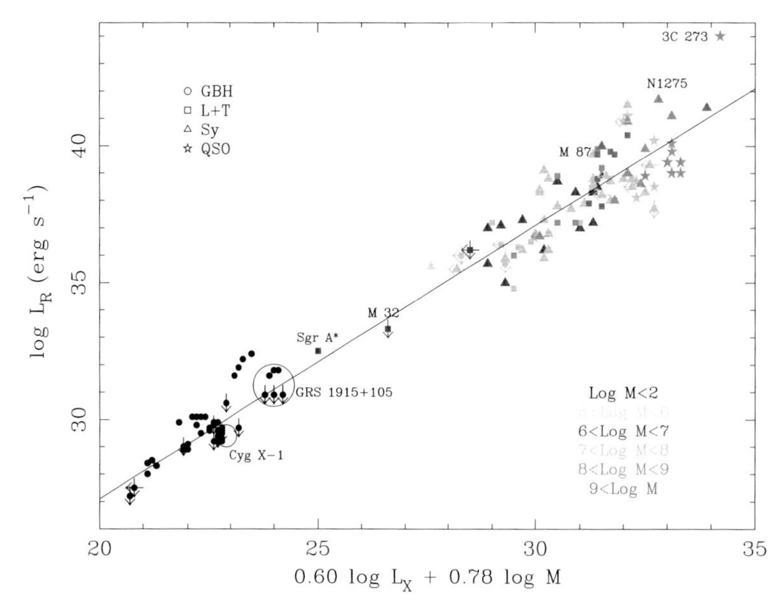

Figure 3.4 The so-called fundamental plane for accreting black holes with which jet outflows are also associated (Merloni *et al.*, 2003). The observable quantities, L_R and L_X are proxies for the jet and disk power, respectively. Objects with mass determinations from other methods were used in constructing this diagram. The tight correlation over many decades of central black hole masses suggests a commonality in the physics underlying the disk-accretion and jet launching mechanism across object subclasses. For a color version of this figure, please see the color plates at the beginning of the book.

of readily measurable emission lines. These can emanate from hot gas in the nuclear region or from the velocity dispersion associated with gas in the bulge.

A recent development has revealed another relationship between observable AGN properties and the central black hole mass. Heinz and Sunyaev (2003) first demonstrated that accepting the assumption that the jet formation process is qualitatively equivalent among the massive black holes in AGN and stellar-mass galactic black holes, one can derive a universal scaling between the radio luminosity, which serves as a proxy for the jet power, and both mass and mass-accretion rate. This relation is independent of details of the jet model. Merloni *et al.* (2003) used X-ray luminosity as a proxy for the disk luminosity and used a sample of objects with known black hole mass to construct the diagram shown in Figure 3.4. In essence this represents jet power versus a mass-adjusted disk luminosity. The tightness of this correlation over nearly seven decades in black hole mass is highly suggestive of a common underlying physics for the broad class of radio-loud AGN and the BHXRBs. It is now commonly referred to as the *fundamental plane*.

3.1.3 Gravitational Field Near a Black Hole: the Schwarzschild Metric

The gravitational field near a nonrotating black hole was first described in the context of Einstein's theory of General Relativity by the German astrophysicist Karl Schwarzschild in 1916. The solution Schwarzschild derived involves a spherical surface centered on the black hole known as the event horizon. This surface, defined by the radius $R_S = 2GM_{BH}/c^2$ and referred to as the "Schwarzschild radius," is the distance from the black hole at which the escape velocity is equal to the speed of light (Figure 3.5). Thus, no matter or electromagnetic radiation contained within can escape. The space-time surrounding the black hole event horizon is described by the space-time line element or "metric" as

$$\mathrm{d}s^2 = \left(1 - \frac{R_S}{r}\right)^{-1} \mathrm{d}r^2 + r^2(\mathrm{d}\theta^2 + \sin^2\theta\, \mathrm{d}\phi^2) - \left(1 - \frac{R_S}{r}\right) c^2 \mathrm{d}t^2 \qquad (3.4)$$

One consequence of the space time solution described by the Schwarzschild metric, is that the proper time, that is, the a time interval measured by an observer at a radial distance r from the black hole is $(1 - R_S/r)^{1/2}\mathrm{d}t$. This is less that the time interval $\mathrm{d}t$ that a distant observer would infer for the same event. Thus a distant observer will observe the clock of an observer at radius r to run more slowly than the distant observer's own clock, by a factor $(1 - R_S/r)^{1/2}$. This time dilation factor tends to zero as r approaches the Schwarzschild radius R_S. This implies that an event occurring at the Schwarzschild radius will appear to freeze to a stop, as observed by anyone outside the Schwarzschild radius. This gravitational time dilation also leads to a gravitational redshift of electromagnetic radiation. That is, a distant observer will observe radiation emitted from the vicinity of the black hole event horizon to be redshifted to lower frequencies, or equivalently to longer wavelengths.

At large distances from the black hole the R_S/r terms effectively vanish and the metric becomes indistinguishable from the Newtonian case. However, in the study of AGN there are regimes where the general relativistic effects are important, notably in calculations used to model accretion disks and the emergent line and continuum radiation they produce. In particular, calculations of the structure and thermodynamic properties of the innermost regions of an accretion disk, from which the highest-energy radiation emitted by the disk emanates, require a relativistic treatment.

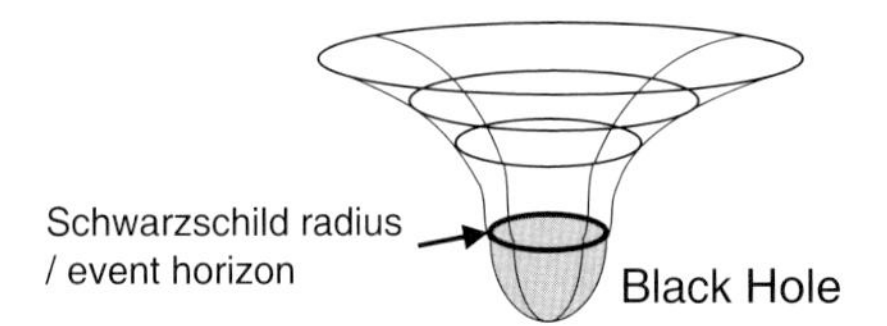

Figure 3.5 Schematic representation of the gravitational potential of a nonrotating black hole.

3.1.4
Rotating Black Holes: the Kerr Metric

In general relativity theory, the Kerr metric describes the space time geometry in the vicinity of a rotating massive object such as the central black hole of an AGN. Roy Kerr derived this solution in the early 1960s. Rotating black holes are formed in the gravitational collapse of a collection of stars or gas which would almost certainly have nonzero angular momentum. Thus it is reasonable to expect that most AGN black holes, or astrophysical black holes in general, are of the Kerr type. There are several implications of this in the context of our efforts to understand AGN. For one thing, it is possible to extract large amounts of energy from the rotating black hole at the expense of its rotational energy through the so-called Penrose process. This is suspected to play a role in the launching of AGN jets, which exhibit enormous energetic outputs. A detailed discussion of the Penrose mechanism is beyond the scope of this book, but it involves the spinning black hole's so-called ergosphere; an ellipsoidal region just outside its event horizon. Matter within the ergosphere corotates with the black hole, but because that region is outside the event horizon the matter can escape and carry energy drawn from the black hole with it. For details see for example Misner *et al.* (1973); Penrose (1969). Although the detailed physics of jet formation and propagation remain very much a work in progress, the suspicion that spinning black holes are the underlying energy source has persisted as a central theme in modeling effort for decades.

Another important aspect of Kerr black holes pertaining to AGN research has to do with the accretion disk's inner radius, believed to be related to the innermost stable Keplerian circular orbit. This radius is smaller for a spinning black hole of a given mass compared to the disk surrounding a nonspinning black hole of the same mass. This effects both the continuum emission from the disk and the profiles of any emission lines formed in the inner region of the disk. The issue is that the inner edge of the accretion disk reaches closer to the black hole in the case of the Kerr black hole than for the Schwarzschild case. Specifically, if expressed in units of the terms of the Schwarzschild radius $R_S = 2GM_{BH}/c^2$, the inner disk radius lies at $r = 1.23R_S$ for a maximally rotating black hole and $r = 6R_S$ for a zero-angular momentum black hole. Since the local temperature within the disk is predicted by basic theory to scale radially as $T \propto r^{-3/4}$ this leads to a different spectral energy distribution and intensity for the emergent radiation from the disk. In AGN the peak-disk temperature is usually in the unobservable far-ultraviolet spectral region, but efforts to rigorously model the emission need to take the choice of metric (and thus the radial extent of the inner disk) into consideration. For example, multiepoch observations of the quasar Q0056-363 revealed dramatic variation in the near UV and soft X-ray intensity as well as changes in the spectral energy distribution (SED) from a thermal-like to a more power-law-like configuration. This was interpreted by Matt *et al.* (2005) as a truncation of the accretion disk inner edge. We note that this effect is more dramatically evident in the SED of galactic X-ray black hole binaries, which have much hotter accretion disks, peaking in the $\sim$ 1 keV region where the continuum can be accurately measured. In fact, models of the X-ray continuum from those objects have been applied to observational data to estimate the black hole specific angular momentum (e.g., McClintock *et al.*, 2011).

More profoundly for AGN is the effect of black hole angular momentum, or spin, on the shapes of certain emission line profiles. In particular, the fluorescent Fe K_α line at 6.4 keV is believed to be formed in the inner accretion disk in response to irradiation by Compton scattered X-rays. It has long been known that lines formed in a disk far from the black hole have a double-peaked profile resulting from the Doppler effect. Closer to the black hole, the symmetric double-peaked profile appears distorted due to relativistic effects. The shape of the iron line profile is determined by the black hole mass and angular momentum, and thus the radial extent of the disk. It also depends on the emissivity profile of the disk, and the viewing geometry, specifically the disk inclination angle relative to the observer's line of sight. A schematic representation of the effect is shown in Figure 3.6.

A complete model for the iron line must also take into account the effects of the Doppler shift, relativistic beaming, and general the effects of gravitational redshift. Thus, the black hole spin is critical to a precise model calculation. In fact, there have now been many attempts to invert the problem so to speak, and use line profile models to estimate AGN black hole spins (e.g., Brenneman and Reynolds, 2009). The first indication that AGN K_α line profiles can be relativistically modified came from observations of MCG-6-30-15 obtained using the ASCA satellite (Tanaka *et al.*, 1995). The emission line was found to be extremely broad, with a width indicating velocities of about $0.3c$. They observed a marked asymmetry towards energies lower than the 6.4 keV rest energy of the line. It was asserted that this asymmetry was most likely caused by gravitational and relativistic Doppler shifts near the central black hole at the center of the galaxy. Since then, many observations using more sensitive X-ray telescopes seem to have confirmed this interpretation, although skepticism persists (Miller *et al.*, 2009). We will come back to this topic in Section 5.5 when discussing the appearance of AGN in the X-ray domain.

3.2 Accretion Processes

As we have discussed in previous sections, the accretion of matter onto a central massive black hole has long been postulated to power the enormous radiative output of AGN. Details of the exact nature of this accretion flow remain to be worked out, however, disk accretion scenarios are strongly favored by theoretical arguments (for a recent review of disk accretion the authors recommend King (2008)). Here we offer an overview of some possibilities that have been explored in the literature since the discovery of the AGN phenomenon.

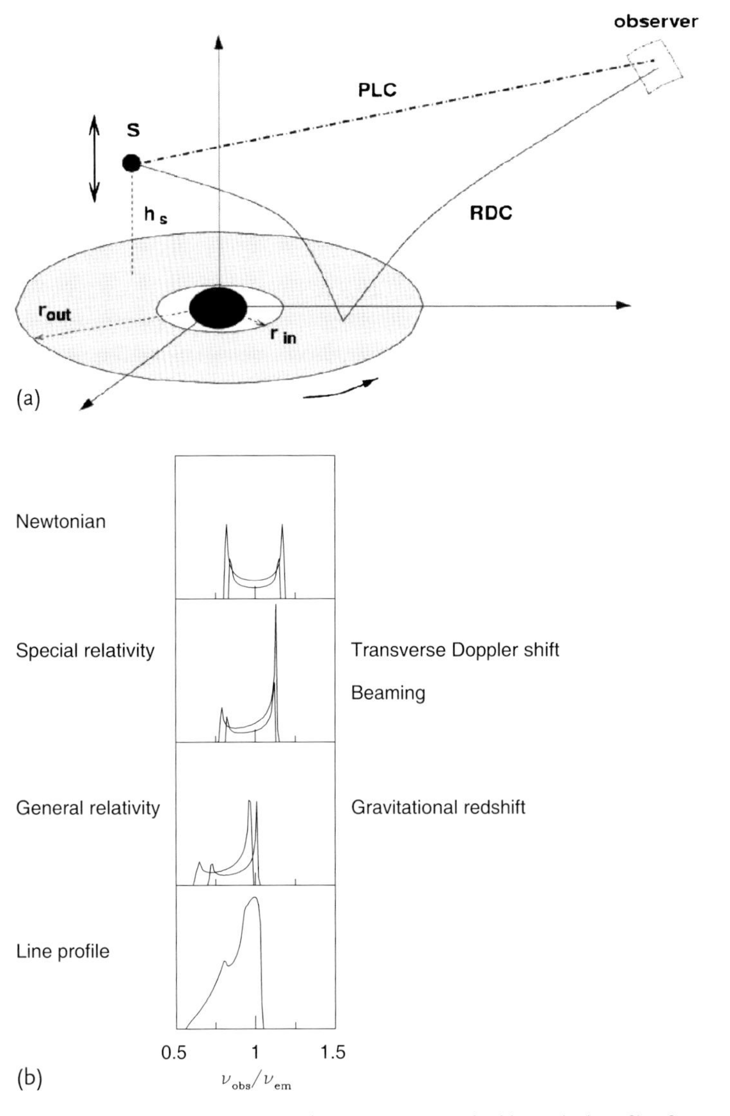

Figure 3.6 As illustrated in this schematic (reproduced from Fabian, 2008), the profile of the broad iron line is caused by the interplay of Doppler and transverse Doppler shifts, relativistic beaming and gravitational redshifting. The light path from the source S at height *h* close to the black hole causes the symmetric double-peaked profiles from two narrow annuli on a nonrelativistic disk (a). The direct power-law component (PLC) and the reflection-dominated component (RDC) are indicated. The effects of transverse Doppler shifting and relativistic beaming have been included, as well as gravitational effects (b).

3.2.1
Accretion Basics: Bondi Accretion and the Eddington Limit

As pointed out in the previous section, the basic mechanism underlying the AGN central engine is accretion. Matter falls onto a compact object – a black hole with a mass in the range of a million to a billion times that of the sun – leading to the conversion of gravitational potential energy to electromagnetic radiation. The simplest configuration one might consider is an approximately spherically symmetric accretion flow onto the black hole. This could occur for example from a uniform ambient wind (an analogous situation could occur when a neutron star or black hole of a few stellar masses move through a uniform interstellar medium). The black hole will accrete matter at an approximate rate

$$\dot{M} = \pi r^2 \rho v \tag{3.5}$$

where ρ and v are the wind density and velocity. This type of "spherical accretion" is often referred to as Bondi accretion (also called *Bondi–Hoyle* accretion) as the first detailed calculations were worked out by the physicist Herman Bondi in the early 1950s (Bondi, 1952). The effective capture radius for accretion is then approximately equal to the escape velocity for a particle at distance R from the black hole to the wind velocity: $\sqrt{2GM_{\mathrm{BH}}/R} = V$ or $R = 2GM_{\mathrm{BH}}/V^2$. The accretion rate is then

$$\dot{M} = \frac{4\pi\rho G^2 M_{\mathrm{BH}}^2}{V^3} \tag{3.6}$$

This is an approximate relation for Bondi's more detailed formulation, but it is sufficient to provide physical insight. Ultimately, accretion onto a compact object is limited by the effects of the radiation pressure experienced by the in-falling plasma. This limit, first pointed out by Arthur Eddington in the 1920s depends on the mass of the compact object and the mean opacity of the in-falling material. Often in astrophysical environments it suffices to use the opacity of ionized hydrogen. The observable quantity corresponding to the critical mass-accretion rate for a source at a known distance is its luminosity, commonly referred to as the *Eddington luminosity*. This quantity is obtained by equating the pressure gradient of the in-falling matter,

$$\frac{\mathrm{d}P}{\mathrm{d}r} = \frac{-GM\rho}{r^2} \tag{3.7}$$

to the radiation pressure

$$\frac{\mathrm{d}P}{\mathrm{d}r} = \frac{-\sigma_{\mathrm{T}}\rho}{m_{\mathrm{p}}c} \frac{L}{4\pi r^2} \tag{3.8}$$

where M is the central object mass, $\sigma_{\mathrm{T}} \simeq 6.65 \times 10^{-25}\,\mathrm{cm}^{-2}$ is the electron or *Thomson* scattering cross-section (Eq. (2.3)) and m_{p} is the proton mass (here we have assumed that the in-falling plasma is predominantly ionized hydrogen). This leads to:

$$L_{\mathrm{Edd}} = \frac{4\pi G M m_{\mathrm{p}} c}{\sigma_{\mathrm{T}}} \simeq 1.3 \times 10^{38} \frac{M}{M_{\odot}} \,\mathrm{erg\,s^{-1}} \tag{3.9}$$

One then defines the *Eddington ratio* or *Eddington rate* λ in terms of the bolometric luminosity and the Eddington luminosity:

$$\lambda_{\mathrm{Edd}} = L_{\mathrm{bol}}/L_{\mathrm{Edd}} \tag{3.10}$$

The *Eddington limit* is reached when $\lambda = 1$. The Bondi accretion process is unlikely to be significant in powering AGN. The main reason for this is that the efficiency with which gravitational potential energy can be converted into emergent radiative energy is low. Radiation in an accretion flow is thermal in nature and is produced by the viscous heating of the plasma. The basic problem is that in the absence of angular momentum, that is, for pure Bondi accretion, the plasma falls onto the central compact object before it has time to radiate the bulk of its thermal energy. If the plasma is collapsed into a disk-shaped structure, as is expected to occur from basic dynamical arguments as long as it has some angular momentum, it can radiate far more efficiently.

It should be noted that while Bondi accretion is unlikely to power AGN, it should in principle produce an observable signature in nearby nonactive galaxies or the center of our own galaxy which harbors the $M_{\mathrm{BH}} \simeq 4 \times 10^{6}\,M_{\odot}$ black hole associated with Sgr A*. The central gas densities of these external galaxies are known to reasonable accuracy, as are in many cases the central black hole mass. However, the observed central X-ray luminosities are found to be orders of magnitude smaller than what would be predicted based on the Bondi accretion rate and assuming a $\sim$ 10% radiative efficiency. The same is true for Sgr A*, for which the surrounding region is readily resolvable with the Chandra X-ray observatory. These results imply that the radiative efficiency of the in-falling gas is exceedingly low and/or that much less material is accreted by the black hole than predicted by the Bondi mechanism.

Recently, Wong *et al.* (2011) have searched for a signature of Bondi accretion in the nearby ($\sim$ 10 Mpc), nonactive S0 galaxy NGC 3115. This galaxy is known from dynamical studies of its bulge region to contain a massive, $> 10^{9}\,M_{\odot}$ black hole and it is close enough to resolve the expected temperature profile within the Bondi capture radius using Chandra. They find their data to be consistent with a Bondi radius of about 4–5 arcsec (corresponding to 188–235 pc). This in turn suggests a black hole mass of about $2 \times 10^{9}\,M_{\odot}$, which is consistent with the values inferred from dynamical studies, and a mass accretion rate of $\dot{M} = 0.022\,M_{\odot}\,\mathrm{yr}^{-1}$. However, the measured X-ray luminosity is about 6 orders of magnitude less than expected for the case of a 10% radiative efficiency. Possible explanations include radiatively inefficient accretion flows such as the ADAF model discussed in Section 3.2.4.

3.2.2
Accretion and Viscous Dissipation in a Thin Disk

Dynamical arguments suggest that matter orbiting a central massive object will settle into a flattened structure (note the rings of Saturn), which herein we refer to as a disk, or if the matter spirals inwards onto the massive object an "accretion disk." In the case of AGN the central object is the massive black hole. Its gravitational field draws the material inwards, leading to viscous heating and thus in principle to an observable signature in the emitted radiation. But early models of disk accretion faced a basic problem: that the orbiting material must, through some unknown microphysics, lose angular momentum. The total angular momentum of the system is conserved, thus, the angular momentum lost due to matter falling onto the center has to be offset by an angular momentum gain of matter far from the center. This process is referred to as angular momentum "transport."

The basic physics underlying this angular momentum transport remains an active field of study to the present day. However, a remarkably useful approximate solution to the problem, applicable to geometrically thin but optically thick accretion disks within a constant rate of accretion, was proposed in the early 1970s (Shakura and Sunyaev, 1973). The resulting class of models are often referred to in the literature as "alpha disk" models as we discuss in what follows.

A basic assumption in the alpha-disk scenarios is that the energy from accreted material is dissipated within a small region at its radius r. Furthermore, since the disk media is optically thick, the emergent spectrum is that of a black body with a temperature $T(r)$. If we start with the simplistic assumption of a region of the disk releasing gravitational energy at a rate $G M \dot{M}/r$ we can apply the virial theorem to relate this to the local temperature. Specifically, one half of this should go into the kinetic (thermal) energy of the gas. For a local equilibrium to be maintained the other half is radiated away, thus

$$L = \frac{G M \dot{M}}{2r} = 2\pi r^2 \sigma T^4 \tag{3.11}$$

Here σT^4 is the black body radiation formula for the energy per unit area and $2\pi r^2$ is the disk area (top plus bottom). This leads to a temperature $T(M, \dot{M}, r)$ satisfying

$$T \propto (M \dot{M})^{1/4} r^{-3/4} \tag{3.12}$$

This turns out to be the same result one obtains following a more rigorous treatment in terms of the radial dependence of the characteristic disk temperature.

As we have alluded to, the early models of disk accretion faced a basic problem: that the orbiting material must, through some unknown process, lose angular momentum. The total angular momentum of the system must be conserved, thus, the angular momentum lost due to matter falling onto the center has to be offset by angular momentum gain of matter far from the center (thus angular momentum "transport" must occur). Shakura and Sunyaev (1973) proposed turbulence in the gaseous material of the disk as the source of increased viscosity enabling this process. Assuming subsonic turbulence and the disk height as an upper limit for the linear scale of the putative turbulent structures, the disk viscosity ν can be approximated as $\nu = \alpha c_s h$, where c_s is the speed of sound within the medium and h is the scale height of the disk. In this simplified prescription, the parameter α, is a dimensionless quantity between 0 and 1, respectively corresponding to no accretion and a maximum rate of accretion. The introduction of this free parameter allows for the disk structure to be calculated by using the equation of hydrostatic equilibrium, combined with conservation of angular momentum with the assumption that the disk is thin.

Applying the alpha prescription for the account for the viscous dissipation of energy within the disk medium one obtains

$$T(r) = \left[\frac{3GM\dot{M}}{8\pi\sigma r^3}\left(1 - \sqrt{\frac{R_{\text{in}}}{r}}\right)\right]^{1/4} \tag{3.13}$$

In the limiting case where $r \gg R_{\text{in}}$, where R_{in} is the radius of the inner edge of the disk, this reduces to

$$T(r) = \left(\frac{3GM\dot{M}}{8\pi\sigma R_S^3}\right)^{1/4}\left(\frac{r}{R_S}\right)^{-3/4} \tag{3.14}$$

where R_S is the Schwarzschild radius (Eq. (3.1)). This relation for $T(r, M, \dot{M})$ has the same parameter dependencies as the expression derived from simple arguments.

The emergent spectrum can be calculated as well. Each annular segment of the disk has a temperature and emits as a black body with a luminosity $2\pi r \delta r \sigma T^4$, with a spectral energy distribution described by the Planck function,

$$B_\nu(T) \propto \nu^3\left[\exp\left(\frac{h\nu}{kT}\right) - 1\right]^{-1} \tag{3.15}$$

The flux at a given frequency is then given by the integral:

$$S_\nu \propto \int_{R_{\text{in}}}^{R_{\text{out}}} B_\nu(T) 2\pi r dr \tag{3.16}$$

This results in the form of the spectral energy distribution depicted in Figure 3.7. For AGN, the observable disk emission is primarily in the upper, relatively flat portion of this distribution.

For typical AGN physical parameters, for example for a few times $10^8\ M_\odot$ black hole accreting at about 10% of its Eddington rate, the peak temperature of the disk should correspond to photon energies of ~ 50 eV, which is in the extreme ultraviolet (or very soft X-ray) portion of the spectrum at zero redshift. That makes the peak temperature difficult to obtain observationally since the galaxy becomes opaque in

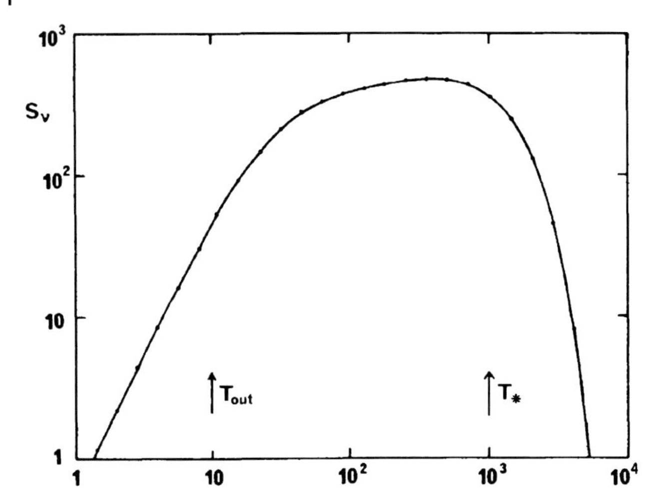

Figure 3.7 Spectral energy distribution for a geometrically thin optically thick steady-state accretion disk based on the "α-disk" model. The units are arbitrary. The vertical arrows indicate the frequencies corresponding to the inner- and outer-most disk annuli. For an AGN with a mass of $\sim 10^8\ M_\odot$ accreting near the Eddington limit the inner disk temperature corresponds to frequencies of a few 10^6 Hz, in the far ultraviolet. Thus for low-redshift objects, the disk emission is most prominent in the ultraviolet. The central, relatively flat portion of the distribution is characterized by a $\nu^{1/3}$ slope, which is generally found to be consistent with observations.

that spectral domain. Another prediction however is that the spectral shape of the disk continuum emission is in principle observable. The optical-UV continuum is expected to be approximately $f_\nu \propto \nu^{1/3}$ (see Figure 3.8). In practice, there can be dust reddening which can alter the continuum shape and the blue bump continuum must also be deconvolved from the underlying broad AGN continuum as well as possible stellar light contribution from the host galaxy. In practice, this "α-disk" prescription has been applied to a wide range of accretion-powered phenomenon in astrophysics including AGN (e.g., Sun and Malkan, 1989; Laor, 1991).

A complete discussion of basic accretion disk physics is beyond the scope of this book. Some good references include the book by Frank *et al.* (1992) and the review article by Pringle (1981).

The angular momentum transport problem is now believed to be qualitatively understood in the context of magnetohydrodynamical calculations. The basic idea, worked out by Balbus and Hawley (1992) involves a magnetic field with a component in the axial direction of the disk; the so-called "magnetohydrodynamic instability." Radially adjacent fluid elements within the disk will interact in a manner analogous to two mass points connected by a spring, the spring tension being related to the magnetic tension of the perturbed field. The inner fluid element would begin orbiting more rapidly than the outer element, causing the hypothetical spring to stretch. The inner fluid element then slows down due to the spring action, reducing its angular momentum causing it to move to a smaller orbit. The outer fluid element being pulled forward will speed up, increasing its angular momen-

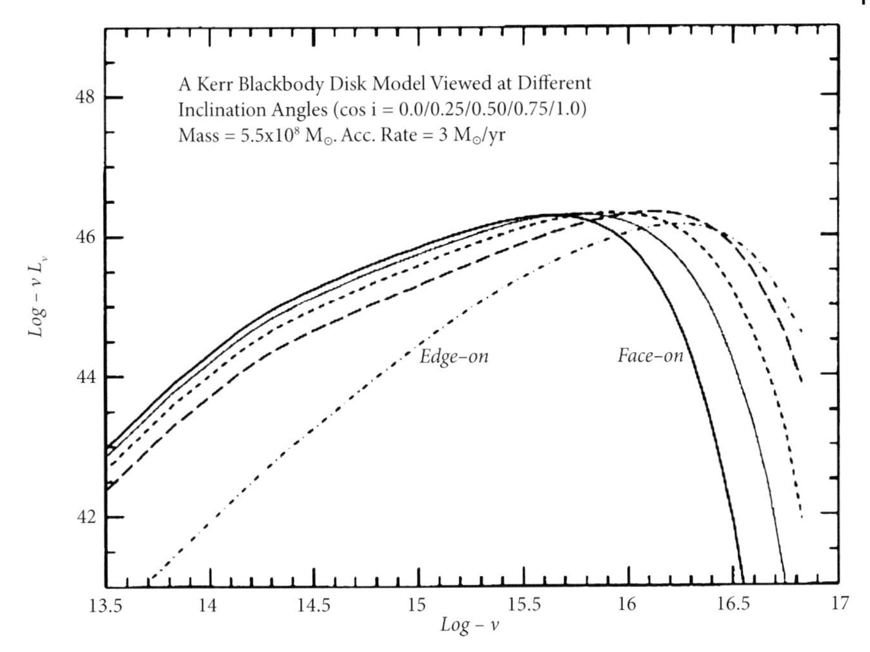

Figure 3.8 These calculations from Sun and Malkan (1989) illustrate possible optical-UV spectral energy distributions assuming the big blue bump component is due to an "alpha" accretion disk. Effects of viewing angle are indicated, and in this case, a Kerr black hole (Section 3.1.3) is assumed. Fitting these types of models to actual data requires additional assumptions regarding the underlying nonthermal continuum. Another practical problem is that for typical Seyfert 1 AGN with redshifts $\sim$ 0.1 the emergent disk flux peaks in the unobservable far UV spectral region. Furthermore, attempts to extend these models into the soft X-ray range have been problematic. Also, the optical-UV and X-ray variability time scales are very different.

tum and move to a larger radius orbit. This process, also called the Balbus–Hawley mechanism, can escalate, leading to a dynamically unstable situation. Numerical calculations employing these basic ideas (Krolik *et al.*, 2005) have led to plausible models of real astrophysical disks.

3.2.3
Accretion in Thick Disks

In AGN accretion disks there may be parameter-space regimes where the thin-disk scenarios break down. If for example, the accretion rate significantly exceeds the Eddington value, or the cooling of the disk becomes highly inefficient (corresponding to high values of the viscosity parameter α), the flow cannot be vertically confined and the standard α-disk model is not sustainable. Toroidal, or "thick-disk" geometries may then be required to model the accretion flow, depending on whether the pressure is dominated by radiation or by hot gas (e.g., Abramowicz, 1988). Observational determination of the frequency at which the thermal UV component peaks is very uncertain but it is crucial in constraining the models. The emergent radiation from a thick-disk photosphere has a characteristic black body spectrum, with a peak temperature and frequency that scales with $M^{-1/4}$ and $\dot{M}^{1/4}$ in a manner similar to geometrically thin disks. Those maxima are reached at radii of about $R \simeq 5 R_S$.

If the big blue bump extends into the soft X-rays, the required inner-disk temperatures could be $\sim 10^6$ K which would favor super-Eddington accretion rates, thus necessitating thick disk scenarios. However, observational evidence, in particular multiwavelength variability studies (Nandra *et al.*, 2000) do not support the idea that the X-rays simply are the hot tail of the big blue bump.

Comptonizing sources for the more energetic nonthermal radiation, seen observationally as power-law spectra, are additional AGN characteristics. The optically thin and heated corona of the disk, jets or winds are likely candidates for these Comptonizing media, and thick disks may facilitate the production of these components.

Another possible argument for the presence of thick disks is the observed presence of double-peaked emission lines, most commonly Hα (e.g., Eracleous and Halpern, 1994). The most common interpretation is that these features emanate from the disk with the red and blue peaks resulting from the Keplerian dynamics. However, if the lines are from photoionized material in a thin disk, the illumination geometry becomes problematic (assuming the ionizing radiation innermost disk region). Schemes to scatter the radiation from above the disk, preferentially illuminating the inner regions must be invoked. Furthermore, it has been questioned whether or not there is sufficient mass density and thermal energy in the outer disk for production of the observed lines. Launching of winds or jets could also be difficult in the context of thin accretion disks. Thick-disk scenarios could possibly facilitate solutions to these problems.

3.2.4
Advection-Dominated Accretion Flows

While the basic alpha-disk model and variants thereof have been successful in interpreting a number of observed AGN properties some issues remain unresolved. For example, some low-luminosity AGN do not seem to have a big blue bump (BBB) spectral component (e.g., Nemmen *et al.*, 2010). We note that in other cases, specifically objects where relativistic beaming is predominant the BBB is not seen, but that is assumed to be a contrast issue. The lack of a BBB component is even more profound in "quiescent" galactic nuclei, notably the 10^6 solar mass black hole associated with the Sgr A* source at the center of our Galaxy. For example, Sgr A* is believed to have a mass supply rate of about $10^{-5} M_\odot \text{ yr}^{-1} \simeq 10^{-4} \dot{M}_{\text{Edd}}$ based on a very detailed dynamical study of stars and gas in the region. On the other hand, its luminosity is $\sim 10^{-9} L_{\text{Edd}}$. Since our Galaxy is not believed to be atypical among its local group neighbors there are likely to be similar situations in the nuclei of a majority of galaxies (e.g., Fabian and Canizares, 1988).

Radio-loud and radio-quiet

The definition, what is a radio-loud and what is a radio-quiet AGN, has not always been the same over the years. It has to be pointed out that radio-quiet is not the same as radio silent. At a certain flux level probably all AGN emit radio waves, and their detection depends on the sensitivity of the telescope which is used. The term radio-loud is usually defined as flux ratio in the radio band versus the optical band. Sramek and Weedman (1980) used *k*-corrected flux values, that is, values which are corrected for the redshift

$$z = \frac{\lambda_{\text{observed}} - \lambda_{\text{emitted}}}{\lambda_{\text{emitted}}} = \frac{\nu_{\text{emitted}} - \nu_{\text{observed}}}{\nu_{\text{observed}}} \tag{3.17}$$

of the observed versus emitted energy band, defining the radio-loudness as:

$$R^* = \left(\frac{f_{5\,\text{GHz}}}{f_{2500\,\text{Å}}}\right) \tag{3.18}$$

Here one also has to be aware that the flux measurements can depend on the resolution of the instrumentation and whether it includes just the core of the AGN, the whole galaxy, or, in case of radio galaxies, also the radio lobes. Sometimes a source is called radio-loud when the simple flux ratio is larger than 10, or if the radio luminosity is larger than $L_r > 10^{33}\,\text{erg}\,\text{s}^{-1}\,\text{Hz}^{-1}$ (Stocke *et al.*, 1991). As definition for radio-quiet for example (Peterson, 1997) used $0.1 < R^* < 1$. Often the definition of a spectral slope between the optical and radio band is used for the distinction. This can be defined as

$$\alpha_{\text{ro}} = \frac{\log\left(\frac{L_{5\,\text{GHz}}}{L_{2500\,\text{Å}}}\right)}{\log\frac{\nu_r}{\nu_o}} = \frac{\log R^*}{5.38} \tag{3.19}$$

Here, sources with $\alpha_{\text{ro}} > 0.35$ ($R^* \gtrsim 76$) are called radio-loud (Della Ceca *et al.*, 1994).

Another issue is that AGN are broadly separated into two categories (Section 4.6): radio-loud and radio-quiet, the former exhibiting collimated jets. The radio-loud objects are additionally subclassified into FR-I and FR-II objects. Other astrophysical objects believed to be driven by disk accretion, galactic X-ray binaries, are also characterized by distinct observational spectral states and intermittent jet activity. All of this suggests that accretion disks undergo transitions from one type of configuration – for example one reasonably well-approximated by the alpha-disk prescription – to another which is radiatively inefficient and perhaps accommodating of outflows. The advection-dominated accretion flow, or ADAF (e.g., Narayan and Yi, 1995), scenarios may address this in a natural manner (see Figure 3.9). Basically, accreting gas is heated viscously and cooled radiatively. Any excess heat is stored in the gas and then transported in the flow. This process represents an "advection" mechanism for the transport of thermal energy. The conditions for an ADAF to

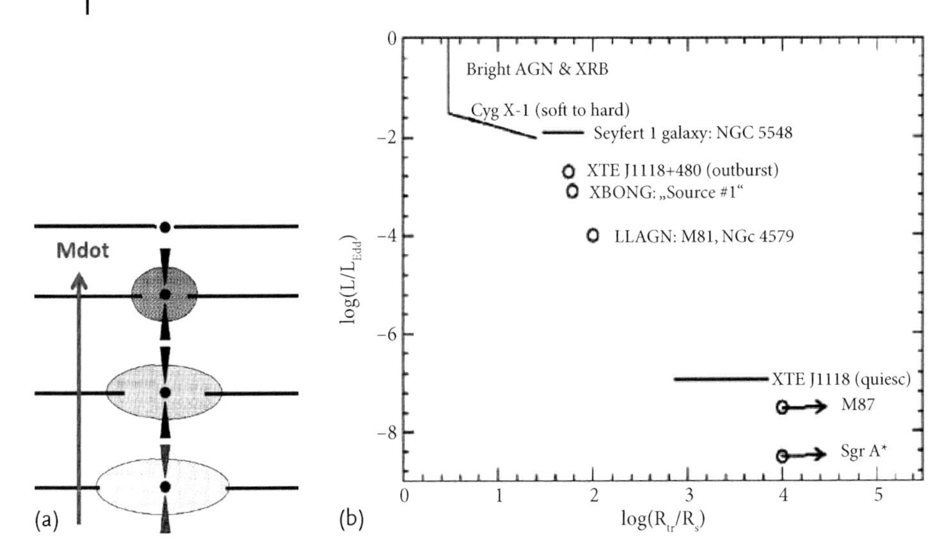

Figure 3.9 Shown schematically, the ADAF plus thin-disk configuration that may underlie the observed X-ray binary spectral states and the AGN subtypes delineated by luminosity, radio properties and the spectral energy distribution (SED), in particular the presence or absence of a big blue bump component (a). Yuan and Narayan (2004) illustrate how AGN are delineated in terms of their luminosity (specifically L/L_{Edd}) and the disk truncation radius derived from fitting the ADAF plus thin-disk model to the observed SED (b).

form are a low (sub-Eddington) accretion rate and a gas opacity which is very low. An element of the gas is unable to radiate its thermal energy in less time than it takes it to be transported through the disk onto the black hole.

ADAFs may be associated with jets. The observed radiation is a combination of emission from the ADAF and the jet with the radiation from thermal electrons likely to be from the ADAF rather than from nonthermal electrons that comprise the jet. The thin disk to ADAF boundary apparently occurs at luminosities $L \sim 0.01{-}0.1 L_{\text{Edd}}$, or $\dot{M}_{\text{crit}} \sim 0.01{-}0.1 \dot{M}_{\text{Edd}}$. This boundary is consistent with AGN transition from bright quasars mode to low-luminosity LINER (Lasota *et al.*, 1996; Quataert *et al.*, 1999) and it is also approximately where galactic X-ray binaries switch from the well known high-soft to low-hard spectral states.

Unlike galactic binaries, where these state transitions occur on observable time scales, the AGN analogs must be addressed through the compilation of statistical ensembles of AGN subclasses. With that though comes the possibility of selection biases or incomplete sampling, so the BHXRB-AGN analogies remain open issue.

3.3 Absorption Close to the Black Hole

In addition to the multiple velocity component gaseous outflows evident from optical and UV spectroscopy, improved instrumentation in the X-ray spectral domain has also revealed a wealth of new AGN phenomenology. In particular, X-ray CCD and dispersive grating spectroscopy has revealed multiple absorption zones characterizing the AGN source population as a whole and covering a range of ionization. Variability on relatively short time scales may also be in some cases indicative of dynamical absorption media in close proximity to the central black hole. The gas responsible for the X-ray absorption is not only clearly distinct from the UV absorbing gas, but a stratification of ionization structure (e.g., Krolik and Kriss, 2001) or alternative geometrical structure (Behar *et al.*, 2003) is required to model the X-ray absorbing media alone.

The ionization levels involved and the inferred temperatures lead to the designation as "warm absorber." This terminology was, to our knowledge, first introduced by Halpern (1984) to denote the fact that the inferred electron temperatures, typically $\lesssim 10^5$ K, were much lower than those of collisionally ionized thermal plasmas at a similar level of ionization. We should note that for certain geometrical configurations, the warm absorber gas can alternatively give rise to observable narrow emission lines. In the literature, this type of configuration is on occasion referred to as the "warm emitter."

The warm absorber gas resides close to the AGN central engine and as such its thermodynamic properties are determined by the radiation field of the central engine. The ionization state of the gas can in principle be modeled with photoionization calculations. This is not a simple problem though as it is thus a function of the luminosity and spectral energy distribution of the central engine continuum, the distance between the illuminated surface of the warm-absorber cloud media and the central engine, thermodynamic and geometric stratification within the cloud media, the elemental abundances, and the radiative transfer between the central engine and the warm absorber cloud if there is significant intervening matter. In addition, the ionizing continuum is likely to vary in intensity and shape, thus the problem is not a static one. Nor is the warm absorber media components likely to be static as they are dynamically tied to the potential well of the AGN central black hole. Thus even with very accurate and detailed observational data, complex calculations and some simplifying assumptions are in order in any attempt to model the warm absorber.

Figure 3.10 illustrates a model calculation of the emergent spectra of an AGN (Turner and Miller, 2009). The inherent complexity of this type of modeling is evident. Early studies were limited by the available spectral resolution of the X-ray detectors applied to the problem, for example $\Delta E/E \sim 1$–5 for gas proportional counters or 10–50 for CCD detectors (see, e.g., Komossa, 1998). Realistically, the only detectable features were bound-free absorption "edges," primarily those due to helium- and hydrogen-like oxygen, which appeared as residual structures in efforts to accurately model the continuum emission. The depths of the features varied from object to object and sometimes from epoch to epoch, presumably due to different line of sight gas structures. Nonetheless, a consensus developed that the basic warm-absorber paradigm of a photoionized $T \lesssim 10^5$ K gas close in to the AGN core provided the most viable interpretation of the observations.

More recent observations, utilizing the much improved spectral resolution available with the dispersive spectrographs on-board the Chandra and XMM-Newton

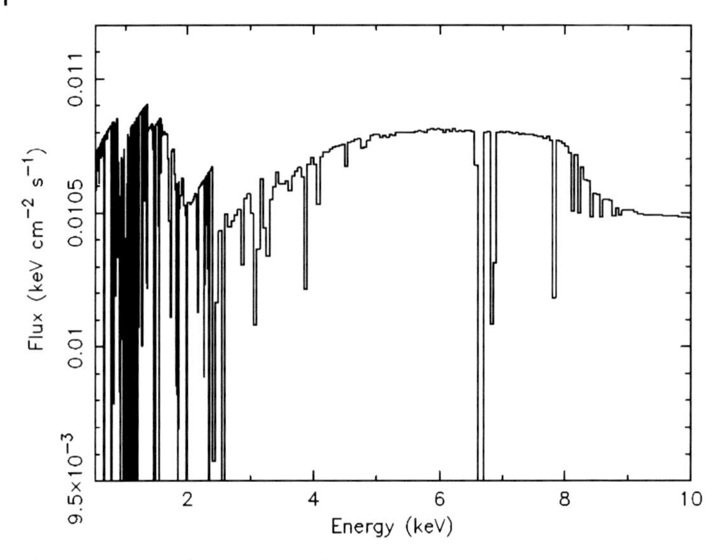

Figure 3.10 Example of a model calculation for the emergent X-ray spectra of an AGN (reproduced from Turner and Miller, 2009). In this case, a gas column density of $N_H \sim 5 \times 10^{22}$ cm^{-2} and a gas ionization parameter $\xi \simeq 3$ erg cm s^{-1} were assumed (note that here the ionization parameter is defined as $\xi = L/nr^2$). The input spectrum was a power law. The curvature is due to *K*-shell edges from high-ionization Mg, Si, S and *L*- and *K*-shell edges of Fe.

observatories have begun to reveal the complex nature of the warm absorber gas. These dispersive spectra allowed separation of lines that had been unresolved with the CCD observations. Multiple gas components involving different temperatures and ionization states were found to be needed to understand the observations, for example for the case of NGC 3783 (Netzer *et al.*, 2003); also refer to Figure 3.14 for an example of a high-resolution X-ray spectrum of that same object.

Additional structures such as pseudocontinua due to radiative recombination can be identified in detailed modeling. Spectral components of certain He-like triplets, for example of oxygen and neon can in some cases be resolved and their intensity ratios used as gas temperature and density diagnostics. The ability to identify individual lines provided improved determinations of the gas kinematics, showing outflow velocities covering the range 10^2–10^3 km s^{-1} (e.g., Kaspi *et al.*, 2002).

This is a field which has continued to evolve, not only as a result of the improved observational capabilities but also due to improved computational tools (such as the XSTAR photoionization code), computational capabilities and atomic physics databases. There may also be new information forthcoming from X-ray polarization measurements from the GEMS satellite planned for launch sometime in the coming decade. Calculations suggest a distinct signature for the warm absorber gas when viewed in polarized X-rays (Dorodnitsyn and Kallman, 2011). For details we refer the reader to a number of excellent review articles, for example, Turner and Miller (2009); Komossa and Hasinger (2003); Crenshaw *et al.* (2003).

3.3.1
The Torus Model

Radio-quiet AGN unification models generally invoke an obscuring dust structure with a toroidal geometry surrounding the central accretion disk (see Figure 3.11). In the context of these scenarios, the Seyfert 1 objects represent cases where the observer has a relatively unobstructed view of the active nucleus and the broad-line region. In Seyfert 2s on the other hand, it is observed through an obscuring structure, which largely obscures the optical-UV emission. Given the compact nature of the broad-line region, as inferred from reverberation mapping studies and photoionization calculations, the obscuring dust is expected to be located well within the region dominated by the gravitational influence of the black hole.

Dynamical studies that have been carried out tend to favor a clumpy structure rather than a uniform toroid (e.g., Elitzur and Shlosman, 2006). One basic difference between smooth and clumpy density distributions is that for the latter, a large range of dust temperatures can coexist at similar distances from the source (Nenkova *et al.*, 2008). This potentially explains certain observations such as the low dust temperatures found near the nucleus in NGC 1068. While, as noted, it obscures viewing of the broad-line region in the UV and optical, the IR emission can be partially transmitted and in some cases can reveal the presence of the broad-line region (BLR). This is the essence of the radio-quiet AGN unification scenario described in Section 4. The main point is that distinct AGN subtypes can comprise intrinsically similar phenomena but can appear very different just because of the angle between the disk-torus structure and the observer's line of sight. Statistical compilations of Seyfert 1 and 2 galaxy observations, suggest a torus scale height-radius relationship of $H/R \sim 1$ (Schmitt *et al.*, 2001).

How the torus structure is formed is another question to be answered. The origin could, for example, involve matter coming off the relatively cool, outer regions, of the accretion disk. That matter could form a clumped wind structure about the disk's perimeter. Alternatively, matter could be accreted from ambient matter from

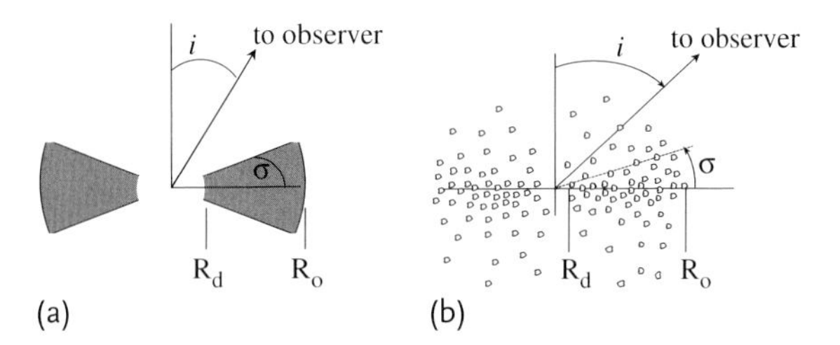

Figure 3.11 Schematic depiction of possible AGN torus configurations, smooth or clumped (from Elitzur, 2007). The classification of AGN into types 1 and 2 is based on the visibility of the nuclear region. The viewing angle and the torus geometry, parameterized here by σ, are together the determining factor for a smooth density torus (a). For a clumped torus, the delineation between types 1 and 2 is on the other hand not strictly due to orientation but rather to the of probability of the observer having direct view of the AGN nucleus (b).

within the host galaxy. For example, Krolik and Begelman (1988) have considered a hydrostatic model whereby a toroidal structure populated by molecular clouds can be accreted from the host galaxy. A problem was quickly pointed out though regarding the H/R constraint (e.g., Davies *et al.*, 2006). Specifically, the origin of vertical motions capable of sustaining the clouds in a hydrostatic structure with $H \sim R$ was recognized to be problematic for those models. Alternative scenarios, involving outflowing clouds from the disk embedded in a hydromagnetically driven disk wind have also been proposed.

While the torus model is still central to the AGN type 1–type 2 unification scenario, a problem was pointed out in 1991 by Lawrence (1991). He noted that the fraction of type 2 AGN was a decreasing function of luminosity in optical and IR-selected samples. It was also noted that X-ray and radio-selected samples indicated a subsample of type 2 objects which were heavily reddened, and that these were increasingly common towards lower luminosities. These observations suggested that a simple unified unification model involving only the orientation of a molecular torus was untenable. It suggested instead that the tori must have a range of geometrical thickness as well as locations of their inner walls as determined by the radial distance from the central engine at which the dust reaches its sublimation temperature. In more luminous objects this distance is expected to be larger. If the height of the torus remains constant, the critical angle at which the observers line of sight to the central region is obstructed decreases, thereby leading to a luminosity dependence in the observed type 1 AGN fraction. This is known as the "receding torus" model.

We note that this issue has additional implications regarding the AGN contribution to the cosmic X-ray background (CXB). The shape of the CXB spectrum suggests that the type 2 AGN are being undercounted and thus the receding torus model impacts proposed resolutions to that issue.

The receding torus problem was recently revisited by Simpson (2005) who used the complete magnitude-limited sample derived from the Sloan Digitized Sky Survey (SDSS; see box in Section 4.6) to address the type 1–type 2 fractions with respect to luminosity. He finds that the fraction of type 1 AGN increases with the luminosity of the isotropically emitted narrow-line [O III] flux (Figure 3.12). This is again generally consistent with the receding torus model, but they are able to construct luminosity functions for both types, and consider the problem in greater detail. They find that the luminosity functions have different shapes along with the expected increase in type 1 fraction with luminosity. They derive a modification to the basic receding torus whereby in addition to the inner torus wall radios, the height of the torus has a specific luminosity dependence of $h \propto L^{0.23}$. With this modification, the receding torus model which seems to provide a reasonable correction factor to account for the "missing" type 2 AGN in surveys.

There have been suggestions that AGN torus clouds can be directly observed by virtue of the apparent presence of water masers in certain objects. For example, Kondratko *et al.* (2005) have observed molecular cloud structures in NGC 3079. It has been suggested that these clouds have resulted directly from vertical wind ejection from the accretion disk. Improvements to our understanding of the AGN

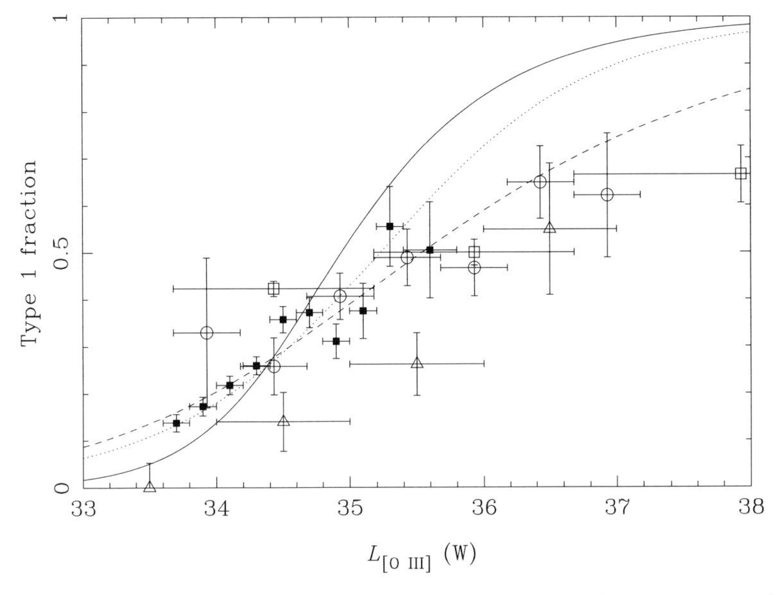

Figure 3.12 The type 1 AGN fraction as a function of luminosity in the isotropically emitted [O III] narrow emission line (reproduced from Simpson, 2005). The filled squares are based on the optically selected SDSS sample, and the open squares and triangles are derived from X-ray surveys plus the SDSS. The solid line curve is the best-fit standard receding torus model and the dashed line is the best-fit model where the torus height is allowed to vary with luminosity.

torus model can be expected as new observational capabilities are brought to bear on the problem. Current (Herschel Observatory) and future (James Webb Space Telescope) space-based IR observatories and enhanced radio interferometric techniques, notably involving the use of embedded maser sources, will add significantly to the available knowledge base.

3.3.2
Mass Loss in AGN

It has long been recognized that mass outflows occur in a large fraction of AGN. This is evident from the presence of blueshifted absorption lines in the UV and X-ray spectra at different widths and velocities. Multiple outflowing components, characterized by distinct velocities relative to the systemic AGN velocity are often observed. Roughly 50% of Seyfert 1 exhibit UV absorption troughs of gas outflowing at inferred velocities $v \lesssim 1000\,\mathrm{km\,s^{-1}}$ (e.g., Crenshaw *et al.*, 1999). An example for multiple absorbers in the UV spectrum of an AGN is shown in Figure 3.13. A similar fraction ($\simeq$ 50%) of AGN also exhibit absorption in their X-ray spectra (George *et al.*, 1998), with X-ray absorption features present in each case exhibiting UV absorption but not necessarily of the same properties (velocity, ionization parameter). This raises the issue of the relation between the X-ray and UV absorbers.

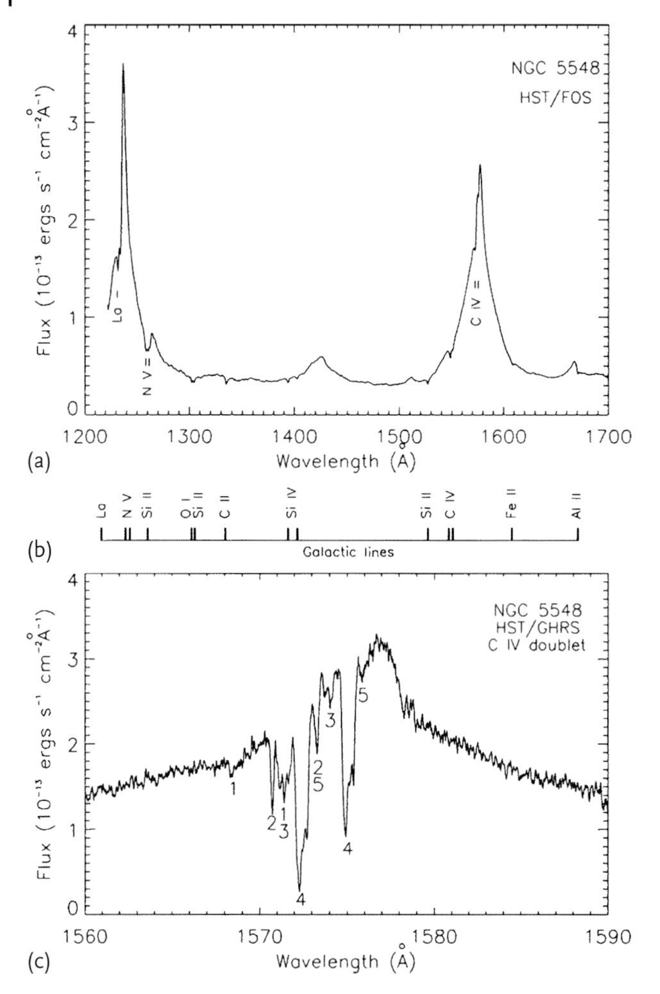

Figure 3.13 Example of multiple velocity kinematic components in the UV spectra of an AGN, in this case NGC 5548 (Crenshaw *et al.*, 2003). A low-resolution spectrum from the HST FOS (a), interstellar absorptions (b), and the C IV region at higher resolution, and identifies 5 separate kinematic components by virtue of absorption in the C IV doublet (c).

An example of the richness of X-ray absorption line systems is illustrated in Figure 3.14. Generally speaking, the inferred mass-loss rates can be comparable to the mass accretion rates. In many cases, the absorbing material exhibits time-varying column densities of particular ions due presumably to both the gas dynamics and variations in the ionizing flux.

Recent studies of AGN absorption systems that make use of the spectroscopic monitoring observations, photoionization modeling, and magnetohydrodynamic calculations are leading to new constraints on the physical and dynamical conditions. The source of the outflowing material, relative to the compact central engine, which may be near the inner nucleus ($\sim$ 0.01 pc) or as remote as the galactic disk

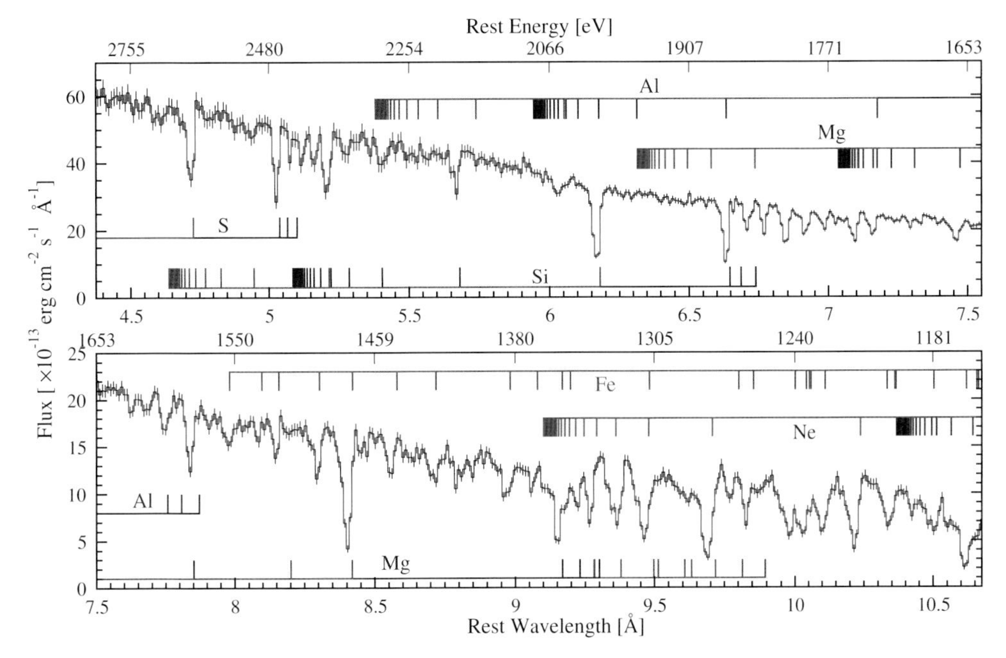

Figure 3.14 Since the late 1990s, dispersive X-ray spectroscopy has provided sufficient spectral resolution ($R \sim 10^2 - 10^3$) to identify and study outflowing gas from more highly ionized regions that are presumably closer to the AGN central engine than in the optical or UV. In this figure, adopted from Kaspi *et al.* (2002), Chandra high- and medium-energy grating spectra of NGC 3783 are illustrated. The source frame hydrogen and helium-like lines of Si and Ne are annotated, as is a forest of Fe features. Many of the lines are clearly detected, and can be interpreted in the context of wind models.

or halo ($\sim$ 10 kpc) can also be probed. Models that are based on thermal winds, radiation pressure driven winds, as well as magnetohydrodynamic winds have all been applied and are leading to an improved understanding.

The intense AGN radiation field presents an obvious agent for driving the observed outflowing winds. There may be problems however. Detailed models of radiatively driven winds, both semianalytical (Murray *et al.*, 1995) and fully numerical (Proga *et al.*, 2000), have shown the importance of preventing the X-ray ionization of the wind materials. The point is that ions with UV resonance or bound-free lines must be present for the radiation pressure to significantly accelerate the gas. This mechanism is believed to be of particular importance in driving the high-velocity material evident in the absorption features of broad absorption line quasars (BAL QSOs). The BAL QSOs it should be noted are generally characterized by low X-ray-to-optical luminosity ratios, and as such, the aforementioned problem may be minimal. An altogether different class of models is based on MHD driven outflows (e.g., Blandford and Payne, 1982; Fukumura *et al.*, 2010). In these scenarios, a centrifugally driven outflow of matter from the disk can occur depending on the magnetic field configuration, in particular poloidal component and its angle relative to the disk. A fraction of the accreting plasma could then be launched for example, from a thin-disk surface with a Keplerian rotational velocity profile and then be magnetocentrifugally accelerated along magnetic field lines threading the disk. X-ray emission from the central inner most region of the disk, which is effectively a point source at the wind location, irradiates the outflowing material. The X-rays are progressively absorbed as the radiation propagates through the line of sight plasma. Combining a hydrodynamic model of the wind – that is, its geometry and its density and velocity profiles – with photoionization modeling one can interpret the observed absorption lines and determine useful physical information such as its ionization structure. This in turn can impact studies of AGN populations, since the different AGN types are believed to be delineated by the environment ambient to the central engine and their orientation relative to our line of sight.

In addition to offering the potential to probe the kinematics, thermodynamics, geometry, and chemical abundances within the active nucleus and its host galaxy, absorption studies may provide insight into the feedback mechanism (see Section 3.1.2) which has been proposed as a mechanism to regulate the growth of the central massive black holes and provide a natural explanation for the M–σ relation. Because the outflow absorbing gas is directly in the line of sight to the continuum source, the blueshifted absorption provides unambiguous evidence of outflow and in many cases, both radial and transverse velocities can be obtained for the absorbers. They can thus be used as tracers of the dynamical forces such as radiation pressure, magnetic fields and gravity in AGN.

Recently, Faucher-Giguère *et al.* (2012) have applied this type of analysis to a subset of the BAL quasars for which low-ionization species of iron, specifically Fe II have been used to estimate the mass outflow rates and the corresponding kinematic luminosities of these objects. They find that the kinematic luminosities can be in the range 1–5% of the bolometric luminosity, which is approaching values needed by the feedback models to explain the M–σ relation normalization (e.g., Di Matteo *et al.*, 2005). They also find that the outflow properties of these objects are comparable to those recently inferred for molecular outflows from some ultraluminous infrared galaxies, suggesting that active galactic nuclei are capable of driving those outflows as well. Giustini *et al.* (2011) studied the time variability of X-ray absorption in the BAL QSO PG 1126-041. Apparently the wind varies significantly on time scales of months down to hours, causing the observed spectral slope to vary significantly on short time scales. The authors also detect differences in outflow speed between the faster X-ray wind and the slower matter flow observable in the UV. Thus it seems the higher ionized matter responsible for the flux changes in X-rays is more variable than the moderately ionized phase, which varies rather on monthly time scales. This can be understood by considering that the hot and highly ionized wind emanates closer to the central engine than the UV-detected wind.

3.4 Photoionization Modeling

Prominent, broad emission lines that can vary significantly in intensity on time scales of days to years are a defining characteristic of AGN. Also, the narrow, constant intensity, lower ionization lines are ubiquitous in low-luminosity AGN as well. The study of these AGN emission lines is of importance in regard to a number of issues. The lines provide information pertaining to the central engine, as they are expected to be broadened, for example, as a result of either virialized or orbital bulk motion. The resulting emission line widths can be applied to estimates of the central black hole mass. The resulting black hole mass estimates in turn can be used in other domains of astronomy; for example in studies of the coevolution of the black holes and central regions of their host galaxies.

It was quickly realized that the photoionization of gas in the nuclear regions, presumably by UV and X-ray emission from the AGN central engine was fundamental to the modeling of AGN emission lines. By the time AGN were discovered significant progress had been made on the understanding of gaseous nebulae in the galaxy in which photoionization calculations had matured (see, e.g., Osterbrock, 1989). The temperatures inferred in narrow-line regions of Seyfert galaxies are of order a few 10^4 K. Collisional or shock heating, which is the only other plausible means of exciting the gas, involves higher temperatures, for example [O III] lines would be radiated at temperatures $T > 5 \times 10^4$ K. It was also clear that a wide range of ionization stages were present suggesting that the source of ionizing radiation had to extend to higher energies – UV to X-rays – than was available from hot stars that powered the galactic H II regions and planetary nebulae. The observed, approximate power-law continua of AGN would seem to serve this purpose. The SED of the radiation incident upon the line-emitting gas is not known. However, reasonable assumptions can be employed in calculations based on knowledge of the emergent SED and inferred filtering effects of the media ambient to the central engine. This input spectrum usually must be extrapolated into the unobservable extreme UV as well. The basic results are generally supportive of the photoionization hypothesis. For example, the Hα line luminosity versus monochromatic blue continuum luminosity is linear over some six orders of magnitude for a variety of radio-quiet AGN; see for example Figure 3.15.

Since early calculations, photoionization models have improved, driven in large part by the improvements in the observational data. For example, increased sensitivity and spectral resolution have revealed a variety of line profiles indicating dynamical structure more complex that had been assumed. It had been assumed that a single region or "zone" was sufficient to model the gas applying a single ionization parameter and gas density. In that scenario, the line emission can be calculated from the ionization parameter

$$\xi = \int \frac{L_\nu}{h\nu 4\pi r^2 n c} \mathrm{d}\nu \tag{3.20}$$

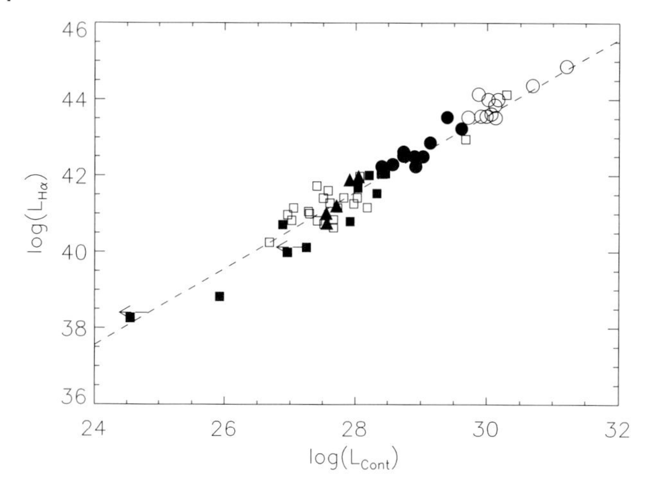

Figure 3.15 Luminosity in the Hα emission line versus monochromatic continuum luminosity (at λ4800) for different AGN types: QSOs (open circles), Seyfert 1 galaxies (filled circles) and Seyfert 2 galaxies (open squares). The dashed line shows the predicted relationship for a photoionization model with an input spectrum $L_\nu = C\nu^{-1.05}$ (from Osterbrock, 1989).

which describes the balance between the ionization rate as driven by the incident photon flux and the rate of recombination which is directly dependent on the density n. Typical values were $\log \xi \simeq -1.5$ and densities $n \sim 4 \times 10^9\ \mathrm{cm}^{-3}$ for the BLR (Kwan and Krolik, 1981). Limits on the gas density were estimated from the widths and the presence or absence of forbidden and/or semiforbidden lines. The covering fraction of the BLR gas was inferred from comparing the observed Lyα equivalent width to the value computed. The results, $\sim$ 10%, suggested that the BLR gas was nonuniformly distributed. Instead it seemed to be clumped into discrete filaments or clouds.

The reverberation mapping campaigns of the 1980s (Section 3.5) led to an improved understanding of AGN broad-line cloud structure, but initially posed problems for the photoionization models. The observed lag times between continuum flux variations and broad-line response to those variations were shorter than anticipated based on the photoionization models. The immediate implication was that the line emission was occurring much nearer to the central black hole than had been inferred from the calculations. This also implied that the gas was denser than previously believed. It was also found that lines of different ions exhibited different lag times. For example, in NGC 5548 it was noted that [C III] lagged the continuum by 3–4 weeks, whereas C IV and Lyα, ions expected to be present at up to higher densities than [C III], lagged by $\sim$ 1 week (Ferland *et al.*, 1992). Thus it seemed that stratification of the broad-line emitting gas in AGN was a needed feature for more realistic models. This led to refinements invoking multiple zones and subsequently weighted averages over gas densities and ionization parameter values (e.g., Baldwin *et al.*, 1995).

With advances in X-ray spectroscopic observing capabilities, in particular dispersive spectrometers flown on the Chandra and XMM-Newton observatories, this work has been extended to higher energies and a wider range of ionization levels (e.g., Kallman, 2010). Fitting photoionized plasma models to the spectra obtained with these instruments can provide insight into the nature of the so-called warm absorbers (Section 5.5) and related structures. Issues to be addressed include the degree of ionization, density, geometry, composition and kinematics. However, these efforts are still in their early stages of development relative to optical and UV studies, and thus the models are likely still incomplete in some areas. Typical X-ray grating spectra, with the best available statistics, most often do not yield truly acceptable fits in a statistical sense to standard models; typically, reduced chi-square values per degree of freedom of $\chi^2_\nu \sim 2$ or greater are obtained. This may be a result of, for example, missing lines in the available atomic databases, incorrect treatment of line broadening, incorrect ionization balance, idealized assumptions (such as ionization equilibrium), inaccurate treatments of radiative transfer or geometrical effects.

Nonetheless, significant results have been obtained. In Figure 3.16, model fits to a Chandra grating spectrum of NGC 3783 illustrate the potential benefits to AGN studies. In this example, Kallman (2010) and collaborators were able to obtain these fits by invoking revised dielectric recombination rates for iron, suggested previously (Badnell, 2006) but not observationally confirmed. The next generation of X-ray instruments, based on a calorimetry principle, are expected to obtain dramatically improved X-ray spectra in terms of resolution and sensitivity. In parallel, improvements in laboratory data and computational methods will be needed to realize the full potential of the observations.

A comprehensive review of photoionization modeling of AGN is beyond the scope of this book. For more details, see for example Krolik (1999), Kallman and Bautista (2001), Leighly and Casebeer (2007), and Kallman (2010).

3.5 Narrow and Broad-Line Regions

One of the first distinguishing observational properties of AGN is the presence of redshifted, time variable intensity emission lines with Doppler widths of order of 10^3 to a few 10^4 km s^{-1} (see Figure 3.17). The most prominent of these lines are the hydrogen Balmer series lines H$\alpha\lambda$6563, H$\beta\lambda$4861, and H$\gamma\lambda$4340, hydrogen Ly$\alpha\lambda$1216. Also common are lines of ions Mg II λ2798, [C III] λ1909, and C IV λ1549. While ubiquitous, depending on the redshift of a given AGN, they may not be observable for a spectrograph of a given bandpass. These lines are in addition in many cases, much narrower (a few hundred kilometers per second) forbidden or semiforbidden emission lines. The narrower widths and lack of variability of the latter led early on (circa 1970s), to the conclusion that they emanated from a region that was much larger and kinematically separate from that of the broad lines. This led to their respective designations as the broad-line region (BLR) and narrow-line

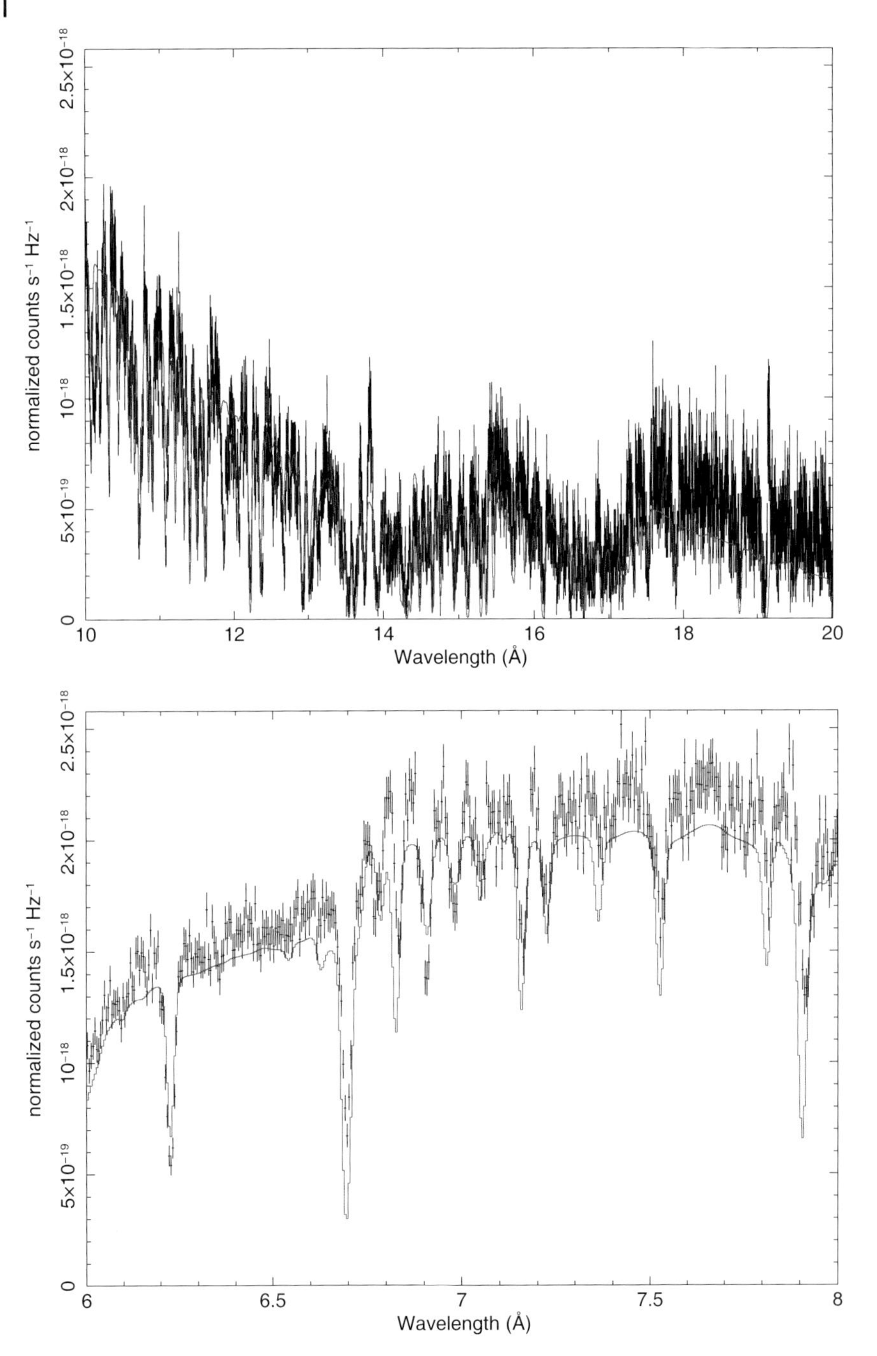

Figure 3.16 Examples of photoionization models of ionized plasma fitted to dispersive X-ray spectra obtained with the Chandra X-ray observatory (Kallman, 2010). The fits led to refinements in iron recombination rates and revealed multiple gas velocity components. For a color version of this figure, please see the color plates at the beginning of the book.

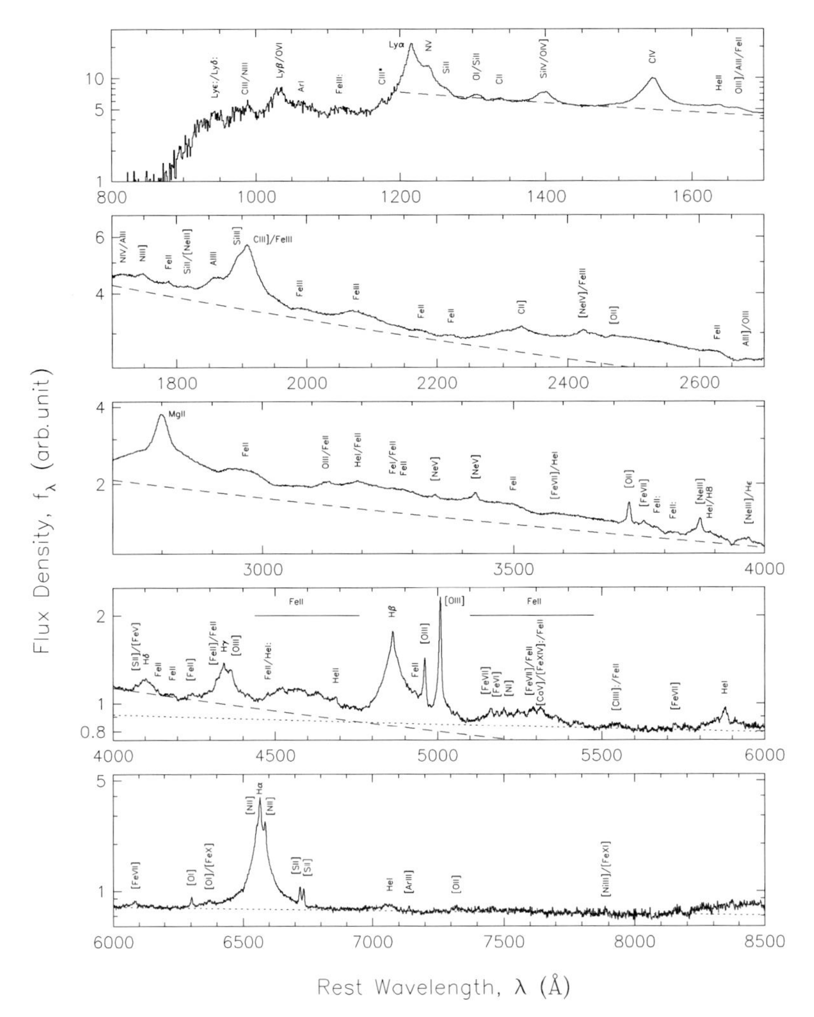

Figure 3.17 Median quasar composite spectrum derived from the SDSS. Labels ending with a colon (:) are uncertain identifications. Two power-law continuum fits are shown by dashed and dotted lines. For the rest frame wavelengths of the lines see Table 4.1. From Vanden Berk *et al.* (2001).

region (NLR). The physical properties of the BLR have been assessed on the basis of photoionization modeling. For a recent review of the BLR see Gaskell (2009). The BLR clouds are assumed to be in photoionization equilibrium, that is, the rate of photoionization is balanced by the rate of recombination. Mathematically, this can be expressed in terms of an "ionization parameter" ξ defined in Eq. (3.20) in the previous section. For a plasma cloud satisfying this equilibrium condition one can infer physical parameters. If two or more lines of different ionization states are observed, it is possible to determine the relative number densities of the different ion stages and the gas temperature. This leads commonly to temperatures of order

10^4 K and densities $n_e \sim 10^9\,\text{cm}^{-3}$. It was also evident from early studies that the BLR is coalesced into filaments or individual clouds. The argument for this being the case is that a homogeneously filled ambient gas would be optically thick blueward of the Lyman limit (e.g., Ferland and Mushotzky, 1982). Since temperatures $T \sim 10^4$ K correspond to thermal line widths of order $10\,\text{km}\,\text{s}^{-1}$, it is evident that the BLR gas or individual clouds exhibit some bulk, supersonic motion. This could be an orbital motion about the central compact object, however, net in-falling or outflowing motions are plausible as well. In any case, the large Doppler widths are strongly suggestive that they reside deep inside the gravitational potential well, as is largely corroborated by reverberation mapping studies; thus, the BLR provides important insight into the central engine.

The inferred BLR scales are light days to months for low-luminosity AGN, but they are larger, of order light years perhaps, for higher luminosity quasars. In those latter cases measurements are sparse, as the required monitoring campaigns become more difficult to carry out. Also, real long-term variations in the BLR properties could be occurring concurrently on the required observational time scales. In practical terms, these types of photoionization calculations are difficult to carry out and interpret and can be computationally intensive. Modern astronomers benefit from the availability of much faster and less expensive computers than those available to their predecessors who pioneered these types of calculations. Also, a number of photoionization codes have been produced by specialized groups and made available to the general public. These include Cloudy (Ferland *et al.*, 1998), ION (Netzer, 1990), and XSTAR (Kallman and Bautista, 2001).

The conventional "cartoon" picture of the BLR has for many decades consisted of a central, isotropically emitting source of ionizing radiation surrounded by a roughly spherical arrangement of cloudlets. This is depicted in Figure 3.18a, reproduced

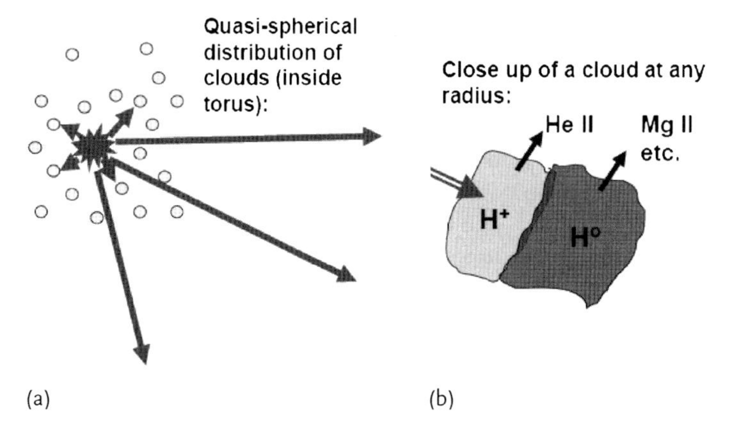

Figure 3.18 Schematic illustration of the an AGN broad-line region (BLR) reproduced from Gaskell (2009). A possible geometrical configuration, albeit the most simplistic of possibilities (a), and the side of the cloud facing the central engine (front side) will be more highly ionized than the back side (b). Reverberation mapping campaigns are gradually leading to refinements of this standard picture.

from Gaskell (2009). Individual clouds, depending on their size, will have a structure such as that illustrated schematically in Figure 3.18b. The side of the cloud facing the central engine (front side) will be highly ionized. Depending on its column density, it will be less ionized or mostly neutral on the opposite (back) side. Observationally, this manifests as the front emitting high-ionization lines from ions such as He II, He I, O VI, N V, and C IV, while the back emits low-ionization lines from Mg II, Ca II, O I, and Fe II. All these lines are well-established in AGN spectra, and in principle, this basic geometrical scenario can be assessed observationally with the reverberation mapping technique.

An important property of AGN nuclear emission and the BLR is the relation between the emission line strength, characterized by the equivalent width (EW, see Figure 4.6), and continuum luminosity was first established by Baldwin (1977). It was thus established that the regions emitting these two spectral components are associated. Baldwin found the EW of C IV $\lambda 1549$ Å, to be inversely correlated with the monochromatic luminosity at $\lambda = 1450$ Å, namely,

$$\log \mathrm{EW(C\ IV)} = -K \log L_{1450\,\text{Å}} + \mathrm{const} \tag{3.21}$$

This correlation became known as the "Baldwin effect," a designation now widely used in the astronomical literature. Baldwin's original work was based on only a few tens of quasars.

Subsequent surveys, including the use of space-based UV instrumentation, have enabled investigation of this relation with wider luminosity and redshift ranges revealing that similar correlations exist in other broad emission lines such as Lyα, [C III] $\lambda 1908$, and Mg II $\lambda 2798$ Å. An analogous X-ray Baldwin effect has also been identified (see, e.g., Wu *et al.*, 2009, and references therein), whereby the EW of the narrow Fe K_α line at 6.4 keV is anticorrelated with X-ray luminosity, $L_{2\,\mathrm{keV}}$. However, this X-ray Baldwin effect may be a separate phenomenon from the optical-UV case, since the Fe K_α equivalent width does not seem to exhibit any correlations with C IV or with the X-ray-to-optical luminosity ratios among observed samples.

The presence of lower ionization forbidden emission lines in Seyfert galaxies and QSOs are indicative of a spatially distinct component from which the broad high-ionization lines emanate. For one thing the lack of observable flux variations in the lines suggests a much larger region. Also, similarities to galactic nebulae and analysis of the underlying atomic physics reveal very different thermodynamic conditions than those which characterize the broad lines. Thus the terminology broad- and narrow-line regions (BLR and NLR) has long been part of the AGN lexicon. While there has been debate in the past about the possible role of shocks or perhaps radio jet/cloud interactions, extensive optical/UV studies (e.g., Kraemer and Crenshaw, 2000), as well as X-ray observations (e.g., Kallman and Bautista, 2001) in parallel with advances in computational methods have solidified the picture. It is now well-established that the NLR gas is photoionized by the UV-X-ray continuum radiation emitted by the central source. That ionizing flux, which must extend to far UV or X-ray energies, emanates from the central engine, but is filtered by its transmission through ambient gas. Some basic physical characteristics of the NLR can be summarized as follows.

The FWHM (Figure 4.6) of the emission lines are typically $\sim$ 400–500 km s^{-1}, thus an order of magnitude less than typical BLR line widths. The presence of forbidden as well as semiforbidden and permitted lines of, for example, oxygen, neon, magnesium and sulfur, are indicative of gas densities of $\sim 10^3$–10^5 cm^{-3}, as compared to densities of $\sim 10^9$ cm^{-3} or greater inferred for the BLR. The size of the NLR can provide a probe of the distribution of dust and gas in the central regions of AGN. It is believed to extend to hundreds of parsecs and to scale with the luminosity in some of the prominent forbidden lines. For example, Bennert *et al.* (2004)) find that the nominal NLR radius R scales as $R \propto L^{\alpha}$, where L is the luminosity in the [O III] λ5007 line, and $\alpha = 0.55(0.32)$, respectively, for type 1 and type 2 AGN (see Figure 3.19). This is analogous to the so-called Strömgren law for galactic H II regions. The dichotomy between type 1 and type 2 AGN is likely related to unification and specifically the collimating and filtering effects of the torus and on viewing geometry. There could also be greater complexity regarding the filtering of the ionizing flux as well, for example from disk winds.

The total mass of the NLR gas is estimated to be $\sim 10^6\ M_\odot$. Structures of this scale and surface brightness can now be definitively resolved and their basic morphologies revealed in some Seyferts (e.g., Kraemer *et al.* (2008) Figure 3.20). Those

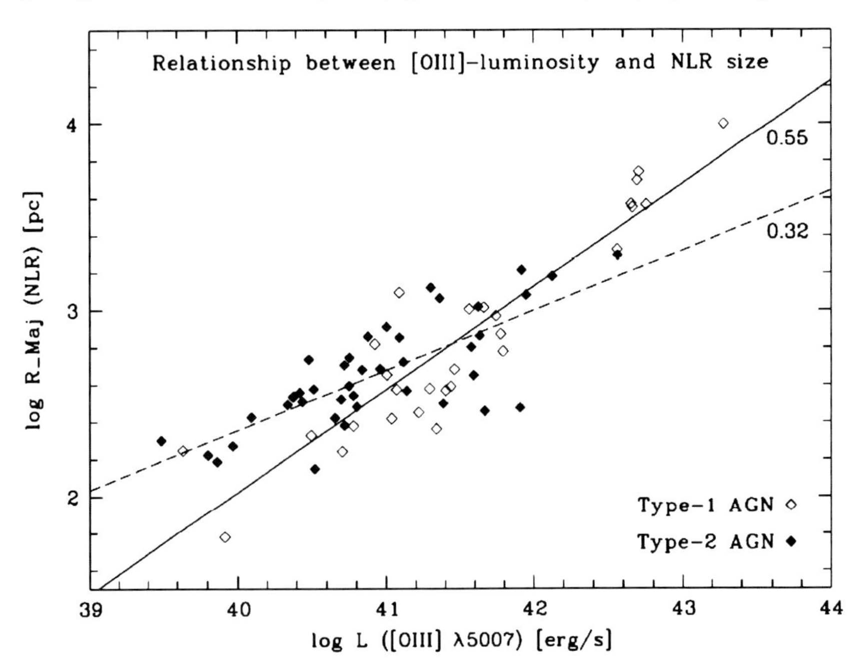

Figure 3.19 The size of AGN narrow-line region for a sample of objects based on narrow-band imaging plotted as a function of the [O III] λ5007 luminosity. The difference between the type 1 and type 2 AGN may be related to the effects of the putative dust torus surrounding the central black hole and accretion disk. (Reproduced from Bennert *et al.*, 2004, with permission by Cambridge University Press.)

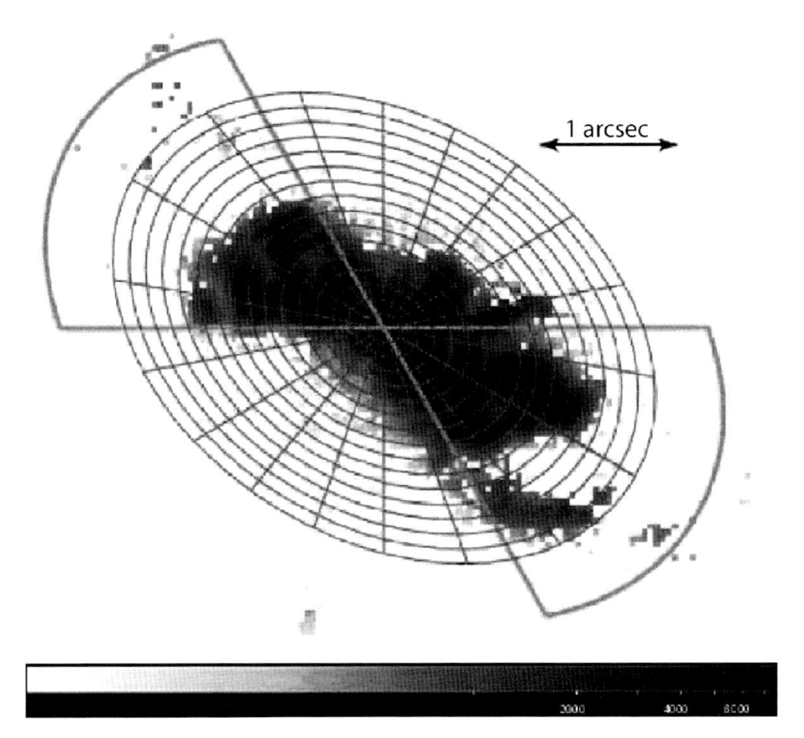

Figure 3.20 Narrow-band HST [O III] image of the central region of NGC 4151, with the biconical morphological structure depicted (Kraemer *et al.*, 2008). Those authors analyzed data within each of the overlaid annuli and compared line ratio measurements to the results of their photoionization calculations. Their results suggest that the NLR structure is due to filtering effects of the ionizing flux.

observations support the influence of a jet or collimated outflow on the NLR structure. The simplest explanation is that the putative dust torus collimates the ionizing radiation field leading to the observed roughly conical morphology.

The narrow line emission from AGN has recently been utilized as a tool for the study of a problem of longstanding interest. Binary supermassive black holes (SMBHs) have been proposed theoretically (e.g., Begelman *et al.*, 1980). In addition, many examples of galaxy mergers are now documented and it is also evident that most bulge-dominated merging galaxies harbor central SMBHs. The formation of binary SMBHs from galaxy mergers would thus seem inevitable, and in at least some cases these mergers could take place during (or trigger) active phases (or starburst activity) in one or both galaxies. However, observationally confirmed binary SMBHs are surprisingly rare. Only about 0.1% of quasars are found to be in physical pairs with projected separations of tens to hundreds of kpc (e.g., Hennawi *et al.*, 2010). In the Chandra deep X-ray observation, Schawinski *et al.* (2011b) found a triple-AGN using HST direct imaging and spectroscopic data. The three

AGN cores reside in a clumpy galaxy at redshift $z = 1.35$ and have black hole masses in the range of only $M_{\rm BH} = 3–10 \times 10^6\ M_\odot$. The Eddington ratios of these cores are quite different, with $L/L_{\rm Edd} = 1.4$, 0.3, and 0.09, with a combined X-ray luminosity of only $L_{0.5-8\,\rm keV} = 1.4 \times 10^{42}\ \rm erg\,s^{-1}$. The X-ray emission appears to be absorbed ($N_{\rm H} = 5 \times 10^{22}\ \rm cm^{-2}$), but not Compton-thick. This case might indeed represent a situation where one observes recently formed black holes at a comparably late epoch, about 4.8 Gyr after the Big Bang. Whether this is a particular source or whether there are many black holes forming late in the Universe is under debate.

Recently, Shen *et al.* (2011) have attempted to address the issue of binary black holes using a sample of type 2 AGN-selected from the Sloan Digitized Sky Survey (SDSS) that exhibit evidence for double-peaked [O III] emission lines. The basic idea is that the NLR is relatively compact about the central black hole, $R_{\rm NLR} \sim 1\ \rm kpc$ scales, and the lines are narrow relative to the anticipated dynamical wavelength shifts. They performed follow up near IR imaging and spectroscopy in attempt to identify binary SMBH candidates. They found that $\simeq 10\%$ of their objects are best explained by binary AGNs with projected kpc-scale separations, where two stellar components with spatially coincident NLRs are seen. An example is reproduced from their paper in Figure 3.21.

Here a two-dimensional spectrum of the AGN SDSS J1108+0659 clearly reveals the spatial plus spectroscopic separation of the [O III] and Hβ lines. Direct imaging observations (not shown) also spatially resolved the galactic nucleus into two components separated by $0.5''$.

It should be kept in mind, though, that there are other ways than a supermassive binary to produce a double-peaked emission line. Fischer *et al.* (2011) studied the optical spectra of the nearby Seyfert 2 galaxy Mrk 78 which shows a double-peaked [O III] line. Using HST data, the authors found that the specific line structure can be reproduced when assuming an asymmetric distribution of outflowing gas in the NLR. Because the outflow is not homogeneous but rather consists of single knots or clumps, one does not observe one broad line from the NLR, but instead the approaching and receding parts of the outflow produce the characteristic double-peaked structure. Fischer *et al.* (2011) argue that this scenario might explain also other double-peaked emission line profiles.

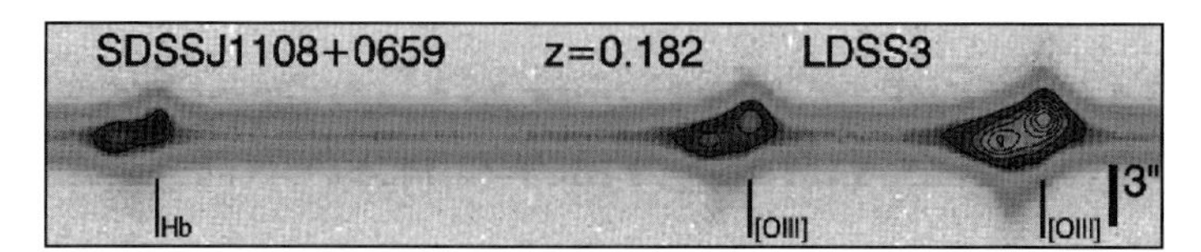

Figure 3.21 Two-dimensional spectrum for the Hβ-[O III] region of the type 2 AGN SDSS 1108+0659, which is at a redshift of $z = 0.182$ (reproduced from Shen *et al.* (2011)). The corresponding systemic line positions are marked. The two velocity components are spatially offset by 0.9 arcsec. The underlying stellar continua are not separated in the spectrum due to the limited seeing and the proximity of the two stellar nuclei. The nuclear region is also spatially resolved in a recent NIR adaptive optics direct imaging observation (not reproduced here; see Fu *et al.*, 2011).

3.6
Reverberation Mapping: Probing the Scale of the BLR

The basic idea underlying reverberation mapping is as follows (for a more complete discussion see, e.g., Peterson, 2008). Qualitatively, the observed AGN spectra consist of a strong continuum superposed by broad emission lines. The continuum emission, particularly towards the ultraviolet region of the spectrum, is believed to be due to accretion onto the central black hole. The simple geometrical picture then has that production at the center of the system surrounded by the gas producing the broad emission lines. Large Doppler motions in the gas that result from its proximity to the black hole and the strong gravitational field lead to the broadened line profiles. Since the lines are produced by photoionization or excitation from the continuum radiation, variations in the continuum should lead to changes in the line emission delayed by a time $\tau = R_{\mathrm{BLR}}/c$. The mass of the black hole can then be determined, following for example Peterson and Horne (2004) as

$$M_{\mathrm{BH}} \simeq f R_{\mathrm{BLR}} \frac{\Delta v^2}{G} = \frac{\tau c}{G} f \Delta v^2 \tag{3.22}$$

where f is a scale factor that depends on the geometry of the system. While this is in principle a powerful method, it requires in practice a large carefully orchestrated and thus "expensive" observing program. The lag times are on the order of weeks to months and the continuum and line variations occur sporadically. This necessitates a long baseline observing campaign, that is, one which spans many lag cycles. Furthermore, the preferred time series analysis method used in searching for characteristic time scales is the cross-correlation function (CCF) which presumes uniformly sampled data. Thus, the telescope scheduling problem becomes increasingly complicated (and political) as other observing programs will inevitably become interrupted or compromised. Finally, since as noted, the vacuum ultraviolet is the preferred spectral region in which one can cleanly sample the accretion-disk continuum and the high-excitation broad emission lines in nearby (and thus typically bright) AGN, the campaigns are best carried out from space. Notwithstanding these difficulties, a number of campaigns have now been carried out leading, collectively, to major advances in our knowledge (Peterson and Wandel, 2000; Bentz *et al.*, 2009b). One of the first successful reverberation mapping campaigns occurred in the late 1980s involving the Seyfert 1 AGN NGC 5548. Observations every 4 days were carried out over an 8-month period using the International Ultraviolet Explorer (IUE) satellite which covered the $\sim$ 1200–3300 Å range. Line-to-continuum light curve lags were clearly detected, with characteristic time scales that depended on ionization (Clavel *et al.*, 1991).

Subsequent observing campaigns using the Hubble Space Telescope (HST) which covers the UV with much greater sensitivity than IUE and from the ground using the Balmer lines and expanding samples to include higher-luminosity objects have been carried out over the subsequent decades. There are currently over 40 AGN with reverberation mapping black hole mass determinations (Bentz *et al.*, 2009b). The resulting distribution is depicted in Figure 3.22. These results can in

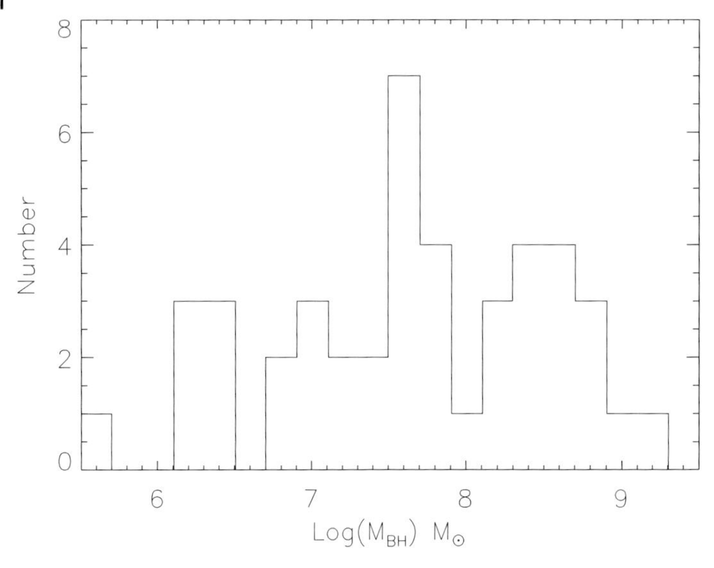

Figure 3.22 The distribution of black hole mass determinations resulting from reverberation mapping campaigns. This figure was reproduced from Bentz *et al.* (2009b).

some cases then be compared to black hole mass determinations obtained by other methods in both as a self-consistency check and to cross-calibrate the different methods.

For example, Nelson (2000) have investigated the relationship between narrow-line widths, specifically [O III]λ5007 Å and the reverberation mapping determined black hole masses. The narrow-line widths are a plausible proxy for the bulge velocity dispersion, which is often not accurately known in AGN which in turn is known to be related to the central black hole mass via the M–σ relationship (Section 3.1.2). They find that there is a relationship, but that there is considerably greater scatter than is the case for the M–σ relation. Nonetheless, such a relationship can facilitate compilation of a much larger database of AGN black hole mass estimates, albeit with larger uncertainties. Nelson (2000) also make black hole mass estimates in a subset of the reverberation mapped sample based on stellar dynamics. They find that there is general consistency between the mass estimates obtained through both methods, thus further substantiating the reverberation mapping technique.

The black hole mass is a fundamental parameter in any effort to model AGN. Thus, the importance of the reverberation mapping campaigns which provide the most reliable estimates has had a profound impact on the field. For example one of the most basic questions is at what fraction of the Eddington luminosity is an object radiating? Figure 3.23 illustrates the black hole masses for about 30 AGN as a function of their luminosities. The diagonal lines represent the Eddington luminosity for a given mass and the corresponding 1 and 10% $L_{\rm Edd}$ luminosities. This supports the idea that a typical AGN radiates at about 10% of Eddington, but that there is considerable scatter. The open circles represent the subset of reverberation-mapped AGN that are among the NLS1 subclass (Section 4.1.4). The position of

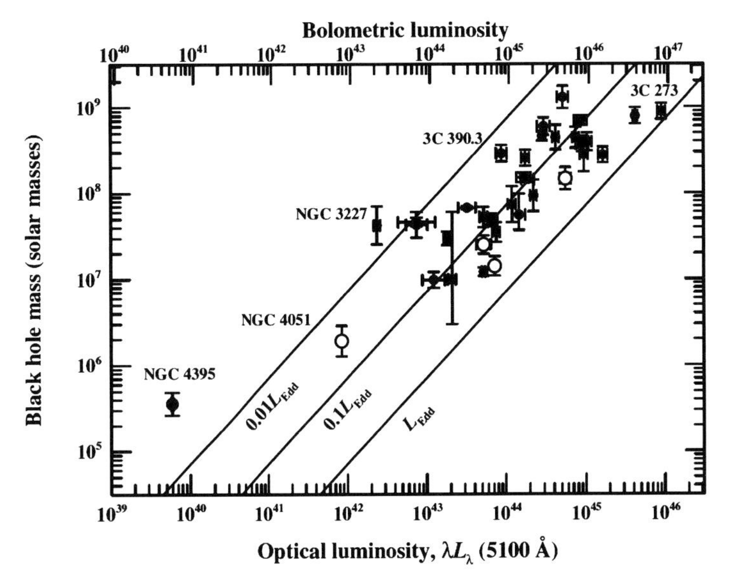

Figure 3.23 The black hole mass-luminosity relationship for reverberation mapped AGNs reproduced from Peterson (2007). The open circles identify NLS1s, which are generally believed to be high-Eddington ratio objects (see Section 4.1.4). The bolometric luminosity scale on top horizontal axis assumes the scaling $L_{bol} = 9 \times L_{5100}$, that is, a factor of 9 times the monochromatic 5100 Å luminosity.

those objects on this mass-luminosity diagram seems to be consistent with the notion that the NLS1s tend to radiate at a higher fraction of the Eddington rate than their broad-line counterparts.

A variant reverberation mapping technique has recently been extended into the X-ray domain (Miller and Turner, 2011; Papadakis *et al.*, 2001). In this case reverberation signals involve the cross-correlation analysis of time series from two separate energy intervals within the overall bandpass of a single X-ray detector. These datasets could result from a single long (relative to typical variability time scales) observation or from a well sampled monitoring sequence. Although AGN variability time scales in X-rays are typically shorter than in the optical-UV domain, these observation campaigns are nonetheless "expensive" as they most be done from space using heavily subscribed facilities.

The Size of an AGN

We now look at the different components which we consider to be part of the AGN phenomenon. At its center, the black hole, of which we know the size as soon as we know its mass. It is more difficult to estimate the size of the other components, the accretion disk, the broad- and narrow-line region, and the extension of the absorber, because these are model-dependent. Finally, the size of the jet is easier to determine as one is able to observe it directly in the radio domain. To summarize what we know about the size of an AGN, we give in the following table the size range and formulas or methods which are used to estimate the extension. Obviously, these are approximate values and are only meant to give a rough idea of the extent of the various aspects of an AGN.

Component	Approximate size	Estimator/method
BH event horizon	$R_S \sim 0.01-10\,\mathrm{AU}$	$R_S = 2GM_{BH}c^{-2}$
Inner disk radius	$R_{in} \sim 0.01-60\,\mathrm{AU}$	$R_{in} = 1.2 - 6R_S$
Outer disk radius	$\sim 1-1000\,\mathrm{AU}$	$\log R \simeq 15.8 + 0.8\log(M_{BH}/10^9\,M_\odot)$ (R in (cm); see Morgan *et al.*, 2010)
Broad-line region	$R_{BLR} \sim 0.01-1\,\mathrm{pc}$	$R_{BLR} \propto L^{0.6}$ (see, e.g., Bennert *et al.*, 2004)
Molecular torus	Parsec scale	Time delay measurement
Narrow-line region	$R_{NLR} \sim 10^2-10^4\,\mathrm{pc}$	$R_{NLR,type\ 1} \propto L_{O\,III}^{0.6}$ $R_{NLR,type\ 2} \propto L_{O\,III}^{0.3}$
Jet	Up to $\sim$ 100s kpc length	Radio imaging

The time lags between the different time series which may originate predominantly from separate spectral components can be used to address issues such as the importance of Compton reflection in AGN and physical scales involved. Hard lags, that is, hard-band photons trailing soft-band photons are generally expected since the former can Compton scatter with circumnuclear material much more efficiently than the latter, which tend to be partially absorbed. As an example of the technique example, Miller and Turner (2011) have used long observations of three NLS1 AGN made with the Suzaku Observatory to explore the scale and geometrical configuration of scattering media in the vicinity of the nuclei. They have computed the cross spectrum (see Section 6.2), which provides the hard-soft band lag times as a function of frequency.

Hard lags are what Miller and Turner (2011) observed at low frequencies, but the lags tend towards zero at higher ($\sim 10^{-3}$ Hz) frequencies, and in one case, 1H 0707-495 they become negative, as shown in Figure 3.24. They further interpret the hard-to-soft transition frequency to be indicative of a Compton up-scattering corona with size of order a few thousand light seconds. They argue against a compact, $\sim 1R_S$ reflector as suggested in the literature in the context of the interpretation of broad K and L shell transitions in AGN (e.g., Fabian and Miniutti, 2009). While this issue may continue to be debated, the viability of the technique for probing the size and AGN inner regions is evident.

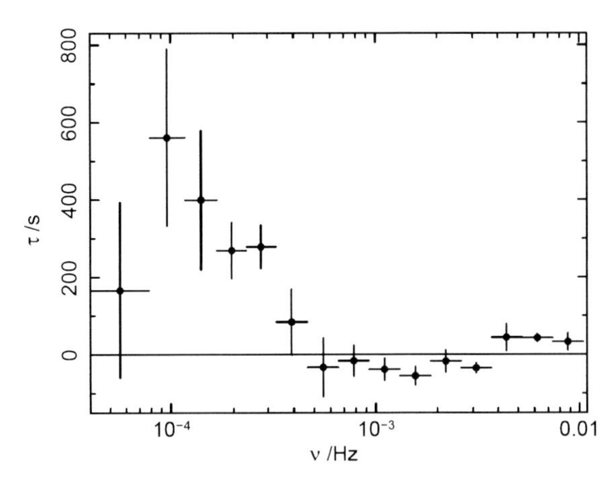

Figure 3.24 Fourier cross-spectrum, or lag spectrum, of the NLS1 1H 0707-495 derived from 0.4–1 and 4–7.5 keV time series obtained with the XMM-Newton observatory. Note the statistically significant negative lags at a few 10^{-3} Hz and the transition from positive to near zero lag new 10^{-3} Hz. Reproduced from Miller and Turner (2011).

3.7 AGN Jets: Emission, Dynamics and Morphologies

A notable feature of certain subclasses of AGN is the presence of jets emanating from the region near the active nucleus. The first example noted in the literature was M87 (e.g., Curtis, 1918) in the Virgo Cluster. These are now believed to be extremely energetic and highly collimated outflowing plasma structures launched from an accretion disk. They can extend up to hundreds of kiloparsecs or even megaparsecs into space often retaining a remarkably high degree of collimation. Apparent propagation velocities can be up to 40 times superluminal. We will explain the phenomenon of superluminal motion in Section 3.7.3. The jets are most prominent in radio frequency observations, where in particular long baseline interferometry can resolve many AGN jets in remarkable detail. Jets occur in about 10% of AGN, the so-called radio-loud objects. Those AGN are further categorized by a broad, nonthermal continuum attributed to synchrotron radiation from the charged particle population that presumably comprises the jets. The observational appearance of a compact jet in a radio-loud AGN consists of a core, which is often marginally resolved or unresolved at VLBA resolution.

We show the example of the inner jet of M87 and what the recent X-ray, optical and radio observatories can achieve with respect to resolution power, in Figure 3.25. The protruding structures can exhibit a variety of appearances due to both projection effects and intrinsic properties. They can appear short and irregular, long and straight or curved. They can be relatively smooth morphologically, or can be dominated by brightness enhancements known as knots. The knots are often superluminal, although they can be subluminal or even stationary in some instances (see, e.g., Kellermann *et al.*, 2004). Separate morphological substructures of a given jet

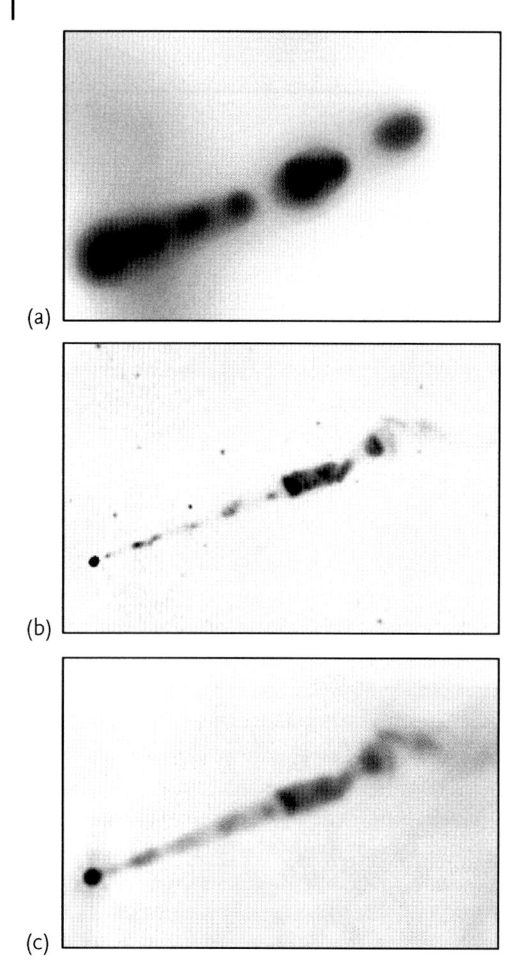

Figure 3.25 This composite shows the M87 jet as observed by the Chandra X-ray Observatory (a, X-ray), the Hubble Space Telescope (b, optical), and the Very Large Array (c, radio). At about 17 Mpc distance, M87 provides one of the best opportunities to study an AGN jet in detail. Credit: X-ray: NASA/CXC/MIT/H. Marshall *et al.*, Radio: F. Zhou, F. Owen (NRAO), J. Biretta (STScI), Optical: NASA/STScI/UMBC/E. Perlman *et al.*

appear to move with a common characteristic velocity. This presumably represents an underlying continuous flow, but as noted, stationary, slower moving and even apparently reverse-moving features are sometimes seen. There is evidence for a systematic decrease in apparent velocity with increasing wavelength, but this could be a result of the observations sampling different parts of the jet.

Another issue yet to be resolved is the composition of AGN jets. Although current blazar studies, in particular analyses of correlated X-ray and TeV gamma-ray flares, favor leptonic jet models involving efficient acceleration of relativistic electrons (e.g., Mastichiadis and Kirk, 1997), models involving hadronic jets cannot be ruled out (e.g., Mücke and Protheroe, 2001). We note as a point of clarification that the presence of hadrons in the flow does not necessarily imply that they significantly contribute to the observed emission. It may be the case that the essential aspects of leptonic emission models are valid both for pure leptonic or mixed lepton-hadron composition. The hadronic emission models on the other hand necessarily require ultrarelativistic hadrons in the plasma. The present consensus is that AGN jets are likely composed of a mixed proton and lepton plasma. The protons would comprise most of the kinetic energy with the leptons being responsible for the entire emission we observe. A useful description of these ideas can be found in Sikora *et al.* (2009).

One argument for hadronic blazar jets, put forth by Aharonian (2000) was based on the fact that the reported shape of the TeV spectrum which, during strong flares of the source, remains stable despite dramatic variations of the absolute gamma-ray flux. He argued that this type of source behavior could be explained if the TeV emission was a result of synchrotron radiation of extremely high-energy ($E \geq 10^{19}$ eV) protons in highly magnetized ($B \sim 30$–100 G) compact regions of the jet with typical size $R \sim 10^{15}$–10^{16} cm and Doppler factor $\delta \simeq 10$–30 (here δ is related to the bulk Lorentz factor γ by the expression $\delta = [\gamma(1 - \beta \cos\theta)]^{-1}$). This model may face problems however in accommodating the rapid TeV gamma-ray variability seen in some instances (see Section 6.5).

There could be other observable consequences of hadron acceleration resulting from interactions with ambient material or photon fields as well. These could include for example the high-energy electromagnetic radiation from the secondary leptons and γ-rays (e.g., Atoyan and Dermer, 2004). Another possibility is the production of neutrinos although a neutrino signature associated with individual AGN is unlikely to be observable with current technologies. The production of collimated beams of ultrahigh-energy ($E > 10^{15}$ eV) neutrons and gamma rays formed in the same process of hadronic interactions is also a possibility and that could in principle explain the apparent transport of energy from AGN cores to multikiloparsec distances. Observations at other wavelengths, that is radio and X-rays, could in principle test this idea. For example, for a given initial energy, the neutron decay process would correspond to a predictable angular scale on the sky which could be compared to observations.

Attempts to delineate between leptonic and hadronic jets have produced ambiguous and sometimes conflicting results (e.g., Wardle *et al.*, 1998; Ruszkowski and Begelman, 2002). The former cite circularly polarized radio emission as a signature of electron–positron pairs, while the latter group argues that turbulent magnetic fields could produce similar effects in normal electron-ion plasmas. One definitive signature of electron–positron pair plasmas would be the detection of 511 keV annihilation radiation, but no such detections have been reported in the literature to our knowledge. It is worth noting once again though that there is one observationally definitive example of a hadronic jet associated with the galactic *microquasar* SS 433.

3.7.1
Raising the Jet

How AGN jets form and the nature of their composition and mechanical configuration is uncertain. The conventional picture involves magnetic fields threading the accretion roughly parallel to the disk axis. The twisting of these magnetic fields through their interaction with ionized material comprising the disk leads to collimation of the outflow along general direction of the rotation axis of the central object. Thus, under suitable conditions a jet will emerge from each face of the accretion disk, i.e., a *bipolar* structure. If the jet line of sight is oriented approximately towards our line of sight from Earth, relativistic beaming effects will change its apparent brightness and the relative prominence of the approaching and receding components. The detailed mechanism underlying the creation of the jets and their material composition is still a highly debated issue within the astronomical community.

We do not know the detailed physics or the precise geometrical configuration of the accretion flow in the vicinity of the black hole. The presumption is that it is a thin disk that continues down to the innermost stable circular orbit. On the other hand the flow could become advection-dominated and comprise a more nearly spherical geometry. The accretion disk emits thermally in the ultraviolet, optical, and infrared bands due to viscous dissipation. There is also a putative hot corona – either a disk "atmosphere" or perhaps the base of the jet – whose electrons Compton scatter some of these photons to X-ray energies. Differential rotation of the magnetic field within the disk, or perhaps within the black hole ergosphere is then invoked in most modeling scenarios to launch the jet.

The most common theory for powering the jet involves the Blandford–Znajek mechanism (Blandford and Znajek, 1977), which allows for extraction of energy from the central black hole. It requires a spinning black hole. The magnetic fields threading the accretion disk or in the vicinity of the ergosphere are "dragged" by the spin of the black hole. The relativistic material can then be launched by the reconfiguration, or tightening, of the field lines.

Recently, numerical simulations invoking three-dimensional relativistic magnetohydrodynamics by Tchekhovskoy *et al.* (2011) support this scenario. Their calculations lead to powerful outflows in the case of a spinning black hole. Furthermore, the fraction of the available energy increases with the presumed black hole angular momentum (or spin). If the black hole is spun up to its nearly maximal rate characterized by spin parameter $j = 0.99$ (see Eq. (5.8)), they find that an amount of energy equivalent to 140% of that available from the accretion flow onto the black hole directed into the jet. Thus, clearly an additional energy source is required, and they conclude that the only plausible source is the net extraction of energy from the spinning black hole via the Blandford–Znajek mechanism.

One common theme among the magnetic launching models is that the regularity of the magnetic field geometry. Specifically, in those scenarios it is expected to form a helical configuration through its interaction with the disk. A dilemma thus arises, as many observations seem to be consistent with the presence of disordered fields and turbulent plasma (e.g., Lister and Homan, 2005). Apparently, the helical structures exist (see, e.g., Gabuzda *et al.*, 2008), but they may often only occur on scales compact enough to elude observation at VLBI resolution, e.g., Sikora *et al.* (2005).

In any case, further advances in our understanding of the basic problem of launching and powering the jets are likely to involve increasingly large and detailed numerical simulations. Advances in computing capabilities as well as in the codes applied have led to slow, but consistent progress. For example, see McKinney (2006) for a recent discussion of a robust set of simulations and interpretive discussion.

Another unresolved issue is whether the jet is accelerated to a high bulk Lorentz factor in the immediate vicinity of the central black hole, for example within tens of Schwarzschild radii. Alternatively, the jet acceleration could occur gradually over distances of hundreds or thousands of Schwarzschild radii. One argument for the latter is that the dense radiation field in the inner region would lead to a "Compton drag" effect, that is the jet particles would experience substantial energy loss due to Compton scattering with the photon field ambient to the central engine. Other calculations invoking magnetic or focusing or gas dynamics generally accelerate the jet over longer distances, but discrepancies between different models persist in the literature. For recent reviews of AGN jet models see Meier (2003) and Worrall (2009).

Recent advances in the study of jet formation and propagation are emerging owing to intensive observing campaigns and enhanced facilities. Notably, the Fermi Gamma-Ray Space Telescope launched by NASA in 2008 provides coverage of the full sky from about 100 MeV–300 GeV on hourly time scales. In its first two years of operations, Fermi detected nearly 900 AGN a majority of which are identified as blazars. Roughly half of these are BL Lacs and half are flat-spectrum radio quasars. The energy budgets of these classes of objects often peaks in the gamma-ray spectral domain covered by Fermi. Contemporaneous with Fermi, VLBA campaigns such as the MOJAVE collaboration are sampling the formation and propagation of jet structures in many of the same objects at submilliarcsecond scales. There are also spaced-based X-ray observatories, notably the Rossi X-ray Timing Explorer (RXTE) which operated until early 2012 and the Swift satellite. Those observatories each have flexible scheduling capabilities and thus naturally facilitate the study of time variable and transient phenomenon such as blazar AGN.

The gamma rays are widely believed to result from Compton scattering within the jet – either through the synchrotron self-Compton mechanism or from upscattering of photons from a source external to the jet. A persistent question since gamma-ray emission from blazars was discovered in the 1990s though has been the where are the gamma rays are formed? Two basic constraints that need to be considered in addressing this question are the associated variability time scales, that is, shorter time scales favoring more compact sites, and photon–photon opacity effects which can constrain the compactness of the emission sites.

Regarding the photon–photon opacity, two scenarios should be considered. Strong opacity is expected for sources in which the environment, for example the

BLR perhaps within $\sim$ 1 pc scales, surrounds the jet and produces a dense isotropic UV-optical photon field. This is likely for example in the case of the FSRQ. In this case the source of the opacity is ambient to jet, such that photons leaving the jet within the BLR region have to propagate in this environment, which is characterized by very large opacities (e.g., Donea and Protheroe, 2003). In this case, the opacity does not depend on the jet parameters such as the Doppler factors, but rather it is a property of this ambient environment.

Recently, Poutanen and Stern (2011) have interpreted the apparent $\sim$ GeV spectral breaks in some gamma-ray emitting FSRQ in terms of photon–photon opacity effects within the BLR radius that result from the intense UV radiation field therein. The compact nature would also provide a natural explanation for the observed short time scale flux variability. However, observational evidence for gamma-ray emission sites well beyond the central parsec-scale region has emerged from multiwavelength plus VLBI campaigns.

A different source of opacity (e.g., Dondi and Ghisellini, 1995) is expected due to soft photons produced inside the source, that is, the synchrotron photon field. Gamma rays must propagate inside the jet radiation field and can be absorbed before they escape. This type of absorption, of course, depend on the jet's intrinsic properties, in particular the jet Doppler factor. This then imposes a limit on the "compactness" of the gamma-ray emitting region for a given Doppler beaming factor (e.g., Maraschi *et al.*, 1992).

Jets that bear some resemblance to AGN jets, though on a much smaller scale, are now known to be associated with the accretion disks of galactic neutron star and black hole X-ray binaries. These binary systems are often referred to as microquasars. The first discovered and famous example is SS 433, whose well-observed jet has a velocity of $0.23c$. It is also the only instance where a baryonic matter component is directly observed in an astrophysical jet; this by virtue of red and blueshifted atomic emission lines in optical-UV-X-ray spectroscopy. There are other microquasars which appear to have higher-velocity jets. GRO J1655-40 for example has much exhibited radio jets in 1994 interpreted by Hjellming and Rupen (1995) to be propagating at about $\simeq 0.9c$, but it is a transient source, and has never duplicated this behavior subsequently.

One important breakthrough in studies of the galactic microquasars, which may have significance towards understanding AGN jets as well, is the observation that the jet formation seems to occur during a certain accretion state of the system. This is characterized by its evolutionary track in a hardness-intensity diagram as shown in Figure 4.18. The jet forms in the low-hard state where radio intensity increases roughly linearly with X-ray intensity. Superluminal knot ejection occurs at the transition to the high-soft state, after which the jet apparently turns off. This evolution in the accretion flow typically occurs on time scales of days to weeks in the microquasars. The analogous changes in AGN would presumably scale with the black hole mass in some manner and not be discernible on observable time scales. However, if AGN disk accretion follows analogous evolutionary tracks one should be able to sample this by observing an ensemble of many objects.

Recently, Chatterjee *et al.* (2009) presented results from a 5-year radio-interferometric – X-ray monitoring program on the radio-loud Seyfert galaxy 3C 120 using the VLBA and the RXTE satellite. Over the course of their 5 year of observation baseline, significant dips in the X-ray light curves were detected and quantified. Their analysis revealed ejections of bright superluminal knots in the VLBA images occurring subsequently to these X-ray dips. The striking correlation between the time of the X-ray dips and the knot ejections is shown in Figure 3.26. Additionally, the X-ray flux and the 37 GHz flux are anticorrelated with X-ray leading the radio variations. Furthermore, they found that the total energy radiated in a radio flare was related to the equivalent width of the corresponding X-ray dip. This behavior is strongly suggestive of a disk-jet connection analogous to the aforementioned patterns in the galactic microquasars.

AGN jets are now routinely detected and imaged in the X-rays as well as in radio and optical; see Massaro *et al.* (2011) or Harris and Krawczynski (2006) for report on a compilations of X-ray observations and their interpretations for over 200 AGN jets. Massaro *et al.* (2011) consider the X-ray to radio flux ratios in the jet knots finding similarity between quasars and FRI objects within their sample. This is difficult to explain if the X-rays, as is commonly believed, result from inverse Compton scattering of jet electrons with the CMB.

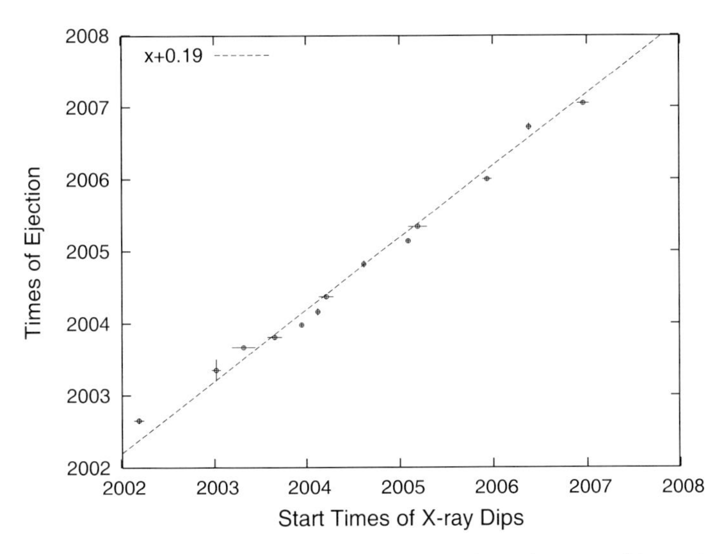

Figure 3.26 This plot, from Chatterjee *et al.* (2009), results from a compilation of joint VLBA and X-ray flux monitoring of 3C 120. The *y* axis is the estimated time of plasma ejection leading to jet propagation gleaned from the time-sequenced VLBA images. The *x* axis shows the times of dips in the X-ray light curves. The remarkably tight correlation thus derived is suggestive of correlated disk-jet behavior analogous to that now well-established in galactic microquasars.

3.7.2 Shocks and Knots

Superluminal knots, manifest as localized intensity enhancements in radio interferometric images of AGN jets, are generally attributed to the presence of shocks in the collimated plasma outflow (e.g., Blandford and Konigl, 1979). VLBI observations provide us with the unique tool to explore the jets with milliarcsecond angular resolution corresponding to parsec-scale resolution in linear size (see Section 5.1). A particularly striking example of the current VLBI capabilities is illustrated in Figure 3.28, where the inner-jet structure of Cen A, the nearest AGN, has been resolved at linear scales of a few hundredths of a parsec. Typically, parsec-scale flows are characterized by a one-sided knotty structure with sometimes pronounced curvature, rapid variations of flux density, and detection of superluminal motions (Section 3.7.3). Shocks are present in a wide variety of astrophysical phenomena, from our own solar system up to galaxy cluster scales. In AGN the ejection of a relativistic plasma "blob" could propagate through the outflowing plasma stream leading to a trail of shocks. This basic idea is supported by detection of strong polarization and enhancement of the magnetic field in the brightest jet features as determined from frequency-dependent imaging (e.g., Pushkarev *et al.*, 2008, and references therein). The polarization directions measured in many AGN are consistent with a shock front propagation transverse to the jet direction (e.g., Lister and Homan, 2005), although in some cases the measurements suggest turbulent magnetic fields.

In any case, the physics of how to generate shocks in highly relativistic plasmas at speeds is precarious (see Figure 3.27). It requires a relative fluid velocity that exceeds the local sound speed. In practical terms a factor of ~ 2 in the relative speeds is likely required to produce the shocks. The superluminal knots thus formed by shocks are thus likely to arise from activity close to the AGN central engine near to putative origin of the jet and models are driven by that paradigm. Realistic models must be able to account for, qualitatively at least, the time variability of the flux, polarization, and for the jet contribution (or domination in the case of blazars) to the continuum spectral energy distributions. There has been considerable success with numerical modeling nonetheless (e.g., Marscher and Gear, 1985; Polko *et al.*, 2010).

In many of these modeling scenarios, electrons are accelerated at the shock front and then radiate losing energy. The plasma also expands transverse the jet flow lagging behind the front. This leads to stratification in the energy density, with the highest energy electrons occupying a narrow region behind the shock front. The synchrotron emission is correspondingly stratified in terms of its emergent spectrum. Variability, in particular flaring events, can be applied to these models and information extracted if the observed spectral changes can be reproduced.[1] Alternative models to those invoking shocks attempt to explain the knots as discrete plasma density enhancements, often referred to in the astronomical literature as

1) From Perlman *et al.* (2011), images from HST shows polarization in M87 knot HST-1 being variable between $\Pi \simeq 20 \rightarrow 40\%$.

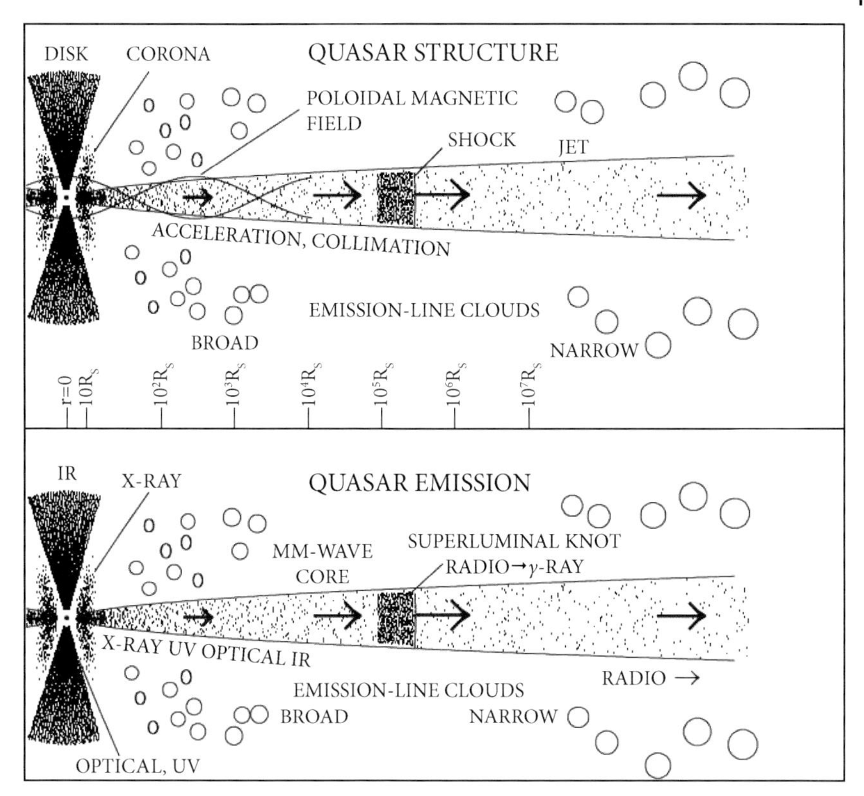

Figure 3.27 Schematic diagram of a disk-jet structure in a radio-loud active galaxy. The density of dots corresponds to the intensity of the emission. The jet emission is relativistically beamed. It is unclear whether the emission from the ambient jet between the black hole and the core is visible. The length of the arrows along the jet axis indicates the Lorentz factor of the flow. Note that the distance scale is logarithmic and is labeled in units of Schwarzschild radii. Graphic from Marscher (2005).

"blobs." These blobs would then propagate along with the jet flow, but they could expand longitudinally as well as laterally. They would also have to have internal magnetic field configurations to explain the polarization data. Most astronomers currently favor models based on shocks as the more likely explanation for appearance of knots in the high-resolution radio images.

3.7.3
Superluminal Motion

Observations of processes where matter seems to travel at velocities exceeding the speed of light seem to contradict special relativity. In fact, the apparent superluminal motion is based on a geometric effect, as shown, e.g., in Blandford *et al.* (1977)

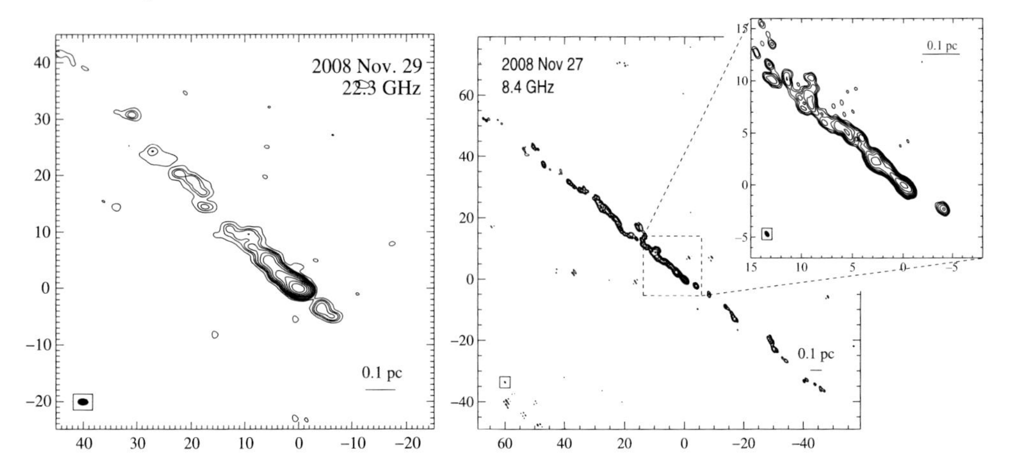

Figure 3.28 High-resolution VLBI image at 22.3 and 8.4 GHz of the inner jet of the radio galaxy Cen A. All axes are in units of milliarcseconds relative to the phase center. Knot structures on linear scales of a few hundredths of a parsec are clearly resolved into discrete components. From Müller *et al.* (2011).

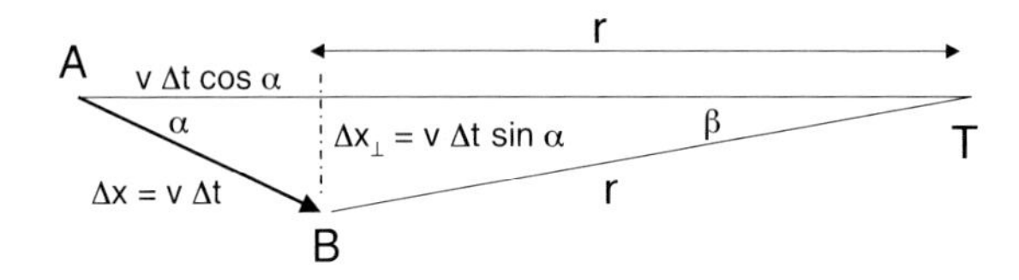

Figure 3.29 Schematic representation to explain superluminal motion. The AGN core at point A emits a blob of matter at time t_1, which reaches point B at $t_2 = t_1 + \Delta t$. The observer is located at point T with a distance r to the emitted matter at point B.

and we follow the explanation of Peterson (1997). Figure 3.29 shows a schematic representation of the problem. Consider that we observe an AGN at point A. This AGN emits matter, for example in a jet, with speed v with an angle to the line of sight α at a given time t_1. Because the light travels with c, we observe this event at a later time t_1' when the signal has covered the distance from point A to point T, where the telescope is set up. After a time interval Δt the matter in the jet will have traveled the distance $\Delta x = v\Delta t$ and is now in point B at time $t_2 = t_1 + \Delta t$. The projected (observed) distance of this movement is $\Delta x_{\text{obs}} = v\Delta t \sin\alpha$. The angle under which we see this movement is

$$\beta = \frac{\Delta x_\perp}{r} = \frac{v\Delta t \sin\alpha}{r} \tag{3.23}$$

with r being the distance to the emitted matter at time t_2. Since $r \gg \Delta x$ and β is small, the projection of the position B′ into the line of sight towards the AGN core at B thus defined at distance r. We as observer detect the time of emission of the

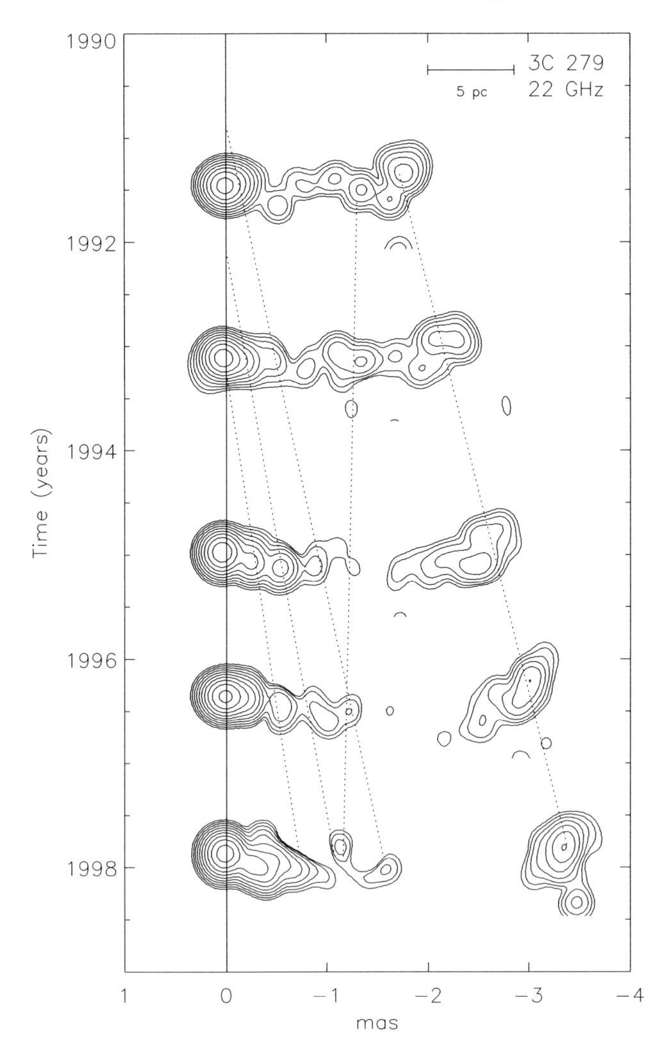

Figure 3.30 Time series of radio observations of the core and jet in the blazar 3C 279. The 22 GHz VLBI data show the apparent superluminal motion with respect to the stationary core of the AGN, which is shown on the left. For example, the component on the right appears to have a speed of $(7.5 \pm 0.2)c$ with respect to the central engine. Dotted lines indicate the assumed movement of the various jet components. From Wehrle *et al.* (2001).

event at the core of the AGN at

$$t_1' = t_1 + \frac{r + v\Delta t \cos\alpha}{c} \tag{3.24}$$

and we see the matter which arrived at point B at $t_2 = t_1 + \Delta t$ finally at

$$t_2' = t_2 + \frac{r}{c} \tag{3.25}$$

Thus, for the observer at point T the time interval between emission of the matter at point A and arrival at point B is

$$\begin{aligned}\Delta t' = t_2' - t_1' &= t_2 + \frac{r}{c} - t_1 - \frac{r + v\Delta t \cos\alpha}{c} \\ &= \Delta t - \frac{v\Delta t \cos\alpha}{c} = \Delta t \left(1 - \frac{v}{c}\cos\alpha\right)\end{aligned} \tag{3.26}$$

We observe a perpendicular movement with the speed $v_\perp$ to the line of sight of the emitted matter of

$$\frac{v_\perp}{c} = \frac{1}{c}\frac{\Delta x_\perp}{\Delta t'} = \frac{1}{c}\frac{v\Delta t \sin\alpha}{\Delta t'} = \frac{v \sin\alpha}{c\left(1 - \frac{v}{c}\cos\alpha\right)} \tag{3.27}$$

As shown in Peterson (1997), this function has a maximum at an emission angle $\alpha_{\max}$ to the line of

$$\alpha_{\max} = \frac{1}{\cos\frac{v}{c}} \tag{3.28}$$

The maximum apparent perpendicular speed one can observe then is

$$v_\perp^{\max} = v\frac{1}{\sqrt{1 - \frac{v^2}{c^2}}} = v\Gamma \tag{3.29}$$

where Γ is the Lorentz factor of the emitted matter. Because the Lorentz factor can be $\Gamma \gg 1$, the apparent motion can be superluminal already for mildly relativistic cases when the jet emission is close to the line of sight.

Figure 3.30 shows a prominent example of apparent superluminal motion in the blazar jet of 3C 279 at redshift $z = 0.536$. The high observed velocity of 7.5 times the speed of light implies a low pitch angle helical motion (see Section 2.2.4), and an emission of the jet within $\alpha \lesssim 2°$ of the line of sight (Wehrle *et al.*, 2001).

4
AGN Types and Unification

AGN comprise a broad class including of a large variety of subtypes that collectively occupy a vast parameter space. For example, the mass of the central black hole can range from a few times $10^5\,M_\odot$ in the case of NGC 4395, up to $10^{10}\,M_\odot$ for objects like the quasar 3C 273. They range in bolometric luminosity from some $L_\text{bol} \sim 10^{41}$–$10^{48}\,\text{erg}\,\text{s}^{-1}$. Some AGN undergo large intensity outbursts, while others appear as persistent sources. The optical-UV spectra of some AGN exhibit strong emission lines, while in others lines seem to be absent.

One has to keep in mind that the AGN phenomenon was first defined based on observational characteristics. In 1943 Carl Seyfert observed emission line galaxies in which the lines were redshifted. Relating the redshift to a distance, these galaxies had to be very luminous, thus, something special had to be at work in the center of these objects that distinguished them from the nonactive spiral or elliptical galaxies. With the advent of powerful radio telescopes in the 1960s, strong radio sources were discovered whose optical counterparts often appeared point-like, thus the designation "quasi-stellar objects" (QSOs) was introduced. Assuming that the point-like appearance is indicative of large distances, one was led to the conclusion that these QSOs had to be rather extraordinary extragalactic sources.

Thus, a large number of different flavors of AGN were discovered. From our current perspective, in which one postulates a common type of central engine at the core of all these different AGN, an effort is being made to try to unify these disparate subclasses. The task is to find the most simple model possible to explain the different appearances. We will first take a look at the different classes and subclasses of AGN which are known and then will consider scenarios for their unification in Section 4.8. A concise review on the topic of classification and unification of AGN has been provided by Tadhunter (2008).

4.1
Seyfert Galaxies

Seyfert galaxies are the most common class of AGN we observe in the local Universe. Owing to their relative proximity we have the best available spectra and images for them, allowing us to study in detail the physical processes at work. Many

Active Galactic Nuclei, First Edition. Volker Beckmann and Chris Shrader.

of the main characteristics we discussed in the previous chapter, have been discovered in Seyfert galaxies. This is not true though for the jet phenomenon, because the emission of Seyfert galaxies is evidently not beamed. In the following we will discuss first the optical classification of Seyfert galaxies, because their definition as a class originates from optical spectroscopy. Because H II regions can be easily misidentified as Seyfert 2 galaxies we have a closer look at them in Section 4.1.2. X-ray observations are currently a common tool to discover and study AGN, and classification can be based on those data, as we will see in Section 4.1.3. A special subtype of Seyfert galaxies are the narrow-line Seyfert 1 galaxies (NLS1), and Section 4.1.4 is dedicated to them.

4.1.1 Optical Classification

Seyfert galaxies were the first AGN identified as such. As mentioned above, Carl Seyfert (1943) observed galaxies using the 60-in and 100-in telescopes on Mount Wilson. The 100-in telescope, inaugurated in 1917, was the world's largest telescope until 1948, and it had already served Edwin Hubble for his extragalactic studies, and also Albert Michelson who had worked on interferometry at Mount Wilson.

What distinguishes Seyfert galaxies from nonactive galaxies in direct images is the bright, central, point-like core. With the advent of larger telescopes and improved instrumentation, it became straightforward to obtain spectra of the unresolved core and from the host galaxy, and it was discovered that the highly ionized emission lines are indeed emitted from the central core, while the hosts had typical galactic spectra. These host galaxy spectra are interpreted as a synthesis of the stellar populations and thus they display photospheric absorption lines imprinted on their continuum. Emission lines from low-ionized gas may be present depending on the type of host galaxy. We will come back to the relation of AGN to their host galaxies in Section 7.1.

The identification as a Seyfert galaxy is nowadays based on the spectral signature of the AGN core. In case this spectrum shows highly ionized emission lines, it qualifies as a Seyfert. Studying the spectra of Seyfert galaxies Khachikian and Weedman found in the early 1970s two generally distinct types of optical spectra. While all spectra had narrow and unresolved emission lines from highly ionized material, only some of the Seyferts also exhibited broad lines. They separated the Seyfert galaxies into classes according to their relative widths of narrow (forbidden) lines and Balmer lines (Khachikian and Weedman, 1974). In the Seyfert class 1 the Balmer lines, mainly Hα, Hβ, and Hγ, would appear broader than the forbidden lines, like the oxygen [O II] and [O III] lines, and the nitrogen and neon [N II], [Ne III], and [Ne IV] lines. Table 4.1 gives an overview of the most prominent lines observed in AGN spectra and their rest frame wavelength λ_0. In the type 2 objects, both the forbidden and the Balmer lines show the same narrow width. A closer inspection of the spectra shows that actually the Seyfert 1 galaxies have a broad and a narrow component in the Balmer lines. Another indication of a type 1 AGN is the presence of a strong Fe [II] line at 4570 Å.

Table 4.1 Typical lines observed in spectra of AGN. Note that most of these lines can appear in emission and in absorption. The G-band, Na I-D, Ca II-H and Ca II-K are absorption lines of the host galaxy and the latter two mark the calcium break observable in many AGN spectra. The calcium break and the Mg II absorption line are often used to identify the redshift of otherwise featureless blazar spectra. Figure 3.17 shows a quasar spectrum with the lines indicated here.

Line	λ_0 (Å)	Line	λ_0 (Å)	Line	λ_0 (Å)
Lyβ	1026	Hη	3835	[O III]	4959
O VI	1032	[Ne III]	3869	[O III]	5007
Lyα	1216	Hζ	3889	[Mg I]	5174
N V	1240	Ca II-H	3935	[Ca V]	5309
Si IV	1400	Ca II-K	3970	[Fe VII]	5721
C IV	1549	Hϵ	3970	He I	5876
He II	1640	[S II]	4071	Na I-D	5894
[C III]	1909	Hδ	4101	[O I]	6300
[C II]	2326	G band	4306	[N II]	6548
[Mg II]	2798	Hγ	4340	Hα	6563
[Ne V]	3346	[O III]	4363	[N II]	6584
[Ne V]	3425	[He II]	4471	[S II]	6716
[He I]	3587	[Fe II]	4570	[S II]	6731
[O II]	3727	[He II]	4686	[Ar III]	7136
[Fe VII]	3760	Hβ	4861		

The narrow lines are characteristic for low-density gas with electron densities of $n_e \simeq 10^3 - 10^6\ \mathrm{cm}^{-3}$ and typical velocities of a few hundred kilometers per second. In contrast to this, the broad Balmer lines arise from dense matter ($n_e \geq 10^9\ \mathrm{cm}^{-3}$) and the line widths indicate velocities of $10^3 - 10^4\ \mathrm{km\,s}^{-1}$ (see Section 3.5).

In the continuum emission, one observes a superposition of the host galaxy and the AGN core. In Seyfert 2 galaxies the AGN core is usually less dominant with respect to the surrounding galaxy than in the Seyfert 1 objects. This is one reason why it is generally more difficult to find Seyfert 2 galaxies based on their optical spectra: often longer exposure times are needed in order to separate the weak emission line spectrum superposed on the synthetic stellar spectrum of the galaxy. On the contrary, Seyfert 1 galaxies often exhibit a strong continuum which cannot be attributed to the surrounding galaxy. This continuum emission appears to be featureless, that is, without the characteristic stellar absorption lines. These featureless spectra often exhibit nonthermal continua which further distinguish them from the stellar spectra. This can be a somewhat misleading term, if we recall that the continuum emission in the AGN core can be understood as the signature of the accretion disk, and thus a superposition of many black body spectra of a continuous range of temperatures, as outlined in Section 3.2. On the other hand, in some radio-loud Seyfert 1 objects, emission can indeed be influenced by nonthermal synchrotron radiation from a weak jet, which clearly contributes to the optical emission. Generally, one must study the broader spectral energy distribution, as it is otherwise difficult to distinguish between a multicolor black body arising from an accretion disk which has a characteristic $\nu^{1/3}$ spectral shape (Section 3.2), from truly nonthermal processes. Figures 4.1 and 4.2 show examples for optical spectra of a Seyfert 1 and Seyfert 2, respectively.

In spite of the clear distinction between Seyfert 1 and Seyfert 2, intermediate and borderline objects are observed. These objects are sorted into subclasses Seyfert 1.2, 1.5, 1.8, and 1.9 according to their Balmer line characteristics, following the scheme introduced by Osterbrock (1977). The subtypes are based on the relative width of the Hβ line: a Seyfert 1.2 is a Seyfert 1 galaxy with typical broad Balmer lines, but with a slightly less broad Hβ line. A Seyfert 1.5 (like NGC 4151) has an Hβ line in which the broad component is as strong as the narrow one. Seyfert 1.8 and 1.9 show some broad component in the Balmer lines, with the Seyfert 1.9 having only a broad Hα line, while the Hβ line is narrow. It is obvious that the subdivision into these Seyfert classes is somewhat arbitrary. Often the optical spectra lack sufficient signal-to-noise to measure precisely the line widths. Therefore, when objects have been classified in the past based on low resolution spectra, they have often been separated simply into the Seyfert 1 and 2 classes, while better studied, bright objects are over-represented in the subclasses. A study by Trippe *et al.* (2010) seems to indicate that as many as 50% of the Seyfert 1.8 and 1.9 are indeed misidentified

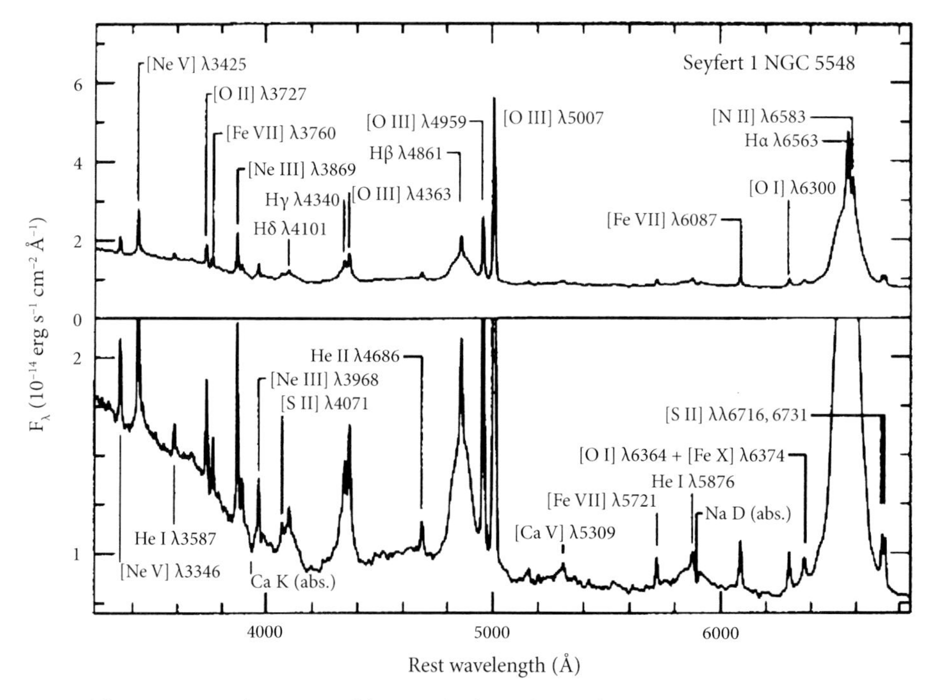

Figure 4.1 Optical spectrum of the X-ray bright Seyfert 1 galaxy NGC 5548. The spectrum shows the broad Balmer lines Hα, Hβ, Hγ, and Hδ and the narrow forbidden lines like [O III], [O II], and [Ne III]. Figure from Peterson (1997).

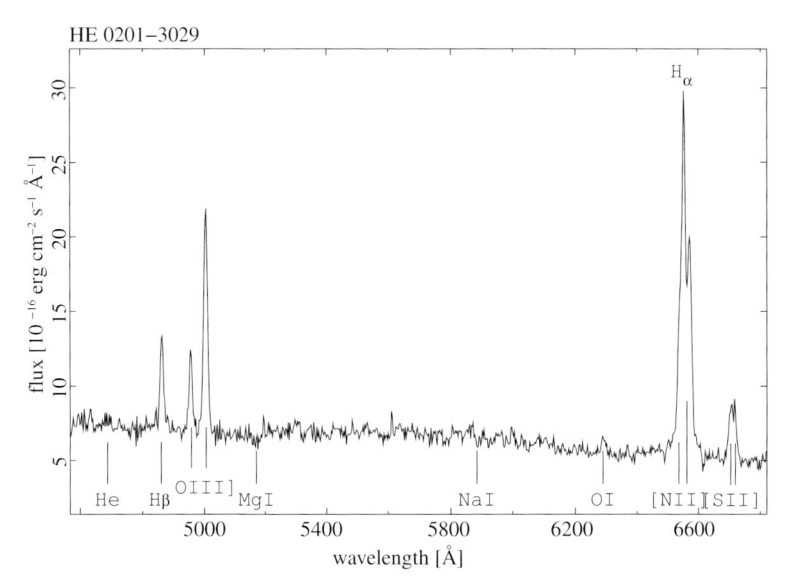

Figure 4.2 Optical slit spectrum of the Seyfert 2 galaxy HE 0201-3029 ($z = 0.036$). The wavelengths indicated are rest frame values. Note the absence of broad-line components and the strong forbidden oxygen ([O III] $\lambda\lambda$4959, 5007, [O I] λ6300), nitrogen ([N II] $\lambda\lambda$6548, 6583), and silicon ([S II] $\lambda\lambda$6716, 6731) lines, characteristics which are typical for type 2 objects. Note that the [N II] lines are strongly blended with the Hα line at $\lambda = 6563$ Å. The spectrum was taken with the Danish 1.54 m telescope on La Silla (Chile).

Seyfert 2 galaxies, in which the flux of the [N II] line has not been correctly taken into account, leading to the false conclusion that the Hα line is broad.

Starting from a spectrum, one can follow for example Caccianiga *et al.* (2008) who presented a flow chart allowing to distinguish between type 1 AGN (Seyfert 1–1.5, NLS1, type 1 quasars, BLRG), type 2 AGN (Seyfert 1.8–2, type 2 QSO, NLRG), and blazars.

4.1.2
HII Regions

Spectroscopic screening of AGN candidates often reveals galaxies with emission line spectra similar to those of Seyfert 2 galaxies. These are often produced by massive gas clouds of ionized hydrogen, so-called HII regions. They are detected preferentially in spiral galaxies and often exhibit strong forbidden oxygen lines in comparison to the Balmer lines with [O III] $\lambda 5007/\mathrm{H}\beta > 3$. In order to correctly classify an AGN spectroscopically, it is, therefore, necessary to examine its ionization level through studies of the various emission lines. Veilleux and Osterbrock (1987) presented several classification diagrams in which the different object classes can be distinguished based on measured line ratios.

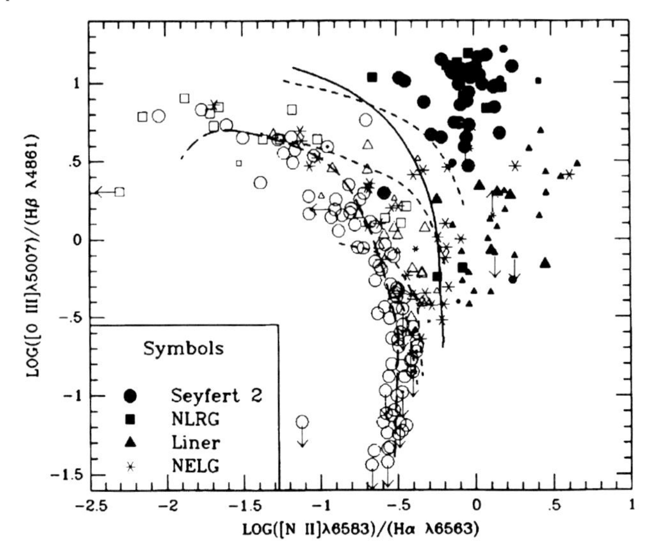

Figure 4.3 Classification diagram using reddening-corrected [O III]λ5007/Hβ vs. [N II]λ6583/Hα. Open circles indicate H II regions, open squares are H II galaxies, and open triangles are starburst galaxies. The other symbols represent AGN as indicated in the plot. The long-dashed curve represents a H II region model, the solid curve divides AGN from H II region-like objects (Veilleux and Osterbrock, 1987). (Reproduced by permission of the AAS.)

In Figure 4.3 we show the diagram for the [O III]/Hβ versus [N II]/Hα. It can be seen that although the different object types mostly occupy separated regions in this diagram, there is certainly some ambiguity. Therefore, for a correct classification, other line ratios have also been considered such as [S II]/Hα and [O I]/Hα. In the case the spectral resolution does not allow measurement of the weaker emission lines, a rough classification can also be based on the [O III]/Hβ versus Hβ diagram (see, e.g., Caccianiga *et al.*, 2008). The type 2 objects will settle at [O III]/H$\beta > 3$. And although type 1 objects have similar [O III]/Hβ ratios as H II regions and starburst galaxies, their broader Hβ line (FWHM $\gtrsim 1000\,\text{km s}^{-1}$) distinguishes them.

Figure 4.4 shows an example of the optical spectrum of an H II region. Another distinctive characteristic of these objects is the smaller range of ionization levels compared to the spectra of Seyfert galaxies. Electron temperatures within the H II regions are in the range 5000–20 000 K, with the cooler objects showing higher abundances of heavy elements. Electron densities are typically of the order of a few dozen to a few hundred per cm^3. Similar to the case of an AGN, the line-emitting region is assumed to be illuminated by a central region with an ionizing continuum. Unlike in AGN, the ionizing sources are hot, massive O and B stars embedded in star-forming regions. For a review on the properties of extragalactic H II regions we refer to Shields (1990).

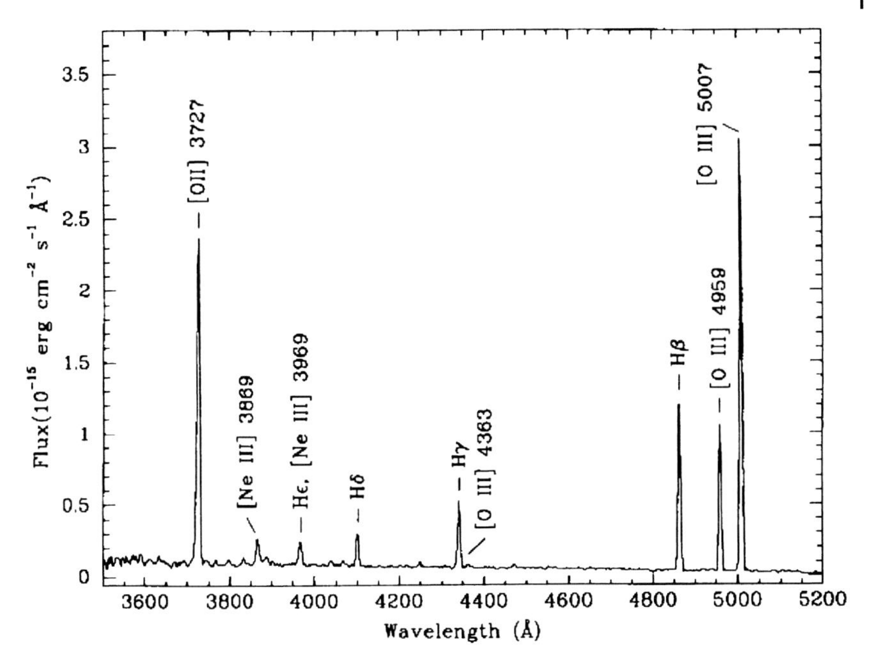

Figure 4.4 Spectrum of an H II region in the spiral galaxy NGC 2541 (Zaritsky *et al.*, 1994). (Reproduced by permission of the AAS.)

4.1.3 X-ray Classification

Up to now we have distinguished Seyfert types based on the optical spectrum, separating them into type 1 and type 2. A similar distinction is made in the X-rays, based on the *intrinsic* absorption measurement in the soft ($E \ll 5$ keV) X-ray band. Here, intrinsic absorption indicates matter which is close to the central engine of the AGN, whereas other absorption of line of sight photons, for example in our own Galaxy, has been accounted for. The absorption is measured as a column density of hydrogen $N_{\rm H}$ in the line of sight in atoms per cm^2. The unification scheme for AGN, which will be described in Section 4.8, assumes that the differences between type 1 and type 2 Seyferts result from the amount of absorbing material close to the central engine. The most strongly absorbed sources are often referred to as type 2, and those with lower absorption are labeled type 1. As a dividing line a hydrogen column density of $N_{\rm H} = 10^{22}\,{\rm cm}^{-2}$ has been chosen. Most, but not all, Seyfert galaxies which have an inferred intrinsic absorption of $N_{\rm H} < 10^{22}\,{\rm cm}^{-2}$ are classified spectroscopically as Seyfert 1 or Seyfert 1.2, while most, but again not all, Seyfert galaxies with $N_{\rm H} > 10^{22}\,{\rm cm}^{-2}$ are optical Seyfert 1.8, 1.9, or 2.

It is likely that the transition between absorbed and unabsorbed sources is smooth, and that not all type 1 AGN exhibit low absorption (Awaki *et al.*, 1991). In addition, Seyfert 2 galaxies like NGC 3147 and NGC 4698 show no intrinsic absorption (Pappa *et al.*, 2001). A catalog of hard X-ray-selected AGN contains typically around 10% of objects in which the optical classification does not match the one based on intrinsic absorption measurements, as seen for example in the AGN catalogs based on data from the hard X-ray satellites Swift (Tueller *et al.*, 2008) and INTEGRAL (Beckmann *et al.*, 2009). Thus, it is important to know for each study whether the classification of an object has been based on X-ray or optical measurements. In addition, in some objects a variation of the intrinsic absorption with time has been observed, which causes objects to be unabsorbed in one X-ray measurement, while they are of absorbed type in another.

XBONGs

X-ray satellites picked up extragalactic sources which apparently have no optical emission lines but are also not blazars, like NGC 4156 and NGC 3862 (Elvis *et al.*, 1981). With ongoing X-ray surveys, more objects like these were found, and called "X-ray bright optically inactive galaxies" (XBONG). Most of these puzzling objects have by now been identified either as diluted sources, in which the starburst outshines the emission line region and therefore the AGN lines appear weak, or the X-rays originate from galaxy groups or non-obvious clusters of galaxies. In the latter case, the X-rays are produced in the hot gas in between the galaxies of the group or cluster, and their spectra can be represented by bremsstrahlung or thermal emission rather than by a power-law model. The name XBONG will likely intrigue the community for a while but ultimately fade into oblivion.

4.1.4 Narrow-Line Seyfert 1 Galaxies

A subset of the Seyfert-1 galaxies, the so-called narrow emission line Seyfert 1 galaxies (NLS1) are strong X-ray emitters, but while their Hα lines are broad, their Hβ line is narrow (FWHM(Hβ) $< 2000\,\mathrm{km\,s^{-1}}$), similarly as in Seyfert 1.9 (Osterbrock and Pogge, 1985). Figure 4.5 shows an example of an optical spectrum of a NLS1. In the UV regime NLS1 also show narrower lines than Seyfert 1 and they have weaker [C IV] and [C III] emission. Therefore, they seem to be reddened Seyfert 1 galaxies, where the absorption is effective only at wavelengths $\lambda \ll \lambda(\mathrm{H}\alpha) = 6562.8\,\text{Å}$.

Equivalent width

The strength of emission and absorption lines is often expressed with respect to the underlying continuum. The *equivalent width* gives the line flux in terms of how much continuum flux one has to integrate in order to get the same flux value. In practice this means to draw a rectangle with the height of the continuum and the same surface of the line. The width of this rectangle is then the equivalent width EW. If we assume that the continuum intensity f_c is constant over the wavelength range λ_1 to λ_2 and $f(\lambda)$ is the total flux (line plus continuum) at wavelength λ, we can write

$$\text{EW} = \int_{\lambda_1}^{\lambda_2} \frac{f(\lambda) - f_c}{f_c} d\lambda \tag{4.1}$$

Figure 4.6 gives a schematic drawing of the definition of the equivalent width.

Fe II emission with respect to the Hβ line (Fe II/Hβ) is about twice as strong in NLS1s as compared to other Seyferts, and the equivalent widths of these two lines are strongly correlated with the optical luminosity (Zhou *et al.*, 2006). When studying the masses of the central engine in NLS1s it appears that they are on average less massive than other Seyferts, in the range of $M_{\text{BH,NLS1}} \sim 10^5\text{–}10^7\ M_\odot$, yet their bolometric luminosities are comparable. This led to the conclusion that NLS1s are accreting at a higher rate, i.e., that these objects are closer to the Eddington limit than other Seyferts (Collin and Kawaguchi, 2004). The steep X-ray spectra of NLS1s first seen in observations made using the ROSAT satellite suggested that these sources might be the analog of galactic black holes seen in their high-soft emission state (Boller *et al.*, 1996). A description of the spectral states of galactic black holes is shown in Section 4.8.4 and summarized in Figure 4.18.

The hypothesis that NLS1 are accreting near the Eddington limit has been a topic of frequent discussion. Some investigations apparently indicate that the black hole masses and therefore the accretion efficiencies are similar compared to other

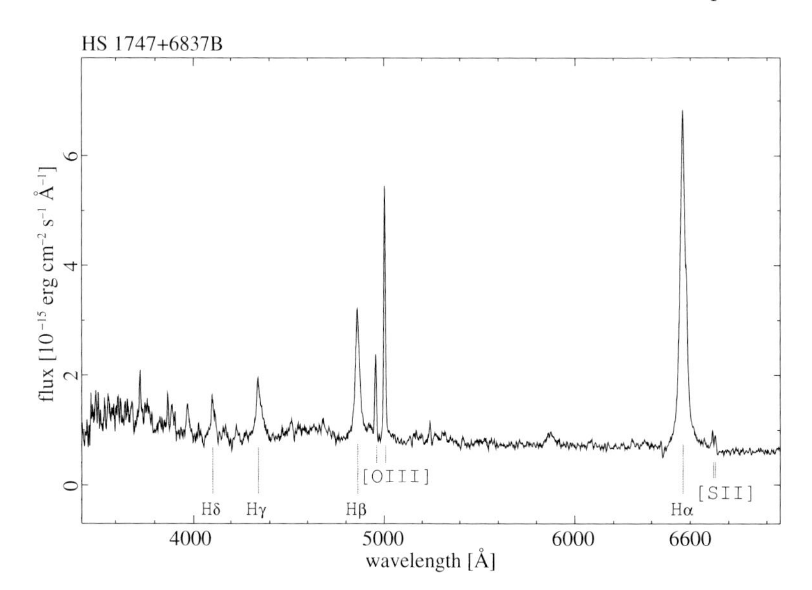

Figure 4.5 Optical spectrum of the narrow-line Seyfert 1 galaxy HS 1747+6837B taken in the course of the *Hamburg Quasar Survey*. Spectrum courtesy of Ralf Keil.

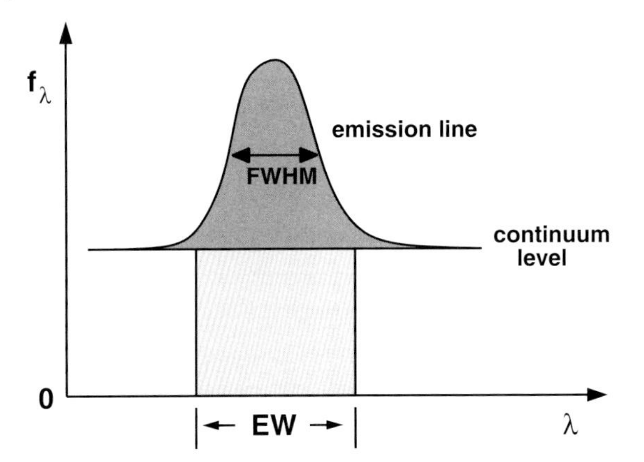

Figure 4.6 The equivalent width (EW) gives an integral of continuum flux with the same surface as the emission (or absorption) line. FWHM is the full-width at half-maximum, as measured from the continuum of a line in Å.

Seyferts (Marconi *et al.*, 2008). Recent studies confirm that the NLS1 have steeper low-energy X-ray spectra than broad-line Seyfert 1. These studies further suggest that they are fainter in the UV and hard X-ray bands (Grupe *et al.*, 2010). Although NLS1s show strong X-ray variability, they vary only marginally in the UV. This could indicate that the accretion disk, as seen in the UV, is rather stable, while in the X-rays nonthermal beamed emission can be responsible for the variability. Optical studies using the HST seem to indicate that the black holes in NLS1 are still growing and the growth is governed by secular processes rather than by mergers (Mathur *et al.*, 2012). On the other hand, the peculiar properties of NLS1 can be partially explained by low inclination with respect to the observer, that is by a near pole-on view of these AGN (Peterson, 2011).

In recent years the class of NLS1 drew new attention, as some of these AGN are detectable in the gamma rays by the Fermi satellite. This association with gamma-ray sources indicates that some NLS1 might contain a misaligned jet, leading to strong nonthermal emission (Foschini, 2011, and see Section 5.6). Multifrequency campaigns have shown that in the case of PMN J0948+0022, a NLS1 at redshift $z = 0.585$, there is indeed a jet present with a power comparable to that in blazars: during an outburst in 2010, the source flux indicated a luminosity of up to $\sim 10^{48}\,\mathrm{erg\,s^{-1}}$. The latter value is assuming isotropic (i.e., nonbeamed) emission, thus, if the emission was highly collimated, the true luminosity was lower. If most of the NLS1 hosted a jet, one would expect them to be strong and variable radio sources, where the synchrotron emission of the relativistic outflow should dominate. But strong radio variability is seen only in about 7% of the sources (Komossa *et al.*, 2006). Studying a sample of 76 NLS1 galaxies, among which 46 are radio-loud ($R^* > 10$; see the definition in the box of Section 3.2.4), the others radio-quiet (but not radio-silent), Foschini (2011) found seven to be gamma-ray emitters detected by Fermi/LAT above 100 MeV. Thus, displaying a jet remains the exception rather than the rule for NLS1, and gamma-ray detectable and/or radio-loud objects will remain rare objects in this AGN class.

Another indication that NLS1 are a distinct class of Seyfert galaxies is the discovery of the first known quasi-periodic oscillation (QPO) in the NLS1 RE J1034+396 (Gierlinski *et al.*, 2008); also see Section 6.4. In this source a strong peak for a periodicity of 3730 ± 130 s, with slight changes over the observation was found, indicating that the feature wanders in phase, amplitude and/or frequency, as seen in QPOs of galactic black hole binaries (Remillard and McClintock, 2006).

4.2 Low-Luminosity AGN

Since the discovery that our own galaxy hosts a supermassive black hole at the position of the radio source Sgr A* with about $4 \times 10^6\ M_\odot$, it has become a common view that the majority of normal galaxies might harbor a massive black hole similar to that in active galactic nuclei. As in the Milky Way, there seem to be cases where the black hole is starved or undergoing radiatively inefficient accretion, and thus do not appear as an AGN (Section 3.2.4). These may have experienced an active phase in the past. It may be the case that supermassive black holes are present in the majority of the galaxy population. Greene and Ho (2007) estimate that the duty cycle of relevant accretion onto the central massive black hole in AGN is as low as 10^{-2}, implying that the AGN spends 99% of its life time in a quiescent state.

Considering its low bolometric luminosity, less than $L_{\rm bol} = 10^{37}\ {\rm erg\,s^{-1}}$, Sgr A* would not be detected if even moved to a neighboring galaxy. It should be noted that evidence for past periods of enhanced activity in Sgr A* have been reported in the literature but these are considered speculative in nature. On the other hand, we do see AGN with luminosities of $L_{\rm bol} > 10^{40}\ {\rm erg\,s^{-1}}$. It can be expected that there should exist objects intermediate in luminosity, which may represent a link between Sgr A* and Seyfert galaxies.

A class of object which seems to fill in this luminosity gap, are the low-ionization nuclear emission-line regions (LINER). These show faint core luminosities and strong emission lines originating from low ionized gas. Expected line widths are $200{-}400\ {\rm km\,s^{-1}}$ and their properties are very similar to the Seyfert 2 galaxies, but, although they are overall of lower luminosity, LINER do have stronger forbidden lines. LINER seem to mark the low-luminosity end of the AGN phenomenon, and they may be accreting with low radiative efficiency or at a low rate compared to Seyferts.

LINER are further assumed to represent the link between H II regions and Seyfert galaxies, as can be seen in Figure 4.7. Different than Seyfert galaxies, LINER can show [O III] $\lambda 5007/{\rm H}\beta < 3$. And in comparison to H II regions, LINER have stronger [N II] lines when compared to the Hα line flux, as can be seen in Figure 4.3. Graves *et al.* (2007) pointed out that there seems to be a LINER core at the heart of many early-type galaxies. Galaxies with strong LINER-like emission seem to be younger than those without this component. This is also consistent

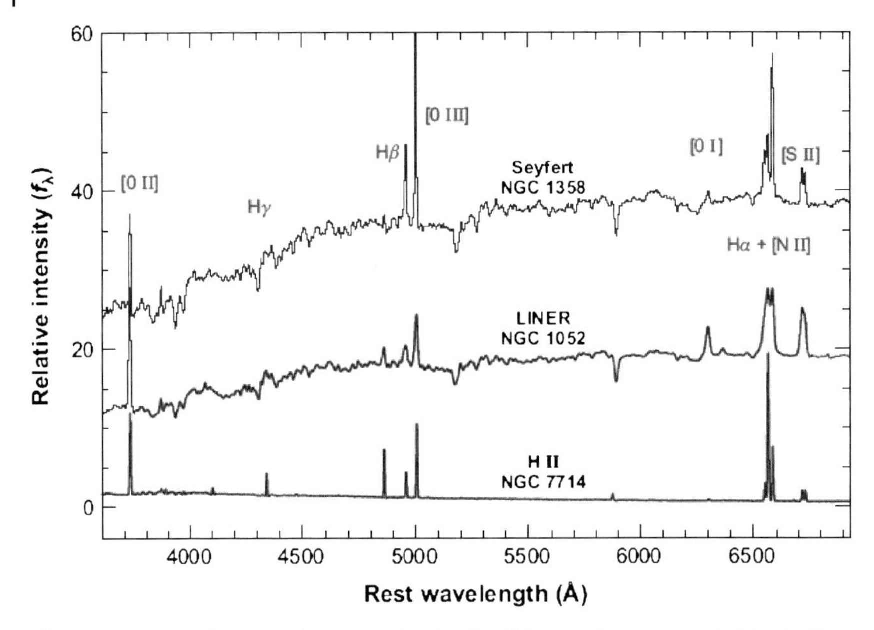

Figure 4.7 Three optical example spectra showing the differences between a typical Seyfert 2 galaxy, a low-ionization nuclear emission line region (LINER), and an H II galaxy. (Reproduced from Ho (2008) with permission of Annual Reviews via Copyright Clearance Center.)

with the observation that the X-ray spectra of LINER are dominated by starburst emission rather than by the AGN core. Absorption seems to play an important role in the case of LINER, as there are more highly absorbed Compton-thick[1] sources among them in comparison to Seyfert galaxies. If there is indeed a strong absorption component around the LINER core, the viewing geometry will play an important role in our characterization of these sources. In fact it seems that the strength of the narrow lines varies by a factor of ~ 2 depending on the viewing angle (Fine *et al.*, 2011).

As for other low luminosity AGN, also in LINER it is often difficult to distinguish between the AGN core emission and the host galaxy contribution. Starburst activity and binaries can also dilute the core emission. Cid Fernandes *et al.* (2010) even argue that most objects which have been classified as LINER are indeed consistent with being evolved galaxies, whose ionizing photons are produced by an aging star population. Although a supermassive black hole could still reside in its center, it would be dormant and thus these galaxies would not be classified as AGN. Capetti (2011) estimate that only 10% of the proclaimed low-luminosity AGN (LLAGN) are indeed AGN, the others being merely emission-line galaxies. On the other hand, in the UV band LINER seem to be dominated by an AGN core, as evidenced through strong variability and a blue, nonthermal continua, something which a "normal" nonactive galaxy will not produce. González-Martín *et al.* (2009) assume that 60%

1) Absorption is called *Compton-thick* when the column density in the line of sight exceeds a value of $N_\mathrm{H} = 1.5 \times 10^{24}\,\mathrm{cm}^{-2}$, that is, the inverse of the cross-section for Thomson scattering.

of the LINER could host AGN cores which are emitting X-rays and that they are intrinsically Seyfert 2 galaxies with an additional extinction component which suppresses the lines in the optical domain.

Let us assume that the key parameter which determines the brightness and activity of an AGN core is indeed the accretion rate. In relative units this can be expressed in terms of the Eddington ratio

$$\lambda_{\text{Edd}} = \frac{L_{\text{bol}}}{L_{\text{Edd}}} \tag{4.2}$$

with the Eddington luminosity L_{Edd} as described in Eq. (3.9) and the bolometric luminosity L_{bol}. Following this idea, the low luminosity AGN are likely not emitting through the same processes as the more active Seyfert and quasar cores. In fact at such low bolometric luminosity, and therefore low accretion rates, that is, lower than $\dot{M} \simeq 10^{-2}\,\dot{M}_{\text{Edd}}$, the in-falling matter will be optically thin, unable to form an optically thick disk which is able to cool efficiently. Therefore, energy will have to be transported outwards not by electromagnetic radiation, but rather through flow of matter, in processes such as winds, advection, or convection. Thus, the emission of low-luminosity AGN such as LINER might be better described by models like the Advection Dominated Accretion Flow (ADAF) described in Section 3.2.4.

An overview on low-luminosity AGN has been given in the review article of Ho (2008).

4.3
Ultraluminous X-ray Sources

Ultraluminous X-ray sources (ULX) were first discovered in nearby galaxies by the Einstein observatory (see Section 5.5.1). They appear brighter than expected from X-ray binary systems by exceeding $L_X > 10^{39}\,\text{erg}\,\text{s}^{-1}$ (Long, 1982). This is about $10\times$ super-Eddington for galactic X-ray binaries. At the same time, the ULX are not located at the dynamical center of their host galaxies and are thus not typical AGN cores. The high-resolution Chandra X-ray telescope has revealed numerous ULX. Swartz *et al.* (2004) found 154 ULX within 82 galaxies observed with Chandra. They also pointed out that ULX appear preferentially in young galaxies with a strong star-formation component. At least some of the ULX appear to be associated with extended radio emission (see, e.g., Mushotzky, 2004). ULX are variable sources, and for some of them there have been reports of periodic and quasiperiodic oscillations in their light curves. In the X-rays, the spectra appear either to be simple power laws or a combination of power law and a thermal emission component as is expected from an accretion disk. In some cases, however, the inferred temperatures are difficult to reconcile with black hole accretion.

So far there is no consensus on what ULX really are. Several possibilities have been brought forward. The reason why we include ULX in this chapter is that they might be in fact down-scaled versions of AGN, an idea which has been proposed by Colbert and Mushotzky (1999). Here they would fill the gap between stellar mass, that is from 3 to up to several tens of $M_\odot$, and the supermassive black holes observed in AGN with $M_{\rm BH} \gtrsim 10^5\ M_\odot$. These putative *intermediate*-mass black holes (IMBH) are also of interest in that they could serve as the seeds for supermassive black holes growing in the local Universe. Before we come back to the AGN interpretation, let's have a look at alternative explanations for the ULX phenomenon.

One idea explains the existence of ultraluminous, off-center X-ray sources as simply the bright end of the luminosity function of supernovae and X-ray binaries (e.g., Grimm *et al.*, 2003). This would not require a new class of objects, but it stretches the distribution of luminosities we observe in our own Galaxy in order to match objects at $L_{\rm X} > 10^{39}\ {\rm erg\,s^{-1}}$. In this model the ULX appear outside the center and bulge of the galaxies, simply because they follow the star formation in the spiral arms and are therefore off-center. The correlation with star formation also appears natural: massive binary systems, as required to form the bright population, are short-lived and therefore require an environment of high star formation. In this context, the very bright end of the luminosity function of X-ray sources needed to provide the brightness of ULX could be derived by metal-poor star evolution. In a low-metal environment massive stars could conceivably grow to large masses, avoid supernova explosions, and collapse directly into a massive black hole with some 25–80 $M_\odot$ (Fryer *et al.*, 2001; Mapelli *et al.*, 2011). An X-ray binary system hosting such a massive compact object should be able to produce the required output without exceeding the Eddington limit. For a black hole with 80 $M_\odot$ the luminosity at the Eddington limit is $L \sim 10^{40}\ {\rm erg\,s^{-1}}$ (see Eq. (3.9)). If one allows for short periods of super-Eddington accretion, accordingly smaller masses would help to explain the ULX population. Also, the fact that these binary systems are short-lived is consistent with the fact that ULX are mainly found in star-forming regions.

High apparent luminosities can also be reached if the emission is collimated or "beamed" toward the observer. This possible explanation for ULX has been outlined concisely in King *et al.* (2001). This anisotropic emission would require an enhancement by b^{-1}, with $b = \Omega/4\pi$, where Ω is the solid angle of emission. Already mild beaming with $b \sim 0.1-0.01$ would be sufficient to explain ULX by X-ray binaries with masses observed in our own Galaxy. In this scenario, the ULX could be higher-mass higher-luminosity analogs to the galactic low-mass X-ray binaries known as microquasars, examples of which include GRS 1915+105 and SS 433 (see Section 3.7).

An alternative scenario, also involving massive stars, is the possibility that ULX are the remnants of a putative subclass of extremely luminous supernovae called hypernovae. The remnants of these events could exceed luminosities of $L > 10^{40}\ {\rm erg\,s^{-1}}$, as suggested for example from the cases of SN 1988Z and SN 1995N (Immler and Lewin, 2003) as well as the possible hypernovae remnants in M101 (Wang, 1999). In support of this idea there are ionized nebula observed near some of the ULX. For example, a He II emission line nebula can be seen near the ULX in the galaxy Holmberg II (Pakull and Mirioni, 2003). Apparently, the ULX has ionized the matter in a large spherical bubble around it, which is an argument against a beamed emission from the ULX, at least in this case. But because the X-ray spectra of supernova remnants are well approximated by models of thermalized atomic plasma, they are not likely to be responsible for the majority of ULX.

As mentioned above, accreting intermediate-mass black holes (IMBH) are also viable candidates for being the counterparts of ULX (how these objects could be formed is another issue that is presently unresolved). With respect to the other theories, an accreting IMBH explanation has the advantage of providing the observed spectral features, like the power law-like continuum and soft X-ray excess (Section 5.3). The latter appears at higher energies than expected. Solar-mass black hole binaries have characteristic disk temperatures of about $\sim 1\,\text{keV}$ while AGN disks are much cooler, peaking in the extreme UV or soft X-ray bands. As shown in Section 3.2, this has to do with the $R^{-3/4}$ dependence of the disk peak temperature and the much larger (in proportion to their Schwarzschild radii and thus their black hole masses) inner-disk radius of the AGN. Viable IMBH disk models should have characteristic temperatures intermediate to these two cases.

The observation of iron line emission which is seen in some ULX spectra would also be naturally explained in the context of an AGN. In addition, the extended radio emission sometimes seen in ULX is consistent rather with an AGN-like origin. An AGN in the center could also explain why there are large photoionized nebula ambient to certain objects. Recently, Pakull *et al.* (2010) observed what looks like a small radio galaxy inside NGC 7793. The ULX with $L_X \sim 10^{40}\,\text{erg}\,\text{s}^{-1}$ appears to have ionized a large nebula, indicating that the jets are about 10 000 times more energetic than the ULX core itself.

One of the best candidates for an IMBH has been found in the edge-on spiral galaxy ESO 243-49 with a variable X-ray luminosity extending to $10^{42}\,\text{erg}\,\text{s}^{-1}$, implying a lower limit on the mass of $M_{\text{BH}} \sim 500\,M_{\odot}$ (Farrell *et al.*, 2009). We show the optical image of the galaxy and the bright X-ray source in Figure 4.8. This specific ULX has drawn further attention as it appears to exhibit transient episodes with a recurrence time of about a year (Lasota *et al.*, 2011). However, only two outbursts fit this picture precisely and a third must be extrapolated backwards in time for full consistency. The upcoming years will show whether this periodicity is robust, supporting the hypothesis of an eccentric binary system. Further support for the IMBH explanation comes from the *fundamental plane* considerations we will discuss in view of the unification of AGN in Section (4.12).

A major challenge of the IMBH theory is to plausibly explain how these objects are formed. Such massive black holes would require a long time to grow to their current size of $M_{\text{BH}} \sim 100\text{–}1000\,M_{\odot}$, perhaps a substantial fraction of the ages of their host galaxies. They should thus preferentially be found in the older parts of the host galaxies, where star formation occurred long ago, like in the halo of galaxies. They should arguably also appear preferentially in elliptical galaxies rather than in spirals. But as mentioned above, ULX follow the areas with ongoing star formation and seem to be also more abundant in spirals than in ellipticals.

As with many unsolved problems in astrophysics, the ultimate explanation for the nature of ULX might not be a single one. Winter *et al.* (2006) used XMM-Newton data of 100 ULX candidates in 32 nearby galaxies, searching for low-state

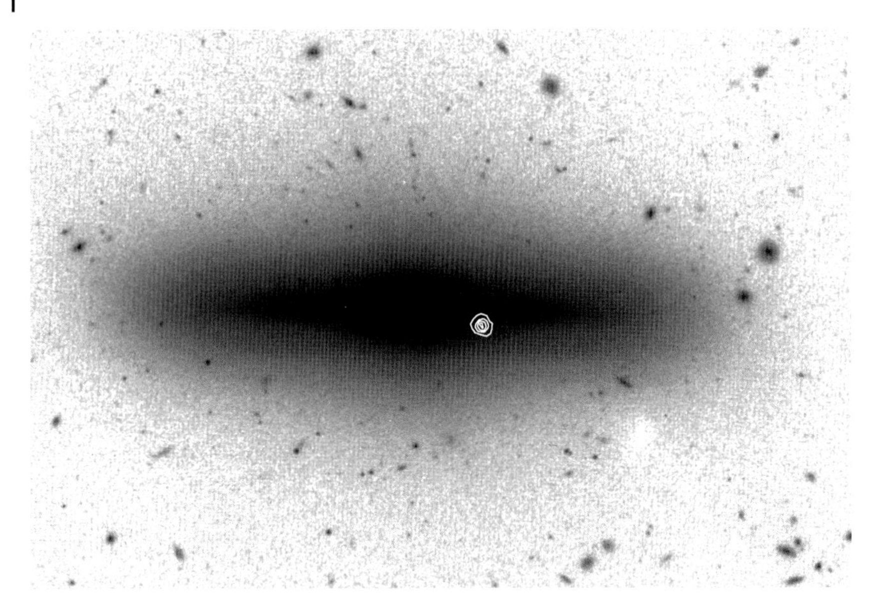

Figure 4.8 Optical image of the galaxy ESO 243-49, as derived from an 800 s long exposure of HST/WFC3, overlaid with X-ray contours, based on a 10 ks Chandra/ACIS-S observation, both taken in September 2010. The X-ray source is ultraluminous, not located in the center of the host galaxy and is one of the best candidates to date for an intermediate-mass black hole (IMBH). (The HST data were obtained from the Mikulski Archive for Space Telescopes (MAST). STScI is operated by the Association of Universities for Research in Astronomy, Inc., under NASA contract NAS5-26555.)

and high-state objects with properties similar to those of galactic X-ray binaries. They then divide the high-state objects into those which have black body temperatures similar to X-ray binaries, that is $kT \sim 1\,\text{keV}$ and $L_X < 2 \times 10^{39}\,\text{erg}\,\text{s}^{-1}$, and those with lower temperatures of $kT \sim 0.1\,\text{keV}$ and higher luminosity, which are rather IMBH candidates. The conclusion of Winter *et al.* (2006) is that 75% of the ULX found in the 0.3–10 keV band are IMBHs, while one quarter are X-ray binaries. Thus it is likely that at least the fainter end of the ULX population is formed by massive binary systems, and beaming can push the limit significantly. Although hypernovae cannot account for all ULX, in some cases supernova remnants might appear as ultraluminous sources. Additionally, the possible AGN-ULX connection needs further study. As we have seen in the case of low-luminosity AGN, distinguishing them from other, nonactive object classes is often difficult. Revealing the true nature of the sources observed as ULX today will require further high-resolution and high-sensitivity observations, not only in the X-ray but also in the radio domain.

A concise summary of most of these arguments around the nature of ULX can be found in Mushotzky (2004), and a more recent review has been presented by Feng and Soria (2011).

4.4 Ultraluminous Infrared Galaxies – ULIRGs

Galaxies are infrared emitters due to their stellar population and their dust content. In particular, old, massive galaxies, like ellipticals, show a strong infrared component. But observations in this energy range reveal a fraction of galaxies that seems to exceed their expected infrared emission significantly. These extremely infrared bright sources, called luminous infrared galaxies (LIGs) are defined by a far infrared ($\lambda > 10\,\mu m$) luminosity exceeding $L_{IR} > 10^{11}\,L_\odot$. The complete census of the sky in the far infrared by the Infrared Astronomical Satellite (IRAS) led to the discovery of many of those bright galaxies, and also of a substantial population of galaxies with $L_{IR} > 10^{12}\,L_\odot$, the so-called ultraluminous infrared galaxies (ULIRGs). While infrared galaxies with $L_{IR} < 10^{11}\,L_\odot$ are mainly single, gas-rich and star-forming galaxies, this picture changes dramatically toward the brighter end of the population. The more dominant the far-infrared component, the more likely it is to find that the galaxy has been involved in a merging process with another galaxy. Among the ULIRGs, 95% of the sources are mergers (Sanders and Mirabel, 1996). Observations by ESA's Infrared Space Observatory (ISO; 1995–1998) seem to confirm the close connection of ULIRGs and mergers.

At the same time, one finds strong indication of AGN activity, with about 70% of the ULIRGs displaying a Seyfert or LINER core. This would mean that while in normal infrared galaxies the dust is heated by star-formation processes, in the case of ULIRGs an important contribution to the dust heating comes from the AGN at the heart of the galaxy. But as Genzel *et al.* (1998) pointed out, apparently even for the ULIRGs more than half of the bolometric luminosity originates from starburst activity, at ages of 10^7–10^8 years. Thus, star formation and AGN activity seem to go closely together in these galaxies. The AGN core dominates stronger in the more luminous ULIRGs. ISO observations showed that at the low luminosity end of the ULIRG population ($L \sim 10^{12}\,L_\odot$) the sources are clearly dominated by star formation, while at luminosities $L > 3 \times 10^{12}\,L_\odot$ the AGN is the dominant component. There are exceptions to this rule. In fact the advent of NASA's Spitzer space telescope, launched in 2003, revealed ULIRGs with exceptional brightness, like MIPS J142824.0+352619, with $L_{IR} \simeq 3 \times 10^{13}\,L_\odot$ but no trace of AGN activity (Borys *et al.*, 2006). Spitzer opened the view onto ULIRGs out to redshifts $z > 2$. In this high-redshift regime, it is even more unclear as to what drives these sources: an AGN core, sometimes hidden from direct observations, or a large amount of star formation. A comprehensive summary of Spitzer results on extragalactic sources has been presented in Soifer *et al.* (2008). Another important aspect here seems to be the merging of galaxies. As we will see in Section 7.3, there appears to be a close connection between the amalgamation of two galaxies and their AGN and star-forming activity. Sanders (1999) thus puts the ULIRGs as a stage or episode between gas-rich merging disk galaxies and the resulting ellipticals. He argues that ULIRGs might be the precursors of the more luminous and less absorbed quasars. Genzel *et al.* (2001) pointed out that although ULIRGs might become giant elliptical galaxies in the future, they generally involve a more dynamic host galaxy with moderate rotation, with decoupled gas dynamics and stellar dynamics. Thus, they rather resemble intermediate-mass elliptical galaxies than old and heavy giant ellipticals. On the other hand, the matter in the host of ULIRGs is already more relaxed than in an ongoing merger, placing the ULIRGs right in between merging and giant ellipticals.

As an AGN class the ULIRGs seem to have been more prominent at higher redshift, as pointed out by Goto *et al.* (2011a). While in the local Universe, their total energy output does not seem to contribute significantly to the overall emission of AGN (i.e., they do not dominate the local luminosity function), they seem to be more dominant and important at $z > 1$. As we will see in Chapter 9 this might indicate that in the past the sources in the Universe were on average more strongly obscured. The obscuring matter would then reprocess the higher-frequency emission of the central engine into the infrared range.

ESA's Herschel space observatory is going to provide a deep look into the infrared universe, both through pointed and survey observations. This is likely to boost our understanding of the starburst vs. AGN dominance in ULIRGs, especially at high redshifts, and to clarify the importance of massive molecular outflows (e.g., Sturm *et al.* (2011); see also Section 5.2).

4.5 Radio Galaxies

When the central region of a quasar is hidden but the object produces bright radio jets and large radio luminosities, the existence of an AGN core is assumed. A commonly used morphological classification scheme uses the appearance of the extended radio emission. The two subgroups are the low-luminosity Fanaroff–Riley class I (FR-I) galaxies which show a rather compact emission arising from close to the core, and the high-luminosity FR-II objects, in which the structure is dominated by the radio lobes and most of the emission appears to come from the far end of the extended emission (Fanaroff and Riley, 1974). As a dividing line a luminosity of $L_R = 10^{32}\,\mathrm{erg\,s^{-1}\,Hz^{-1}\,sr^{-1}}$ at a frequency of 175 MHz has been established.

Figure 4.9 shows the apparently brightest extragalactic radio source, the FR-II radio galaxy Cygnus A. At a redshift of $z = 0.057$, the radio structure of Cygnus A has an angular extent of about 100 arcsec, corresponding to 100 kpc with a total radio luminosity of $L_R = 10^{45}\,\mathrm{erg\,s^{-1}}$. Thus, the radio lobes seen in Figure 4.9 extend far outside of the host, a giant elliptical galaxy with a diameter of about 70 kpc which fits in between the two radio lobes (Carilli and Barthel, 1996). The radio image also shows a confined jet reaching from the central engine, located at the bright spot, out into the lobes.

The jet's emission is polarized and nonthermal, and thus it is thought to be synchrotron emission. Although the image suggests that the jet is originating at the core and then terminating in the large lobes with a termination shock at its outermost edge, one cannot directly confirm this picture, as the outflow speed cannot be directly measured. The radio spectrum of the core, the jet and the lobes is feature-

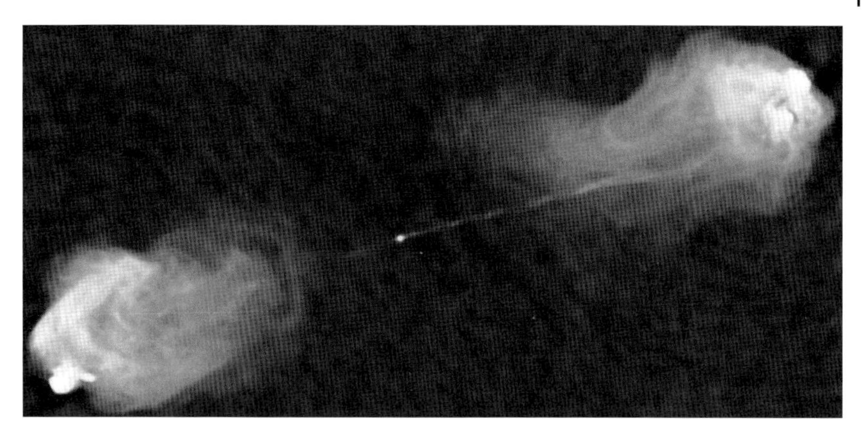

Figure 4.9 The FR-II type radio galaxy Cygnus A is the brightest extragalactic radio source. The picture shows the 5 GHz image taken with the VLA telescope array with 0.4″ resolution (Carilli and Barthel, 1996). The AGN core is located at the bright spot at the center, the radio lobes extend out to about 50 kpc from the core, far beyond the host galaxy which is not visible in the radio domain. For a color version of this figure, please see the color plates at the beginning of the book.

less, thus no Doppler shift can be measured. One can determine outflow speeds and energetics only through indirect and thus questionable methods. In only a small number of cases and only close to the AGN core, the observation of blobs traveling down the jet can give us an idea about the starting speed of the outflow (Figure 3.30).

In some cases, asymmetric morphologies in FR-II are observed, for example objects in which the two radio lobes seem to have equivalent brightness, but the jet is visible only on one side. An example is displayed in Figure 4.10, showing the radio galaxy 3C 175. This source is located at $z = 0.77$ and hosts a Seyfert 1.2 core. The radio structure extends over $\sim$ 212 kpc. This kind of single-sided jet can be used to estimate the jet speed, following several assumptions. First, we assume that the morphology of the FR-II is intrinsically symmetric. In addition, also the physical parameters, like speed, density, power, describing the jets on both sides of the central engine are supposed to be the same. Then, the only difference would be caused by different Doppler enhancement of the emission toward the observer. This would lead to an amplification of the observed flux of the jet approaching us (f_+), and a dimming for the receding jet with flux f_-. The ratio can be derived from the Doppler factors, following Eq. (5.13):

$$\frac{f_+(\nu)}{f_-(\nu)} = \left(\frac{1 + \beta \cdot \cos\theta}{1 - \beta \cdot \cos\theta}\right)^{2+\alpha} \tag{4.3}$$

Here, θ is the angle of the jet with respect to the line of sight, $\beta = v/c$ and α is the spectral energy index of the observed continuum emission. In the case of a single-sided jet, an upper limit can be applied for f_- of the receding jet, the angle θ can be assumed from the observed geometry, f_+ and α are observables, and thus β

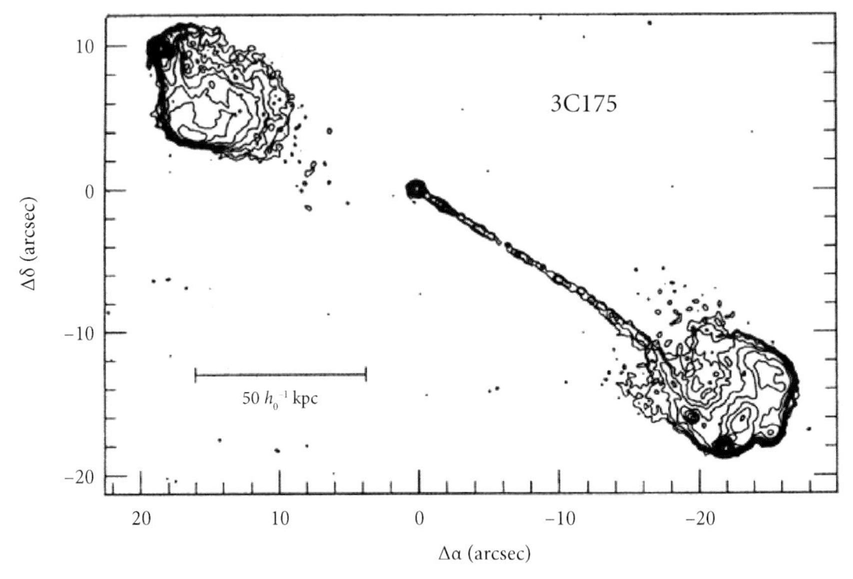

Figure 4.10 The FR-II type radio galaxy 3C 175 has a symmetric lobe structure with hot spots at both ends, but a single-sided jet. The reason might be that the visible jet component is Doppler boosted toward the observer, while the nonvisible part is pointing away from us. 5 GHz image taken with the VLA (Bridle *et al.*, 1994).

can be determined. The caveats of this method are the assumptions going into it. Especially the assumption that the radio morphology is intrinsically symmetric.

Some models have been developed to explain the single-sided jet by alternating jet emission. In this concept, the jet would be emitted to one side for a period of time, then switching to the opposite direction. Rudnick and Edgar (1984) found several radio galaxies which showed gaps in their radio jets. Whenever such a gap on one side would occur, the corresponding counter jet would be visible at the same distance from the core: apparently the jet alternated in time, causing the complementary appearance. Theoretical models have been developed in order to explain these phenomena (Icke, 1983) by assuming interaction of the central engine with a surrounding medium or its periodic movement within a larger gravitational potential which causes the formation of constraining walls, which determine the direction of the outflow.

The expansion speed of the radio lobes and thus of the hot spots can be estimated applying various models. This includes a description of the hot spots as the terminating shock of the beamed jet in a Mach disk (see box), where the jet is slowed down rapidly by the ambient medium causing electron acceleration through the Fermi process.

Mach disk

The Mach disk is one way to get a planar shock in a jet (see, e.g., Hartigan, 1989). In order to form such a shock, the pressure in the jet has to be lower than the ambient medium (or, in general, significantly different than in the surrounding matter). This jet can expand freely, it widens until the pressure inside the jet is equal to the ambient pressure. The expansion bounces back, the jet contracts and reaches a point of high inner pressure, where a standing shock is created. The Mach disk will appear at the same position in the jet as long as the outflow is steady and the general characteristics do not change. At the Mach disk, the speed of the outflowing material is significantly reduced. Mach disks are a commonly observed feature in the exhaust of jet engines and rockets.

The density of relativistic electrons (or charged particles, to be more general) and the magnetic flux density in the jet are large, which causes the bright hot spots downstream of the Mach disk. The relative brightness of hot spot, radio lobe, and jet can be used to estimate their relative speeds. The lobes are usually considered to expand at nonrelativistic speeds with $v_{\rm lobe} \ll 0.3c$, and more likely $v_{\rm lobe} < 0.1c$. A relativistic forward movement of the lobes is also unlikely for more simple reasons: if the lobe plowed into the surrounding medium at relativistic speed, a similar appearance as for the hot spot would be expected, that is the lobe should have a sharp and well constrained front edge. As can be seen also in the examples shown in this section, this is generally not the case. If one assumes a speed of the hot spots of about $0.1c$, the jet flow has to have a speed of $v \simeq 0.5c$. From measurements of blobs moving along the jet near the central engine, relativistic outflow speeds have been inferred. This also indicates that the jet has to slow down substantially on its way toward the radio lobes and before causing the hot spots.

All these unknowns about the physical conditions in the various parts of radio galaxies also play a major role in the discussion about several open issues. How are the jets so well collimated very close to the central engine? How is the jet confined over a distance of, in some cases, several 100 kpc? What particles dominate the jet – they have to be charged particles, as they emit synchrotron radiation, but are they electrons, positrons, protons, or a mix of them? While electrons and positrons can be accelerated more easily than protons simply because of their lower mass, a heavier jet is more likely to be stable and confined out to large distances. As we can see in Figure 4.9, the jet emission also seems to be stable in time. The jet extending out to $\sim$ 100 kpc must have been active for at least half a million years, even when assuming the outflowing speed being close to the speed of light. The lobes in addition indicate that the process has been stable on time scales an order of magnitude larger. We will face these questions again in the discussion of the blazars, which are also dominated by jet emission.

Figure 4.11 shows an example of M84, a FR-I radio galaxy inside the Virgo Galaxy Cluster at a distance of 17.4 Mpc. When comparing this image to the one of Cygnus A, it is apparent that in M84 strong radio emission arises from the jet,

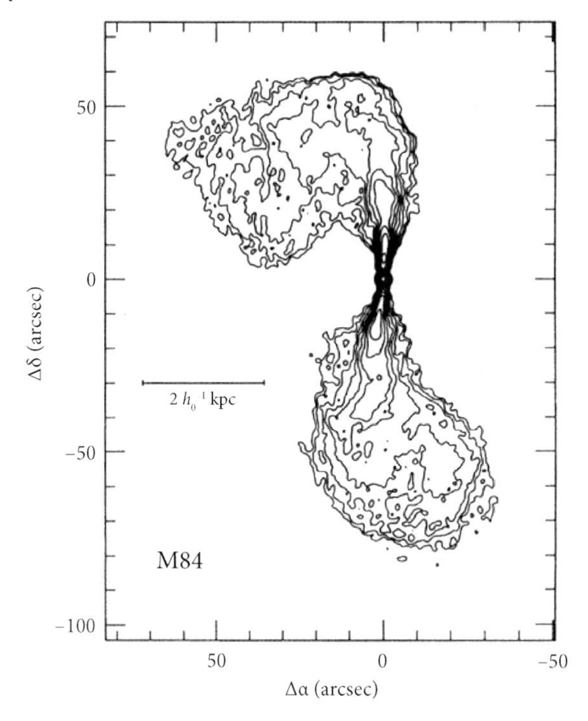

Figure 4.11 The FR-I radio galaxy M84 at 5 GHz as seen by the VLA (Laing and Bridle, 1987). The FR-I objects show a jet-dominated radio structure, in contrast to the FR-II, which have a lobe-dominated radio morphology.

whereas in Cygnus A the hot spots in the lobes are dominant. It is intriguing to hypothesize that the jet is aligned closer to the line of sight in the case of the FR-I galaxies. But again, as we cannot measure the outflow speed of the jet directly, we do not know which part is flowing toward us or whether the whole radio jet and lobe structure is aligned perpendicular to the line of sight. However, if we assume that the jet is more closely aligned to the line of sight in these cases, why is there no hot spot seen in at least one of the radio lobes? A reason might be that the jets in FR-I are not as collimated and/or powerful as in the FR-II. Not all FR-II show hot spots. There seems to be a correlation between the Eddington luminosity and the appearance of hot spots. Apparently only those FR-II that accrete at a high Eddington ratio seem to be able to produce jets powerful enough to cause hot spots in the radio lobes (Koziel-Wierzbowska and Stasinska, 2011). Another issue is the fact that most FR-I have symmetric jets. If one jet is pointing toward the observer, an enhancement is expected, similar to that observed in single-sided sources like 3C 175. The difference might be again intrinsic, that is, the jet might be less powerful in FR-I objects. Alternatively the different appearances could be caused by properties of the ambient media of individual objects. It may also be a mixture of both effects.

When studying polarization of the radio lobes, Laing (1988) and Garrington *et al.* (1988) noticed that FR-II with single-sided jets show in many cases lower polarization in the lobe lacking the jet component. In general the lobe with lower depolarization seems to be on the same side as the stronger jet and/or hot spot component. The explanation of this Laing–Garrington effect is that the depolarization is caused through Faraday rotation as the polarized emission travels through an ionized medium. If the radio galaxy is enclosed in a large ionized cloud, the emission from the radio lobe on the far side would be more strongly depolarized than the one closer to the observer. This is another indication for relativistic jets in FR-II. Although this effect is also observed in the less luminous FR-I sources, it appears to be much less prominent here. This points, in turn, to slower jet speeds in FR-I when compared to FR-II.

So far, we have been focusing on the extended emission from radio galaxies, as this is the defining aspect for this group. When studying the AGN core in radio galaxies, one finds similarities to the cores of Seyfert galaxies. One finds both, Seyfert 1 and Seyfert 2 optical spectral types in radio galaxies, and also the intermediate Seyfert types. For example, the bright FR-II radio galaxy Cygnus A shown in Figure 4.9 has a Seyfert 2 core, the central engine of the FR-II 3C 175 (Figure 4.10) has been identified as a Seyfert 1.2, and the FR-I M84 (Figure 4.11) has a LINER core. Radio galaxies with a type 2 AGN are referred to as narrow-line radio galaxies (NLRG), while those with broad optical-UV lines are called broad-line radio galaxies (BLRG). In addition, some radio galaxies are classified as weak-line radio galaxies (WLRG), commonly counted into the type 2 classification. WLRG show very weak [O III] emission lines and are often optically classified as LINER, although they are strong radio sources, usually of the powerful FR-I type. The emission line properties and diagnostics of WLRG have been discussed for example in Lewis *et al.* (2003). There is no correlation detectable between the FR classification and the Seyfert type of the central engine. Thus, when discussing a radio galaxy of a certain Fanaroff–Riley type, we do not know *a priori* whether it has broad, only narrow, or weak lines in the optical spectra.

Radio galaxies are observed out to high redshifts. These high-redshift radio galaxies (HzRGs) have been found out to redshifts $z \simeq 5.2$, thanks to the follow-up observations of radio sources by optical (e.g., HST) and infrared (e.g., Spitzer) space telescopes. On average, these are heavy, giant elliptical galaxies, often residing in galaxy clusters or in protoclusters. With larger distance it becomes more and more difficult to resolve the morphology of these sources, and eventually these objects appear point-like and fall into the category of quasars, which are discussed in the next section.

4.6 Quasars

In the late 1950s and early 1960s, large surveys of the sky using radio telescopes were performed. An early milestone was the compilation of the third Cambridge catalog, containing 471 sources detected at 159 MHz with a flux density larger than 8 Jy (Bennett, 1962). Optical follow-up observations of these sources discovered, as expected, a large number of radio galaxies and supernova remnants. But many of the sources could not be easily identified, and appeared in optical images like blue stars. These sources were therefore called *quasi-stellar radio sources*, or quasars. Optical spectroscopy then revealed strong emission lines in these sources, and Caltech astronomer Maarten Schmidt was the first to realize that these lines in the case of 3C 273 are indeed the Balmer lines, but redshifted by 16% (Schmidt, 1963). Spectroscopy of other 3C sources then led to the discovery of a number of quasars with significant redshift. It was soon suspected that these sources are indeed the more distant equivalents of nearby Seyfert galaxies. But the telescopes used at that time in combination with the photographic plates, which were yet to be superseded by electronic detectors, were insufficient to image the low-surface brightness host galaxies of distant quasars. Thus research on the two classes of Seyfert galaxies and quasars developed in parallel until it became clear that they were different manifestations underlying the same phenomenon. The advent of space-based UV observations in the 1970s and 1980s solidified this picture (Section 5.4).

As more sensitive observations were obtained, the gap between Seyfert galaxies and quasars was filled with intermediate objects. In order to separate the two groups of AGN, an arbitrary dividing line was introduced, and Seyfert galaxies with absolute magnitudes brighter than $M_{\rm B} < 23$ mag are now referred to as quasars (Schmidt and Green, 1983).

Another way to search for quasars is their distinct appearance in the optical and UV range. This was already reflected in the early days of discovery that quasars appear to be very blue objects compared to galaxies for example. The reason for this is the big blue bump in the spectra of quasars. This bump is often thought to be caused by the accretion disk as described in Section 3.2. In fact, while the spectra of solar-type stars peak around $\lambda = 500$ nm, those of O-stars at around $\lambda = 350$ nm, the blue bump appears at $\lambda = 100$–400 nm reaching even shorter wavelengths and peaking at frequencies $\nu = 10^{15}$–10^{16} Hz. Therefore this feature cannot have its origin in the starlight of the host galaxy.

Even though for many years both type 1 Seyfert galaxies and quasars with broad optical emission lines have been observed, one puzzling fact was that until the late 1990s no narrow line quasars could be found, that is, sources that are bright and at large cosmological distance, but show an optical spectrum similar to that of a Seyfert 2. Indeed these objects are hard to find, as they lack significant radio emission and, at the same time, the optical emission lines are inherently weaker than in type 1 objects. At high redshifts, type 2 AGN will thus easily escape detection. The discovery of type 2 quasars was finally realized by the advent of X-ray surveys. Because all Seyferts and quasars are potentially bright X-ray sources, the follow up of sources detected by ROSAT and ASCA unveiled a number of type 2 quasars, for example RX J13434+0001 (Almaini *et al.*, 1995). Currently data from large optical surveys like the Sloan Digital Sky Survey (SDSS, York *et al.*, 2000, see box below) seem to indicate that bright type 2 quasars are as numerous as type 1, at least in the nearby ($z \lesssim 0.8$) Universe (Reyes *et al.*, 2008), with 15% ± 5% of these objects being radio-loud (Lal and Ho, 2010).

The Sloan Digital Sky Survey (SDSS)

The SDSS (York *et al.*, 2000) provides us with the largest optical sample of AGN to date. The SDSS uses a 2.5 m telescope in the northern hemisphere with a wide field of view (1.5 $\deg^2$) in order to take deep images in five optical bands (ugirz). Then, spectra of selected objects are taken with the same telescope using a customized aperture plate and fiber optics, allowing for spectroscopy of as many as 600 objects within a single observation. Operational since 2000, the SDSS now provides photometric measurements for $\sim 5 \times 10^8$ stars, galaxies, and quasars, and spectra for nearly two million of them.

As quasars are the most luminous AGN, they are also the ones we see far across the Universe. To date, the highest-redshift quasars reach redshift up to $z = 7.085$ (Mortlock *et al.*, 2011), which puts these massive objects at times as early as 800 million years after the Big Bang. In total, there are more than 40 quasars known at redshifts $z > 6$ (Willott *et al.*, 2010), mainly thanks to spectroscopic surveys such as the Canada-France High-z Quasar Survey (CFHQS) and the SDSS. The question, how quasars can have formed in such a short time, will be discussed in Chapter 9. The redshift distribution of known AGN at the time of writing is shown in Figure 4.12, based on 168 000 AGN which are included in the catalog by Véron-Cetty and Véron (2010). Although this is a compilation of AGN rather than a complete sample by any means, one can see clearly the peak in the redshift distribution around $z \simeq 1.8$. This means, that at earlier times in the Universe there have been many more quasars per unit volume than today. We will return to this point in Chapter 9.

4.6.1 Radio-Quiet Quasars

We described in Section 3.7 that radio emission is a signature of a relativistic outflow in the form of a jet, as it produces synchrotron radiation. Quasars lacking this kind of strong radio emission are more difficult to find than their radio-loud counterparts. Large optical spectroscopic surveys were necessary in order to demonstrate that most quasars are not radio bright and do not produce a jet. About 90% of all quasars are radio-quiet, and unlike the separation into type 1 and type 2 objects, there seems to be no smooth transition. This could be understood in a simple scenario in which sources either have a jet, and are therefore radio-loud, or they do not. As described in the box of Section 3.2.4, the definition of radio-loudness arises from the comparison of the radio flux to the optical flux. One definition calls an

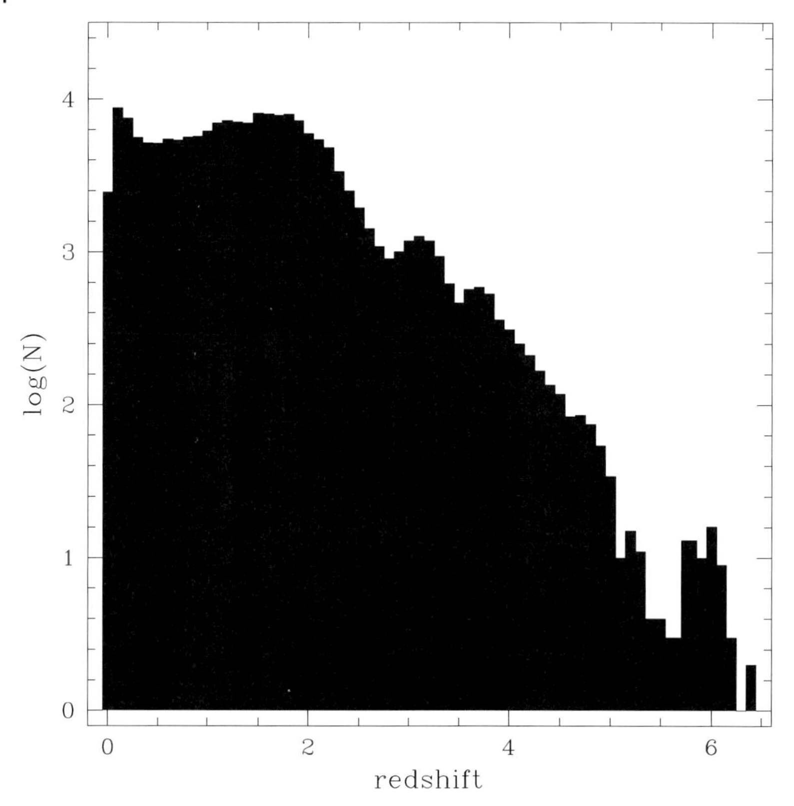

Figure 4.12 Histogram of redshifts of all 168 000 AGN in the thirteenth version of the catalog by Véron-Cetty and Véron (2010). Data have been binned in redshift in $\Delta z = 0.1$ intervals. Note that the y axis is in logarithmic scale. A significant peak of the quasar population appears in the range $1.5 < z < 2.0$. It has to be taken into account, though, that this is a compilation of all published AGN and not a complete sample.

object radio-quiet when

$$R^* = \log \frac{f_{\text{radio}}}{f_{\text{B}}} < 1 \tag{4.4}$$

Thus, an object which is radio-quiet is not necessarily radio-silent: it can still show a certain amount of radio emission. The mechanism to produce the radio flux in radio-quiet AGN is not well understood yet. It certainly is synchrotron emission as in the radio-loud objects. One solution might be a small-scale jet, which is simply much less powerful than in the radio-loud case, as proposed for example by Ulvestad *et al.* (2005a). This is supported by the detection of compact nuclear radio components moving at nonrelativistic speeds observed in a few radio-quiet AGN like the Seyfert 1.5 galaxy NGC 4151 (Ulvestad *et al.*, 2005b; Middelberg *et al.*, 2004). However, this scenario does not account for the clear distinction between radio-loud and radio-quiet objects, and one would expect a smooth transition between the two classes which is so far not observed. Laor and Behar (2008) proposed as an alternative for the emission mechanism magnetic heating, similar to what is taking place in the coronae of active stars. A third possibility might be the production of synchrotron emission through shocks which occur in the accretion flow, as explained by Ishibashi and Courvoisier (2011). One limitation of this model is that one needs to know the strength of the magnetic field, the maximum Lorentz factors of the electrons and their density in order to calculate the flux arising from the synchrotron radiation. As each of these three parameters is hard to measure, the shock model is left with large uncertainties.

4.6.2 Radio-Loud Quasars

As described above, radio-loud quasars were the first to be discovered. Because of high luminosity, a strong radio emission, and distinct spectroscopic features, they are easier to detect than radio-quiet objects. As previously pointed out, only 10% of the quasars are indeed radio-loud. They show similar characteristics as the radio-quiet objects, like a variable continuum flux, a strong UV component, and broad emission lines. But their host galaxies tend to be more massive than those of radio-quiet AGN (e.g., Dunlop *et al.*, 2003) and they produce a powerful jet component which provides significant radio emission through synchrotron radiation.

Radio-loud quasars are further classified as radio bright Flat Spectrum Radio Quasars (FSRQ), and Steep Radio Spectrum Quasars (SRSQ). The latter ones are dominated by radio lobe emission of the host galaxy, the former have a compact radio structure. FSRQ are often also referred to as blazars; see the next Section.

While these sources have radio spectra which are either flat or rising toward higher frequencies, there is a large class of radio detected sources which have inverted spectra because they appear to have a spectrum which peaks in the GHz range. These sources are therefore called inverted spectrum sources or GigaHertz Peaked Spectrum sources (GPS). One striking feature of these sources is that they do not seem to produce a significantly large radio jet. A review on this subject has been presented by O'Dea (1998). Several theories have been brought forward to explain what these sources are and how they fit into the larger class of radio galaxies. Apparently, there are both nonactive GPS and quasar-like GPS. For the galaxy-type GPS sources three main explanations have been brought forward. They could represent very young sources which will eventually evolve into FR-I/II radio galaxies, that is, the jet could have been recently formed and will continue to grow with time. In some cases radio features are seen in the vicinity of GPS sources, which led to the assumption that these sources might produce jets for short periods of time which then shut off again. This possibility is called the recurrent scenario. A less favored explanation, the so-called frustrated scenario, is that the jets are suppressed by a dense ambient medium. It has been argued, for example by Stanghellini *et al.* (2005) that quasars and non-AGN GPS are in fact different, and they would appear similar only in their GHz peaked spectrum and the nondetectability of a large jet. Support for this interpretation comes from high-resolution radio observations using the VLBA of GPS quasars by Orienti *et al.* (2006). They find complex radio structures and the superposition of small lobes and hot spots might indeed mimic a spectrum which peaks in the GHz range, but the intrinsic AGN spectrum in these cases would indeed be flat. Thus, while there is a class of GPS sources, they are not related to the AGN phenomenon, and when a GPS galaxy contains a quasar, this AGN remains a FSRQ or SRSQ.

4.7 Blazars

Blazars are a special subclass of quasars. The common model for blazar emission is that these sources are quasars in which a relativistic jet is pointing at the observer, or very close to the observer's line of sight. Therefore, this class is dominated by high variability. Blazars are prominent emitters from the radio frequencies all the way up to the Very High Energies (VHE) above 1 TeV (see Section 5.7). The blazars are subdivided into BL Lac objects and FSRQ, which are again divided into the Optical Violent Variables (OVV), and the Highly Polarized Quasars (HPQ). BL Lac objects do not show prominent features in the optical spectrum, as seen in the example of RX J1211.9+2242 in Figure 4.13. A criterion used to distinguish between FSRQ and BL Lacs is therefore the equivalent width of the emission lines, and a dividing line of EW = 5 Å is applied, with the BL Lac objects having only lines below that value. This dividing line is somewhat arbitrary, because the continuum emission of blazars is variable and so is the equivalent width of the lines.

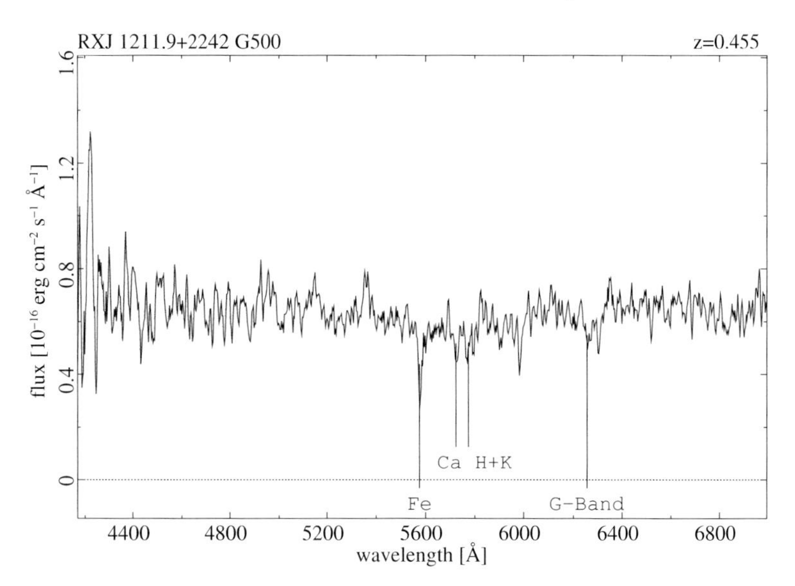

Figure 4.13 Slit spectrum of the blazar RX J1211.9+2242. Only weak absorption lines caused by the jet emission being absorbed by the host galaxy are detectable.

Although their line emission is weak, BL Lacs seem in some cases to possess a narrow-line and a broad-line region as well. The fact that lines are usually not seen in their spectra is due to the dominance of the underlying nonthermal continuum. Stocke *et al.* (2011) showed that some bright BL Lac objects do display weak and narrow Lyα lines. The FSRQ blazar subtypes, like OVV and HPQ, show stronger broad emission lines. Additionally, HPQ show polarization in their continua.

Blazars show dramatic variations in their emission throughout the electromagnetic spectrum. This was the first property to find and identify them. Other ways to find a blazar is to identify sources with strong (and polarized) radio emission (radio-selected blazars, RBL), or sources with a high flux in the X-ray band compared to their optical emission (X-ray-selected blazars, XBL). As these selection criteria are strongly dependent on the kind of observation and the single source characteristics, the different types of blazars are currently distinguished based on their overall spectral energy distribution. Here, sources which are dominant in the X-ray peak at higher energies than the BL Lac objects with relatively strong radio emission. Thus, one calls the X-ray-selected blazars rather "high-frequency peaked BL Lacs" (HBL) and the ones with strong radio flux "low-frequency peaked BL Lacs" (LBL). For a discussion of the spectral energy distribution of blazars see Section 5.8.1.

Variations in blazars are reported on time scales from years down to less than a day, the so-called intraday variability (IDV). In the radio band very high amplitudes ($\Delta f_r / f_r \sim 1$) on hourly time scales are observed. There has been some debate whether this variability is intrinsic, for example due to intensity variations of the jet, or extrinsic, like differing amounts of absorption in the vicinity of the emission region, or even close to the telescope. Lovell *et al.* (2008) found in a study of 443 FSRQ that the short term variability at 5 GHz is likely to be caused by scintillation in the turbulent, ionized interstellar medium of our own Galaxy. They also observe that the amplitude of the IDV in the radio depends on the galactic latitude. Therefore, at least in the low frequency radio band, the observed short-term variability is not related to the source itself.

The optical band is well studied and variations down to minute time scales are found with amplitudes up to 20% (Wagner and Witzel, 1995). Here, scintillation cannot be the responsible mechanism. Typically X-ray-selected BL Lac objects spend most of their time in a quiescent state, which is interrupted by large outbursts. The fraction of time in which the BL Lac is variable, the so-called "duty cycle," depends strongly on the overall spectral type of the source. The X-ray dominated HBL show a duty cycle of $\lesssim 0.4$ while LBL have duty cycles of ~ 0.8 and also show larger variability. In fact, while radio-selected blazars show variabilities up to $\sim 30\%$ within one day, this value is $< 5\%$ for the HBLs.

Strong ($\Pi > 3\%$) and variable polarization is seen in blazars in the radio and in the optical domain. For radio-selected blazars one finds polarization with average values $\Pi \sim 5\%$ (Mantovani *et al.*, 2011) with some objects reaching up to $\sim 40\%$. For these sources the polarization can vary significantly in strength and orientation, while the HBL have a maximum of $\Pi_{\max} \simeq 15\%$ and do not exhibit strong variability. The duty cycles[2] of polarization seem to be similar, though, with $\sim$ 80% of the LBL and $\sim$ 70% of the HBL showing polarization (Heidt and Nilsson, 2011). There is also evidence for a correlation between the peak frequency of the synchrotron emission and the degree of polarization in a sense that more X-ray dominated objects show less polarization in the optical region, confirming earlier results. At the same time a correlation of polarization with luminosity has not been found.

As mentioned above, FSRQ candidates can be identified as AGN by the presence of broad permitted Balmer lines on an otherwise nonthermal continuum. In direct images of blazars the strong, beamed core emission very often outshines the host galaxy, making its identification a more demanding task. The jet not only outshines the host galaxy, but also the broad-line region emission, such that the Balmer lines appear less prominent than in other radio-quiet quasars.

The criteria to identify a BL Lac object as such have been mostly determined by practical observing considerations rather than real physical distinctions between different types of objects. To distinguish the BL Lac galaxies from nonactive elliptical galaxies, a criterion was applied to the strength of the calcium break in the optical continuum emission at 4000 Å. A nonactive elliptical galaxy has a break strength of $\sim$ 40%. Therefore Stocke *et al.* (1991) used a criterion of a break $\leq$ 25% for BL Lac objects found among the X-ray sources detected by the Einstein Medium Sensitivity Survey (EMSS). In fact, there are no objects within their candidates with a break value of 25% $\leq \mathrm{Ca}_{\mathrm{break}} \leq$ 40%, thus a clear distinction could be made between a nonactive elliptical galaxy and a blazar. However, at a later time Marchã *et al.* (1996) found several transition objects, which could be identified as BL Lac due to their radio properties. Therefore, the existence of a break $\geq$ 25% in BL Lac objects appears to be more frequent in radio-selected samples and LBL than among the X-ray-selected HBL objects. The distinction between FSRQ and BL Lac as objects with and without broad lines can be defined more quantitatively. Sbarrato *et al.* (2012) found that the Eddington luminosity of the broad lines is a good indicator to distinguish both classes, although the transition between them appears to be smooth. Here, the BL Lac objects show low Eddington ratios with $L_{\mathrm{Balmer}}/L_{\mathrm{Edd}} < 5 \times 10^{-4}$, whereas objects with brighter Balmer lines can be considered to be FSRQ. Based on modeling of the entire spectral energy distribution, one can derive further insight into the physical parameters of the emission region in blazars, as we will see in Section 5.8. Based on these models, Ghisellini (2011) describes the difference between FSRQ and BL Lac objects which have been detected in the gamma ray in terms of jet power, magnetic field strength, Eddington ratio, and inferred accretion disk and jet luminosities (Table 4.2). We see that although the black hole masses, distance of the emission region from the black hole, and the speed of the jet as expressed in the bulk Lorentz factor are similar for FSRQ and BL Lacs, the applied magnetic field strength and accretion flow are lower in BL Lacs, resulting in the observed lower luminosities.

2) In this case duty cycle is the fraction of time an object spends with a degree of polarization > 4%.

Table 4.2 Average physical properties of γ-ray bright blazars as presented by Ghisellini *et al.* (2010). Here, R_{diss}/R_S is the distance of the dissipation region in units of the Schwarzschild radius (Eq. (3.1)) from the black hole with mass M_{BH}. Γ is the bulk Lorentz factor of the jet, B the magnetic field strength, L_{disk} the accretion disk luminosity as derived from the emission lines, and λ the Eddington ratio (Eq. (3.10)) with respect to L_{disk}. P_e is the power injected by the relativistic electrons, and P_{jet} is the total jet power, assuming that each electron is accompanied by a proton.

	$\log M_{BH}$ ($M_\odot$)	R_{diss}/R_S	Γ	B (G)	$\log L_{disk}$ (erg s^{-1})	λ	$\log P_e$ (erg s^{-1})	$\log P_{jet}$ (erg s^{-1})
BL Lacs	8–9	300–1000	10–20	0.1–2	42–44	< 0.01	41–43	43.5–45
FSRQ	8–9.5	300–3000	10–16	1–10	44–46.5	> 0.01	42.5–44	45–48

As compared to FSRQ with emission lines, the BL Lac objects with generally featureless spectra render redshift determination observationally challenging. In BL Lac objects, at distances where the host galaxy cannot be cleanly separated from the AGN core, it is often impossible to determine the spectrum of the host galaxy. Therefore, only few BL Lacs have measured redshifts beyond $z = 0.5$. Apart from measuring the absorption lines of the host galaxy in order to determine the redshift, one can also use the apparent magnitude of the galaxy itself as an indicator of the distance. This method assumes that the host galaxies of BL Lac objects, which are mainly giant elliptical galaxies, are similar in size and brightness, and the relation is calibrated by using BL Lac objects with known redshifts. Sbarufatti *et al.* (2005) found a correlation between redshift z of the host galaxy and apparent optical R-band brightness m_R of

$$\log(1 + z) = \left(0.293 m_R^2 - 7.19 m_R + 45.1\right) \times 10^{-2} \tag{4.5}$$

Although the average difference between spectroscopic and photometric redshift for a sample of 64 BL Lacs is only $\langle \Delta z \rangle = 0.01 \pm 0.05$ (rms), the difference for the single object can be large, up to $\Delta z/z \sim 0.5$.

The situation is slightly more favorable for the FSRQ, where the AGN core itself with its broad-line region provides the means to measure its distance. However, here also at redshifts above $z = 2$ the beamed emission usually outshines in most cases the BLR, leaving the observer with no handle to determine its distance. This can be seen in the redshift distribution of blazars known to date as shown in Figure 4.14. Distinct from the quasar redshift distribution shown in Figure 4.12, only 25% of the blazars with measured redshift are at $z > 0.5$, and only a handful of blazars above redshift $z = 2$ have been identified, while 80% of all AGN are at redshifts $z > 0.5$ and 20% are at $z > 2$. Apart from the difficulty to measure the emission or absorption lines in blazars, several selection effects might contribute to the lack of known high-redshift objects. When, for example, in a high-redshift survey a source is found and an emission line can be measured, there is often no distinction made between a quasar and a blazar. Thus, these objects are not necessarily classified beyond being identified as an AGN. One aspect which also limits the redshift distribution of blazars is the fact that most of those sources are found

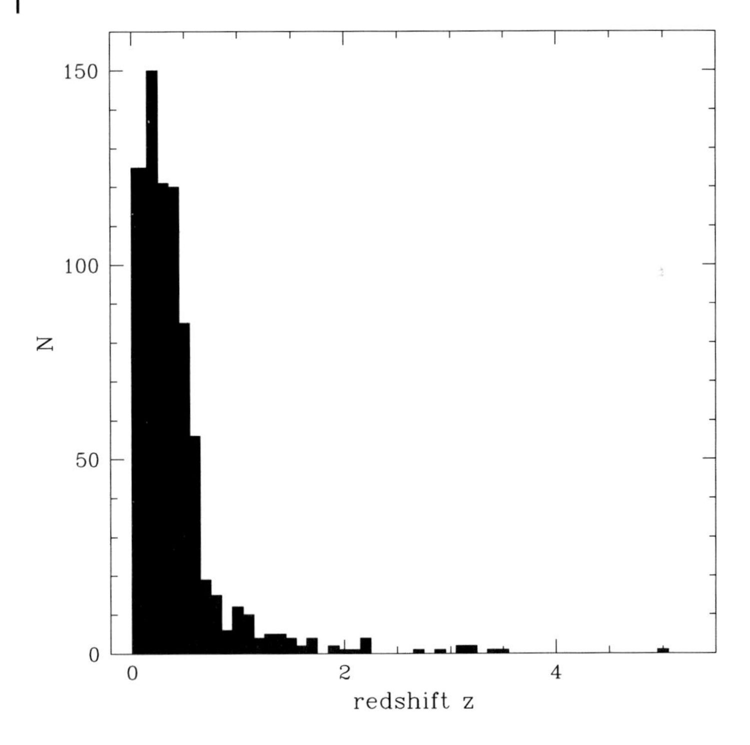

Figure 4.14 Histogram of the 760 blazars with known redshift in the catalog by Véron-Cetty and Véron (2010).

by analyzing radio or X-ray sources, and surveys in both bands are on average less deep than the high-redshift optical surveys.

Determining the mass of the central black hole also turns out to be more difficult in the case of blazars than for the unbeamed sources. As the jet outshines in many cases the host galaxy, estimates using the width of the lines originating in the galaxy as a proxy for the mass of the central engine are often impossible. Mass determination is therefore biased by the inclination angle of the jet with respect to the line of sight – the more we look directly into the jet, the less likely it is to give a reasonable estimate of the black hole mass. A recent study found that all BL Lac objects seem to have approximately the same central mass, $M_{\rm BH} \simeq 10^{8.5}\,M_\odot$ (Plotkin *et al.*, 2011). But as this study is based on the detectability of the lines of the host galaxy, there might be indeed a selection effect occurring.

4.8 Unification of AGN

A fundamental question in AGN research is, whether all these distinct appearances of the AGN phenomenon can be explained by a common underlying model, or whether the different classes are intrinsically distinct. It was pointed out early on that a Seyfert galaxy is in fact in most cases a spiral galaxy to which a faint quasar is added in the center (Weedman, 1973; Penston *et al.*, 1974). In addition, Kristian (1973) showed that the fainter quasars indeed appear to have an extended form rather than being point-like, indicating that they reside in galaxies. Rowan-Robinson (1977) made an attempt to unify Seyfert galaxies and radio sources. While he correctly assumed that absorption by dust is important in order to explain the differences in infrared emission, he did not take into account beaming effects which are an important ingredient when trying to understand radio-bright AGN.

At a 1978 BL Lac conference in Pittsburgh the foundations for the beaming unification were outlined (Blandford and Rees, 1978), a concept which is still believed to be true. In this picture, if an AGN appears to be a blazar it is because the emission is beamed along the symmetry axis of the AGN towards the observer (Figure 4.15). In a next step, Scheuer and Readhead (1979) proposed that the radio core-dominated quasars could be unified with the radio-quiet quasars by assuming the former ones are beamed towards the observer, similar to the case of blazars. This implies that all radio-quiet quasars also host a relativistic radio jet, but they are only FSRQ when the jet is along the line of sight. This concept turned out to have a problem though. As Orr and Browne (1982) pointed out, the core-dominated and radio-loud quasars indeed showed extended radio emission in MERLIN and VLA observations. Therefore, radio-quiet AGN could not simply be misaligned radio quasars. Later studies explained the differences by two effects: difference in orientation, and difference in obscuration (Barthel, 1989). A still valid and rather complete overview of the problem of AGN unification was given by Antonucci (1993). In the most simplified picture, there are basically two types of AGN: radio-quiet and radio-loud. For each type a range of luminosities is observed, leading for example to the Fanaroff–Riley classes as well as to the distinction between Seyfert and quasar. All other observed

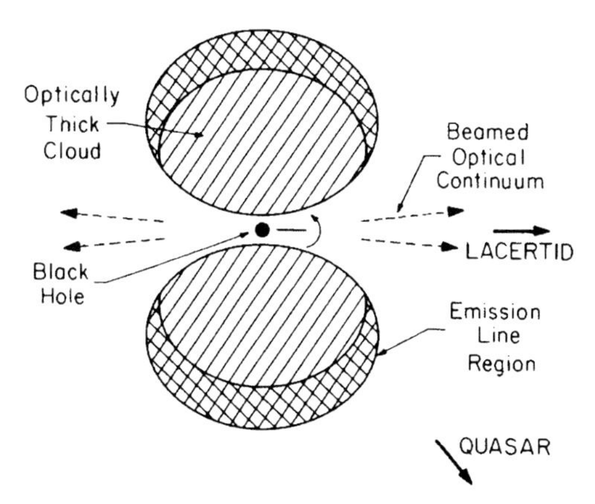

Figure 4.15 Schematic representation of a geometrical interpretation of the BL Lac phenomenon by Blandford and Rees (1978). If the optical continuum is beamed along the symmetry axis, then the emission lines may be suppressed when the source is viewed from this direction. In this figure Lacertid stands for blazar.

differences would be explained by orientation effects. In this scenario all objects which show a quasi-stellar radio core and blazars would emit beamed radiation towards us, with a closer alignment to the line of sight in the case of the blazars. Radio galaxies, in this picture, emit their jet at large off-axis angles with respect to the line of sight.

Antonucci pointed out that the existence of an optically thick torus surrounding the central regions of an AGN on scales of 1–100 pc would lead to the absence of broad emission lines in the case of Seyfert 2 if they were observed edge-on, as their broad-line region would be hidden, compared to Seyfert 1 objects, which are mostly observed face-on. As the narrow-line region lies further away from the central black hole than the broad-line region, the NLR would still be observable when the BLR is obscured by the torus. In addition, Antonucci was also aware of the shortcomings of this simple model and that it left open the question, what is the intrinsic difference between radio-loud and radio-quiet AGN, and why do radio-loud AGN mainly reside in elliptical galaxies, while radio-quiet AGN are hosted by spiral galaxies.

A subsequent review by Urry and Padovani (1995) explained the unification of the subgroup of radio-loud AGN. The aim was to study whether the low radio luminous FR I can be the parent population of the BL Lac objects, while FSRQ would be a subset of the FR II galaxies. Urry and Padovani (1995) pointed out the difference in evolutionary behavior between BL Lacs and FSRQ, and considered the suggestion that FSRQ evolve into BL Lac objects, becoming weak-lined objects by virtue of increased beaming of the continuum, that is, of a Lorentz factor increasing with cosmic time (decreasing with redshift; Vagnetti *et al.*, 1991). A problem became evident though, as the luminosity functions of the two object groups could not be connected smoothly, for example, because of very different radio power and line luminosity at comparable redshifts. In the following section we will look at unification scenarios involving absorption only and then investigate whether radio bright and radio faint AGN can indeed be unified.

4.8.1
Absorbed versus Unabsorbed AGN

If we put aside for the time being the differences between radio-loud versus radio-quiet AGN, the unified model predicts a distinction between the various types of sources based solely on the viewing angle. The anisotropy of the AGN population is then assumed to be caused by the different level of absorption in the line of sight. This concept leads to certain predictions which can be tested through observations. All the intrinsic properties, that is, the appearance of the AGN when absorption effects are not relevant or when they have been modeled out, should be similar for all Seyfert types on one side and all radio emitting sources on the other. On the other hand, we should be able to observe a consistent set of differences which can be explained by the absorption, and which should correlate with the optical depth of the absorber.

A strong test for the unification model is the prediction that the broad-line region lies at smaller radii than the absorbing material, whereas the narrow-line region should be visible in all Seyfert types as the emitting material resides further out. This part of the model was based on the observational fact that Seyfert 1 galaxies have broad and narrow emission lines whereas Seyfert 2 have only narrow lines. The distinction between type 1 and type 2 should disappear if we find a way to exclude the influence of absorption in our observation of AGN spectra. One method to investigate the intrinsic line width in optical spectra is to study the polarized light. Although the broad-line region (BLR) is hidden by obscuring material in the case of type 2 AGN, the light of the BLR can escape in directions where no material hinders the view to the central engine and its surroundings. If the BLR's light then hits, for example, electrons, it can be scattered into the line of sight of the observer and thus still reach us. This scattering of photons follows the rules of Thomson scattering and in Section 2.1.1 we have seen that the scattered light can be linearly polarized. Thus, by looking at the polarized emission only, the broad component in Seyfert 2 can still be visible, and the scattering material acts like a mirror which enables us to look behind the absorbing matter.

A first proof of this concept was given by Antonucci and Miller (1985), who showed that the Balmer lines in the Seyfert 2 NGC 1068 are broad when the AGN is observed in polarized light. They also showed that the nonthermal continuum emission of the central engine has the same level of polarization as the Balmer and Fe II emission lines, in the case of NGC 1068 about $\Pi \simeq 15\%$. Subsequent spectropolarimetric observations of highly polarized Seyfert 2 galaxies discovered further hidden broad-line regions (HBLR), for example in Mrk 3, Mrk 348, Mrk 463E, Mrk 477, Mrk 1210, NGC 7212, NGC 7674, and Was 49b (Miller and Goodrich, 1990; Tran *et al.*, 1992). The observations confirmed that continuum and broad Balmer lines show the same degree of polarization, which however can differ substantially from object to object. At the same time, the narrow forbidden lines do show little or no polarization at all, confirming that the narrow-line region is observed directly. This discovery was certainly a strong argument in favor of the unified model. On the other hand, there are numerous type 2 AGN which do not show any broad-line component in polarized light. In total, only about 40% of the Seyfert 2 galaxies are seen to have an HBLR (Wu *et al.*, 2011b). Gu and Huang (2002) found that those which have polarized broad lines are mainly the AGN which have a powerful central engine and thus a high accretion rate. A similar result was reached by Trump *et al.* (2011) when studying accretion rates in a large sample of several hundred AGN with multiwavelength data, ranging from the infrared up to the X-ray band. They defined the intrinsic luminosity $L_{\rm int}$ of these AGN as the sum of the disk, as measured in the optical and UV band, of the corona on top of the accretion disk, as detectable in the X-ray band, and of the reprocessed emission as visible in the infrared. In this study, a strong dependence of the broad-line detectability on the accretion rate was evident. Broad emission lines would only be created in objects with a high Eddington ratio $\lambda = L_{\rm int}/L_{\rm Edd} > 0.01$ (Eq. (3.9)). The broad-line region in these objects will be detectable, either directly in the unobscured Seyfert 1, or in the polarized light in the type 2 AGN. At low accretion rates with $\lambda < 0.01$, narrow-line AGN are observed which do not show strong absorption. Thus, in these cases, neither in normal light nor in polarized spectroscopic observation, is a broad-line region detectable. The nondetectability of the broad-line region would not necessarily break the unification, although it adds another dependency, the strength of the AGN activity. We will come back to this point when discussing the potential problems of the unified model in Section 4.8.3.

Another way of testing the unified model is to look at the average intrinsic, unobscured luminosity of samples of AGN. If the central engines of different types are the same, then the average unabsorbed luminosity should be the same as well. As we saw in Section 4.1, the [O III] line at $\lambda = 5007$ Å is a strong indicator of AGN power. At the same time, this line is a narrow forbidden line that should be produced outside of the absorbing material. In addition, the strength of the NLR should not be affected by the existence or absence of a jet. For example, FSRQ and radio galaxies of similar power should then have similar [O III] or [O II] line strengths. Jackson and Browne (1990, 1991) studied spectra of narrow-line radio galaxies and of FSRQ. They compared sources of the same extended radio luminosity and found that the [O III] luminosities of quasars exceed those of NLRG by a factor of 5–10. This would indeed be a strong indication against the unified model. Jackson and Browne (1990) therefore suggested that part of the narrow-line region is absorbed by the obscuring matter, adding another complication to the unification scheme.

More recent studies using the Spitzer Space Telescope come to a different conclusion. Here, the Spitzer mid-infrared data are used as an indicator of the overall AGN power. The idea is that the circumnuclear dust which is dominating this energy range acts as a bolometer for the central engine. Although reprocessed, the Spitzer data might thus give a good proxy for the bolometric luminosity of the AGN, hidden or not. Comparing the mid-infrared luminosity of 46 radio-loud AGN from a complete sample with $f_{2.7\,\mathrm{GHz}} > 2\,\mathrm{Jy}$ with [O III] line luminosities, Dicken *et al.* (2009) find no major difference between quasars, narrow-line and weak-line radio galaxies, nor between FR-I and FR-II. Other studies call into question the claim that the [O III] line indeed represents the bolometric luminosity of the AGN.

Moving further into the infrared, where absorption should play a lesser role, one finds for example the forbidden narrow line [O IV] at $\lambda = 25.9\,\mu\mathrm{m}$. Kraemer *et al.* (2011) studied a sample of 40 Seyfert galaxies and found that the ratio of [O III]/[O IV] is lower for the less luminous sources and for the Seyfert 2 objects. This indicates that the [O III] luminosity might after all be affected by absorption. On the other hand, Kraemer *et al.* (2011) found that in the Seyfert 1.5 galaxy NGC 4151 only one third of the [O III] emission seems to arise from the inner ~ 30 pc. Thus, in order for Seyfert 2 galaxies to have the same [O III] profiles as the unobscured Seyfert 1, the absorbing dust must extend out to large radial distances.

In summary, the results of these tests using the narrow emission lines seem to be inconclusive. The problem here is to find a reliable measurement of the intrinsic luminosity. In addition, if the NLR is not fully visible, further corrections to the [O III] line luminosity have to be applied and further dilute the results. An extensive review on the emission line spectrum of AGN and the implications on the unification scheme has been provided by Véron-Cetty and Véron (2000).

Another way to study differences and similarities between different AGN types is to look at their spectral shape. Absorption also has an influence on the continuum emission of AGN. The appearance of the underlying continuum will be changed and in case one can measure the absorption, the intrinsic spectrum can be recovered. The effect on the intrinsic spectrum will obviously be less pronounced for lower absorption, and should also be less strong when observing at wavelengths less affected by absorbing material. The optical domain is strongly affected by material in the line of sight. For example, in the V-band, for a hydrogen column density N_H, measured in atoms per unit area in the line of sight, Predehl and Schmitt (1995) showed that the extinction is

$$A_\mathrm{V} = \frac{N_\mathrm{H}}{1.79 \times 10^{21}\,\mathrm{cm}^{-2}}\,\mathrm{mag} \tag{4.6}$$

Already the galactic hydrogen column density at high-galactic latitude ($|b| > 20°$) is in the range $N_\mathrm{H,Gal} = 10^{20}$–$10^{21}\,\mathrm{cm}^{-2}$ and thus can lead to an extinction of the optical flux of $A_\mathrm{V} = 0.5$ mag. Through the galactic plane, observation of extragalactic objects in the optical domain is very difficult to impossible. Moving into the infrared regime, absorption becomes less efficient. In the near infrared bands J ($\lambda = 1.3\,\mu\mathrm{m}$), H ($\lambda = 1.7\,\mu\mathrm{m}$), K ($\lambda = 2.2\,\mu\mathrm{m}$), and L′ ($\lambda = 3.8\,\mu\mathrm{m}$), the extinction with respect to the V-band extinction is $A_\mathrm{J} = 0.276 A_\mathrm{V}$, $A_\mathrm{H} = 0.176 A_\mathrm{V}$, $A_\mathrm{K} = 0.112 A_\mathrm{V}$, and $A_{\mathrm{L}'} = 0.047 A_\mathrm{V}$ (Schlegel *et al.*, 1998).

Holes in the galactic absorption

In order to study the spectrum and spectral energy distribution of AGN, those areas in the sky where our Galaxy has little gas and dust in the line of sight, are most interesting. Not only is the optical domain strongly affected by absorption, but in the UV band the effect is even worse, and the soft X-rays, below a few keV, are also efficiently absorbed. Lockman *et al.* (1986) discovered a "hole" in the galactic hydrogen column density in the northern hemisphere centered at RA = 10h52m and DEC = $+57°22'$ (in J2000.0 coordinates). The so-called *Lockman hole* covers an area of about 15 deg^2 with a minimum of only $N_\mathrm{H} = 4.5 \times 10^{19}\,\mathrm{cm}^{-2}$ and an average of $N_\mathrm{H} = 5.7 \times 10^{19}\,\mathrm{cm}^{-2}$. This sky area has been used extensively especially for X-ray observations, in order to get an unobscured view on the AGN population. In the southern hemisphere, the Chandra X-ray Observatory used a field around RA = 3h32m and DEC = $-27°48'$ with a column density of $N_\mathrm{H} \simeq 7 \times 10^{19}\,\mathrm{cm}^{-2}$ to perform deep extragalactic survey studies.

The problem with recovering the intrinsic spectrum in the infrared to optical domain is that in this energy range many different components contribute. Not only is the underlying continuum of the central engine visible, but also the dust and stars in the bulge of the host galaxy, and the thermal emission of the accretion disk. An energy range better suited to study the intrinsic spectrum in type 1 and type 2 objects and to determine whether it has the same shape in the X-ray and hard X-ray domain. The X-rays below 10 keV are significantly affected by absorption in the line of sight. Although some AGN show some contribution of the surrounding starburst activities and/or some additional excess below 2 keV, the continuum as seen in the $\sim$ 3–10 keV is dominated by the emission of the central engine (see Section 5.5). The only diversion of the continuum from a single power law, as expected for Comptonization processes, is then due to the absorption in the line of sight. Therefore, the X-ray range is very well suited for measuring the column density of the absorber.

X-ray data show that most but not all AGN unabsorbed in the X-ray are Seyfert 1 type, and most but not all AGN that are absorbed belong to the Seyfert 2 group (e.g., Awaki *et al.*, 1991). The distinction between absorbed and unabsorbed appears around $N_{\mathrm{H}} = 10^{22}\,\mathrm{cm}^{-2}$ as a dividing line. At energies above 10 keV, absorption will have very little effect on the emission, unless the absorber is Compton-thick, that is when the column density in the line of sight significantly exceeds a value of $N_{\mathrm{H}} = 1.5 \times 10^{24}\,\mathrm{cm}^{-2}$. When studying a sample of 25 Seyfert 2 galaxies, Risaliti *et al.* (2002) found that 90% show significant variations of their X-ray absorption column density. This cannot be explained by the simple torus model for the absorber, instead one has to assume some clumpiness in the absorbing matter. The most prominent of those "changing look" AGN is the Seyfert 1.8 NGC 1365. In the X-rays, this source changes from a Compton-thick to a Compton-thin absorber and back on a monthly time scale, as shown by Risaliti *et al.* (2005). The component in the X-ray spectrum which is thought to arise from reflection on the absorbing medium, does not seem to vary; thus, the assumption that there is indeed a massive, clumpy absorber at some distance ($\sim$ 1 kpc) from the central engine seems to be valid. Type 2 sources are not the only ones that show strong variable column densities in the X-ray. Seyfert 1 and Seyfert 1.5, such as NGC 4151, MCG-6-30-15, and NGC 3227 do as well (Risaliti, 2010), as does the narrow-line Seyfert 1 galaxy Mrk 766 (Risaliti *et al.*, 2011).

The combination of X-ray results should tell us whether type 1 and type 2 AGN are intrinsically the same: in the softer X-rays we can measure the absorption strength, and at the hardest X-rays we can determine the true intrinsic spectral shape. Early X-ray surveys seemed to indicate that there is indeed a difference in the intrinsic continuum spectrum of Seyfert 1 and Seyfert 2, in the sense that the spectra of the absorbed sources ($N_{\mathrm{H}} > 10^{22}\,\mathrm{cm}^{-2}$) appeared flatter than those of type 1 AGN. This has been noticed by Zdziarski *et al.* (1995) based on data taken in the 2–10 and 50–300 keV band by Ginga and CGRO/OSSE, respectively. The same discrepancy between the spectra of Seyfert 1 and 2 was later confirmed for example by Gondek *et al.* (1996) using combined EXOSAT, Ginga, HEAO-1, and CGRO/OSSE spectra, and also by INTEGRAL at hard X-rays above 20 keV, where absorption should not play a role (Beckmann *et al.*, 2006a). A study using data from another hard X-ray experiment covering the 15–200 keV band (BeppoSAX/PDS) using spectra of 45 Seyfert galaxies has come to a similar conclusion, although the spectra of Seyfert 2 appeared to be steeper when considering a possible cutoff in the spectra of Seyfert 1 galaxies (Deluit and Courvoisier, 2003).

A problem in measuring the true spectral shape is the complex nature of the intrinsic hard X-ray spectrum. As we will see in Section 5.5, the inverse Compton emission spectrum, which cuts off exponentially at around $\sim 100\,\text{keV}$, can also be altered by the reflection from cooler material leading to a Compton reflection "hump" around 30 keV. Therefore high-quality data are necessary in order to disentangle the different components. Data from the hard X-ray missions BeppoSAX, INTEGRAL, and Swift seem to indicate now that the underlying continuum has the same spectral slope when all components are taken care of correctly. Analysis of a sample of 105 Seyfert galaxies using the spectra collected with BeppoSAX in the 2–200 keV band (Dadina, 2008) provided no evidence of any spectral slope difference when applying more complex model fitting including a reflection component. The INTEGRAL data show consistent slopes for the spectra of unabsorbed/type 1 and absorbed/type 2 objects already when a simple power-law model is used. When applying more complex models with geometrical dependencies, the underlying spectral slope seems to be fully consistent over different inclination angles (Beckmann *et al.*, 2009). Other studies seem to indicate that the spectral slope is the same, but the reflection component is of different strength when comparing Seyfert 1 and 2 galaxies. Ricci *et al.* (2011) find in an analysis of hard X-ray spectra of 165 Seyfert galaxies that the strongest reflection is originating from Seyfert 2 galaxies with intrinsic absorption of $10^{23}\,\text{cm}^{-2} \leq N_\text{H} \leq 10^{24}\,\text{cm}^{-2}$, whereas objects with more or less absorption do not show this feature strongly. In the unified model this is difficult to explain and requires a complex absorption geometry, in which the objects with a strong reflection component would have to have an absorber which covers a high fraction of the X-ray source. Clouds of matter of different sizes could be a possible solution, in the sense that the strongly reflecting sources display smaller matter clumps in the vicinity of the X-ray source than the other Seyfert galaxies. The smaller clump size would lead to a larger surface available for reflection, assuming that the total absorber mass is about the same.

Concerning the accretion activity, the INTEGRAL-selected sources seem to indicate that the mass of hard X-ray-selected Seyfert galaxies does not depend on source type and is on average $\sim 10^8\,M_\odot$. But at the same time the average luminosity of type 1 AGN is higher than that of Seyfert 2, and thus also the Eddington ratios of Seyfert 1 ($\lambda \simeq 0.06$) appear higher than those of Seyfert 2 with $\lambda \simeq 0.02$ (Middleton *et al.*, 2008) in the local Universe. These values have to be treated with caution, as the black hole masses were determined using different methods. In addition, the X-ray luminosity was used as a proxy for the bolometric luminosity with $L_\text{bol} = 2 \times L_{3-1000\,\text{keV}}$, assuming that the radio to optical branch of the AGN emits as much as the inverse Compton branch.

More general approximations for determining the bolometric luminosity based on measurements in some bands are described in the box below. Gallo *et al.* (2010) found that the accretion rate of AGN appears to be a function of the black hole mass. In their study the Eddington ratio seems to be anticorrelated with M_BH, another indication for a violation of the unified model.

Bolometric approximation

Using the X-ray luminosity as an indicator for the overall emission of the AGN is a common approach, for example Marconi *et al.* (2004) used the 2–10 keV luminosity multiplied by ten in order to derive a bolometric correction for low luminosity AGN. In fact, the correction factor is a function of the bolometric luminosity itself. The larger the bolometric luminosity, the higher the factor to be applied to the X-ray measurement. Marconi *et al.* (2004) derived luminosity-dependent correction factors based on templates of spectral energy distributions of AGN. These can be used to estimate bolometric luminosities based on X-ray or optical measurements. The corrections are

$$\log \frac{L_{\text{bol}}}{L_{2-10\,\text{keV}}} = 1.54 + 0.24\mathcal{L}_{\text{bol}} + 0.012\mathcal{L}_{\text{bol}}^2 - 0.0015\mathcal{L}_{\text{bol}}^3 \tag{4.7}$$

$$\log \frac{L_{\text{bol}}}{L_{0.5-2\,\text{keV}}} = 1.65 + 0.22\mathcal{L}_{\text{bol}} + 0.012\mathcal{L}_{\text{bol}}^2 - 0.0015\mathcal{L}_{\text{bol}}^3 \tag{4.8}$$

$$\log \frac{L_{\text{bol}}}{\nu_{\text{B}} L_{\nu_{\text{B}}}} = 0.80 - 0.067\mathcal{L}_{\text{bol}} + 0.017\mathcal{L}_{\text{bol}}^2 - 0.0023\mathcal{L}_{\text{bol}}^3 \tag{4.9}$$

Here luminosities are given in solar units ($L_\odot = 3.83 \times 10^{33}\,\text{erg}\,\text{s}^{-1}$), $\mathcal{L}_{\text{bol}} = (\log L_{\text{bol}} - 12)$, and the optical approximation is given for the B-band with $\nu_{\text{B}} = 6.8 \times 10^{14}$ Hz.

Numerous investigations have tried to explain the differences between the Seyfert types, which cannot be covered in the unified model by differences in the geometry or physical properties of the absorber. Ramos Almeida *et al.* (2011) conclude based on a small sample that the absorbing tori in Seyfert 2 have a larger covering factor, a *lower* optical depth, and are more clumpy than those in Seyfert 1. In addition, if we assume that the absorbing medium is not homogeneous, but rather clumpy, observing a type 1 or type 2 AGN is rather given by the probability of the light of the broad-line region shining through it. In a clumpy absorber model, a small inclination angle object in which we observe the AGN disk face-on can also appear as a Seyfert 2.

Here we also might face a situation where the overall picture agrees with the unified model, but the model needs further adjustment and dependence on other parameters than only orientation and radio-loudness.

4.8.2
Radio-Loud versus Radio-Quiet

As pointed out above, the simple unification scheme which only considers absorption and beaming is not sufficient to answer the question of why some sources are strong radio emitters, and some are not. In other words, what makes the central engine produce a jet. We recall that radio-quiet does not mean that there is no radio emission at all from the AGN, but that the radio to optical flux ratio is low (see box in Section 3.2.4). And as we will see here, a radio bright source is not necessarily radio-loud, and not every AGN which is radio-quiet has to be a faint radio source.

Apparently there is a dichotomy between radio-loud AGN (broad-line radio galaxies, radio-loud quasars, FR-I, FR-II) and radio-quiet AGN (Seyfert galaxies, LINER). Most Seyfert galaxies, although being weak radio emitters, do not seem to harbor a jet. High resolution observations of the radio cores of Seyfert galaxies by Lal *et al.* (2011) do not seem to detect any relativistic beaming, which would be a clear indication for a jet. At the same time, on parsec scales Seyfert 1 and Seyfert 2 appear to be very similar, both appearing to have the same compactness. Also, when comparing the inner parsecs with the extended kpc-scale radio emission, there does not seem to be a difference between the type 1 and type 2 objects, as we would expect from the unified scheme.

A first step to unify the radio-loud objects was made when investigating the properties of the fainter, core dominated FR-I with the brighter, lobe dominated FR-II galaxies. Perley *et al.* (1980) studied compact radio sources and pointed out that although their radio properties are different from the FR-II galaxies, the spectra appear consistent if relativistic beaming effects are considered. Because the jet in FR-I galaxies is closer to the line of sight, the radio core emission would be enhanced in these cases by relativistic beaming. Perley *et al.* (1980) also previously pointed out that the beaming would roughly explain the fraction of core-dominated sources among the radio galaxies. This was further investigated by Orr and Browne (1982), using a simple model for the intrinsic quasar emission consisting of a core which appears relativistically beamed with a radio spectral index of $\alpha_r = 0$ and unbeamed radio lobes with spectral index of $\alpha_r = -1$. The study showed that unification is possible assuming the same average core Lorentz factor of $\gamma \simeq 5$.

A direct correlation of the radio luminosity with black hole mass had been found in several investigations. This connection has roughly the form of $L_r \propto M_{BH}^{2.5}$ as found by Franceschini *et al.* (1998) and confirmed in several other studies. Thus, following this result, strong radio emission, and therefore a powerful relativistic jet is a property connected to the central engines with the highest mass. The same correlation still seems to hold for radio-quiet objects (Nelson, 2000). Another possibly related trend which has been found is the correlation of the mass of the central black hole and the radio-loudness of the AGN. Laor (2000) discovered that nearly all PG quasars with a black hole mass $M_{BH} > 10^9\ M_\odot$ are radio-loud, while quasars with $M_{BH} < 3 \times 10^8\ M_\odot$ are practically all radio-quiet. This led to the assumption that the various types of AGN may be largely set by three basic parameters: M_{BH}, L_{bol}/L_{Edd}, and inclination angle. This extension to the unified model is consistent with what we saw in the previous section when discussing the unified model in view of absorption and inclination angle. It should be pointed out, though, that other studies could not find a simple relation between the radio-loudness and M_{BH}. One also has to keep in mind that the radio-loudness depends strongly on whether the whole radio emission of an object is integrated, or whether only the core flux is measured.

A result which appears to be counterintuitive is the finding that the radio-loudness is anticorrelated with the Eddington ratio λ_{Edd} defined in Eqs. (3.9) and

(3.10). Ho (2002) studied a sample of 80 galaxies including AGN hosting a supermassive black hole of known mass. One has to keep in mind that this study also includes nonactive galaxies, and that most masses have been determined in the local Universe. First of all, the study showed that the relation between the mass of the central engine and its radio luminosity might not be as simple as previously indicated. The $L_r \propto M_{BH}^{2.5}$ law might rather describe an upper envelope than presenting a real correlation, as there are many objects far away but below this line, whereas no object is found above. This means that the relation appears to be rather as $L_r <\propto M_{BH}^{2.5}$ and would just indicate a maximum of radiative jet power possible for a given mass of the central engine. The anticorrelation of radio-loudness and λ_{Edd} might reflect the fact that in objects which are accreting at a low rate, the accretion disk itself is not very prominent. Therefore, a low λ_{Edd} leads to a low thermal disk emission and to a weak blue bump. As both components dominate the optical/UV band, these objects appear as "radio-loud," although they might rather be called "optically quiet." Another explanation for the anticorrelation which indicates that the objects with highest Eddington ratio are the least radio dominated, might be that this rather reflects the anticorrelation between Eddington ratio and M_{BH} as derived by Gallo *et al.* (2010) and discussed earlier. Thus, the intrinsic mechanism could be a result of smaller black holes accreting more efficiently. Larger black holes are more likely to produce a significant jet emission; this would thus appear as an anticorrelation between radio-loudness and Eddington ratio. Sikora *et al.* (2007) extended the study of the anticorrelation of Eddington ratio and radio-loudness to a larger sample of radio-selected AGN, including broad-line radio galaxies and FR-I radio galaxies. Including these objects, the scatter of the correlation of Eddington ratio versus radio-loudness gets much wider. For a given Eddington ratio, the radio-loudness can fall in a range 5 orders of magnitude wide, and vice versa. The $\lambda - R^*$ anticorrelation is then explained by effects of the spin of the central black hole: powerful jets can be produced when rotational energy of the central engine can be extracted via interaction with an external magnetic field, for example from the accretion disk. This is similar to what one observes from galactic black holes when they reside in the so-called *low/hard state* (i.e., low flux and a hard X-ray spectrum).

One has to be aware that Sikora *et al.* (2007) (and many others) use the optical luminosity as a proxy for the bolometric luminosity, simply by performing $L_{bol} = 10 \times L_B$. This makes this bolometric luminosity highly sensitive to the accretion disk's thermal emission profile, that is, the big blue bump, which is different from the bolometric luminosity based on a more complete model. For example, the presence of inverse Compton emission in the X-ray domain is a substantial luminosity component. And, as mentioned in the previous section, absorption can strongly affect the observed optical flux.

Another issue with studying radio-loudness and radio luminosity of radio galaxies is the question whether only the core luminosity should be used or if the total radio emission, including the core and the lobes, gives a better estimate of the overall jet power. The gap between radio-loud and radio-quiet sources appears smaller when using only the core flux (White *et al.*, 2000), and the core dominated sources (FR-I) then show a lower radio-loudness than the lobe dominated ones (Rafter *et al.*, 2011). In a recent study the sample of Sikora *et al.* (2007) has been investigated using only core luminosities. Broderick and Fender (2011) find that the $\lambda - R^*$ anticorrelation becomes less pronounced, as one would expect if the black hole spin is indeed the driving parameter. Instead or in addition to the spin, environmental density or the black hole mass might again be important here. Broderick and Fender (2011) have revisited the radio-loud/radio-quiet dichotomy by using a black hole mass normalized core-only radio luminosity as opposed to the total spatially integrated luminosity. Their investigation was motivated by the knowledge that the bolometric luminosity, jet power and black hole mass are interrelated as demonstrated by the ubiquitous appearance of the black hole fundamental plane. They find that while the bimodal nature of the AGN population sampled is preserved, the magnitude of the separation is significantly reduced. Specifically, they find that FR-I and BLRG are on average more radio-loud than Seyferts and LINER by about 1.6 dex.

Recently a new approach has been applied to try and unify radio-loud and radio-quiet AGN. Garofalo *et al.* (2010) consider the *relative* spin of the central black hole with respect to the accretion disk to be the crucial factor here. In their scenario AGN would start with a black hole which has a very different, even retrograde, spin with respect to the accretion disk, leading to strong interaction with the disk and thus strong jets. As the black hole is spun up in the direction of the accretion disk, the interaction of the rotating black holes with their magnetospheres becomes less efficient and the jet weakens. Thus, the highest prograde black hole spins might be discovered in the least active AGN (Garofalo *et al.*, 2010). This scenario is supported by a theoretical approach of Daly (2009, 2011) determining the black hole spin that is model-independent, but assumes that spin changes only by extraction of the reducible black hole mass. This model applied to a small subset of powerful radio galaxies finds indeed that they harbor low spinning black holes. Further observational support comes from a study of FR-I galaxies, which show low Eddington ratios ($L_{\rm bol}/L_{\rm Edd} < 0.01$) but imply rapidly spinning black holes with $j > 0.9$ (Wu *et al.*, 2011a). Here

$$j = \frac{Jc}{G M_{\rm BH}^2} \tag{4.10}$$

is the dimensionless angular momentum of the black hole and J is its angular momentum. A case with j close to 1 would represent the case of a Kerr black hole, while j close to 0 can be treated as a nonrotating Schwarzschild black hole.

In summary, the separation of AGN into sources with and without a jet might not be as clean as assumed in the past. The gap between radio-loud and radio-quiet appears less pronounced the more high-quality data of AGN we study. The dependency of radio loudness on the Eddington ratio is weak or absent, and the influence of black hole spin, which is difficult to estimate in the first place, might or might not solve the problem of which sources produce jets.

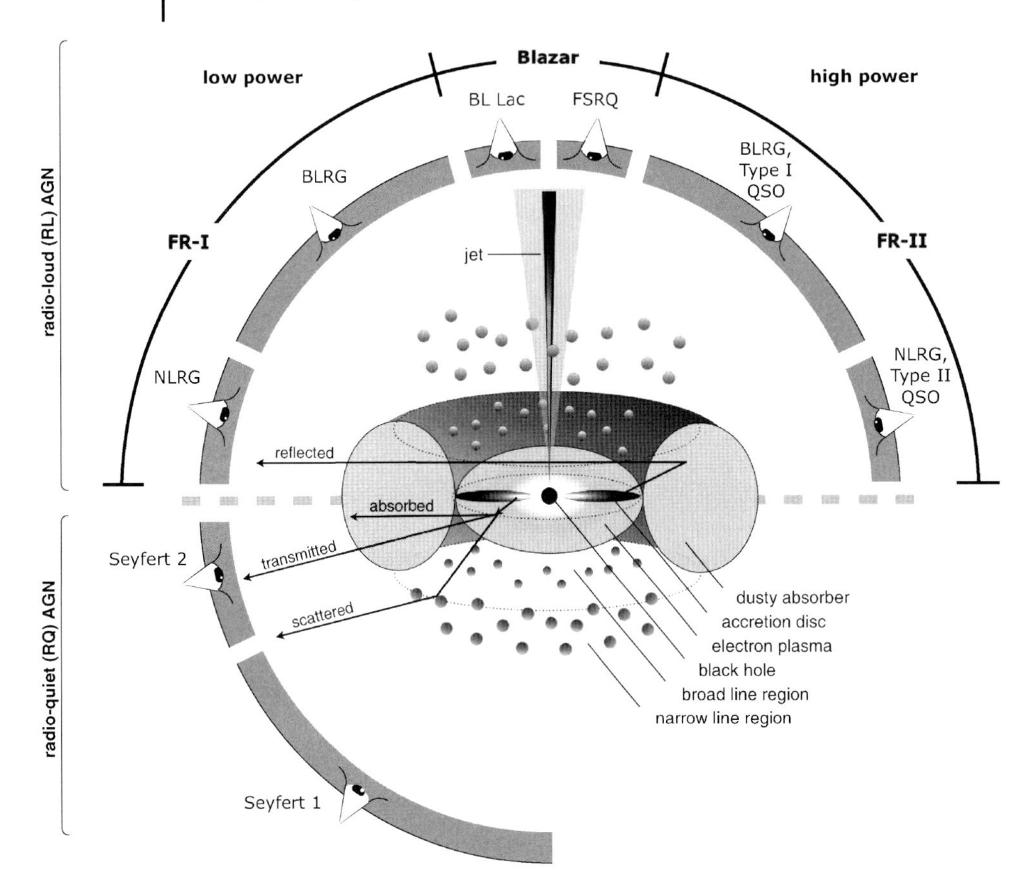

Figure 4.16 Schematic representation of our understanding of the AGN phenomenon in the unified scheme. The type of object we see depends on the viewing angle, whether or not the AGN produces a significant jet emission, and how powerful the central engine is. Note that radio-loud objects are generally thought to display symmetric jet emission. Graphic by Marie-Luise Menzel.

Notwithstanding all of these issues, an overall unification scheme (Tables 4.3 and 4.4) has evolved over the years which is schematically represented in Figure 4.16. It shows the radio-loud AGN which are assumed to display a jet in the upper half, and the radio-quiet objects in the lower part of the figure.. Obviously this is a simplified view if we keep in mind that the transition from radio-quiet to radio-loud may in fact be continuous.

4.8.3 Breaking the Unification

The basic unified model for AGN predicts differences in appearance only based on different orientation toward the observer. This causes different absorption effects intrinsic to the innermost regions of the AGN, as well as geometrical effects re-

Table 4.3 The general unification scheme of AGN, based on the emission lines visible in the optical domain. See also Tadhunter (2008).

Type	Optical lines	Radio-quiet	Radio-loud
Type 1	Broad and narrow lines	Seyfert 1 Seyfert 1.5 NLS1	FSRQ, SSRQ, BLRG
Type 2	Narrow lines only	Seyfert 1.8, 1.9, 2	NLRG, type 2 QSO
	Weak narrow lines	LINER /LLAGN	WLRG
Type 0	No lines	Sgr A*? Dormant AGN[a]	BL Lac, OVV

[a] One might argue here that a dormant or nonactive AGN is indeed by definition not an AGN!

Table 4.4 The unified scheme of AGN, considering various observable parameters. Note that this simplified scheme is true only in general and exceptions are possible in nearly all parameters. For example, nearby radio-quiet AGN can be detected in the gamma-ray domain when a strong starburst component is present.

Radio loud?	AGN type	Subtype	X-ray absorbed? $N_H > 10^{22}$ cm^{-2}	Broad Balmer lines?	Narrow Balmer lines?	Fe$K\alpha$?	γ-rays?
RL							
	Radio galaxy						
		WLRG	Yes	Yes	Yes	No	No
		BLRG	No	Yes	Yes	Yes	Few
		FR I/II	No	Some	Yes	No	No
	Quasar	Type 1	No	Yes	Yes	Yes	Some
	Quasar	Type 2	Yes	No	Yes	Yes	No
	Blazar						
		FSRQ	No	Yes	Yes	Some	Yes
		BL Lacs	No	No	No	No	Yes
RQ							
	Seyfert 1		$< 10\%$	Yes	Yes	Yes	No
	Seyfert 1.5		$\sim 30\%$	Yes	Yes	Yes	No
	Seyfert 2		$> 90\%$	No	Yes	Yes	No
	NLS1		$< 10\%$	Yes	Yes	Yes	Few
	ULIRGs		Yes	Yes	Yes	No	No
	LINER		Yes	No	Yes	Yes	No

garding the beaming of the emission. The previous subsections presented several observational results that do not seem to fit into this scheme. Here we take a look at these open issues and consider how the unified model might be extended to account for these discrepancies.

As stated above, in many but not all Seyfert 2 galaxies one can find a hidden broad-line region when studying the objects in polarized light. The remaining Seyfert 2 objects, which do not show any broad-line region, even when observed in polarized light, might simply have weak BLR emission compared to the underlying continuum, which would make detection difficult. One explanation might be that the power of the central engine is not large enough to sufficiently illuminate the BLR. This might also be used as an explanation, why BL Lac objects, the weakest blazars, do not show any emission lines. What this comes down to is that one needs to add the total power of the central engine as a parameter to the unification model in order to make it work.

If all Seyfert 2 cores are absorbed, this should be observable especially in the X-ray domain, where the emission is supposed to originate from the innermost region around the accreting black hole. But there are examples of type 2 AGN, which indeed show no measurable absorption at soft X-rays, like NGC 3147 and NGC 4698 (Pappa *et al.*, 2001). These cases might represent the same effect as for the Seyfert 2 without a detectable broad-line region even when viewed in polarized light. Also here, there might only be a weak broad-line region because the intrinsic power of the AGN core is low.

If we also invoke luminosity as a parameter in the unified model, we can explain further effects. For example, in hard X-ray surveys of Seyfert galaxies one observes an anticorrelation of the fraction of absorbed sources with luminosity. While for an X-ray luminosity of $L_{20-100\,\mathrm{keV}} = 10^{42}\,\mathrm{erg\,s^{-1}}$ about 65% of the Seyfert galaxies show an intrinsic absorption $N_\mathrm{H} > 10^{22}\,\mathrm{cm^{-2}}$, at $L_{20-100\,\mathrm{keV}} = 10^{45}\,\mathrm{erg\,s^{-1}}$ only 35% are absorbed (Beckmann *et al.*, 2009). An explanation for this coupling can be the scenario of a "receding torus" as proposed by Lawrence (1991). The radiation pressure of the central engine pushes the absorbing material out. If we assume a simple torus as absorber, one can find a correlation between the luminosity of the AGN core and the inner radius of the torus of the form $R_\mathrm{inner} \propto \sqrt{L_\mathrm{bol}}$. Assume that the inner radius is determined by the limit at which the AGN luminosity can evaporate the dust at a temperature of $T = 1000\,\mathrm{K}$. The radius where this is the blackbody equilibrium temperature is roughly at $R_\mathrm{inner} \simeq L_\mathrm{bol} 4 \times 10^{-46}\,\mathrm{erg^{-1}\,s\,pc}$. For a fixed height of the absorbing torus, this will lead to a wider opening angle under which the broad-line region is visible with increasing luminosity. In other words, the fraction of unabsorbed sources we observe increases with luminosity as observed in the X-rays. Thus, these breaks in unification can be explained by an additional dependence on luminosity.

Another challenge for the unified model are the differences found in the luminosity functions of different AGN types. The luminosity function gives a measurement of the density of sources of a given luminosity per unit volume. An in-depth explanation of luminosity functions will be given in Section 9.2.3. In the case of blazars, there appears to be a difference in the luminosity functions of the bright FSRQ, which show broad lines in their spectrum, and the fainter, high-frequency peaked BL Lac objects (HBL). While FSRQ and low frequency peaked BL Lac objects seem to have been more numerous and/or luminous in the past, that is, they show a positive evolution, high-frequency BL Lac objects show either no or slightly negative evolution, making them as numerous and luminous in the local Universe as at redshifts $z \gtrsim 0.3$, or even more abundant now than in the past (e.g., Beckmann *et al.*, 2003). This presents a violation of the unification model, which would predict that the distribution in space does not depend on the AGN type. Again, we can save the unified model by further parameterizing it.

As described by Böttcher and Dermer (2002), one way to unify the blazar classes would be a transformation of FSRQ and LBLs into HBLs as the blazars grow older. In this model, blazars start as powerful FSRQ with jets of high-energy densities. Strong cooling limits the electron energies leading to cutoff frequencies for the synchrotron component at optical wavelengths and for the inverse Compton component in the GeV energy range. By the time the source of the jet gets less powerful the energy density within the jet decreases. The cooling efficiency decreases as well resulting in higher cutoff frequencies for HBLs. The shift of the cutoff frequencies to higher energies is therefore accompanied by decreasing bolometric luminosities, which is evident from the decrease of the luminosities in the radio, near IR, and optical bands. Due to the increasing peak frequencies of the synchrotron branch more energy is released in the X-ray band and the X-ray luminosity increases quite in contrast to the luminosities at shorter frequencies. We will have a closer look at the spectral energy distribution of blazars in Section 5.8.1.

The comparison of Seyfert 1 and Seyfert 2 luminosity functions appears to be difficult. While complete samples of type 1 AGN can be compiled relatively easy, Seyfert 2 samples often lack completeness or include other narrow-line objects, like LINER or H II regions. Turning once more to the hard X-ray band in order to achieve complete samples of AGN in an energy range not affected by absorption, we see that indeed the luminosity functions of absorbed and unabsorbed sources are similar, although there is some indication for Seyfert 2 to dominate at the low luminosity end with $L_X < 10^{43}\,\mathrm{erg\,s^{-1}}$, while Seyfert 1 are contributing stronger to the high-luminosity objects. Again, luminosity (or accretion efficiency in terms of Eddington ratio) would have to be included as a parameter of the unification model in order to explain this discrepancy.

Finally, variability studies can be used in order to verify the unified model. If all sources are intrinsically similar, then they should show variability patterns independent of their source type. Apparently, this is not the case, as the variability seems to be a function of the mass of the black hole. An anticorrelation of X-ray variability with luminosity in AGN is observed in the sense that the more luminous AGN are less variable (e.g., Barr and Mushotzky, 1986; Lawrence and Papadakis, 1993; Beckmann *et al.*, 2007). The same effect has been observed in the UV band and in the optical domain (e.g., de Vries *et al.*, 2005), although narrow-line Seyfert 1 galaxies apparently show the opposite behavior (e.g., Turner *et al.*, 1999). Papadakis (2004) explains this correlation as a connection between luminosity and the mass of the central black hole M_{BH}. This may be explained if more luminous sources are physically larger in size, so that they are actually varying more slowly. Alternatively, they may contain more independently flaring regions and so have a genuinely lower amplitude. The observed correlation might reflect the anticorrelation of variability and black hole mass. This relation has been well studied in the 2–10 keV X-ray band. The X-ray variability amplitude, as measured in the root mean square (RMS) value, is anticorrelated with M_{BH}, as shown in Figure 4.17. The fit shows a correlation of the form $\log(M_{BH}/M_\odot) = (4.85 \pm 0.20) - (1.05 \pm 0.08) \log \sigma^2_{\mathrm{rms}}$.

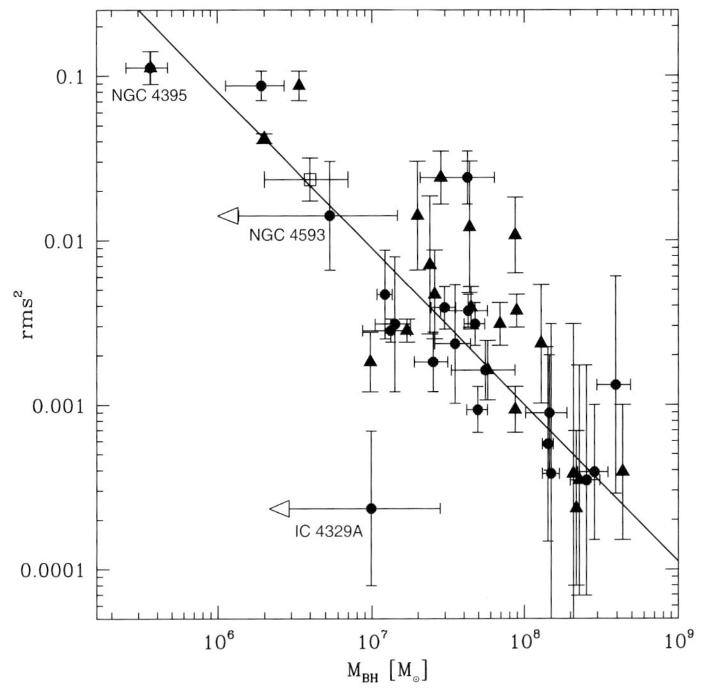

Figure 4.17 Anticorrelation between the X-ray variability amplitude and the black hole mass, with the masses on NGC 4593 and IC 4329A being upper limits. The filled circles denote the objects with the black hole masses measured from reverberation mapping. Triangles are based on stellar velocity dispersion measurements, and the open square denotes the AGN QPO RE J1034+396. The line indicates the best-fit linear relation. The intrinsic dispersion of this fit is 0.2 dex. Data from Zhou *et al.* (2010).

Uttley and McHardy (2004) explained the anticorrelation of variability and $M_{\rm BH}$ by suggesting that the X-rays are presumably produced in optically thin material close to the central black hole, at similar radii (i.e., in Schwarzschild radii, $R_{\rm S}$) in different AGN. As $R_{\rm S} = 2G M_{\rm BH} c^{-2}$, longer time scales for the variability are expected for the more massive central engines. These studies might still be affected in part by absorption. Recent hard X-ray studies of AGN, where absorption should not play a role, do not seem to detect a difference in variability patterns of type 1 and type 2 AGN (Soldi *et al.*, 2010). If this result holds, it would indeed be a strong support for the unified model.

4.8.4
Grand Unification of Black Holes in the Universe

Based on the previous sections it seems that indeed the unified model works, albeit with a need for some refinements. The principle parameters determining the observational appearance of an AGN, are beaming and orientation with respect to the line of sight. Another parameter to be added could be the mass of the black hole, the accretion rate in terms of Eddington ratio, or the bolometric luminosity.

The extragalactic black holes, residing in the hearts of galaxies, accreting matter from a surrounding disk, should be simply scaled-up versions of galactic black hole (GBH) systems. As in AGN, in-falling matter in GBH forms an accretion disks. Furthermore, we observe GBH jets in a number of instances. These objects are called microquasars, indicating that they appear to be analogous to AGN but with black hole masses of only $M_{\mathrm{BH}} < 10\,M_{\odot}$. A few microquasars are known to date, such as SS 433, GRS 1915+105, GRS 1758-258, 1E 1740-2942, and Cygnus X-1.

The quest of the grand unification of black hole activity is to verify whether the same correlations which apply for GBH are also valid for AGN and vice versa. One important aspect in this comparison is the fact that variability time scales should scale with the mass of the central black hole. The equivalent of a flare in a 10 solar mass black hole, which lasts for a couple of days will take several 10^5 years in a supermassive black hole of $10^8\,M_{\odot}$. Extended outbursts of blazars which last over several years, as observed for example in 3C 454.3,[3] would last only a second in a microquasar. The advantage of galactic black holes is that we are able to observe cycles of activity, we see jets appear and disappear, and we can correlate events with flux variations at various energies. Long term observations of microquasars have given us a hint of how accreting black hole systems evolve. Bright flux states are accompanied by steep X-ray spectra. At certain spectral intensity states, jets are observable. In general terms one therefore describes spectral states of GBH systems as "high/soft" and "low/hard." Figure 4.18 shows the common understanding how the coupling between accretion disk, corona, and jet emission works in the case of black hole X-ray binaries. The jet occurs as the inner disk radius increases and the system reaches its lowest flux level at this point (i). The spectrum is now very "hard," that is it shows a flat spectrum with a photon index $\Gamma < 2$. As the jet evolves, the source brightens (ii). After some time the spectrum softens rapidly, the spectral slope becomes steeper, and apparently an internal shock occurs in the jet and the outflow stops. The source enters a high/soft state (iii). The object then faints toward a disk-dominated state (iv), at which point a jet can be started again.

For the reasons given above, we are not able to observe AGN in different states. In spectral terms it seems that AGN display mostly spectra which resemble the low/hard state of GBH. Obviously, on human time scales we do not expect to observe any state changes in AGN as we see in GBH. In addition, the feeding of a GBH differs from that of an AGN. Active GBH are part of binary systems in which a donor star loses matter to the compact object. Nevertheless, by observing many AGN we should be able to find spectral states resembling those in GBH, and also the accretion process itself should basically be the same.

Variability is one of the parameters which should scale simply with the size of the central engine. As we will discuss in Section 6.3, one way to describe the variability of a light curve is to measure the power density spectrum (PDS), that is, the vari-

3) Because of its brightness in gamma rays and its special behavior, 3C 454.3 is sometimes referred to as the *crazy diamond*, especially by the Italian colleagues, following the Pink Floyd song "Shine on you crazy diamond."

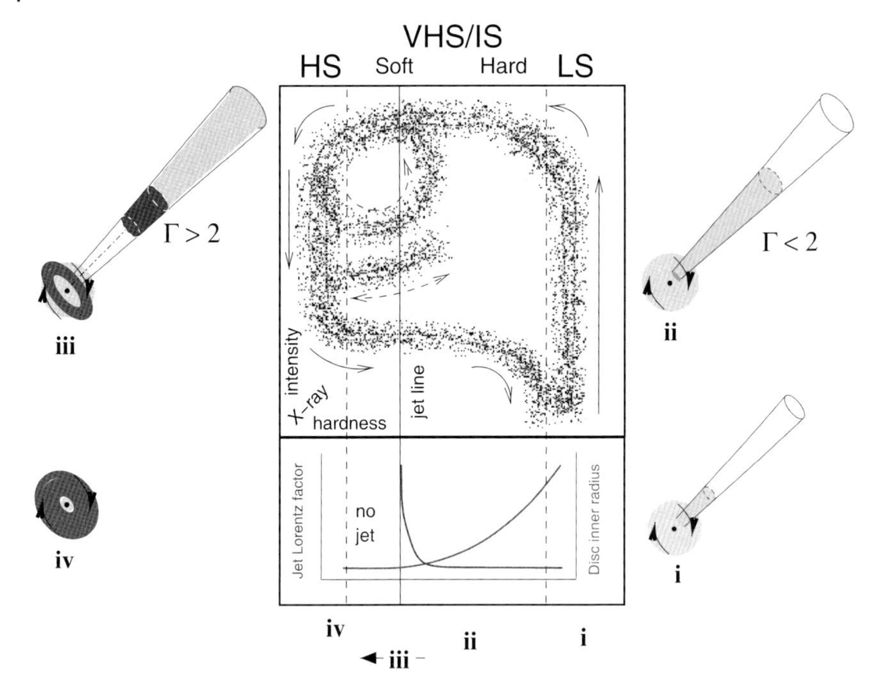

Figure 4.18 Schematic representation of the connection of jet emission, inner disk radius, and spectral state in galactic black hole binaries. The top panel shows the hardness-intensity diagram, in which black holes seem to follow the paths indicated by the arrows. A flat-spectrum radio flux appears and increases with X-ray intensity in the hard state – the right-hand vertical track of the "q" diagram. The radio emission becomes optically thin and the jet appears as the emission transitions leftwards along the upper horizontal track. The jet disappears after the source moves into the high/soft state. In the bottom panel we see the dependence of jet speed and the inner disk radius on the hardness of the spectrum. X-ray states are indicated with HS (high/soft state), VHS/IS (very high and intermediate state), and LS (low/hard state). The sketches around the outside illustrate the concept of the relative contributions of jet, corona (light gray) and accretion disk (dark gray) at these different stages. Graphic from Fender *et al.* (2004). For a color version of this figure, please see the color plates at the beginning of the book.

ability power, $P(\nu)$, at a given frequency, ν. The PDSs of AGN and of galactic black holes show a broken power-law slope, with a slope of $P(\nu) \propto \nu^{-\alpha}$ with $\alpha = -1$ at long time scales, and a steeper slope with $\alpha \geq 2$ at short time scales. The transition at which the PDS steepens can be referred to as the break time scale T_{B}. McHardy *et al.* (2006) discovered that the break time scale T_{B} is a function of the black hole mass M_{BH} and bolometric luminosity L_{bol}, following the same law for both, AGN and GBH, in case there are well sampled light curves available:

$$\log T_{\mathrm{B}} = A \log M_{\mathrm{BH}} - B \log L_{\mathrm{bol}} + C \tag{4.11}$$

The values for A, B, and C turn out to be very similar for AGN and GBH. In the case of AGN, McHardy *et al.* (2006) derived $A = 2.2 \pm 0.3$, $B = 0.9^{+0.3}_{-0.2}$, and

$C = -2.4^{+0.2}_{-0.3}$, and for galactic black holes $A = 2.1 \pm 0.15$, $B = 1.0 \pm 0.15$, and $C = -2.3 \pm 0.2$. When plotting the expected values from this relation versus the actually measured break time scales the relation works indeed for galactic and supermassive black holes over a wide range of M_{BH}, as can be seen in Figure 4.19.

We see here that black hole activity follows a fundamental plane, which is formed by the break time scale, black hole mass, and bolometric luminosity. Other fundamental planes have been found and discussed for black hole activity. Merloni *et al.* (2003) studied the correlation between the radio luminosity L_R, X-ray luminosity L_X and mass of the black hole M_{BH} of a sample of eight galactic and about 100 extragalactic black holes. The correlation found between these properties has the form

$$\log L_R = 0.60 \pm 0.11 \log L_X + 0.78^{+0.11}_{-0.09} \log M_{BH} + 7.33^{+4.05}_{-4.07} \qquad (4.12)$$

The correlation is shown in Figure 3.4. Although the scatter is large ($\sigma_R = 0.88$), the relation holds important implications for our understanding of black hole accretion. The radio luminosity L_R gives a measure of the jet activity in the source,

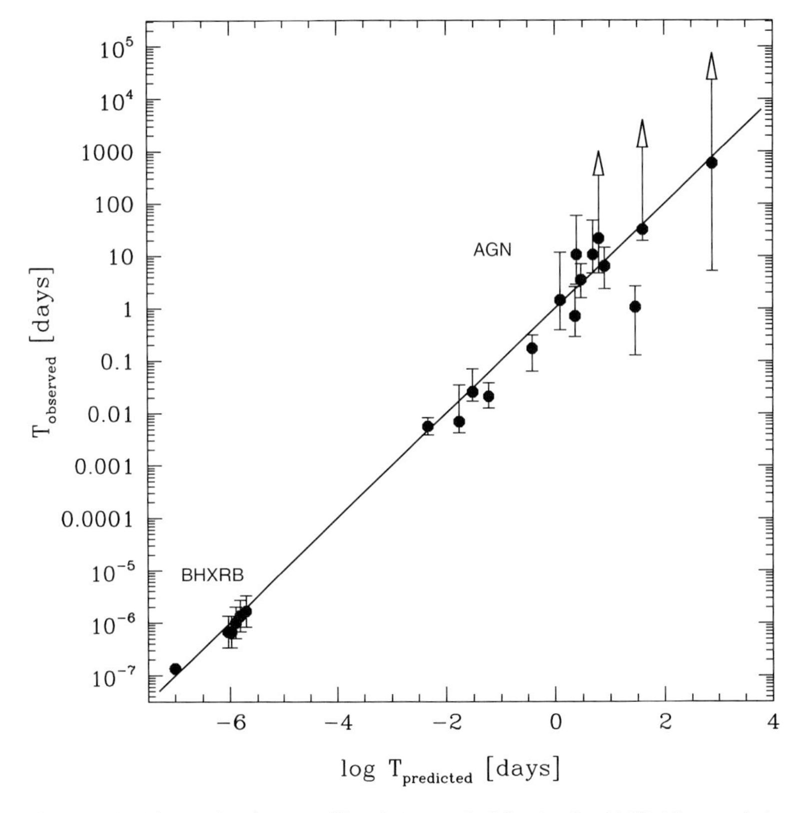

Figure 4.19 Observed and expected break time scale following Eq. (4.11). The correlation, as derived by McHardy *et al.* (2006) holds for galactic black holes and AGN over a black hole mass range from a few to $10^8\ M_\odot$. Data from McHardy (2010).

while L_X is a function of the accretion flow and M_{BH} represents the size of the central object itself. Therefore, this fundamental plane implies a scale invariant correlation of the accretion flow and the central black hole mass with the jet emission. Because the radio emission can be strongly enhanced through beaming effects, the correlation only holds if one either excludes sources with strong relativistic beaming, or if one corrects their values for this effect. Another aspect which has to be considered is the possibility that the nonthermal jet emission also contributes to the observed X-ray flux, and thus L_X might not only represent the accretion, but also the emission process of the AGN.

The fundamental plane of black hole activity opens an interesting possibility to estimate the masses of black holes. For example, in the case of the elusive ultraluminous X-ray sources (ULX; Section 4.3), which might be intermediate-mass black holes, knowing the X-ray and radio luminosity and placing the objects on the fundamental plane described in Eq. (4.12) one can then derive mass estimates. This assumes that the radio flux is a proxy for the jet power (note that the existence of the fundamental plane is believed to be rooted in an intrinsically physical disk-jet relationship).

As an example Merloni *et al.* (2003) compute the mass of an ULX in the nearby dwarf irregular galaxy NGC 5408, deriving a mass of $M_{BH} \simeq 10^{4\pm1}\,M_{\odot}$. As the amount of relativistic enhancement of the flux is unknown in these cases, this value has to be understood as an upper limit of the true mass, and it also implies that ULX are indeed intermediate-mass black holes, which must still be proven.

The fundamental plane as described here indeed presents strong support for the unified model of AGN. As in many correlations discussed in AGN, a major uncertainty is based on the insufficient number of supermassive black holes with high-confidence mass measurements. There is good reason to expect that this problem will be less severe in the near future even within the context of current observational capabilities. We should then be in a position to apply this fundamental plane analysis as a powerful tool for AGN and cosmological research.

5
AGN through the Electromagnetic Spectrum

In addition to the remarkable energetics alluded to in Chapter 1, AGN are prolific emitters of nonthermal radiation that can span more than 16 decades in frequency, from the radio domain all the way up to the TeV energy range. This nonthermal continuum is superposed with a variety of complex structure: the broad and narrow emission lines discussed in Chapter 3, broad thermal excess components in the IR and optical UV bands, and pseudocontinuum structures such as the small blue bump which is comprised of line emission from iron and from the H-Balmer series. In this chapter we will focus on the underlying continua seen in each of the available observational windows and discuss each component's physical origin and their interrelationships. In Section 5.8 we will then have a look at the entire spectral energy distribution of Seyfert galaxies and of beamed sources.

5.1
Radio: Probing the Central Engine

Radio emission was central to the discovery of AGN and to their establishment as a unique astronomical object class. For example, early radio surveys such as the Third Cambridge Survey (3C) carried out in the 1950s and revised in the 1960s, covered northern latitudes (above about declination DEC $> -5°$) to a confusion-limited flux threshold of about 9 Jy at 160 MHz. More surveys with improved spatial resolution were to follow, but the 3C catalog had already yielded a surprising result that would open up a whole new field of study and drastically alter our view of the Universe. While a majority of the high-latitude 3C objects were coincident optically resolvable galaxies, a few seemed to be associated with star-like objects. One of these, known as 3C 273, had also been very accurately positioned based on a lunar-occultation measurement. Its spectrum was studied by Maarten Schmidt and reported on in his now classic 1963 paper (Schmidt, 1963). It was seen to exhibit broad emission lines which were consistent with the hydrogen Balmer series and a few other lines such as Mg II ($\lambda = 2798$ Å) redshifted to the (then) remarkable value of $z = 0.158$, and a new field of study was born. To be fair, Seyfert galaxies had been recognized as unusual objects based on their emission line spectra and

Active Galactic Nuclei, First Edition. Volker Beckmann and Chris Shrader.

nuclear-dominated morphologies, for example by Carl Seyfert in the 1940s (Seyfert, 1943), and even as early as the 1910s by Slipher (1913); but, the association with intense and variable radio emission was what truly opened up the field.

As we discussed in Chapter 4, it is often useful to divide AGN into two subclasses designated as radio-quiet and radio-loud. We now believe that about 10% of all AGN are among the radio-loud AGN subclass. The radio emission from AGN consists of compact and extended components. The spatially extended structure in the radio emission is determined by the interaction between bipolar jets and the external medium. What we observe is modified by the effects of relativistic beaming. The radio emission from both the compact and extended components is synchrotron emission, as is inferred from its very smooth, broad-band character and its strong polarization. The extended emission is produced by the jets (see Chapter 4) with the predominant component involved being the so-called jet lobes or hot spots that result from interaction between the jet and the environment. A compact emission component, often called the core, is unresolved at $\lesssim$ arcsecond scales and is believed to be coincident with a position close to the central black hole, where the jet becomes optically thin. Thus, this position is slightly frequency-dependent, i.e., there is a so-called core shift with frequency: the higher the frequency, the closer the emission region is to the central black hole. Its spectrum is generally flatter than that of the extended emission and it is typically variable in intensity.

The process that launches the plasma jets is not well understood. One possibility involves rotational stressing of magnetic fields in the accretion disk around the massive black hole in the galactic nucleus. This type of disturbance of the fields provides a mechanism to convert gravitational and rotational energy into the kinetic energy of high-velocity bulk outflows perpendicular to the disk. Detailed numerical computations, known as magnetohydrodynamical models, can produce jets, but a basic problem is that they generally require a uniform field threading the disk, whereas observational evidence supports highly tangled or disordered field morphologies.

The synchrotron nature of the AGN radio emission implies that the radio-emitting plasma contains electrons with relativistic speeds (Lorentz factors of $\gamma \sim 10^4$) in an environment characterized by magnetic fields. Since the plas-

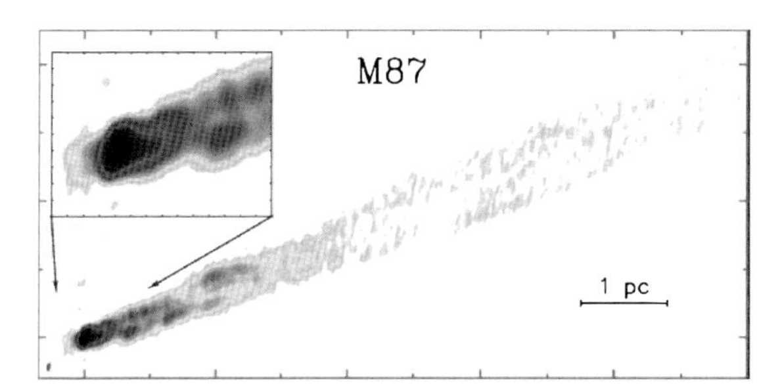

Figure 5.1 VLBA image of the M87 jet taken at $\nu = 15$ GHz; the intensity scale is logarithmic. Graphic adapted from Kovalev *et al.* (2007), reproduced by permission of the AAS. See Figure 3.25 for a comparison with high-resolution optical and X-ray imaging of the jet.

ma must be neutral, it must also contain either protons or positrons, but there is no way of determining the actual composition directly from observations of synchrotron radiation. The resolution of this issue – leptonic versus hadronic jet composition – is central to current studies of AGN jets today.

In the case of the nearby ($d \sim 17$ Mpc) radio galaxy M87 it is now possible to pin down the location of the onset of the jet with respect to the central black hole itself. In contrast to previous measurements, it seems now that the jet is raised in an area not further than some $20 R_S$ away from the core (Hada *et al.*, 2011). The base of the jet has been resolved down to a size of ~ 40 μas, corresponding to $\sim 5 R_S$.

Radio interferometry

A large number of radio telescopes have been built around the world since the first experiments by Karl Jansky in the 1920s and the pioneering work of Grote Reber in the 1930s and 1940s. The fact that radio astronomy provides us with by far the highest resolution images is due to the use of interferometric arrays. Here, several radio telescopes are combined, together providing a telescope with the resolution equivalent to a telescope with the longest base line of the interferometer array as aperture. Worldwide several Very Long Baseline Interferometry (VLBI) arrays have been established. These facilities can be combined globally, providing imaging resolution of the order of some tens of microarcseconds. This allows, for example, to resolve structures of the size of 20 Schwarzschild radii in M87. Figure 5.1 shows the inner jet of M87 as seen by the VLBA. VSOP (VLBI Space Observatory Program) was the successful attempt to install a VLBI in space, by combining HALCA, a satellite based 8 m diameter radio antenna, with ground-based stations, achieving submilliarcsecond resolution at 5 GHz. RadioAstron is the latest space-based VLBI project, with a 10 m antenna on a Russian satellite launched in July 2011. Space-ground VLBI measurements can provide 7 μas resolution at the highest frequency.

The Low Frequency Array (LOFAR) is taking the interferometry idea to the next level and into the 10–250 MHz frequency range – the lowest radio frequencies accessible from the ground before the ionospheric cutoff at 10 MHz. At each station an antenna array consists of very simple omni-directional antennas with the core in the north of the Netherlands and additional stations in Germany, France, Sweden, the United Kingdom (and soon Poland), providing a maximum baseline of 1300 km. LOFAR is able to study radio galaxies up to redshift $z > 6$ due to its high sensitivity. The angular resolution is about 5 arcsec at 10 MHz and about 0.2 arcsec at 200 MHz.

Aside from the material composition of the jets, there is also ambiguity regarding the energy densities in particles relative to that in the magnetic field. This is because the same synchrotron emissivity may be a result of a low-intensity electron beam and a strong field, or a weak field and high-intensity electron beam, or any combination in between these two extremes. Nonetheless, there are some basic principles underlying the synchrotron process from which we can infer physical information of the AGN environment.

Here we summarize some of the basic properties of synchrotron emission. In the simplest case, one can assume a uniformly distributed population of electrons with a power-law energy distribution in a magnetic field of constant intensity and direction. Expressing the electron energy distribution as $n(E)\mathrm{d}E = kE^{-p}\mathrm{d}E$ (Eq. (2.56)) we have shown in Section 2.2.1 that $\alpha = (p-1)/2$ is the spectral index of the corresponding synchrotron radiation spectrum. Note that here and throughout we adopt a convention of implicitly applying a negative sign to quoted spectral indices. For example, $F_\nu \propto \nu^{-\alpha}$, that is, positive values of α indicate fluxes decreasing with frequency or energy. In an optically thin region observations reveal typical lobe emission with power-law indices $\alpha \simeq 0.7$, which imply an electron energy-distribution index of $\simeq 2.4$. In denser regions, particularly in the core, the synchrotron plasma can be optically thick to the synchrotron self-absorption process. In this case, the spectrum can be flat or inverted, that is, having flux density increase with frequency. The turnover frequency is an observable quantity and can potentially provide information on the synchrotron emitting plasma.

The Atacama Large Millimeter/Submillimeter Array (ALMA) consists of fifty-four 12 m and twelve 7 m antennas with flexible configuration setups, allowing for baselines up to 16 km. The telescope ensemble is located in northern Chile at 5000 m altitude with an observational range of 84–720 GHz with a resolution down to 6 mas, depending on the frequency and array setup. The resolution can be approximated by FWHM $\simeq$ 76/maximum baseline/ν, with the full-width half maximum given in arcseconds, the baseline of the setup in kilometers, and the frequency ν at which the observation is performed in GHz. Concerning AGN, ALMA will facilitate mapping of the gas surrounding the core, studying radio jets, and detecting thousands of flat spectrum radio AGN. The complete array is foreseen to be operational in 2013.

One of the most ambitious projects in ground-based astrophysics is the construction of the Square Kilometer Array (SKA) in the southern hemisphere. As the name indicates, this ensemble of radio telescopes will provide a collecting area of about 10^6 m^2. It will consist of different types of telescopes, together covering the 70 MHz to 10 GHz frequency range, and extending out to baselines of up to 3000 km. Although the primary science goals of the SKA are in the cosmology domain, like galaxy formation, dark energy, cosmic magnetic fields, it will also provide a rich database for AGN research. For example, it will be able to measure water masers in AGN with unprecedented sensitivity and thus help also to determine black hole masses (Morganti *et al.*, 2004). The array will allow us to detect and image low-luminosity jets of AGN, clarifying whether Seyfert galaxies and radio-quiet AGN in general produce jets (Bicknell *et al.*, 2004). With the SKA it should be possible to survey about 10^4 faint extragalactic radio sources with extreme low luminosities in the 10^{22}–10^{26} W range at milliarcsecond resolution (Gurvits, 2004), according to current plans. Operations of the SKA are foreseen to commence after 2020.

5.2
Infrared: Dust Near and Far

A major contribution to the bolometric output of various AGN subtypes comes from the infrared (IR) spectral band spanning approximately 2.5 decades from approximately 10^{12}–$10^{14.5}$ Hz, or equivalently about 1–300 μm. Depending on AGN type, the IR emission may consist of thermal and/or nonthermal components. In radio-loud objects, particularly blazars, the same synchrotron emission process producing the radio continuum is the predominant source of IR radiation. In Seyfert galaxies and other low-luminosity AGN subtypes the situation is more complex considering the presence of multiple thermal components.

Historically, the IR band coverage has been relatively limited for example compared to optical and radio. This is due to the fact that only a small subset of that spectral region is accessible from the ground. Even within these atmospheric windows observations can be difficult requiring detector technologies that are different from much more commonly used ones in the optical band. Fortunately, two currently operating space-based observatories, NASA's "Spitzer Space Telescope" and ESA's "Herschel Space Observatory" are collectively compiling an unprecedented legacy of IR imaging and spectroscopic data which will significantly advance this area of study. In this section we will review some of the basic IR observational characteristics and their implications for the AGN paradigm.

In Seyfert galaxies, the IR emission can be roughly understood as comprising three spectral components: (i) thermal radiation from dust in a compact region ambient to the active nucleus. This component, often referred to as a "torus" (see Section 3.3.1), can obscure direct lines of sight to the nucleus leading to distinct observational characteristics. This in turn leads to the unification scheme discussed in Chapter 4. (ii) Thermal dust continuum associated with star formation or perhaps starburst activity. This consists of line emission from macro molecules known as polycyclic aromatic hydrocarbons (PAH) as well as continuum features from heated dust, and (iii) additional line emission emanating from molecular, atomic, and ionic species (see, e.g., Soifer *et al.*, 2008, for additional details). The molecular line spectra are complex due to the large number of oscillatory and rotational modes of the molecular species present. The emission is blended at typical spectral resolutions used in AGN studies. An example of an infrared spectrum of a highly absorbed AGN is shown in Figure 5.2. The Circinus Galaxy, a Compton-thick Seyfert galaxy, was observed by ISO. The average infrared spectrum of 28 quasars as seen by Spitzer is shown in Figure 5.3, presented in Netzer *et al.* (2007).

The dust comprising the torus is heated by the enclosed AGN central engine. This causes it to radiate in the IR band. Its thermal spectral energy distribution thus provides an indirect measure of the AGN luminosity. This information is particularly useful for objects with heavily obscured nuclear regions for which direct measurements may not be possible. If we assume that the infrared emission comes from the dusty torus, we can determine several parameters of the absorbing medium based on the observed spectrum. Here one assumes that we do not have an ideal blackbody but that the emission from the central engine is absorbed and re-

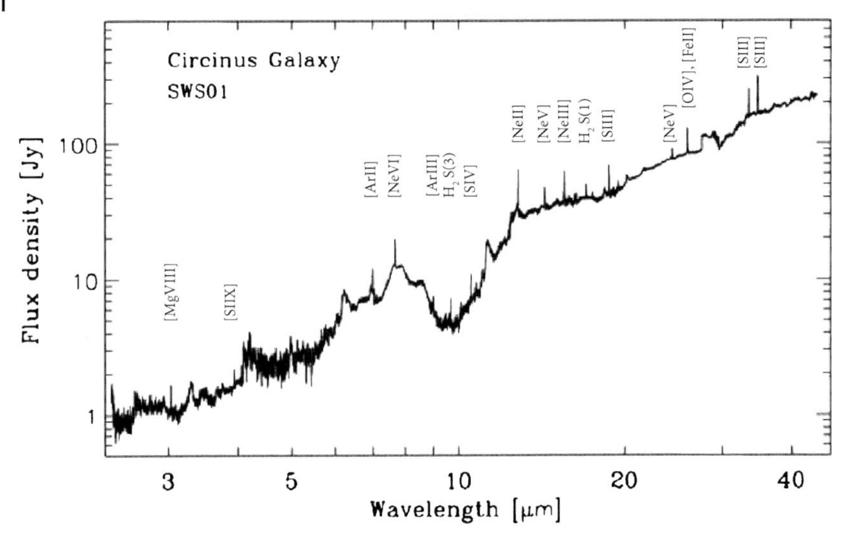

Figure 5.2 Infrared spectrum of the Circinus Galaxy which hosts a Seyfert 1 core, taken by the Infrared Space Observatory (ISO). The spectrum shows emission lines resulting from star formation in the host galaxy at long wavelengths, and from the AGN at higher energies. Graphic from Moorwood *et al.* (1996), reproduced with permission © ESO.

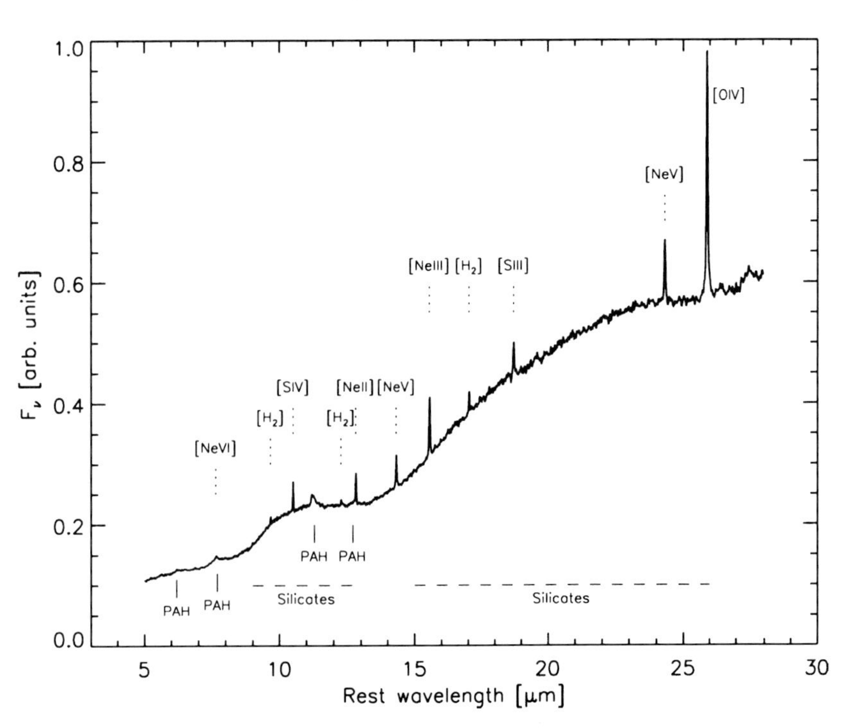

Figure 5.3 Average infrared quasar spectrum based on 28 spectra observed by Spitzer/IRS (Netzer *et al.*, 2007). (Reproduced by permission of the AAS.)

emitted in the outer regions of the torus, which is a medium with an optical depth τ_ν, a function of frequency. If we further assume that the emitting matter has only one temperature, that is, that it is isothermal, we can write the emitted luminosity as (see, e.g., Polletta *et al.*, 2006; Soldi *et al.*, 2008):

$$L_\nu = 4\pi A_{\text{dust}} B_\nu(T)(1 - e^{-\tau_\nu}) \tag{5.1}$$

Here we make the more general assumption that the observer sees a projected surface with area A_{dust}, which does not necessarily have to be a torus. $B_\nu(T)$ is the Planck function of a blackbody of temperature T, see also Eq. (2.71). We can approximate the frequency-dependent optical depth by a power law of the form

$$\tau_\nu = \left(\frac{\nu}{\nu_0}\right)^\beta \tag{5.2}$$

where $\nu_0 = \nu\ (\tau = 1)$ is the frequency at which the dust becomes optically thin. For large frequencies, the observed luminosity will approach therefore that of an ideal black body. β is the dust emissivity index. This index is a function of temperature, frequency, dust composition and geometry. An assumption often used is that of a "gray body," where β is a constant for a given source. In a gray body a common range of values used is $\beta \sim 1.5–2$ (Türler *et al.*, 2006). The observed flux is then

$$f_{\nu,\text{obs}}(\nu_{\text{obs}}) = (1 + z) A_{\text{dust}} d_{\text{L}}^{-2} (1 - e^{-\tau_\nu}) B_\nu(\nu_{\text{em}}, T) \tag{5.3}$$

where d_{L} is the luminosity distance from the observer to the AGN (see Eq. (8.11)). For example, in the case of the blazar 3C 273, Türler *et al.* (2006) assume that $\nu_0 = 1\ \mu\text{m}$ and that the emissivity index is $\beta = 1$, and they fit three absorbers to the data with temperatures of $T_1 \simeq 45$ K, $T_2 \simeq 285$ K and $T_3 \simeq 1300$ K. Assuming spherical absorbers, their respective radii are $r_1 = 8.4$ kpc, $r_2 = 58$ pc, and $r_3 = 1.3$ pc. Although all these numbers are strongly model-dependent, they establish the approximate range of parameters for the absorbing medium, from the large kiloparsec size cool absorbers to the hot, parsec-scale medium. Similar work done by Polletta *et al.* (2000) on IR observations of 19 quasars gave temperatures within a similar range and absorber sizes between 0.1 pc and 9 kpc. It should be pointed out, though, that at sizes larger than ~ 1 kpc this would imply that a significant fraction of the host galaxy itself functions as an absorbing and re-emitting medium, which is probably not realistic.

Studies of emission lines in the IR can often provide valuable diagnostics of the intrinsic SED. This is the case because certain line ratios are sensitive to the shape of the continuum but are less effected by extinction than optical or UV lines of species with comparable ionization energies. Seyfert galaxies, ULIRGs and other infrared galaxies can be distinguished from one another to some degree solely on their infrared emission. A diagnostic plot showing the 9.7 μm silicate strength and the 6.2 μm PAH emission feature is presented in Figure 5.4. Two main branches can be seen in this graphic: a horizontal line of objects with low-silicate strength. Here one finds the Seyfert galaxies and quasars, as well as some ULIRGs and, with the strongest PAH features, starburst galaxies. A diagonal line from high silicate strength and low PAH emission, to low silicate but high PAH, seems to be occupied preferentially by ULIRGs alone. The fact that ULIRGs occupy both branches, that is they do or do not display silicate features, seems to point to differences in the distribution of the dust close to the nucleus of those sources. In the horizontal branch, the dust might be clumpy, not resulting in significant silicate absorption, whereas a more homogeneous distribution of the absorbing matter leads to the strong absorption features. In this context, the objects located in between the two branches might indicate transition from a smooth to a clumpy matter distribution (Spoon *et al.*, 2007).

The PAH lines can also be used in order to estimate the size of the dust grains in the absorber. Generally speaking, the larger the molecules are, the more the emission will be shifted toward longer wavelengths. Using the PAH features at 3.3, 6.2, 7.7, and 11.3 μm, Draine and Li (2007) showed using Spitzer data that the strong 7.7 μm emission is produced mainly by small molecules with less than 1000 carbon atoms. At the same time, the PAH line ratios indicate the ionization level of

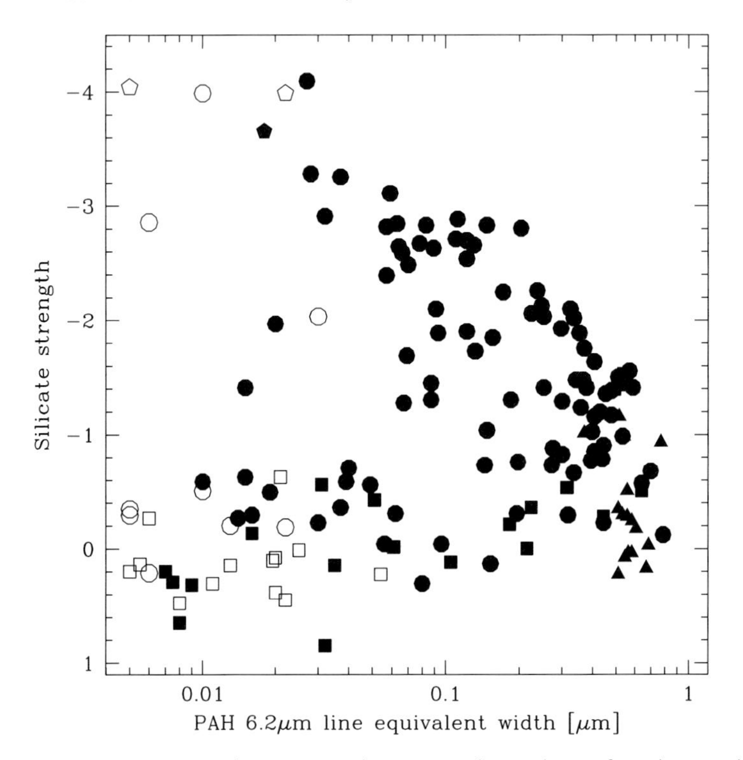

Figure 5.4 Diagnostic diagram using data presented by Spoon *et al.* (2007) using the strength of the silicate absorption feature and the strength of the PAH emission. Galaxy types are distinguished by their plotting symbol: ULIRGs with circles, starburst galaxies with triangles, Seyfert galaxies and quasars with squares, and other infrared galaxies with pentagons. Open symbols represent cases where the PAH line equivalent width measurement is an upper limit only.

the dust. While neutral PAHs have higher values of $f(11.3\ \mu m)/f(7.7\ \mu m)$, ionized PAHs have low ratios.

Infrared space and airborne telescopes

The infrared spectral region is only partly accessible from the ground. Only in the windows spanning 1.1–3.5 and 330–370 μm can observations be done effectively. The majority of the infrared band is accessible only from airborne or space-based observatories.

IRAS was launched in 1983 and performed an all-sky survey at 12, 25, 60, and 100 μm, detecting about 5×10^5 sources. ESA operated its Infrared Space Observatory (ISO) in 1995–1998. ISO provided high-resolution spectroscopy in the 2.5–240 μm band. A higher sensitivity was reached through the Spitzer Space Telescope, launched in 2003 by NASA, carrying a 85 cm telescope. On-board, the IRAC instrument provides images at 3.6, 4.5, 5.8, and 8.0 μm, while the InfraRed Spectrograph (IRS) provides low-resolution slit spectra in the 5.2–38 μm band, and high-resolution spectra ($\lambda/\Delta\lambda \sim 600$) at 10–37 μm. The Multiband Imaging Photometer for Spitzer (MIPS) delivers photometry and imaging at three wavelengths: 24, 70, and 160 μm. Since the coolant expired in May 2009, Spitzer entered the "warm mission" phase, still providing data at the two shortest wavebands provided by IRAC. Contemporaneous with Spitzer, the Wide-field Infrared Survey Explorer (WISE) was a NASA infrared-wavelength astronomical space telescope launched in 2009, and decommissioned in 2011. WISE surveyed the sky in four wavelengths of the infrared band. Its detector arrays have 5-sigma sensitivity limits of 120, 160, 650, and 2600 μJy at 3.3, 4.7, 12, and 23 microns.

The $\sim$ 1–700 μm spectral range begins to become accessible at altitudes of $\sim$ 10 000 km. NASA has flown two major airborne infrared observatories. One is currently operating and its predecessor was operational from the mid-1970s to about 1995.

The Gerard P. Kuiper Airborne Observatory (KAO) involved a highly modified C-141A jet transport aircraft. It had a range of about 11 000 km, capable operations at altitudes of up to 14 km. It began operation in 1974. KAO's telescope was a conventional Cassegrain reflector with a 36-in (91.5 cm) aperture, designed primarily for observations in the 1–500 μm spectral range. Its flight capability allowed it to rise above almost all of the water vapor in the Earth's atmosphere which accounts for much of the infrared opacity. It could travel to almost any point on the Earth's surface for an observation.

The successor to KAO, which recently began operations after achieved its "first light" observation in 2010 is the Stratospheric Observatory for Infrared Astronomy (SOFIA). The telescope is mounted in a Boeing 747 wide-body aircraft that has been modified to accommodate a 2.5 m diameter reflecting telescope and its sophisticated guidance system. This telescope operates at stratospheric altitudes of about 12 km. The optical system uses a Cassegrain reflector design with a parabolic primary mirror and a remotely configurable hyperbolic secondary. It has nine instruments covering 1–655 μm wavelengths providing a complement of imaging, spectroscopic and timing capabilities.

The Japanese Akari satellite has been operational since 2006, allowing observations in the 2–180 μm band and performed an all-sky survey detecting about 9×10^5 sources as well as deep surveys in six bands. In 2009 ESA launched the largest mirror so far deployed in space with a diameter of 3.5 m on the Herschel infrared telescope. It hosts three instruments for spectroscopy and imaging in the 55–672 μm range. The high sensitivity allows for observing molecular mass outflows of AGN. For example, Sturm *et al.* (2011) observed OH gas in ULIRGs with a wide range of velocities relative to the host. The spectra indicate outflow velocities exceeding 1000 km s^{-1} and outflow rates up to $\sim$ 1200 $M_\odot$ yr^{-1}. Apparently, the brighter AGN display higher-outflow velocities, indicating that these molecular winds are directly driven by the active core. The large mass transfer from the center of the host galaxies will have an impact on the host and on the fueling environment of the AGN itself. We will discuss this in a larger context of host galaxies of AGN in Section 7.1.

The next milestone will be the James Webb Space Telescope with a 6.5 m mirror and instruments operating between 0.6 and 27 μm with the main focus on the near infrared.

Some current and future infrared space telescopes are described in the box above. Ground-based observations in the IR are possible in the main window between 8–14 μm, and in addition some narrow bands are observable in the 0.2–5.5 μm range. The absorption in the atmosphere is mainly defined by specific wavebands in which the molecules of water, oxygen, and carbon dioxide can absorb infrared light. The exact transmission of the atmosphere depends also on the location of the telescope and on the water vapor content and temperature at the site.

Most ground-based telescopes provide IR imaging and spectroscopic capabilities in the R, I, J, H, and K photometric bands (0.65–2.4 μm). There are also dedicated telescopes covering the L, M, N, Q, and Z band, covering bands in the 3–40 μm range. One example is the CanariCam instrument at the Gran Telescopio Canarias (GTC), a 10.4 m telescope on La Palma. CanariCam can provide data in the 7.5–25 μm band with a spectroscopic resolution of $R = \lambda/\Delta\lambda \simeq 120$–1300. Together with its large collecting area, the GTC should provide the best measurements for studying absorption in AGN in the near future (Levenson *et al.*, 2008). High spatial resolution is achieved by applying adaptive optics, as is done for example at the Keck observatory or at ESO's Very Large Telescope (VLT) interferometer.

5.3
Optical: Where It All Began

While there is no single observational signature of an AGN, the distinguishing characteristics most commonly noted are found in the optical band. Indeed, it is still the case in the modern era of astronomy with observatories spanning 16 decades in frequency that the "identification" of an object as an AGN comes from its optical properties. Optical spectroscopy is still the main AGN classification tool and also redshift, and thus distance, information is usually measured in this domain. One should bear in mind in the discussion that follows, that the observer's frame optical spectral band may include portions of the source frame ultraviolet. In this section, we will review some of these basic identifying characteristics, again keeping in mind that AGN comprise a broadly diverse class.

For AGN subtypes where we have a direct view of the nucleus continuum emission is the predominant component in the optical band. The optical emission generally has an approximate power-law dependence on wavelength. There is often superimposed a large excess extending into the UV band; the so-called big blue bump component. The big blue bump is believed to be of thermal origin and has been modeled as viscous emission from an accretion disk, while the underlying power law is likely of synchrotron origin. There is often an additional feature superposed on the continuum which is less pronounced than the big blue bump which is accordingly given the name "small blue bump." This is believed to consist of blended Fe emission lines and Balmer continuum. When present and favorably oriented, jets can also contribute to the optical emission. These properties cause AGN to appear "bluer" than nonactive galaxies and most types of foreground galactic objects. This has been useful in cataloging AGN using multicolor photometric surveys. In the past few decades however, multiwavelength methods have proven to be the most viable approach.

Another immediately striking feature of many AGN is the presence of broad emission lines. These are believed to emanate from material close to the central black hole. The lines are Doppler broadened because the emitting material is in gravitationally bound orbits around the black hole. They are associated with hydrogen Lyα and Balmer series lines as well as relatively high-ionization transitions from helium, carbon, and magnesium. Also notable in many objects is the presence of narrow emission lines. These emanate from cooler material than the gas responsible for the broad lines. They also exhibit very little variability consistent with the standard picture in which they lie at much larger distance from the central black hole. These are often associated with forbidden atomic transitions, e.g., [O III] λ5007, indicative of low-density plasmas.

Surveys designed to compile statistically complete samples have a long tradition in the optical domain. Wide field surveys using photographic plates were for a long time the standard approach. The most prominent example was the Palomar Observatory Sky Survey (POSS) which consisted of red- and blue-filtered photographic plates covering the full northern-hemisphere sky. Digitization of some of these surveys has facilitated their wider availability to the astronomical community, for example, the Digitized Sky Survey taken with Schmidt–Cassegrain telescopes in Australia and in the US.[1)]

1) The Digitized Sky Survey is available online under http://archive.stsci.edu/cgi-bin/dss_form, last accessed: 15 May 2012.

Direct imaging however, even in multiple colors, is not sufficient to identify AGN. Because quasars are point-like, their images cannot be distinguished from stars, and nearby galaxies with and without an active core can very much look alike, especially in the case of the low-luminosity objects. One way to identify the AGN, is to correlate two wavebands, for example to look for optical sources with radio counter parts. Obviously, completeness is difficult to achieve when using two wavebands as selection criteria, because the choice already implies an assumption of the spectral energy distribution of the sources. An unbiased sample should give spectral information of all sources down to a well-defined flux limit. Also here, photographic plates have long been the state of the art.

By using Schmidt–Cassegrain telescopes with a prism in the light path, every source in the field of view will crudely produce a spectrum in the focal plane, as shown in the example in Figure 5.5. The larger the angle of the prism, the larger the spectral dispersion. Spectra may also overlap, or be "confused," in crowded fields. This problem becomes more severe the deeper the exposure and the larger the angle of the prism. The Hamburg Quasar Survey (Hagen *et al.*, 1995) and the Hamburg/ESO Survey (Wisotzki *et al.*, 1996) used this technique to perform spectroscopic sky surveys on the northern and southern hemisphere down to a limiting magnitude of $B \sim 18$ mag. Because both surveys were aiming at AGN, the sky below $b = 20°$ was excluded in order to avoid the dense regions. As for the Digitized Sky Survey, the photographic plates were scanned and the resulting low-density spectra were searched for prominent emission lines or screened on the basis of

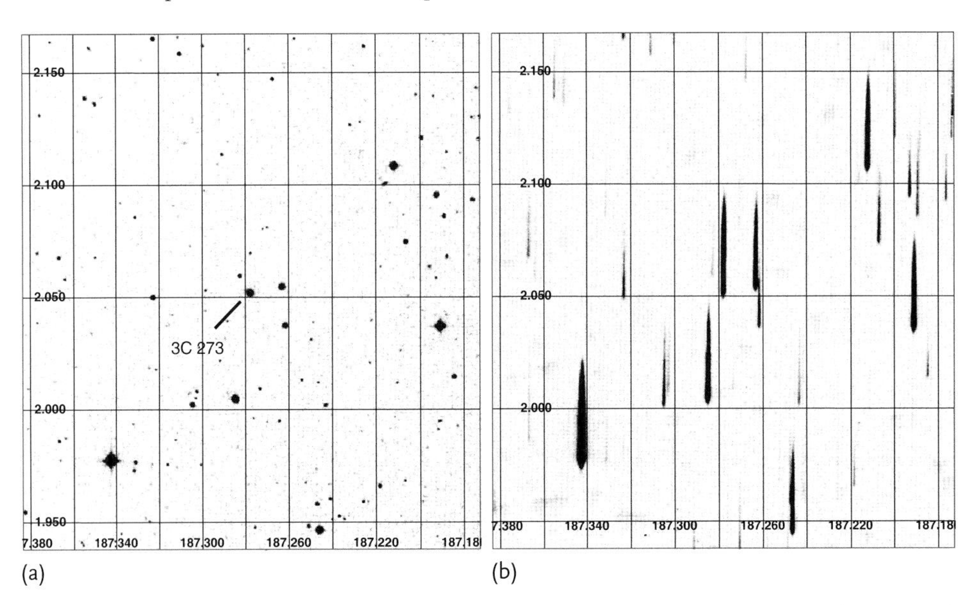

Figure 5.5 Direct image of the sky region around the quasar 3C 273 taken in the blue band with a limiting magnitude of $B \sim 19$ mag (a, image from STScI Digitized Sky Survey). Objective prism plate image of the HQS of the same region (b). Each source produces a spectrum on the photographic plate.

color criteria. The resulting sky catalogs were used for example to identify sources of the ROSAT All-Sky Survey (RASS) performed in the 1990s in the X-rays. Because of overlaps in the objective prism spectra, completeness is difficult to achieve using this technique.

Interestingly, the slitless spectroscopy which was used in the era of photographic plates, is still being applied. ESA's Euclid satellite, which is designed to derive information about the dark energy and dark matter content of the Universe, will use grism slitless spectroscopy in the near infrared. The number of overlaps will be severe. In order to disentangle the source spectra, Euclid will obtain images on each field at two orthogonal position angles, in both red and blue filters. In this manner, overlapping spectra should be identifiable and the source spectra can be reconstructed in most cases. Euclid is expected to be launched around 2020 and its development team anticipates a total of 2×10^8 galaxy redshifts, 10^4 type 2 AGN in the range $0.7 < z < 2$, and about 30 quasars at $z > 8$. Table 5.1 gives the characteristics of some optical AGN surveys.

Another way to obtain a complete sample of AGN is to perform a deep sky observation and to make source identifications through single-object spectroscopy or observations at other wavelengths. The Great Observatories Origins Deep Survey (GOODS; Dickinson *et al.*, 2003) follows this idea. It uses the two deep fields obtained by HST in the northern and southern hemisphere, adding to it X-ray data from Chandra and XMM-Newton, infrared observations from Spitzer, and optical follow-up by telescopes of ESO and NOAO. The total sky coverage is about 320 arcmin^2, but on this small area one finds so far 121 AGN, which gives a density of more than 1000 AGN per square degree at the limiting magnitude of about $B \lesssim 27.8$ mag (Giavalisco *et al.*, 2004). Compared to galaxies, the density of AGN is small – the GOODS fields contain about 60 000 galaxies.

With the advent of powerful and large format CCD arrays and fiber optics, another approach to spectroscopic sky surveys has evolved. Wide field telescopes take direct images identifying each source of interest. A customized aperture plate con-

Table 5.1 Some optical AGN survey samples.

Name	Sky area	Depth (mag)	*N*	z range	Remarks
LBQS	$\sim 454\,\mathrm{deg}^2$	$16 \lesssim B_J \lesssim 18.5$	1055	$0.2 \lesssim z \lesssim 3.4$	Hewett *et al.* (1995)
HQS	$\sim 11\,000\,\mathrm{deg}^2$	$B < 17.5$	$\sim 400^a$	$z < 3.2$	Hagen *et al.* (1995)
HES	$\sim 9000\,\mathrm{deg}^2$	$B < 17$	415^b	$z < 3.2$	Wisotzki *et al.* (2000)
GOODS	$\sim 320\,\mathrm{arcmin}^2$	$B \lesssim 27.8$	121	$z < 7$	Dickinson *et al.* (2003)
2dF	$\sim 2000\,\mathrm{deg}^2$	$B < 19.4$	125	$z > 0.5$	Colless *et al.* (2001)
COMBO-17	$1\,\mathrm{deg}^2$	$R \lesssim 24$	~ 200	$z < 4.8$	Wolf *et al.* (2003)
SDSS	$\sim 10\,000\,\mathrm{deg}^2$	$R \lesssim 23$	$\sim 80\,000$	$z \lesssim 5.2$	York *et al.* (2000)
LSST	$\sim 20\,000\,\mathrm{deg}^2$	$R \lesssim 27.5$	10^7	$z \lesssim 7$	Start 2018
Euclid	$\sim 15\,000\,\mathrm{deg}^2$	$Y/J/H \lesssim 24$	10^7	$z \lesssim 8$	Launch 2020

a only published AGN;
b complete subsample of type 1 AGN.

figured with optical fibers directs light to a spectrometer. This technique has been used for example in conducting the Sloan Digitized Sky Survey (SDSS York *et al.*, 2000). The SDSS uses a 2.5-m modified Ritchey–Chretien telescope in New Mexico, equipped with an array of CCDs and which records the light from a spectrograph up to 1000 fiber channels. In this manner, the SDSS can obtain spectra of $> 10^6$ objects. So far, about 80 000 AGN have been identified.

A different approach is to use optical observations in narrow band filters to determine the nature and even the redshift of sources. Because the redshift information is deduced from broad-band photometry, the derived z values are called "photometric redshifts," or photo-z. The basic idea is that a sharp discontinuity will appear between the photometric bands straddling for example the source frame Lyman-limit, bluewards of which extinction precipitously increases due to neutral hydrogen in the host galaxy. Other less dramatic discontinuities due to the host ISM extinction can in principle be exploited as well. With an increasing number of photometric channels and sufficient precision, this idea can be further exploited to identify differences due to redshifted emission features of individual objects.

An early attempt at a photometric redshift survey was made using the 2.2 m telescope on Calar Alto (Spain) and applying several filters in the optical and near-infrared band. The Calar Alto Deep Imaging Survey (CADIS; Wolf *et al.*, 2001) can be seen as a predecessor to the highly productive project Classifying Objects by Medium-Band Observations in 17 Filters (COMBO-17; Wolf *et al.*, 2003), which applied a similar technique with more filters and used the 2.2 m MPG/ESO-telescope at La Silla in Chile. On an area of 1 deg^2 COMBO-17 detects $\sim 2 \times 10^5$ objects down to magnitudes fainter than $R = 24$ mag. Classification of objects is achieved by comparing the flux measurements in the various bands, which can be seen as a low-resolution spectrum, with samples of spectral energy distributions of the different types of objects expected, such as stars, galaxies, and AGN. Emission lines as well as the calcium break due to the host ISM will leave their imprint on the spectrum and thus can be identified, allowing to approximately determine redshifts. To progress from this kind of deep but small area survey to an all-sky survey is challenging. It requires not only the fabrication of large-format CCD arrays, but additionally poses challenges to current data processing and storage capacities.

Future surveys planning to apply these multiband photometric techniques to obtain classification and redshift of objects, include ESA's Euclid space mission, as well as Pan-STARRS and the Large Synoptic Survey Telescope (LSST) which will be cited in the northern and southern hemisphere, respectively. The ground-based surveys are able to go reach objects as faint as $R \simeq 27.5$ mag in their multiyear program, while the space-based observations will provide ~ 0.2 arcsec resolution and at a depth of better than 25 mag in the optical and near infrared band. The Pan-STARRS project has already begun observations, with Euclid and LSST being currently envisaged to start operations around 2019. These and other similar surveys preparing now will provide order-of-magnitude enhancements in AGN samples relative to what is currently available. Euclid for example is expected to detect and measure the redshift for some 4×10^7 AGN out to redshift $z \simeq 1.8$, and several dozen quasars at $z > 8.1$, LSST is expected to find 10^7 AGN, with 1000 at $z > 6.5$.

The perspective of not only detecting a large number of AGN, but to simultaneously measure their distance without the need for follow-up spectroscopy is truly an exciting step forward. All these projects, as noted, will rely heavily on advances in the state of the art of data processing and storage capabilities.

But how reliable are photometric redshifts? The quality of the photo-z depend strongly on the number of filters which are applied, and on the wavelength range covered. In order to asses the accuracy, photometric redshifts are compared to those derived from slit spectroscopy. Assuming that we know the true redshift z of the objects, one can describe the fractional error Δz of the photometric redshift z_{phot}:

$$\Delta z = \frac{z_{\text{phot}} - z}{1 + z} \tag{5.4}$$

The $(1 + z)$ term is included here because in most cases the errors will be larger for larger redshifts. For a given number N of sources, the systematic error or bias is then the average of the fractional errors $\overline{\Delta z}$. The 1σ dispersion σ_z of the photometric redshifts is determined by

$$\sigma_z^2 = \frac{1}{N} \sum_{i=1}^{N} \Delta z_i^2 = \frac{1}{N} \sum_{i=1}^{N} \left(\frac{z_{\text{phot},i} - z_i}{1 + z_i} \right)^2 \tag{5.5}$$

Another important value in this respect is the number of *catastrophic outliers*, that is, those objects for which the difference between photometric and spectroscopic redshift is larger than the 3σ uncertainty of the former $(z_{\text{phot}} - z) > 3\sigma_z$. In some less strict studies also the fraction of outliers with $(z_{\text{phot}} - z) > 5\sigma_z$ is given when discussing the number (or fraction) of catastrophic outliers. For example, in COMBO-17 the fractional error in the case of AGN is better than $\sigma_z \sim 0.01$ as soon as one ignores the fraction of 10% of catastrophic outliers (Wolf *et al.*, 2008). The relatively large number of cases where the photo-z determination fails is caused by emission lines which are misidentified, for example Hβ and Mg II can be confused. Obviously, the fainter the objects are the higher the failure rate. For example, for faint Seyfert galaxies in COMBO-17 one gets $\sigma_z \sim 0.05$ and a 1σ outlier fraction of $\sim 20\%$ (Wolf *et al.*, 2003). Bordoloi *et al.* (2010) estimate that after the application of data correction techniques the LSST survey should be able to provide photometric redshifts up to $z \sim 3$ as good as $\sigma_z \simeq (1 + z)0.04$ with less than 0.2% of the photometric redshifts deviating more than 5σ away from the actual ones.

For AGN research this means that the results of photometric surveys can be used in a statistical manner, for example to derive luminosity functions, number-flux relations and classifications of source populations. In the individual case the redshift estimate can be indeed very wrong. This does not overly impact the statistical result as long as the redshift bias $\overline{\Delta z}$ is close to zero. But when looking at a particular AGN, slit spectroscopy is still essential to verify the redshift and object type.

The evolution of optical, ground-based telescopes led us from the Hooker 2.5-m mirror, which was the largest optical telescope from its inauguration in 1917 to 1948, when the Hale 5-m took over and reigned supreme until 1976, to the 6-m Bolshoi Teleskop Alt-azimutalnyi (BTA-6) which went operational in 1976 (although it proved to be problematic). At that point, the practical size limit for monolithic mirrors had been reached.

The 1990s saw the advent of the 10 m class telescopes, like the Keck on Mauna Kea (Hawaii) and the four 8 m telescopes of ESO's Very Large Telescope (VLT) in Chile. The largest telescopes of today use segmented mirrors, like the 10.4 m Gran Telescopio Canarias (GTC, Spain). The largest single-mirror telescopes are the Large Binocular Telescope (LBT, USA), the Japanese Subaru telescope on Mauna Kea, and the VLT units.

After the 10 m class, plans are ongoing to reach an aperture of 30 m. This can only be achieved using even lighter segmented mirrors, and naturally issues of engineering difficulties and cost must be overcome. The Giant Magellan Telescope (GMT) will consist of seven 8.4 m mirrors, equivalent to a 21 m single telescope and is foreseen to be built in Chile at the Las Campanas Observatory. At Mauna Kea the next generation will be represented by the Thirty Meter Telescope (TMT) and ESO is preparing for the 40 m European Extremely Large Telescope (E-ELT) near the VLT site in Chile.

5.4 UV: The Obscured Inner Disk

UV light at wavelengths from 100–4000 Å is efficiently absorbed by the Earth's atmosphere even at extreme altitudes, and thus can be observed only using space-based observatories. But even for satellite-based research, the extreme UV (EUV; 100–1210 Å) range is difficult to observe, as the interstellar medium absorbs most of the radiation between the Lyman edge at 912 and 100 Å. Thus, AGN in the EUV are preferentially discovered in directions with low galactic hydrogen column density.

The ultraviolet portion of the electromagnetic spectrum is invaluable to a number of astrophysical studies, from stellar astrophysics, nebulae, to population synthesis of star clusters and external galaxies, and not insignificantly to AGN. This is the case because of the rich atomic emission and absorption line spectra it contains and because of the preponderance of hot 10^4-10^5 K plasma that can be diagnosed through UV line and continuum study. For low- to intermediate-redshift AGN, the UV continuum provides a direct measure of the ionizing nonthermal radiation. The fluxes and spectral profiles of many of the resonance emission lines produced closest to the central engine can be directly measured. In addition, multiple ionization states of various elements can facilitate photoionization modeling of the broad- and narrow-line regions. UV observations also facilitate comparisons between low- and high-redshift AGN samples since for the latter we see the source frame UV in the optical or infrared. The source frame UV is additionally less diluted by stellar light from the host galaxy. However, UV light is extremely sensitive to extinction effects of dust and molecular hydrogen, both in our own galaxy and potentially in AGN hosts.

Early AGN observations in the UV band were made by the Astronomical Netherlands Satellite (ANS) for the brightest objects, notably 3C 273 which was observed in five broad bands between 1550–3300 Å (Wu, 1977) as well as with rocket flights between 1200 and 1700 Å (Davidsen *et al.*, 1977). The first major advances though came with the launch of the International Ultraviolet Explorer (IUE) in 1978. That facility enabled low resolution ($\lambda/\Delta\lambda \sim 200{-}400$) for objects brighter than $m_V \simeq 17$ mag. Its echelle mode, which provided higher resolution, ($\lambda/\Delta\lambda \sim 10\,000$), had insufficient throughput efficiency for all but the brightest AGN, namely 3C 273 and NGC 4151.

One of the first insights was gained from the ability to compare in a statistical sense low redshift, lower-luminosity AGN – Seyferts and radio galaxies – over similar source frame wavelengths bands. Although there were some notable differences, a striking similarity was found between the high-z optical and low-z UV spectra. In particular, many of the prominent emission lines, Lyα (1216 Å), Lyβ at 1026 Å (often blended with O VI at 1032 Å), C IV (1549 Å), [C III] (1909 Å) and [Mg II] (2798 Å) (e.g., Baldwin and Netzer, 1978). The implications of this result are significant, as the range of luminosities involved spans at least 4 orders of magnitude. Thus, the photoionization processes involved are able to accommodate this large range in luminosity by forming clouds at distances from the central engine in such a way that the ionization parameter ξ remains approximately constant (see Section 3.4). The physics underlying this self-similarity must be accounted for in a complete model. Figure 5.6 shows the composite spectrum of AGN observed in the far UV range by the FUSE satellite as an example. As suggested here, this $\simeq 600{-}1100$ Å range, in combination with the $\simeq 1200{-}3300$ Å near UV provides multiple ions of C, O, as well as Lyα and Lyβ in low-redshift objects, which greatly facilitates photoionization modeling efforts.

As an aside note, the high-resolution far-UV coverage of FUSE has been used to exploit AGN as a cosmological tool. Specifically, bright AGN with fortuitous lines of sight can be used as a background light source to obtain the Lyβ absorption line profile of Milky Way galactic halo components. The halo gas is believed to be relatively "pristine," that is free of processing by stars which destroys deuterium. As such, its isotopic deuterium to ^{1}H ratio should approximate the primordial value. The FUSE resolution is sufficient to resolve the Lyβ profiles of the two isotopes as was accomplished using the quasar PG 1259+593. This led to a value of about 22 ppm, which is consistent with estimates of the primordial value and clearly distinct from values obtained from ISM measurements.

The early UV studies also showed that AGN continuum morphologies were qualitatively similar between low-z low-luminosity and high-z high-luminosity subclasses. The flattening of the optical continuum at rest frame UV wavelengths seen in quasars – the big blue bump (Section 1.2) was prominent in the UV continuum of Seyfert 1 AGN as well, and in many cases is the predominant component of their overall energetic output. This facilitated models of the big blue bump as an accretion disk over a larger sample of objects and spanning a large range of luminosities. Figure 5.7 shows an example of the broad-band continuum of the Seyfert 1 galaxy Markarian 335 with the underlying model components considered by those

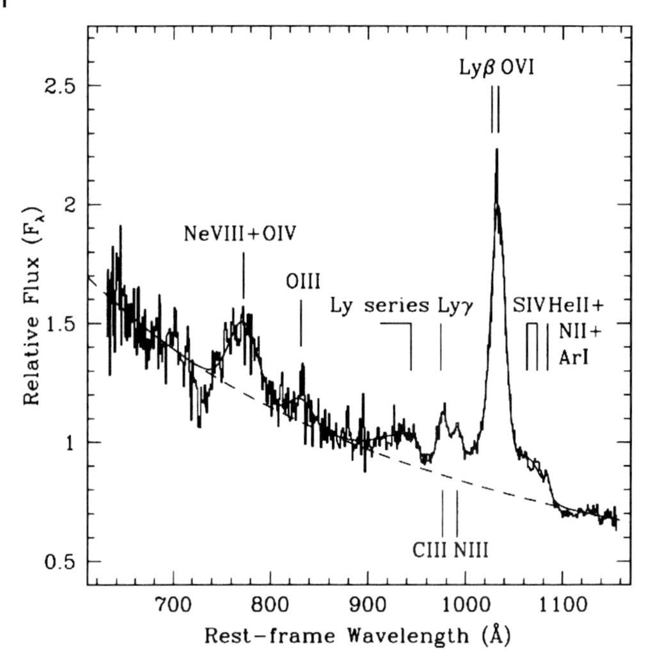

Figure 5.6 Composite AGN spectrum in the extreme UV based on FUSE satellite data as presented by Scott *et al.* (2004), reproduced by permission of the AAS. The dashed line shows the continuum fit, while the solid line represents the fit to continuum and emission lines.

authors. As one can see, the UV spectrum is dominated by a strong rise toward the X-ray domain. This effect was discovered in the 1970s when studying spectra of Seyfert 1 galaxies in the IR-optical-UV range (Malkan and Sargent, 1982b). The feature was called the big blue bump, as it becomes prominent at the blue end of the optical spectrum. It is commonly believed that this maximum in the spectral energy distribution arises from the thermal emission of the accretion disk, as suggested by Shields (1978). Kinney *et al.* (1991) published a catalog of IUE spectra from 69 quasars, blazars and Seyfert 1, showing that the UV emission in AGN is more variable than the optical. This is an extension of the previously noted trend of blueward-increasing variability amplitude in photometric quasar samples. It suggests therefore that the blue bump is a feature not directly related to the underlying continuum we see in adjacent bands. This has lately also been confirmed through observations by the GALEX satellite, showing that the shape of the blue bump is not correlated to the continuum emission (Atlee and Mathur, 2009). We will discuss the accretion disk model for the big blue bump, its implications and alternatives in Section 5.8.2.3 when we have a closer look at the overall spectral energy distribution.

Combined UV and optical observations can also be used to statistically assess the dust content of AGN. For example the source frame monochromatic flux ratios at spectral regions free of emission lines in the optical (e.g., λ4220 and UV λ1770) plotted as a function of the monochromatic luminosity show convergence towards

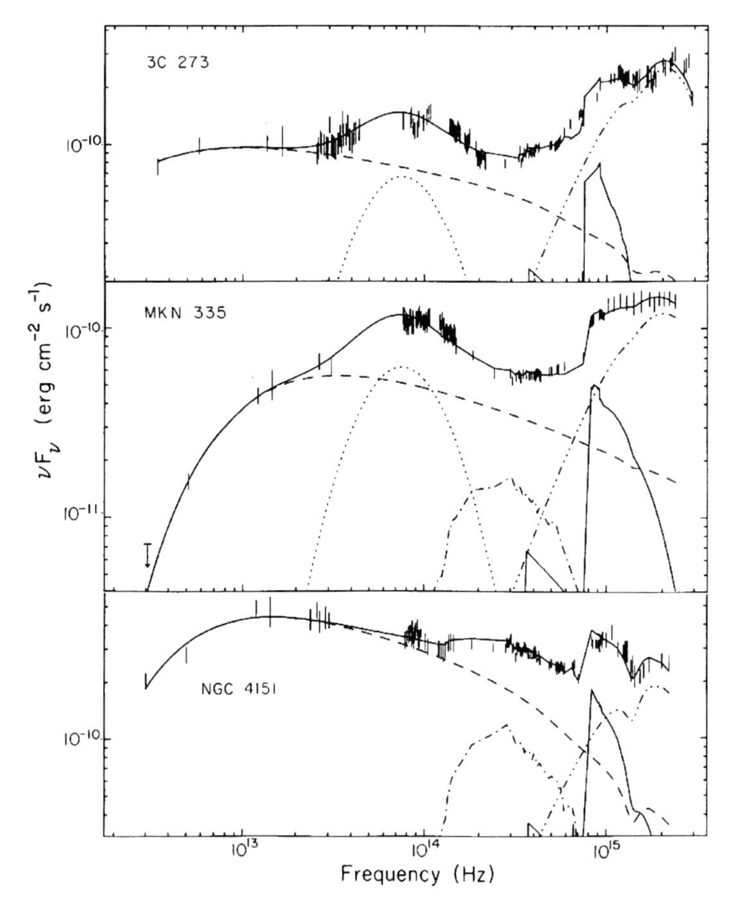

Figure 5.7 IR to UV spectral energy distributions of three prototypical type 1 AGN, with multicomponent models overlayed. The ultraviolet measurements are clearly indicative of a pronounced excess, the big blue bump, relative to the overall broad continuum. In the model decomposition shown it is represented as a single-temperature blackbody. Additional model components such as an IR excess component, Balmer and Paschen continuum are also overlayed. (Reproduced from Edelson and Malkan (1986) by permission of the AAS.)

a constant color ratio with decreasing scatter towards higher luminosities (Malkan, 1984). The scattering envelope is significantly smaller than what would be expected for a Milky Way reddening law. This suggests that dust is more abundant in the least luminous AGN but that it does not survive in the more luminous objects.

More recent data support this idea. For example, by comparing the inferred reddening from type 1 AGN broad-line ratios and the X-ray absorbing column density Maiolino *et al.* (2001) find that the $E_{\rm B-V}/N_{\rm H}$ ratio is nearly always lower than the galactic value by factors ranging from $\simeq$ 3 up to 100. Combined with the lack of prominent absorption features at 2175 Å in the spectra of reddened Seyfert 1s, this suggests that dust in the circumnuclear region of AGNs may be different from that in our galaxy. In Figure 5.8 the UV spectra of five type 1 AGN obtained with

HST are compared to an average AGN continuum that has been reddened using a galactic extinction law. This supports the mass fraction of dust or the nature of the dust itself is different in AGN environments. One possibility is that the dust composition in the circumnuclear region of AGNs could be dominated by large grains,

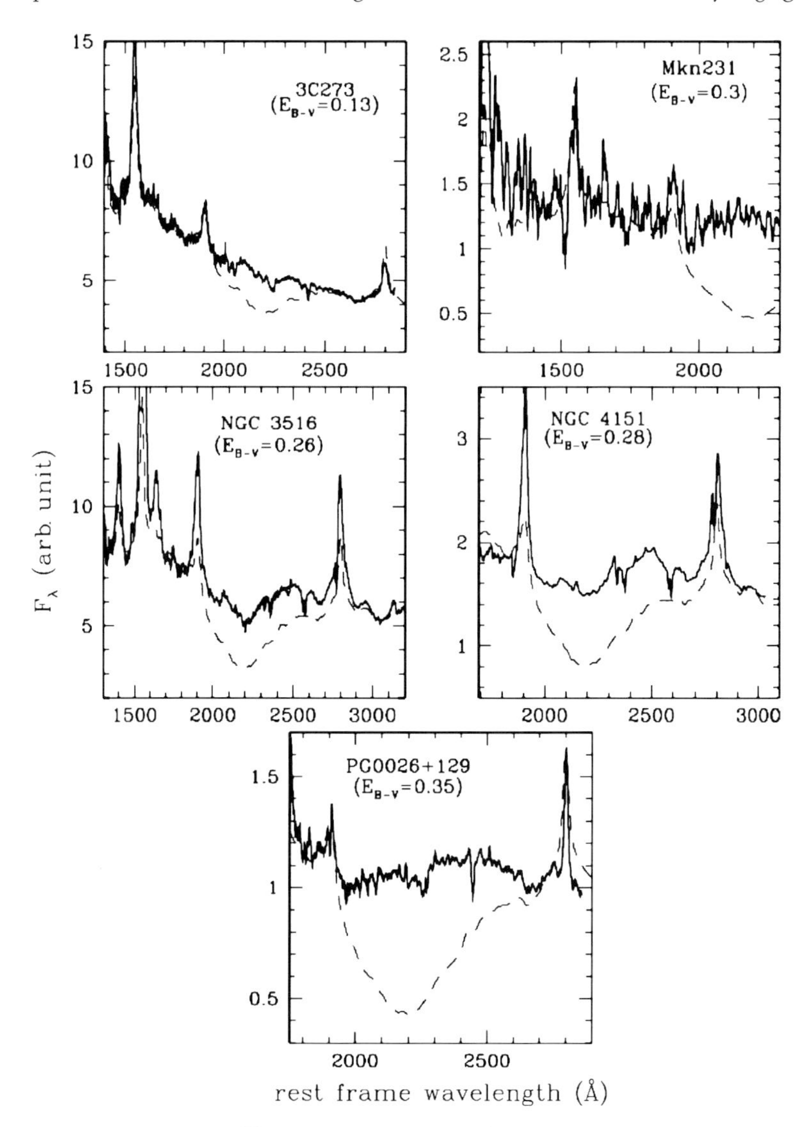

Figure 5.8 UV spectra of five type 1 AGN obtained with HST, reproduced from Maiolino *et al.* (2001). The broad-line ratios and continuum indicate the presence of dust absorption. The thin dashed line is an averaged type 1 AGN spectrum reddened with a standard galactic extinction curve with an E_{B-V} value consistent with that inferred from the broad-line ratios. Notably, the galactic extinction curve always predicts significantly greater absorption than what is observed.

which make the extinction curve flatter and featureless. In any case, whatever the physical origin of this effect may be, the reduced dust absorption with respect to what is expected from the gaseous column density should be carefully considered when using optical and X-ray measurements for making AGN classifications.

A number of other areas of AGN study benefit from UV observations. UV spectroscopy has provided the highest precision dynamical measurements in the inner regions of AGN, specifically resolving multiple component mass outflow structures as we discussed in Section 3.3.2; in particular see Figure 3.13. It has also greatly facilitated reverberation mapping studies expanding the accessible range of luminosities and ionization levels (see Section 3.1.2). High-resolution UV spectroscopy of bright AGN can reveal absorption-line systems associated with the galactic halo or intervening galaxies or in the intergalactic medium (e.g., Kriss *et al.*, 2001; Tripp *et al.*, 1998), and broad absorption line quasars studies are greatly enhanced through the inclusion of ions such as C IV (1549 Å) and Mg II (2798 Å) in low-redshift objects. Imaging of galaxies in the UV, notably with the GALEX satellite, have facilitated studies of recent star-formation histories of early-type galaxies in the local Universe. This can in turn enable study of their possible evolutionary connection with AGN activity.

UV telescopes

The era of space-based UV astronomy began, practically speaking, in the 1970s. The Copernicus satellite (OAO-3) was launched in 1972 and performed high-resolution UV spectral scans in the 900–1560 and 1650–3150 Å band. Spectra were obtained for 551 objects, most primarily bright stars. Soon thereafter, the Astronomical Netherlands Satellite (ANS), an X-ray and UV mission launched in 1974, carried out observations between 1500 and 330 Å within five bands and ran for 20 months, detecting about 400 objects.

A breakthrough for UV astronomy was the International Ultraviolet Explorer (IUE), which operated from 1978 until 1996. Its instruments provided spectroscopic coverage at 1150 and 3200 Å. It had a low-resolution mode, $R \sim 200$–400 and an echelle mode, $R \sim 10^4$. IUE took thousands of spectra of more than 100 AGN.

The NASA explorer mission Extreme Ultraviolet Explorer (EUVE, 1992–2001), at 70–760 Å, conducted an all sky survey, discovering 801 objects, and detected the first extragalactic object in the extreme UV, the BL Lac PKS 2155-304. NASA continued its UV program with the Far Ultraviolet Spectroscopic Explorer (FUSE), which operated from 1999 to 2007 in the 900–1200 Å band, observing hundreds of low-redshift AGN in the local Universe ($z < 0.1$), and dozens at larger distances.

Today, the Galaxy Evolution Explorer (GALEX) performs grism spectroscopy, starting in 2003, in the near and far UV (1771–2831 and 1344–1786 Å) allowing to study hundreds of Lyα emission line sources at $0.2 < z < 0.44$ and $0.65 < z < 1.25$.

The Hubble Space Telescope (HST), launched in 1990 has had a succession of UV spectrometers; the FOS, GHRS and currently two operational instruments, the Space Telescope Imaging Spectrograph (STIS) and the Cosmic Origins spectrograph (COS). STIS failed in 2004, but was repaired on the last HST servicing mission in 2009. It uses two-dimensional detectors operating from the ultraviolet to the near infrared providing both grism spectroscopy as well as an echelle mode. It covers the UV to the near infrared band (1150–10 000 Å). The echelle gratings provide the highest spectral resolution in the near UV thus far, reaching $R \sim 10^5$. The COS UV spectrograph covers the 1100–3200 Å range with two instrumental bands. It has a maximum resolving power of $R \sim 40\,000$, and its unprecedented low-instrumental background make it the most sensitive UV spectrometer flown to date.

5.5 X-rays: Absorption, Reflection, and Relativistically Altered Line Profiles

With increasing frequency we are getting closer to the central engine of the AGN. Electromagnetic radiation between ~ 120 eV and ~ 120 keV is referred to as X-rays. As is the case in the UV range, X-rays are similarly unable to penetrate the Earth's atmosphere. It is thus necessary to fly detectors at high altitudes, for example using balloons or rockets. For longer exposure times and thus higher sensitivity, spaceborne telescopes are required. In the 1950s only the sun was a detectable source of X-rays, specifically, its corona. Other stars would be at distances too large to be relevant. Neutron stars and black holes were still theoretical constructions rather than a demonstrable physical reality, and the concept of AGN was still to come.

The X-ray domain has seen an enormous evolution over recent decades, and is currently one of the key energy ranges to study AGN. As such, we include here a brief historical review how X-ray astrophysical studies of AGN became such an important tool.

5.5.1 AGN in the X-ray from 1965 to the 1990s

The first X-ray measurements of AGN were made with detectors on-board an Aerobee rocket in April 1965, which provided evidence for high-energy emission from Cygnus A and M87 (Byram *et al.*, 1966). A flight in 1969 then provided the first detection of the radio galaxy Cen A and of the quasar 3C 273, at a significance level of 3.0σ and 3.9σ, respectively. Rocket flights however did not provide sufficient exposure times to further advance the field once the brightest sources had been detected. Uhuru, the first X-ray telescope on an orbiting satellite was then launched in December 1970. The mission was equipped with two large area proportional counter detector systems, with $840\,\mathrm{cm}^2$ effective area each. It performed the first sky survey leading to a catalog of 339 sources in the 2–6 keV range. The predominant fraction of these sources were found to be compact, mass-exchange binaries, thereafter referred to as X-ray binaries. Examples included Cyg X-3, Her X-1, and Vela X-1 named after their host constellations and the order of their discovery. These objects provided a general confirmation of theoretical models that suggested accretion onto a compact object can lead to X-ray radiation.

Uhuru also provided the first detection of the Seyfert galaxies NGC 4151 and of NGC 1275, and confirmed the earlier X-ray detection of Cen A, Cygnus A, M87, and of 3C 273. In total, 15 Seyfert galaxies were detected, all of them being Seyfert 1 or Seyfert 1.5. The largest class of extragalactic objects though were the galaxy clusters, of which Uhuru detected 45.

A succession of two missions as follow up to the highly successful Uhuru mission was planned by NASA. In the years 1977–1979, the first High-Energy Astronomy Observatory (HEAO-1), which carried four experiment packages was flown. Two of those experiments, A1 and A2, consisted of nonimaging X-ray proportional counters covering the nominal 0.2–25 keV X-ray band. The A1 Large Area Sky Survey experiment included seven proportional counter modules of 10^3 cm^2 collecting area. It cataloged over 800 sources, including 90 identified as AGN (Wood *et al.*, 1984). The A2 experiment also conducted a survey ($f_{2-10\,\text{keV}} > 3.1 \times 10^{-11}\,\text{erg}\,\text{s}^{-1}\,\text{cm}^{-2}$) at galactic high-latitude ($|b| > 20°$) containing 85 sources, out of which 61 were extragalactic. It also provided a precise measurement of the broad-band X-ray extragalactic background radiation in the 3–250 keV band, revealing its peak around 30 keV (Marshall *et al.*, 1980).

Proportional counters did not allow direct imaging of the X-ray sky though, thus the application of mirrors was the next challenge in this field. Hans Wolter had studied the use of mirrors at grazing incidence angles as a way to focus X-rays (Wolter, 1952), and the use of Wolter-type mirrors had been proposed before the first detection of an extrasolar X-ray source (Giacconi and Rossi, 1960). The first X-ray mirror in space was a 30-cm consisting of two nested mirrors flown on the SKYLAB space station in 1973. The purpose was to perform imaging observations of the sun.

Several satellites hosting X-ray detectors followed, like the Astronomische Nederlandse Satelliet (1974–1977), which observed X-ray emission from stellar coronae and Ariel V (1974–1980), detecting 251 X-ray sources and further establishing Seyfert 1 galaxies as X-ray sources. Based on Ariel V data, Elvis *et al.* (1978) derived the first X-ray luminosity function for Seyfert 1 galaxies and came to the conclusion that the emission originates in the very core of the galaxies and that synchrotron self-Compton radiation is the most plausible emission mechanism in the 2–10 keV band. The 5 yr of data were also used to perform detailed analysis of the light curves of AGN, finding no evidence for a distinction in the X-ray variability of active galaxies with differing morphology or X-ray luminosity (Marshall *et al.*, 1981). Ariel and Uhuru led also to the detection of variable absorption in NGC 4151, which was able to explain the soft X-ray flux variability in this Seyfert 1.5 galaxy. Early on it was noticed that objects which exhibit temperatures able to emit X-ray radiation should also show a prominent line from iron fluorescence at 6.4 keV. First indication of the existence of this line (at a 3.3σ level) was found in data on the X-ray binary Sco X-1 from a 150 s rocket flight observation (Holt *et al.*, 1969) and a few years later also in the supernova remnant Cas A. In AGN the iron fluorescence line was first detected in data of NASA's 8th Orbiting Solar Observatory OSO-8 of Cen A and later in Ariel V data of the Seyfert 1.5 NGC 5548.

A major leap forward occurred with the launch in 1978 of HEAO-2, later named Einstein, which first used Wolter-geometry grazing incidence mirrors on an orbiting X-ray observatory. It hosted four instruments for imaging and spectroscopy which could be rotated into the focal plane, achieving 2$''$ spatial resolution (0.5–4.5 keV) and a spectral resolution of 20% at 6 keV. Einstein performed pointed observations but its high sensitivity lead to a large number of serendipitously detected sources. In the field of extragalactic studies, results included the detection of X-ray jets in Cen A and M87.

Prior to this mission, only 5 Seyfert 2 galaxies were known to emit X-rays, NGC 2110, NGC 5506, NGC 7582, NGC 2992, and MCG-5-23-16, but with the Einstein data it was now possible to conduct statistically meaningful tests on the differences of type 1 and type 2 objects in the X-ray domain. A well-defined sample based on Einstein data is the Einstein Observatory Extended Medium Sensitivity Survey (EMSS; Gioia *et al.*, 1990; Stocke *et al.*, 1991) on a sky area of 778 deg^2, containing 835 X-ray sources in the flux range 5×10^{-14}–10^{-11} erg cm^{-2} s^{-1}, including 427 Seyfert galaxies and 36 blazars.

From then on, population studies based on X-ray-selected samples' data could be rigorously applied. In addition, the main ingredients of AGN in the X-rays were established: a supermassive black hole in the center of the host galaxy, which emits throughout the electromagnetic spectrum through accretion processes, absorbing material in the vicinity of the central engine affecting the soft X-ray spectrum, and an iron fluorescence line indicating reflection processes.

The European X-ray Observing Satellite (EXOSAT) was launched in 1983 for a three year mission, carrying on-board a low-energy telescope and a medium-energy proportional counter, allowing observations in the 0.1–10 keV range. Because EXOSAT was launched into a highly eccentric orbit, allowing for uninterrupted exposures of up to 3 days, this mission gave new insights into the X-ray variability of AGN. Green *et al.* (1993) analyzed light curves of 32 AGN, finding variability in 12 of them and discovering short-term variations of less than an hour in a number of sources, indicating that the emission region of X-rays in AGN has to be quite small. EXOSAT data also allowed to further constrain the spectral slope and iron line feature. The spectra of 48 X-ray-selected AGN showed that 50% of them exhibit a soft excess, making this an apparently common feature of Seyfert galaxies (Turner and Pounds, 1989). We will discuss the soft excess in more detail below.

The Japanese Ginga (1987–1991) missions derived AGN spectra from 1.5 up to $\sim$ 30 keV. The energy resolution of about $\Delta E = 1$ keV at $E = 6$ keV showed that iron line features in AGN exist in emission and in absorption, showing complex reprocessing processes in the central engine of Seyfert galaxies.

The Japanese science satellite program

The Japanese community has been very active in X-ray satellite-based astronomy since the early 1980s. The Japanese missions are called Astro-A, -B, -C, and so forth, but are given a less prosaic name shortly after launch. Astro-A (Hinotori) was launched in 1981 to observe the sun in the 10–40 keV band. Astro-B (Tenma, 1983–1988) performed high-resolution X-ray spectroscopy of stars, Astro-C (Ginga, 1987–1991) and Astro-D (ASCA) are discussed in the text of this section, and Astro-E was lost at launch in 2000 but the mission was reborn as Astro-E2 (Suzaku) with a launch in 2005 and further information on this X-ray satellite is also included in the text. Astro-F (AKARI) is an infrared astronomy satellite launched in 2006, and Astro-G (VSOP-2) is a radio satellite with possible mission start in 2013. Astro-H is an X-ray mission with a long focal length allowing for the first time to perform imaging and spectroscopic observations applying mirrors at hard X-rays, and covering a band from 0.3–600 keV. Up to 80 keV it should be about 100 times more sensitive than Suzaku. This mission will also provide a high-energy resolution microcalorimeter with a FWHM of 5 eV. The launch might be as early as 2014.

Arguably, the 1990s represented a *golden age* of X-ray astronomy. ROSAT provided spectroscopy in the 0.1–2.4 keV energy band and performed an all-sky survey, the RASS. This survey revealed 10^5 X-ray sources, opening the X-ray band for large population studies. In addition, the ROSAT/HRI instrument provided high-resolution images in the 0.5–2.0 keV band. The Advanced Satellite for Cosmology and Astrophysics (ASCA), a Japanese observatory launched in 1993 and operating until 2000, provided for the first time a powerful combination of imaging and spectroscopic properties at the hard X-ray end toward 10 keV, with spectral resolution of $E/\Delta E \sim 50$ at 6 keV, and ~ 20 at 1.5 keV and a spatial resolution of about 3 arcmin. This allowed accurate, time-resolved studies of the iron line complex. In case of nearby Seyfert galaxies detailed physical models began to be developed and tested. This focused much attention on discussions of the broad iron K_α line profiles in Seyfert galaxies, as we will discuss later.

The Rossi X-ray Timing Explorer (RXTE) was a satellite dedicated to timing studies, decomissioned in January 2012. With the instruments PCA (2–60 keV, 1 keV spectral resolution at 6 keV, 1 μs time resolution) and HEXTE (15–250 keV, 10 keV resolution at 60 keV, 8 μs time resolution) it had moderate sensitivity, nevertheless it provided interesting timing results on AGN (e.g., McHardy *et al.*, 2004), and its large sky coverage was used in order to derive the local luminosity function of AGN in the 3–20 keV band (Sazonov and Revnivtsev, 2004). RXTE data were also used in the spectral analysis of bright AGN, such as 3C 273, Cen A, and NGC 4151.

5.5.2
Today and Future X-ray Missions

We now move on to the X-ray satellites which are taking data today.[2)]

Two major observatories were launched in 1999: NASA's Chandra satellite, and ESA's XMM-Newton mission. Chandra provides the highest spatial resolution X-ray images to-date, with 0.5 arcsec angular resolution. The figure on the book cover shows an X-ray image taken by Chandra/XIS of the radio galaxy Centaurus A. The image reveals a complex X-ray morphology, showing a strong jet and a shorter counterjet. The whole extent of the X-ray structure is about 15 kpc. Another example of Chandra's impressive resolution is given in Figure 5.9, displaying the jet of 3C 273.

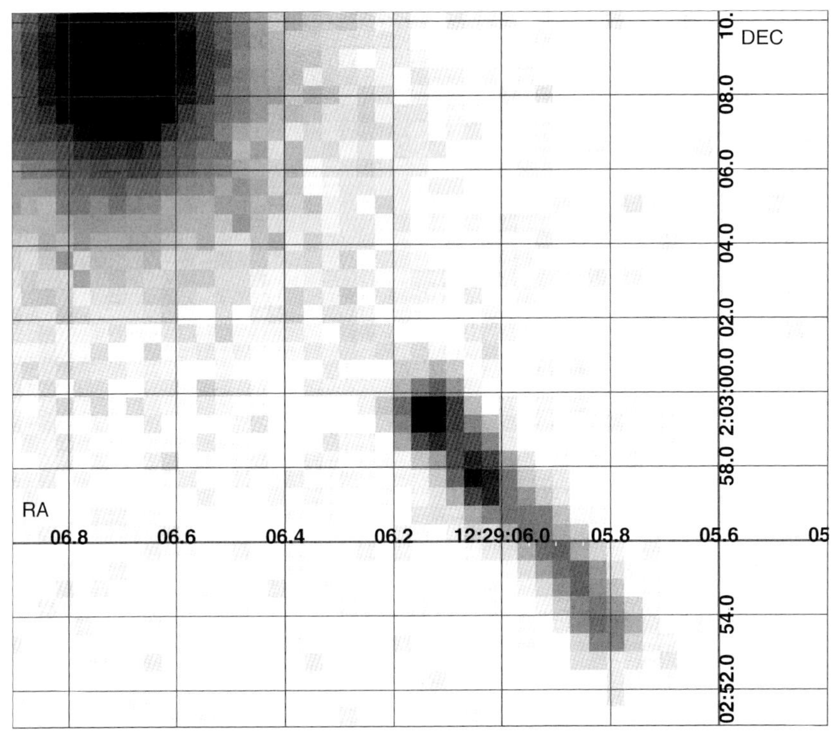

Figure 5.9 X-ray image by the ACIS-S instrument onboard the Chandra observatory of the jet of the blazar 3C 273. The central engine is located on the top left. Although the jet is faintly visible all the way from the core, the bright onset in X-rays appears at 13 arcsec distance from the center, corresponding to a projected distance of about 34 kpc, with the whole jet extending out to about 60 kpc, which is about twice the diameter of our Galaxy. Chandra data obtained from the High Energy Astrophysics Science Archive Research Center (HEASARC), provided by NASA's Goddard Space Flight Center.

2) Table 5.2 presents flux limits for various past, present and future X-ray missions.

Table 5.2 Flux limits of some past, current, and future X-ray and gamma-ray missions. The flux limits are meant to give a general idea. In the individual case, the limit depends on the source spectrum, the absorption in the line of sight, the source density in the field, the instrument mode and other aspects.

Satellite/ instrument	Energy band (keV)	Flux limit (erg cm^{-2} s^{-1})	Exposure time	Remark
ROSAT/PSPC	0.1–2.4	3×10^{-13}	$\sim$ 400 s	RASS
ROSAT/PSPC	0.5–2	2×10^{-15}	140 ks	
XMM-Newton/EPIC	0.5–2	5×10^{-16}	60 ks	XMM Faint Survey
Chandra/ACIS	0.5–2	3×10^{-17}	2 Ms	CDF-N
SRG/eROSITA	0.5–2	10^{-14}	4 yr	All-sky survey
SRG/eROSITA	2–10	10^{-13}	4 yr	All-sky survey
Chandra/ACIS	2–10	8×10^{-16}	200 ks	
NuSTAR	10–30	10^{-14}	1 Ms	
Swift/BAT	14–195	2×10^{-11}	22 months	All-sky survey
INTEGRAL/IBIS/ISGRI	20–60	5×10^{-12}	1 Ms	
INTEGRAL/IBIS/ISGRI	200–1000	3×10^{-10}	1 Ms	
CGRO/COMPTEL	750–30 000	5×10^{-11}	9 yr	All-sky survey
Fermi/LAT	10^5–10^8	4×10^{-12}	2 yr	All-sky survey

XMM-Newton's strength is the large mirror surface with an effective area of 1900 cm^2 for energies up to 150 eV, 1500 cm^2 at 2 keV, and 350 cm^2 at 10 keV. Combined with the high spectral resolution ability allows for the derivation of AGN spectra also from fainter objects in a reasonable amount of observing time.

While Chandra and XMM-Newton operate up to 10 keV, the hardest X-rays up to several hundred keV became accessible through several observatories, like CGRO, BeppoSAX, INTEGRAL, Swift, and Suzaku. Many of these also provide spectroscopy and imaging from the soft X-rays onwards, thus X-ray astronomy has been fully established as a main-stream astronomical discipline over the last 20 years. Rich data archives are now available to the astronomical community, most notably NASA's High-Energy Astrophysics Science Research Archive (HEASARC).

Observations at the hardest X-rays

At the hardest X-ray regime with $E \gg 10$ keV, grazing incident mirror telescopes would need focal lengths of $\gg 10$ m. Thus, in the past mainly two approaches were chosen to observe in this energy range. Nonimaging instruments, which function like a "light-bucket," have been used for example in BeppoSAX/PDS (15–200 keV, 1996–2002), CGRO/OSSE (50–10 MeV, 1991–2000), and Suzaku/HXD (10–600 keV, launched 2005). Another approach has been applied in the case of INTEGRAL (3–1000 keV, launched 2002) and Swift/BAT (15–195 keV, launched 2004). Here coded masks, similar to pin-hole masks but with a pattern of holes allowing for $\sim$ 50% of the light to reach the detector, allow imaging with resolution on the several arcminute scale. All the hard X-ray missions are rather photon-starved and background-dominated, requiring long observation times of 100–1000 ks to observe AGN and to obtain low-resolution spectra. Nevertheless, about 500 AGN have been detected above 20 keV in hard X-rays to date.

One challenge which will be tackled by the upcoming missions in this domain is to focus hard X-ray photons using Wolter-type telescopes. This requires long focal lengths, with a distance between detector and mirror of about 10 m. Several concepts had been laid out for even larger focal length which would require putting the telescope and the detector on two separate spacecrafts and to perform formation flight. Although complex formation flight is technically feasible, as demonstrated for example by *Orbital Express* in 2007, the cost of implementing such a mission has thus far proven prohibitive.

In the single-spacecraft implementation the long focal length is achieved by an expandable boom in the case of NASA's NuSTAR mission (Harrison *et al.*, 2010). This will give an observational window between 5 and 80 keV, with 10″ resolution (FWHM) and a field of view of $10' \times 10'$ at 10 keV. The spectral energy resolution will be only $\Delta E = 0.5$ keV at 10 keV and thus is not likely to lead to major advances in AGN research of the iron line. It does though hold promise for other astronomical endeavors, for example supernova remnant studies for which the nuclear lines like ^{44}Ti at 68 and 78 keV provide a diagnostic of recent galactic supernova history.

The next breakthrough is going to be the German eROSITA survey telescope on-board the Russian Spectrum-Roentgen-Gamma (SRG) satellite to be launched in 2013. eROSITA is going to perform an all-sky survey in the 0.5–10 keV energy range, similar to the ROSAT All-Sky Survey (RASS), but including the harder energies where absorption does not play a major role, and many times deeper than the RASS. The flux limit for the all sky survey will be as deep as $\sim 10^{-14}\,\mathrm{erg\,cm^{-2}\,s^{-1}}$ in the 0.5–2 keV band ($\sim 10^{-13}\,\mathrm{erg\,cm^{-2}\,s^{-1}}$ at 2–10 keV), and it will detect on the order of 10^6 AGN.

A cornerstone X-ray observatory will be the Japanese Astro-H mission. Employing four different detectors, Astro-H will cover the energy range from 0.3–600 keV, with a spectroscopic resolution of $\Delta E = 5$ eV in the 0.3–12 keV band. X-ray imaging will be possible up to 80 keV at a focal length of 12 m, with about 1.7 arcmin resolution. Compared to the 12 arcmin resolution of INTEGRAL IBIS/ISGRI above 20 keV, Astro-H will provide major advances at the hardest X-rays, an essential energy range for AGN research as we will see in the following.

The future of the next large European/American X-ray telescope is less clear. The ATHENA X-ray mission has been proposed as ESA's next large satellite mission. With a focal length of 11.5 m ATHENA is envisioned to carry a wide field imager (WFI) and a microcalorimeter (XMS). Operating at 0.1–40 keV, the mirror would have an effective area of 1 m^2 at 1.25 keV, and still half of that at the iron K_α line at 6.4 keV. The spectral resolution of XMS should be as good as $\Delta E = 3$ eV at this energy, and the angular resolution will be in the range 5–10″ at 0.1–7 keV.

5.5.3
The X-ray Spectrum of AGN

In X-rays, the accretion disk surrounding a black hole is believed to produce a thermal spectrum. The low-energy photons produced from this spectrum are scattered to higher energies by relativistic electrons, residing for example in a corona above the accretion disk, through inverse Compton processes (Haardt and Maraschi, 1993). As the temperature of the disk and the relativistic electrons energy distribution are limited, the resulting inverse Compton spectrum has a high-energy cutoff. The spectrum has an approximate power-law shape with a photon index of $\Gamma \simeq 2$ extending up to a few hundred keV. The soft photons involved in the inverse Compton scattering are believed to originate in the cool thick accretion disk with $kT < 50$ eV, while the relativistic electron gas has a temperature of about $kT \sim 100$ keV. The photons will be up-scattered from their initial energy E_i to the energy

$$E_f = e^y E_i \quad \text{with} \quad y \simeq \frac{4kT}{m_e c^2} \max(\tau, \tau^2) \tag{5.6}$$

with y being the Compton parameter, τ is the optical depth of the electron gas, and T is its temperature. The Compton parameter is the gain ΔE in energy of a photon per scattering event, times the number N of expected scatterings:

$$y = \frac{\Delta E}{E} N \tag{5.7}$$

In the nonrelativistic case with $\tau > 0.01$ this results in $y \ll 10$ and a power law which extends up to the thermal cutoff in the range $E_{cut} \simeq kT$ to $E_{cut} \simeq 3kT$, determined by the cutoff in the thermal distribution of the electrons.

There are a number of additional details involved. For a rigorous discussion of high-energy spectral formation resulting from Comptonizing media in astrophysical environments, we refer interested readers to Sunyaev and Titarchuk (1980).

This spectrum is undergoing reprocessing through absorption. A concise review on the topic of absorption and reflection in the X-ray spectra of AGN has been provided by Turner and Miller (2009). In most cases broad emission lines due to low-ionization stages of iron are visible in the spectra, thus in these objects the absorber is commonly assumed to consist of cold ($T < 10^6$ K) optically thick material (e.g., George and Fabian, 1991). This reprocessing leads to a "hump" in the hard X-ray spectrum, first observed by the Ginga satellite in the Seyfert 1 galaxies NGC 7469 and IC 4329A. The shape and strength of the hump depends on the geometry, chemical composition, and orientation with respect to the line of sight, but has its maximum around 20–30 keV, where the reflection efficiency reaches its maximum. Its measurement has been difficult, as the modeled strength of this component is closely linked to the intrinsic absorption (measured at softer X-rays), spectral slope of the underlying continuum, and the high-energy cutoff, which is not well constrained in many cases. Also for this reason the location and nature of the reflector itself are a question of debate. The reflecting material could reside for example in the outer accretion disk, or at the inner edge of the absorbing gas, or could be located in an outflowing wind. The reflection strength R is normalized so that $R = 1$ represents the case of an isotropic source above an infinite reflecting plane. R is often considered as an estimate of $\Omega/2\pi$ where Ω is the solid angle subtended by the reflector as seen from the isotropic X-ray source.

It has been pointed out that a relative reflection of $R > 1$ represents the seemingly nonphysical situation where more than 100% of the X-ray radiation incident on the reflector is reflected. Stated differently, the reflector as seen by the source subtends a solid angle greater than 4π. This apparent paradox can be explained by variable emission of the central engine and a time delay between the underlying continuum and the reflected component, caused by a large distance of the reflecting material to the primary source (Malzac and Petrucci, 2002). In this explanation R can be larger than unity only for a certain amount of time and has to be balanced by times with $R < 1$ in order to have on the long-term average a reflection component with $\langle R \rangle \leq 1$. Another model which can lead to large R values is based on relativistic effects caused by a dynamic corona moving towards the reflecting disk (Malzac *et al.*, 2001). Other explanations include a special geometry with a high intrinsic covering fraction of the cold disk material (Malzac, 2001) and general relativistic light bending effects caused by a maximally rotating Kerr black hole (Miniutti and Fabian, 2004).

Nevertheless, we caution the reader that the data modeling methods and associated standard software packages in common use can lead to solutions with absurdly large values of R, some of which have found their way into the published literature.

When studying X-ray spectra in the 2–20 keV band taken by the Ginga satellite, Zdziarski *et al.* (1999) observed that the reflection strength R is strongly correlated with the intrinsic spectral slope as measured in the photon index Γ, as shown in Figure 5.10. The correlation is seen not only in AGN, but also in X-ray binaries in their hard state (see Figure 4.18). This means that sources with steeper X-ray spectra show a larger reflection than those with a flat spectrum. An explanation might be that the photons which, together with the relativistic electrons in the corona, are the source for the inverse Compton component observed in the X-rays, are coming from the cool material which is also responsible for the reflection component. If the solid angle under which the reflector is seen by the central engine, is small, less seed photons arrive at the corona and are up-scattered. But in the case that R is large, plenty of photons up-scatter into the X-ray range, and the high density of the photon field leads to efficient cooling: the spectrum is significantly steeper. This is similar to the low-flux/hard spectra versus high-flux/steep spectra behavior observed in galactic X-ray binaries.

The correlation has been confirmed recently in a study including the brightest 27 Seyfert galaxies observed by the INTEGRAL in the hard X-rays. This study, involving also more complex absorption models than previously possible, shows also that the reflection strength $R < 1$ in nearly all cases when considering complex absorption. This would relieve us from accepting questionable explanations for $R \gg 1$. In addition, one seems to find a bimodality of the plasma temperature of the corona as shown in Figure 5.11. This histogram is based on long-term average values,

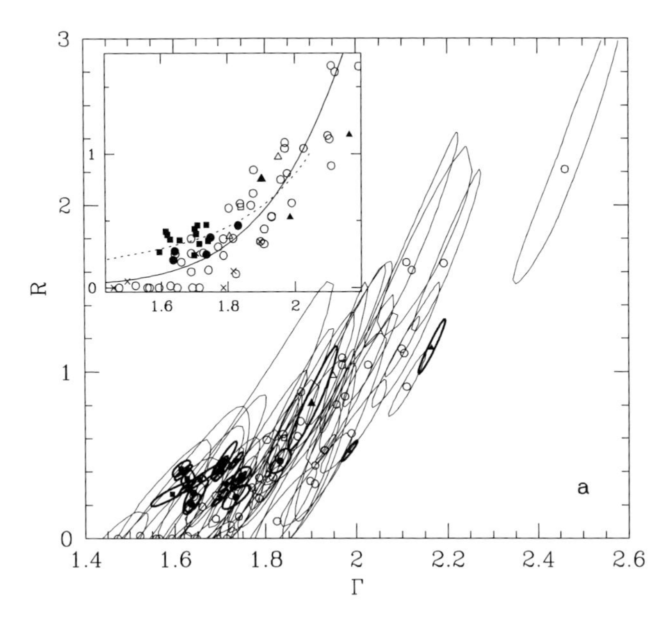

Figure 5.10 The strength of the reflection parameter R the inferred photon index Γ resulting from the application of X-ray spectral models. The analysis is based on 47 datasets for 23 radio-quiet AGN. The correlation shown by the curve has the form $R = u\Gamma^{v}$ with $u = (1.4 \pm 1.2) \times 10^{-4}$ and $v = 12.4 \pm 1.2$. Graphic from Zdziarski *et al.* (1999).

but a similar effect has been observed by Lubiński *et al.* (2010) also in a single AGN which undergoes spectral changes, in NGC 4151, probably the best-studied AGN in hard X-rays. Also here, the source remains either in a state with $kT_e \sim 60$ keV and a corona of high optical depth ($\tau \sim 2$), or the corona is optically thin with $\tau \sim 0.6$ but at a high plasma temperature ($kT_e \sim 200$ keV). If we use these values in order to compute the Compton y parameter (Eq. (5.6)) we see that y appears to be constant. This can imply that the source geometry stays the same, while temperature and optical depth of the corona vary. NGC 4151 seems to spend most of its time in the low-temperature/high-flux state, rather than showing the hard spectrum and a hot corona.

Another signature of the reflection process is the $^{26}\mathrm{Fe}K_{\alpha_1}$ fluorescence line at $E_{K_\alpha} = 6.404$ keV, accompanied by the $^{26}\mathrm{Fe}K_{\alpha_2}$ line at 6.391 keV. The K_α line can be used as a probe for the X-ray reprocessing component and is observable in most Seyfert galaxies with an equivalent width (see box in Section 4.1.4) of on average 50 eV. The latter value has been derived by Shu *et al.* (2010) from a large sample of Seyfert galaxies observed by Chandra. The same line strength was found when stacking about 1000 type 1 AGN with a typical average luminosity $L_X \simeq 1.5 \times 10^{44}$ erg s^{-1} at redshift $z \sim 1.6$ using data of the XMM-Newton COSMOS field (Iwasawa *et al.*, 2012). Earlier studies which came to the conclusion, that is that this feature has on average and equivalent width of 100–200 eV (Matt *et al.*, 1991) seem to have been hampered by insufficient detector resolution and sensitivity. A study of 409 spectra out to redshift $z = 3.5$ observed in the Chandra

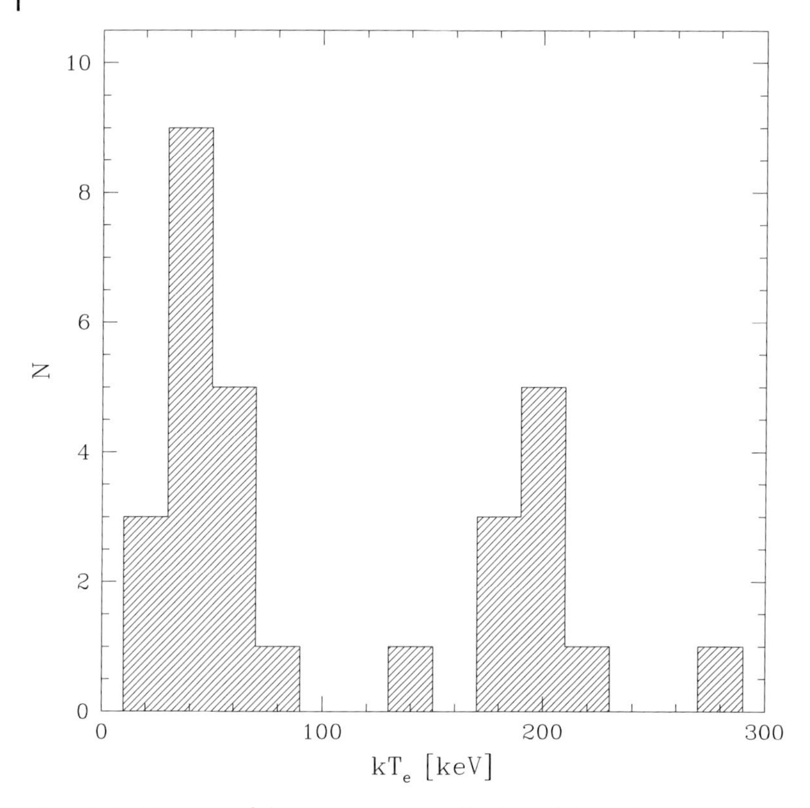

Figure 5.11 Histogram of plasma temperatures of bright Seyfert galaxies as measured using their X-ray spectra. The data seem to indicate a bimodality: the electrons appear to have either a temperature around $kT_e \simeq 50$ or 200 keV. Data courtesy of Piotr Lubiński.

deep fields by Falocco *et al.* (2012) draws a slightly more complicated picture. Absorbed and unabsorbed sources seem to show similar narrow iron line strength of $\text{EW} = 85 \pm 35$ and 73 ± 32 eV, and no significant differences are detected between the high- and low-redshift subsample. Dividing the sample in luminosity, the low-luminous sources with $L_X < 8 \times 10^{43}$ erg s^{-1} seem to show a stronger narrow Fe line ($\text{EW} = 89 \pm 32$ eV) than the luminous AGN ($\text{EW} = 59 \pm 38$ eV).

This shows that the Baldwin effect, which is the anticorrelation between the emission line's equivalent width (EW) and the continuum luminosity is also observed in the X-ray. The proportionality is about $\text{EW} \propto L^{-0.2}$, with L being the bolometric or X-ray luminosity (because the bolometric luminosity goes roughly linear with the X-ray one; see second box in Section 4.8.1). This correlation is sometimes also referred to as the Iwasawa–Tanigushi effect, because Iwasawa and Taniguchi (1993) discovered it based on Ginga data of 37 type 1 AGN. As Shu *et al.* (2010) point out, there might be several factors contributing to the X-ray Baldwin effect. Because the iron emission is likely to be arising from cooler material in the vicinity of the central engine, a more active AGN core might blow this matter away

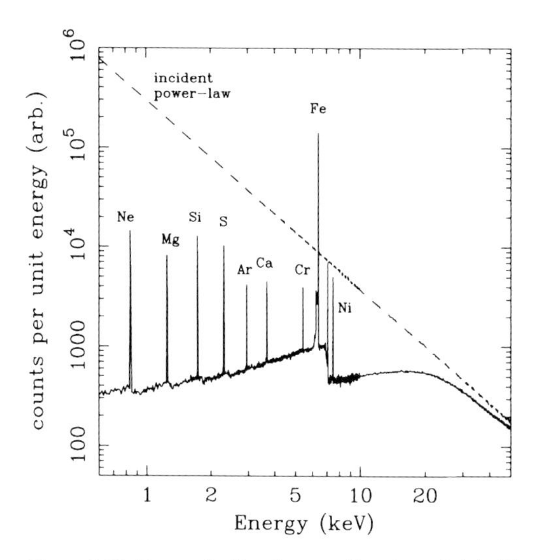

Figure 5.12 X-ray reflection from an illuminated slab. The dashed line shows the incident continuum, and solid line shows the reflected spectrum (integrated over all angles). Figure presented in Fabian *et al.* (2000) is based on Monte Carlo simulations by Reynolds (1996).

through radiation pressure or thermal dissipation. Therefore, the covering factor and/or the column density of the Fe K_α producing gas is likely to decrease with increasing AGN luminosity. In addition, the more active AGN could lead to a higher degree of ionization in the surrounding material, weakening the iron line from low-ionization matter.

Also a K_β line at 7.06 keV is observable, and in good quality data one observes high-ionization lines from Fe XXV at $E = 6.7$ keV and from Fe XXVI at $E = 6.97$ keV, but they are generally much weaker than the iron K_α line. The equivalent width of the iron K_α line is a function of the strength of the reflection component, but also of the iron abundance of the reflecting material, the ionization state of the surface layers of the accretion disk, and of the geometry of the system. The iron lines are not the only emission lines expected in the 0.1–10 keV energy range, but by far the strongest, as can be seen in Figure 5.12.

The physics involved in the line emission process can be modeled assuming an X-ray continuum which illuminates a slab of cold gas (Zdziarski *et al.*, 1994). X-ray photons entering the slab can interact with the reflector in various ways. They can be Compton scattered by free or bound electrons, they can be absorbed, followed by fluorescent line emission, or absorbed followed by Auger de-excitation. The latter effect either destroys the photon, scatters it out of the slab, or reprocesses it into a fluorescent line photon which then escapes the slab (Fabian *et al.*, 2000).

In principle the line profile can be used to examine the movement and gravitational field at the location where the reflection process takes place. Intrinsically, the fluorescence line should be narrow. Thus, line broadening indicates relative movement toward the observer or turbulence in the disk (Pariev and Bromley, 1998). In addition, a nonsymmetrical line broadening can indicate relativistic effects, in which the approaching (blue shifted) line wing is stronger through relativistic beaming than the receding (redshifted) wing. The line profile for a nonrotating Schwarzschild case will look different from the case of a spinning black hole, in which a Kerr metric has to be considered (Laor, 1991; Martocchia *et al.*, 2000). The first example in which a relativistically broadened iron line was observed, was claimed in ASCA data of the Seyfert 1 galaxy MCG-6-30-15 (Tanaka *et al.*, 1995). Subsequent work by Nandra *et al.* (1997b) on other Seyfert 1 galaxies seemed to indicate that the feature is detectable in many cases where the signal-to-noise ratio of the X-ray spectrum is sufficient. An in-depth review on the aspect of the broad iron lines in AGN is given in Fabian *et al.* (2000).

It should be noted however that there is an ongoing debate, regarding how much of the observed broadening can be accounted for by general-relativistic effects, and how much could be due to inaccurate modeling of the underlying X-ray continuum. In the case of MCG-6-30-15 Miller *et al.* (2008, 2009) showed that indeed the spectrum can be modeled in a way which does not require a relativistically broadened iron K_α line, but that the broad line can also be caused by several absorbers partially covering the AGN. The complex absorber makes it difficult to determine the continuum precisely, but when taking it into account, there appears to be no need for a relativistically broadened iron line component. This work was subsequently disputed by Reynolds *et al.* (2009), who claimed that the absorption-dominated model overpredicts the 6.4 keV iron line flux, a claim which was in turn refuted by Yaqoob *et al.* (2010). Support for the importance of absorption in determining the line shape came also from recent Suzaku observations (Miyakawa *et al.*, 2009). They argued that the cross-calibration between the different instruments is not precise enough to allow a firm determination of the black hole spin. Fabian *et al.* (2012) revisited the case of 1H 0707-495 and claimed confirmation of their earlier assertion that light bending effects and X-ray emission close to the central black hole, from within one gravitational radius, were consistent with the data.

Other explanations for the apparently broad iron line have been brought forward. Studying the line profile in cataclysmic variables (CVs) and in neutron stars, Titarchuk *et al.* (2009) explain the asymmetric shape by a wind model. In the case of CVs, an explanation involving general relativistic effects is implausible (the Schwarzschild radius is inside the white dwarf). Because outflows are a common feature of CVs, neutron stars, and also of black holes, the authors conclude that the explanation of the asymmetric line profile is more likely to be found in winds rather than in relativistically blurred reflection.

Yet another way for forming the line shape involves Compton scattering which has its strongest imprint at $\sim$ 30 keV. The shoulder of the Compton hump can reach down into the range of the iron line, affecting the underlying continuum and thus making the line seem to have a red wing (Różańska and Madej, 2008).

Studies by Patrick *et al.* (2011) using Suzaku data again claimed the existence of relativistically broad iron lines in Fairall 9, MCG-6-30-15 and NGC 3783. But fitting the line profiles reveals, although there is a broadened disk line, that the presence of maximally spinning black hole can be ruled out on a 90% confidence level. Other objects which had been claimed to show this feature, like NGC 3516 and NGC 4051,

Table 5.3 AGN in which a spin measurement of the central black hole has been claimed, based on measurements of the iron K_α line, showing the controversy on this topic. Adapted from Brenneman *et al.* (2011).

Name	BH spin j	Reference
MCG-6-30-15	> 0.98	Brenneman and Reynolds (2006),
	$0.5^{+0.2}_{-0.1}$	Patrick *et al.* (2011)
	0.0	Miller *et al.* (2008, 2010) (iron line produced in accretion disk wind and not close to the black hole),
Fairall 9	0.65 ± 0.05	Patrick *et al.* (2011); Schmoll *et al.* (2009)
	0.4 ± 0.2	Anne Lohfink (2011) conference communication
SWIFT J2127.4+5654	0.6 ± 0.2	Miniutti *et al.* (2009)
1H 0707–495	> 0.98	Zoghbi *et al.* (2010); Fabian *et al.* (2012)
	0.0	Miller *et al.* (2008, 2010) (iron line produced in accretion disk wind and not close to the black hole)
Mrk 79	0.7 ± 0.1	Gallo *et al.* (2011)
NGC 3783	> 0.98	Brenneman *et al.* (2011)
	< -0.04	Patrick *et al.* (2011)
Mrk 335	$0.70^{+0.12}_{-0.01}$	Patrick *et al.* (2011)
NGC 7469	0.69 ± 0.09	Patrick *et al.* (2011)

do not seem to require a relativistic line at all. All in all, one has to keep in mind that there have been only eight cases of AGN in which a spinning black hole measurement has been reported. We summarize these cases in Table 5.3. Here the black hole spin

$$j = \frac{cJ}{GM_{\mathrm{BH}}^2} \tag{5.8}$$

is the dimensionless parameter with respect to the maximal spinning Kerr black hole (which has $j = 1$), and J is the angular momentum of the black hole with mass M_{BH}.

Guainazzi *et al.* (2011) argued that the data so far do not rule out the hypothesis of the line being produced close to the event horizon of the black hole, but that also the explanation by partial covering can be true. This ongoing discussion about how much relativistic effects are indeed observable in the X-ray spectra of AGN gathered so far might not be settled before the arrival of higher throughput X-ray telescopes like Astro-H or ATHENA, or with the forthcoming Gravity and Extreme Magnetism Satellite (GEMS) X-ray polarization experiment. Ultimately, it is not possible to determine the black hole spin in AGN in a model-independent manner.

In addition to cold absorbers, warm absorbers are also observed in the X-ray spectra of AGN. The first evidence of absorption due to warm material came from studies of the Einstein X-ray spectrum of the quasar MR 2251-178 (Halpern, 1984), and from AGN spectra taken with ROSAT (Turner *et al.*, 1993). Since then, highly ionized absorbers were observed in about half of the X-ray spectra of type 1 AGN, both Seyfert 1 galaxies and quasars, with column densities up to $N_{\mathrm{H}} \gtrsim 10^{23}\,\mathrm{cm}^{-2}$, and often consisting of several zones of ionized gas.

The column density of the highly ionized material can be indeed large, reaching up to $N_{\mathrm{H}} = 10^{23}$–$10^{24}\,\mathrm{cm}^{-2}$ in the case of Fe XXV ($E = 6.7$ keV) and Fe XXVI ($E = 6.97$ keV). Therefore, although the cold absorber might be moderate, the outflowing and ionized material can be Compton-thick, with significant consequences for the observed continuum. In addition, variations in the outflows with respect to the line of sight can lead to spectral variability. Sim *et al.* (2008, 2010a,b) have performed multidimensional modeling of X-ray spectra of AGN considering disk outflows. When fitting their wind model to the XMM-Newton data of Mrk 766 and PG 1211+143, the study shows that although the Fe K_α line appears to be broad, this can in fact be caused by the warm absorber arising from mass loss of the order of $\dot{M} \sim 0.5\,M_\odot\,\mathrm{yr}^{-1}$. Thus, what might look like a relativistically broadened and red-skewed iron line profile, might be a feature shaped by fluorescence, recombination and scattering around the outflow. This indicates some of the difficulties in verifying the general-relativistic scenario using the iron line shape. Its intrinsic form depends on deconvolving it from the underlying continuum which can be complex. It could for example be heavily affected by ionized absorbers in the line of sight.

Another component observed in AGN is the so-called soft X-ray excess. A soft ($E \lesssim 2$ keV) excess over the power-law component dominant at higher energies has been found in the X-ray spectra of many Seyfert galaxies (Saxton *et al.*, 1993). The origin of the soft excess is still an open issue. In the past the soft excess had often been associated with the high-energy tail of the thermal emission of the disk, but it was recently argued that the temperature of the disk should be nearly constant ($kT \simeq 0.1$–0.2 keV), regardless of the mass and luminosity of the AGN (Gierlinski and Done, 2004). This result implies that some other mechanism is at work, as the temperature of the disk should depend on both the mass of the black hole and the accretion rate. Three competing models have been brought forward in order to explain the soft excess observed in the X-ray domain: an additional Comptonization component (e.g., Dewangan *et al.*, 2007), ionized reflection (e.g., from the disk, see Crummy *et al.*, 2006), or complex and/or ionized absorption (e.g., Gierlinski and Done, 2004; Done, 2007).

For bright AGN grating spectroscopy can be performed providing substantially higher spectral resolution than can be obtained with CCDs or proportional counters, for example $R \sim 10^3$. Figure 5.13 shows an example of a high-resolution soft X-ray spectrum taken by the RGS instrument on-board ESA's XMM-Newton satellite. Absorption and emission lines can be clearly identified, allowing to estimate the temperature of the absorbing material by identification of the ionized lines. In addition, the width of the lines gives information about the velocity of the particles in the absorbing clouds, and their displacement with respect to the laboratory wavelength gives the outflowing (or inflowing) speed.

To summarize, the observed X-ray spectrum of nonbeamed AGN consists of the primary continuum, which is approximated by a cutoff power law, and a reflected component. The spectrum shows signatures of photo-electric absorption, iron fluorescence and Compton scattering. In addition, cold and warm absorption can influence the spectrum, and an additional soft excess is observed in some cases.

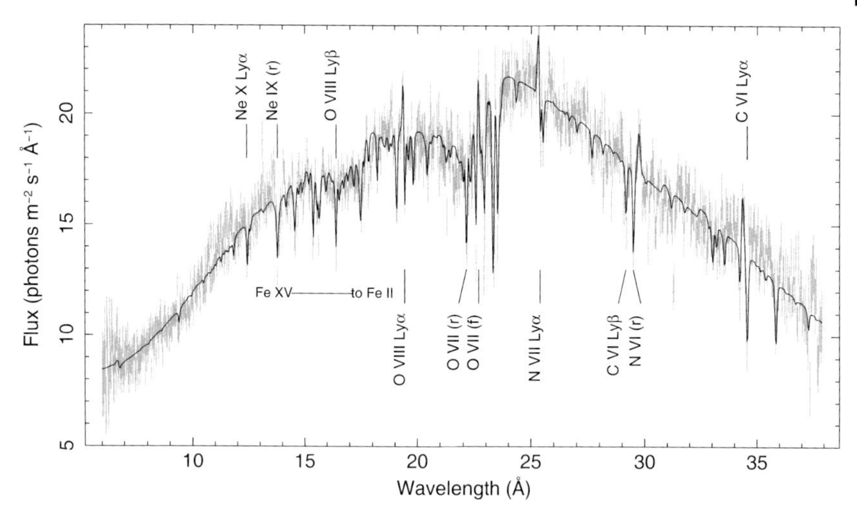

Figure 5.13 Soft X-ray spectrum (0.3–2.1 keV) of the narrow-line Seyfert 1 galaxy Arakelian 564. The data are from XMM-Newton/RGS and the line shows the fit with a power law, blackbody, galactic absorption, three phases of photoionized absorbing gas, and five narrow emission lines due to O VIII Lyα, O VII(f), N VII Lyα, N VI(i) and C VI Lyα (Smith *et al.*, 2008; reproduced with permission of © ESO).

A model spectrum of a type 1 AGN in the X-rays showing the contribution of the various spectral features is shown in Figure 5.14.

For beamed sources, such as blazars, the X-ray spectrum can be represented in most cases by a simple power law, altered significantly only by absorption within our Galaxy. Measurable absorption close to the central engine of blazars has rarely been detected in X-ray spectra. The spectral slope is often close to $\Gamma \simeq 2$, similar as in X-ray spectra of Seyfert galaxies. Thus, classification based solely on X-ray data is often not possible, unless the iron line, reflection hump, and absorption are clearly discernible identifying the object as an isotropic emitter.

5.6
Gamma Rays: the Blazar-Dominated Sky

Above some 120 keV we enter into the gamma-ray domain. In X-rays, grazing incidence mirrors and CCDs can be used in order to detect photons. As we move to higher energies, both techniques fail. At soft gamma rays with $E \ll 10\,\mathrm{MeV}$, solid-state germanium detectors or detectors based on scintillation crystals of for example Na I, Cs I or BGO can be used. Instead of mirrors, mechanical collimation or the coded-mask technique must be applied to spatially locate gamma-ray sources in this energy range.

The coded aperture concept is a two-dimensional generalization of the simple pin-hole camera: a shield with a hole in it is placed at some distance to the detector, and each detected event on the detector plane can be traced back to a position

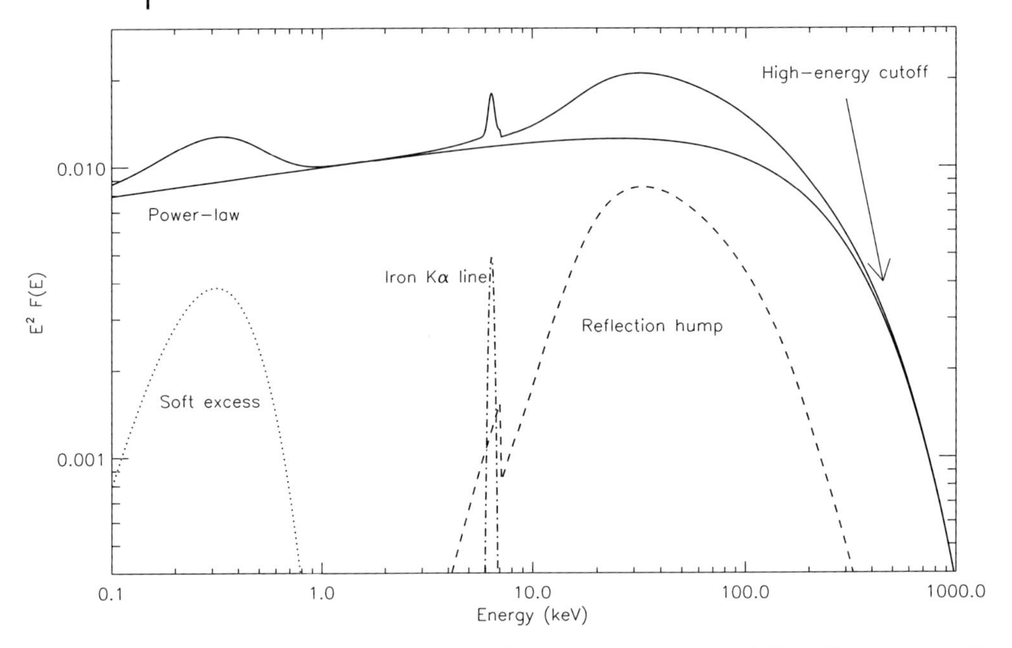

Figure 5.14 Schematic representation of a type 1 AGN spectrum in the X-ray. Figure courtesy of Claudio Ricci.

on the sky. Obviously, the pin-hole mask blocks most of the incoming radiation, and is thus very inefficient. In practice, a coded mask telescope uses a shielding material with about 50% opaque and 50% open elements. Each source on the sky will then cast a shadow of the mask onto the detector plane, and through a mathematical procedure the strength and spatial distribution of the sources in the field of view can be extracted from the sky background. This technique has been used for example on the gamma-ray telescopes SIGMA/GRANAT and INTEGRAL/SPI.

Germanium detectors offer superior spectral resolution and have been applied extensively to other problems, notably nuclear astrophysics. It is difficult for practical reasons to achieve a large detector effective area (relative to the overall instrument mass), and germanium requires cryogenic cooling. As AGN do not show lines in their spectra at energies higher than 7.1 keV and are rather dim sources compared to the bright galactic sources, gamma-ray regime studies in the 200 keV $< E <$ 1000 keV range have yielded only a handful of AGN thus far. Among them are only a few Seyfert galaxies (Cen A, NGC 4151, NGC 4945) and an elusive AGN at $z = 2.4$ (SWIFT J1656.3-3302).

Moving into the MeV range, the coded mask is no longer effective due to the increasing transparency of the mask function, and more fundamentally, to the decreasing cross-section for the photo-electric absorption exploited in the solid-state detectors. However, in the MeV range, Compton scattering becomes the predominant photon–matter interaction and as such an efficient way to detect photons. Compton telescopes usually consist of two detector layers. In the upper layer, the incoming photons lose energy due to Compton scattering processes. If the photon has lost enough energy, it can then be absorbed in the second layer detector. In order to make sure that the scattering processes in the upper detector and the absorption in the lower layer result from the same incoming photon, high timing precision is necessary. This detector concept requires active shielding from events penetrating from the side of the instrument, for example by an anticoincidence system which can send a rejection signal to the processing unit whenever such an event is detected. NASA's Compton Gamma-Ray Observatory (CGRO; 1991–2000) carried such a Compton telescope, COMPTEL, which was sensitive in the 0.75–30 MeV energy band. CGRO/COMPTEL detected 14 blazars and the radio galaxy Cen A. Twelve of these objects were flat spectrum radio quasars and only two BL Lac-type blazars have been detected at low significance, Mrk 421 and S5 0716+714. Seyfert galaxies were not detected, with only an upper limit on a number of sources emerging.

Little progress has been made in the field of MeV astrophysics since the decommissioning of CGRO in June 2000. Although, concepts and laboratory prototypes employing improved technologies have been developed. For example, light silicon detectors would offer the opportunity to build large Compton detectors. The availability of faster electronics allows for more accurate tracking of detector events, thus active shielding, which is again heavy and thus costly on a space mission, might be reduced or excluded in final instrument designs. If successfully implemented, such an instrument would be possible to detect on the order of 0.5 AGN/deg^2, thus several 10^4 objects compared to the 14 sources detected by COMPTEL. Several missions have been proposed to NASA and ESA. The future will show whether the MeV window will be pushed open to reveal the Universe in an energy band rich in extragalactic sources.

At even higher energies, pair production becomes the dominant process through which photons interact with matter. The conversion of an incoming gamma-ray photon in a high-Z material creates an electron/positron pair. The charged particles can then be tracked in a multilayer detector, consisting for example of gas spark chambers or silicon strips. This provides information which can be used to reconstruct the photon track. Downstream from these tracker towers, a calorimeter is installed in order to measure the energy deposition resulting from charged particle ($e^+ e^-$ pair) interactions and thus the energy of the photon can be reconstructed. Also in the case of this kind of tower tracker telescopes, anticoincidence shields are applied in order to reject events entering from the side of the detector.

A succession of such pair-conversion gamma-ray telescopes have been flown since, starting with the SAS-2 satellite in the early 1970s. A major advance came with the CGRO mission of which the Energetic Gamma-Ray Experiment Telescope (EGRET) was one of its four instruments. Its success motivated the Italian satellite AGILE (launched 2007), and NASA's Fermi satellite (launched 2008) with its Large Area Telescope (LAT, Atwood *et al.*, 2009).

Pair-conversion telescopes operate most effectively in the range above 100 MeV. At these photon energies, Seyfert galaxies are not expected to be detectable. Their X-ray emission, which is dominated by inverse Comptonization of UV photons from the accretion disk being up-scattered on relativistic electrons, should fall off rapidly beyond 100 keV. Beamed sources, such as blazars, can radiate much higher photon energies, as the charged particles in the jet providing the energy for the inverse Compton up-scattering can have much higher energies than the thermal electrons close to the accretion disk.

EGRET detected about 300 sources at energies $E > 100\,\mathrm{MeV}$ (Hartman *et al.*, 1999). The precision of the reconstruction of the incident photons trajectory depends on the photon energy (high energies are better). Also lower-intensity sources will ultimately have poorer position determinations than high-intensity ones. Thus, for faint sources the positional uncertainty provided by EGRET was in the degree range. Therefore, many gamma-ray sources found by CGRO could not be identified, as there were simply too many possible counter parts in the error circle. Ultimately, only about half of EGRET catalog sources were unambiguously identified. Most of the identified high-latitude sources turned out to be FSRQ, only a few BL Lacs were found – and the radio galaxy Cen A, the AGN which has been detected in all the energy bands investigated so far. Figure 5.15 shows the gamma-ray spectrum based on CGRO data from three of its four instruments.

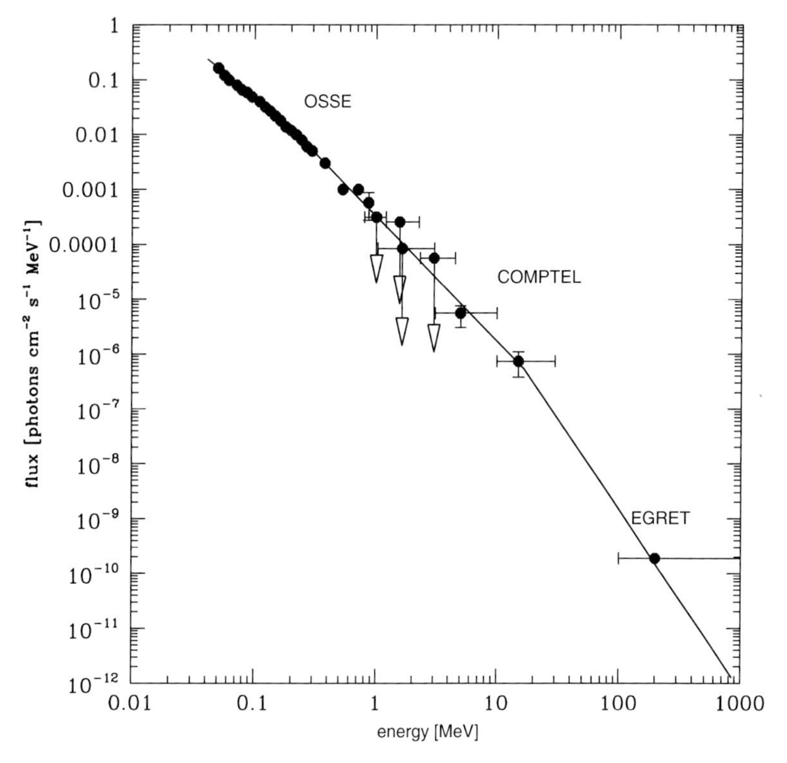

Figure 5.15 Combined spectrum from Centaurus A as measured by CGRO/OSSE (0.05–4 MeV), COMPTEL (0.75–30 MeV), and EGRET (0.1–1 GeV) during one view period during which Cen A was in an intermediate state. The solid line indicates a doubly broken power-law fit to the combined data (Steinle *et al.*, 1998).

With Fermi/LAT, one is now able to look far deeper into the gamma-ray sky. Fermi operates in a scanning mode covering the full sky about every three hours. Thus data are accumulated evenly and a homogeneous and well-defined survey can be achieved for the first time. Its silicon strip tracker technology leads to an enormous improvement over the gas spark chambers of its predecessors. Improvements include more accurate tracks, less (essentially negligible) detector deadtime, and a substantially larger field of view. With the LAT, thousands of blazars have been detected at $E > 100$ MeV. Most of the unidentified EGRET sources, however, could not be identified. These might have been blazars which have been active during the 1990s, and have now fallen back into quiescence. Different than EGRET, Fermi/LAT detects about as many FSRQ as BL Lac objects among the blazars (Ackermann *et al.*, 2011). This is mainly due to the better sensitivity of the LAT compared to EGRET and the fact that the FSRQ are representing the bright tip of the iceberg among the blazar population. In addition, the LAT seems to be able to pick up other extragalactic source types. With NGC 253 and M82 starburst galaxies were detected in gamma-rays. Either these galaxies also host an AGN core with beamed emission, although the spectral energy distribution does not indicate this. Alternatively, the source of gamma-ray emission is cosmic rays, likely to be associated with enhanced star formation. Massive O and B star winds can collide and at the contact edge shock waves might accelerate charged particles to high energies, which then emit gamma-rays through synchrotron radiation. A preponderance of young supernova remnants associated with these massive star populations are also expected cosmic-ray accelerators.

One example of a gamma-ray bright massive star binary in our Galaxy is η Carina (Farnier *et al.*, 2011). Also, the nearby Large and Small Magellanic Clouds (LMC and SMC) and the Andromeda galaxy M31 show gamma-ray emission in the Fermi/LAT data. Recently, Fermi observations have revealed extended gamma-ray emission interpreted as evidence for fresh cosmic ray acceleration in the Cygnus OB2 massive star association. In addition, other star-forming regions in the Milky Way, like W43 and Westerlund 2, are possibly detected in gamma-rays (Lemoine-Goumard *et al.*, 2011).

Apart from Cen A, the Fermi LAT was able to detect additional radio galaxies. While the gamma-ray emission in these sources might be the result of a misaligned jet, the observation of gamma-ray photons from narrow-line Seyfert 1 (see Section 4.1.4) is more difficult to explain. The radio-loud NLS1 seen by the LAT seem to host a relativistic jet in their center, thus also these sources might be hosting a BL Lac core, with the jet pointing in the direction of the line of sight (Abdo *et al.*, 2009b). The NLS1 PMN J0948+0022 even showed a gamma-ray outburst, brightening to $L \sim 10^{48}\ \mathrm{erg\,s^{-1}}$, which is very luminous even when compared to FSRQ (Foschini *et al.*, 2011).

The FSRQ show rather steep spectra in the Fermi/LAT energy band, with photon indices mainly $\Gamma > 2.2$ with a peak around $\bar{\Gamma} \simeq 2.4$ (Abdo *et al.*, 2010). The spectra of the BL Lac objects appear harder, as their peak of the inverse Compton branch usually lies at higher energies. The distribution of the photon indices of BL Lacs is broad though, with the majority of the sources having $1.6 < \Gamma < 2.5$ and an average value of $\bar{\Gamma} \simeq 2.1$. The radio galaxies have on average steep spectra like the FSRQ, thus indicating that their inverse Compton emission also peaks at energies below the energy band observed by the LAT.

Many of the sources are only detected after months of data accumulation. But the bright blazars have been observed to be highly variable in the gamma-ray domain, also on relatively short time scales. AGILE data showed for example that the blazar PKS 1830-211 underwent an outburst lasting for about 4 days during which the flux above 100 MeV was 12 times higher than during quiescence (Donnarumma *et al.*, 2011). Interestingly, the gamma-ray variability was not accompanied by a significant change in X-ray flux.

Radio-quiet AGN do not seem to be significant gamma-ray emitters. Teng *et al.* (2011) searched for counter parts of 491 hard X-ray detected radio-quiet Seyfert galaxies in the first 2 years of data from Fermi/LAT. The sample was taken from the sources detected by Swift/BAT at 14–195 keV with a median redshift of $z \sim 0.031$, thus representing a truly local sample of AGN. Of the 491 sources, only the near-by galaxies NGC 1068 and NGC 4945 were detectable in the gamma rays. However, those two objects also host regions of enhanced star-formation activity, thus the origin of the high-energy emission may not be related to the Seyfert core. A similar result was derived by Ackermann *et al.* (2012), who do not find gamma-ray emission for 120 hard X-ray-selected Seyfert galaxies, with the exception of ESO 323-G077 and NGC 6814, but their detection is not firm yet and might be a coincidence.

5.7 VHE: the Evolving Domain

At the very high-energy (VHE) end of the electromagnetic spectrum, also the tower tracker fails to be a useful instrument. As we have seen in the previous section, the spectra of AGN continue towards higher energies, but they also fall off rapidly. Thus less and less photons arrive per unit area and unit time from AGN and from astrophysical sources in general as we increase the frequency at which we take data. At the same time, it becomes more difficult to stop these high-energy photons. They can simply pass the detector without any interaction.

Thus, in the VHE domain extremely large detectors are necessary if one wants to stand a chance to detect photons from AGN. A solution applied in a number of telescopes is the observation of Cherenkov light by Imaging Atmospheric Cherenkov Telescopes (IACT). This method uses the Earth's atmosphere as a detector. Incoming gamma-ray photons in the GeV to TeV energy range interact with nuclei in the higher atmosphere and produce electron–positron pairs. These particles interact with nuclei through bremsstrahlung, which results in photons which again can cause pair production, and so forth. Thus, the incoming high-energy photon causes a shower of charged particles which travels downward in the atmosphere. The density of particles increases and reaches its maximum at a height of typically 10 km above sea level. Then, as the energy of the photons created in the bremsstrahlung events decreases, pair production decreases. The air shower can be observed from the ground through the emitted light by the charged particles of the shower moving faster than light in air (Cherenkov radiation). Only a small fraction of the air showers observed is caused by gamma-ray photons though. The majority are induced by cosmic ray particles which interact with the atmosphere. Different than photons, the hadronic showers produce mainly pions and can be distinguished by their different appearance in the resulting image of the ground-based detectors. The implementation of the necessary techniques was a difficult undertaking though, and it took nearly two decades before the first source, the Crab Nebula, was significantly detected by a Cherenkov telescope (Weekes *et al.*, 1989).

Crab units

A major problem of X-ray and gamma-ray astronomy is the calibration of the instruments. Precise determination of the sensitivity and response of a detector can often not be achieved, and thus a "standard candle" is used in order to derive flux and spectral calibration and to be able to compare observations from different instruments. The Crab Nebula appears to be a fairly stable source, and its spectrum is described by a simple power law with a photon index of $\Gamma \simeq 2.1$ in the X-rays. 1 Crab is then defined as the intensity of the Crab Nebula in a given band. Thus, for example, 1 Crab corresponds to $f = 2.4 \times 10^{-8}\,\mathrm{erg\,cm^{-2}\,s^{-1}}$ at 2–10 keV, and to $f = 1.5 \times 10^{-8}\,\mathrm{erg\,cm^{-2}\,s^{-1}}$ at 20–100 keV. The situation became more complicated in 2010, when it was discovered that the X-ray and gamma-ray flux of the Crab indeed slightly varies.

It has to be kept in mind that although the original photons are gamma-ray photons, the detected Cherenkov light is in the optical domain. Thus, Cherenkov telescopes operate only at night, and are also severely limited in sensitivity by moon light. Furthermore, they are effective for a fairly limited zenith-angle range. Currently, four major TeV telescopes are operational and collecting data, the Major Atmospheric Gamma-ray Imaging Cherenkov Telescope (MAGIC) and the Very Energetic Radiation Imaging Telescope Array System (VERITAS) in the northern hemisphere, the High-Energy Stereoscopic System (HESS) and the CANGAROO III[3] telescope system in the south. MAGIC consists of two telescopes with 17 m diameter segmented mirrors each, the largest IACT so far, and the system is able to detect a source with a flux of 8 mCrab within 50 h of observing time and provides a spatial resolution of $\sim$ 4 arcmin. One MAGIC observation led to the discovery of an energy-dependent time of arrival of photons from a Mrk 501 flare. During the burst, the higher-energy photons appeared to arrive later than the lower energy ones, with a time delay Δt as a function of energy with $\Delta t = E(0.03 \pm 0.01)\,\mathrm{s\,GeV^{-1}}$, where E is the photon energy in GeV (MAGIC Collaboration *et al.*, 2008). This means that photons with an energy of 1 TeV would arrive on average about half a minute after photons at 100 GeV. Possible explanations include quantum-gravity effects, but the

3) Acronym stands for Collaboration of Australia and Nippon (Japan) for a GAmma Ray Observatory in the Outback.

MAGIC results are still under debate. Alternatively, intrinsic emission from the source could have energy dependencies that we are currently unaware of.

CANGAROO is the oldest and least sensitive IACT array, consisting of four 10 m telescopes since 2004, but only three are still in use, and it has been used mainly for observation of galactic sources. VERITAS is made out of four 12 m telescopes, and so is the HESS array. Located in the southern hemisphere, HESS performed a scan of the galactic plane and center region, detecting the largest number of VHE sources so far. In a second phase, a single telescope with 28 m diameter will be placed in the center of the array, with a maximum sensitivity at $\sim$ 20 GeV. This will make HESS by far the most sensitive IACT and it also provides a significant overlap in energy with the Fermi space telescope.

The first AGN detected in the TeV energy range was the blazar Mrk 421 (Punch *et al.*, 1992) at a redshift of $z = 0.031$. More than one hundred sources are now documented, among them several dozen AGN. In this energy domain above 100 GeV blazars are dominating the extragalactic sky even stronger than in the Fermi/LAT band. Different than at $E \simeq 100$ MeV, the majority of detected blazars are of BL Lac type, as these show the peak of their high-energy emission at higher frequencies than the intrinsically more luminous FSRQ. The only three FSRQ so far are 3C 279, 4C+21.35, and PKS 1510-089. Six nonblazars have been detected: the FR-I radio galaxies Cen A, M87, NGC 1275, IC 310, 3C 66B, and the starburst galaxies NGC 253 and M82. The origin of the VHE emission in the latter object is under discussion, as this source is not only a starburst galaxy, but also hosts a Seyfert 2 core, and from its radio morphology it has been classified as FR-I galaxy. Table 5.4 gives the list of the 46 AGN known to be VHE emitter as of early 2011. The table is based on the TeV catalog website by Wakely and Horan[4] and in the article by Abdo *et al.* (2009a). The last column gives the photon index of a simple power-law model fitted to the VHE spectrum and the associated statistical error. With TeV astronomy being a thriving field and with several Cherenkov telescopes in operation, it can be expected that the number of TeV detected AGN will increase steadily, and the list we show here will be outdated by the time this book is published. We see from Table 5.4 the dominance of high-energy peaked BL Lac objects (HBL) in the TeV domain.

We also notice that nearly all sources detected so far apparently peak at frequencies lower than a few 10^{26} Hz, below the TeV range. This can be deduced from the fact that all but two spectra have a photon index larger than $\Gamma_{\mathrm{VHE}} = 2$. The two objects with $\Gamma_{\mathrm{VHE}} < 2$ were detected in outburst, and the spectral slope of 1ES 0414+009 is consistent with $\Gamma_{\mathrm{VHE}} = 2$ on the 1σ level. Recall that a photon index of 2 corresponds to a horizontal line in a $\log \nu f_\nu$ versus $\log \nu$ diagram. Thus, all of the sources with $\Gamma > 2$ have the maximum of their energy output at frequencies lower than the band in which the spectral slope was measured.

The measured photon spectrum from blazars in the GeV to TeV range might not be the true, intrinsic source spectrum. The reason is that high-energy photons can interact with the background of IR to visible light via pair production process-

4) See website http://tevcat.uchicago.edu/.

Table 5.4 AGN detected in the TeV energy domain. AGN types are indicated as SB = starburst galaxy, LBL = low frequency peaked BL Lac, IBL = intermediate peaked BL Lac, and HBL = high peaked BL Lac. Γ_{VHE} gives the photon index for the VHE spectrum.

Name	RA (J2000.0)	DEC (deg)	Type	Redshift z	Γ_{VHE}
SHBL J001355.9-185406	3.484	−18.902	HBL	0.095	2.0 ± 0.1
NGC 253	11.888	−25.288	SB	0.0008	—
RGB J0152+017	28.165	+1.788	HBL	0.080	3.0 ± 0.4
3C 66A	35.665	+43.035	IBL	0.444 (?)	4.1 ± 0.4
3C 66B	35.798	+42.992	FR-I	0.021	3.1 ± 0.3
1ES 0229+200	38.203	+20.288	HBL	0.140	2.5 ± 0.2
NGC 1275	49.951	+41.512	FR-I	0.0176	—
RBS 0413	49.966	+18.780	HBL	0.19	1.4 ± 0.2
1ES 0347-121	57.347	−11.991	HBL	0.188	3.1 ± 0.2
1ES 0414+009	64.224	+1.083	HBL	0.287	1.8 ± 0.2
PKS 0447-439	72.354	−43.836	HBL	0.2	—
1ES 0502+675	76.984	+67.623	HBL	0.341	—
PKS 0548-322	87.670	−32.272	HBL	0.069	2.8 ± 0.3
RX J0648.7+1516	102.199	+15.274	HBL	0.179	4.4 ± 0.8
RGB J0710+591	107.625	+59.139	HBL	0.125	—
S5 0716+714	110.473	+71.343	LBL	0.31	—
1ES 0806+524	122.455	+52.316	HBL	0.138	3.6 ± 1.0
M82	148.970	+69.679	SB	0.0007	2.5 ± 0.6
1RXS J101015.9-311909	152.566	−31.319	HBL	0.1426	
1ES 1011+496	153.767	+49.434	HBL	0.212	4.0 ± 0.5
1ES 1101-232	165.907	−23.492	HBL	0.186	2.9 ± 0.2
Mrk 421	166.114	+38.209	HBL	0.031	2.31 ± 0.04
Mrk 180	174.110	+70.158	HBL	0.046	3.3 ± 0.7
1ES 1215+303	184.467	+30.117	LBL	0.13	—
1ES 1218+304	185.341	+30.177	HBL	0.182	3.1 ± 0.3
W Comae	185.382	+28.233	IBL	0.102	3.8 ± 0.4
4C+21.35	186.227	+21.380	FSRQ	0.432	3.8 ± 0.3
M87	187.706	+12.391	FR-I	0.004	2.6 ± 0.4
3C 279	194.047	−5.789	FSRQ	0.536	4.1 ± 0.7
1ES 1312-423	198.764	−42.614	HBL	0.108	—
Cen A	201.365	−43.019	FR-I	0.0018	2.7 ± 0.5
PKS 1424+240	216.752	+23.800	IBL	?	3.8 ± 0.5
H 1426+428	217.136	+42.672	HBL	0.129	3.5 ± 0.4
1ES 1440+122	220.701	+12.011	IBL	0.162	—
PKS 1510-089	228.211	−9.100	FSRQ	0.36	—
AP Lib	229.424	−24.372	LBL	0.049	2.1 ± 0.1
PG 1553+113	238.929	+11.190	HBL	?	4.0 ± 0.6
Mrk 501	253.468	+39.760	HBL	0.034	2.22 ± 0.04
HESS J1943+213	295.979	−21.306	HBL	> 0.14	3.1 ± 0.3
1ES 1959+650	299.999	+65.149	HBL	0.048	2.8 ± 0.1
MAGIC J2001+435	300.306	+43.884	HBL	?	—
PKS 2005-489	302.356	−48.832	HBL	0.071	4.0 ± 0.4
PKS 2155-304	329.717	−30.226	HBL	0.117	3.19 ± 0.02
BL Lacertae	330.680	+42.278	LBL	0.069	3.6 ± 0.5
B3 2247+381	342.524	+38.410	HBL	0.1187	—
1ES 2344+514	356.770	+51.705	HBL	0.044	3.0 ± 0.1
H 2356-309	359.783	−30.628	HBL	0.167	3.1 ± 0.2

es and photon–photon collision, as discussed in Section 2.1.5. We derived a large cross-section for frequencies $\nu_1 \simeq 6 \times 10^{40}\,\mathrm{Hz}^2\,\nu_2^{-1}$, which means that the interaction for an IR photon is most probable with a photon at low-TeV energies. The extragalactic background light (EBL) arises from the unresolved radiation from the first cosmic stars and protogalaxies in the early Universe plus a contribution from galaxies throughout the redshift space. The spectrum and strength of the EBL is not well known. It can be determined directly through deep observations of the sky in the optical to IR frequency range. But today it is still not clear whether part of the EBL is truly diffuse emission, or if eventually all of it can be resolved into point sources, and if so, where these sources reside. Figure 5.16 shows the current data on the EBL. These measurements include plenty of upper and lower limits. A very strong constraint comes from the measurements of blazar spectra in the TeV range (e.g., Aharonian *et al.*, 2006). The absorption of the TeV photons has the form

$$f_{\mathrm{obs}} = f_{\mathrm{source}} \times \mathrm{e}^{-\tau(E)} \tag{5.9}$$

with $\tau(E)$ being the optical depth at a given energy E. Based on previous measurements one can assume that $\tau(1\,\mathrm{TeV}) > \tau(0.1\,\mathrm{TeV})$, and thus the effect of the EBL interaction will absorb more high-energy photons, causing the spectrum at VHE to steepen. This effect will be more pronounced in sources with larger redshift z, where the optical depth is larger, as their light will have more opportunities on its way to the observer to interact with the EBL. Thus, if the intrinsic high-energy spectrum of blazars is known, a precise measurement of the EBL's spectral energy distribution, and even of its distribution in space might be achieved. The intrinsic spectral shape might be observed in close by, well-studied blazars, such as Mrk 501 and Mrk 421. As with all blazars, the spectrum is highly variable. Still, the photon index lies within a range of $\Gamma_{\mathrm{VHE}} \simeq 1.5\text{–}2.8$. Thus, if the strength of the EBL is overestimated, the correction for high-redshift blazars will lead to intrinsic spectra much harder than $\Gamma = 1.5$, which seems unrealistic. Together with the assumption that the EBL should have an SED following that of a galaxy, Aharonian *et al.* (2006) determined upper limits for the EBL, as shown in Figure 5.16.

Thus, the observed spectral shape in the VHE range, can give us only some insight about the true intrinsic spectrum, especially at larger redshifts. In addition, recently some claims have been made that our understanding of the high-energy spectrum of blazars might not be correct. For example, Katarzynski *et al.* (2006) demonstrated that a high minimum Lorentz factor $\gamma_{\mathrm{min}} \gg 1$ of the electron energy distribution can explain hard TeV spectra fairly well. Their model even allows for intrinsic photon indices as flat as $\Gamma \sim 2/3$, as demonstrated in the case of the TeV blazar 1ES 0229+200 (Tavecchio *et al.*, 2009). Lefa *et al.* (2011) show that some time-dependent modifications to the simple one-zone SSC model, which we will discuss in Section 5.8.1.1, can lead to inverse Compton branches which have very high-cutoff energies and therefore appear to be flat in the VHE domain. Another way to avoid the strong radiative losses at high energies which would lead to steep spectra is to assume stochastic acceleration models. In this case, the electrons would be accelerated in shocks. As Lefa *et al.* (2011) show, this would lead to steady-state relativistic, Maxwellian-type particle distributions, allowing for very

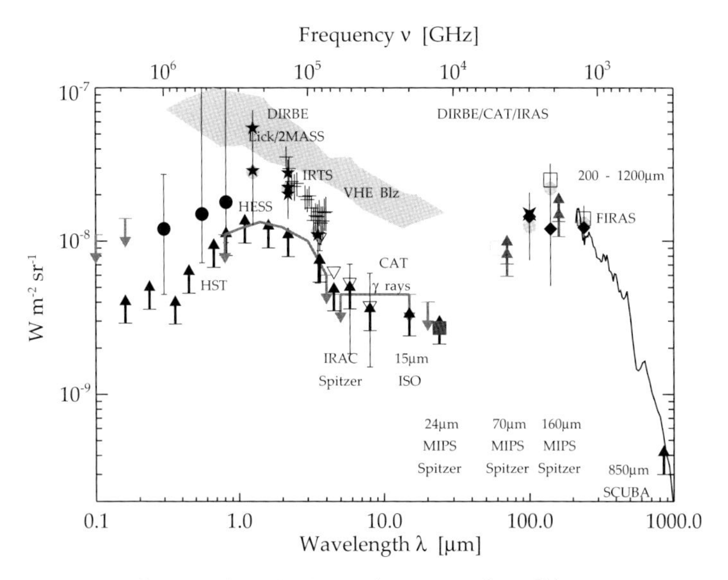

Figure 5.16 The extragalactic background light (EBL) in the optical to infrared range, based on photometric measurements and on indirect techniques (Dole *et al.*, 2006; reproduced with permission © ESO). The HESS TeV data of blazars put strong constraints on the EBL in the range $\lambda = 0.8\text{–}4\ \mu\text{m}$ (indicated by a line). For a color version of this figure, please see the color plates at the beginning of the book.

hard spectra in the TeV range. This shows that a deeper understanding of the intrinsic blazar spectra is necessary if one wants to use the observed VHE spectra in order to constrain the EBL.

VHE astrophysics has made tremendous progress over the last two decades, from the first source detections to a sample of 125 galactic and extragalactic sources today. The TeV domain is expected to see further rapid development in the near future. The Cherenkov Telescope Array (CTA) on the southern hemisphere will consist of some 60 telescopes. Different telescope types ensure detection from some tens of GeV up to several tens of TeV. Four 23 m sized telescopes with a small field of view (4–5°) will cover the low-energy range, the medium range from 0.1–10 TeV will be covered by 23 telescopes of the 12 m-class with a field of view of 7–8°, and at highest energies 32 small 5–6 m telescopes will provide a field of view of $\sim 10°$. The largest telescopes will occupy the center of the array, surrounded by the medium size dishes and with the small type telescopes spread out over a large area. It is expected that the CTA will be able to detect about 1000 VHE sources, giving source numbers comparable to what is provided by Fermi/LAT in the gamma-ray range and by Swift and INTEGRAL in the hardest X-rays (although the source populations are very different). CTA is expected to be operational by the end of this decade. For a recent review see the publication of the CTA Consortium (2011).

5.7.1
The High-Energy End of the Spectrum

At around 50 TeV we reach the high-energy end of the photon spectrum which is still detectable by telescopes available today. It might be possible though to probe even higher-energy emission from AGN. Ultrahigh-energy cosmic rays (UHECR) are observed up to energies of $E \simeq 10\,\text{EeV} = 10^{19}\,\text{eV}$. These charged particles, usually thought to be protons or heavy nuclei, interact with the atmosphere and produce a cascade similar to what has been described for the Cherenkov telescope techniques. The air showers cover a large area though, and at the same time UHE-CR are rare events and thus detectors have to have large sizes. For example, the Auger experiment covers an area of $\simeq 3000\,\text{km}^2$ in the Argentinian Pampa. The origin of the UHECR still being under discussion, some theories claim that they originate in extragalactic sources, possibly in AGN. And, as in other wavelengths, Cen A might turn out to be a source (Fargion, 2008), although the UHECR emission may also be due to the Centaurus cluster of galaxies which lies at a much larger distance but in the same direction. For a recent review on the topic of UHE-CR astrophysics we refer the reader to Kotera and Olinto (2011).

Another way to extend the observed energy range can be the study of neutrinos. Neutrinos have a very small cross-section for interaction with matter, because they only interact by the weak force. Therefore, building a neutrino telescope is a daunting task. Several approaches have been put into place. The IceCube telescope uses the ice of the Antarctic as a detector and applies photo multipliers on kilometer-long lines which are sunk into holes drilled into the ice. Other neutrino telescopes, like ANTARES and Baikal, use water as a detector medium and photo multipliers to detect interactions of neutrinos with matter which produce Cherenkov light. So far, none of the neutrino experiments have been able to identify sources of neutrinos other than the Sun and the supernova SN 1987A. But keeping in mind that other domains of astronomy, such as in the TeV range, took a long time for detection and analysis techniques to mature to the stage where the first sources were seen, there is hope that neutrino and UHECR astronomy might provide exciting insights into AGN physics in the coming years.

5.8
The Whole Picture: the Spectral Energy Distribution

In order to understand the emission processes driving AGN, one has to study each of the different components at work throughout the electromagnetic spectrum. Only by investigating each accessible energy range can one draw a complete and unambiguous empirical picture followed in turn by a viable theory. This is a requirement if our goal is to understand how the central engine works and what physical processes underly its behaviors.

The common method of displaying a spectrum, that is, by plotting the photon flux versus the wavelength, frequency, or energy, is useful in order to study a limited part of the electromagnetic spectrum. For the larger picture, one often reconstructs the Spectral Energy Distribution (SED). We recall that spectra are often of power-law form: the number of photons in an energy band, per unit time and unit area follows then a law $n(E) = n_0 E^{-\Gamma}$, where n_0 is the normalization of the power law and Γ is the photon index. The energy flux is then the photon flux multiplied by the energy of the photons:

$$f(E) = E n(E) = E n_0 E^{-\Gamma} = n_0 E^{-\Gamma+1} = n_0 E^{-\alpha} \tag{5.10}$$

where $\alpha = \Gamma - 1$ is called the energy index. If we want to show the energy output per unit energy in logarithmic scaling, we apply:

$$f(E)\mathrm{d}E = f(E) E \frac{\mathrm{d}E}{E} = E f(E)\mathrm{d}\log E \tag{5.11}$$

Thus, in order to show the energy output of a source, one often plots $E f(E)$ versus the energy E, both in logarithmic scaling. Equivalent are the representations using frequency ν or wavelength λ, because $E f(E) = \nu f(\nu) = \lambda f(\lambda)$. In this representation a spectrum with photon index $\Gamma = 2$ (or energy index $\alpha = 1$) is a horizontal line. Spectra with $\Gamma < 2$ are rising with increasing frequency, and those with $\Gamma > 2$ are falling. The SED is usually shown as $\log \nu f_\nu$ versus the logarithm of the frequency $\log \nu$, or using the luminosity instead of the flux as $\log \nu L_\nu$ vs. $\log \nu$. The latter plot differs only by an offset on the ordinate axis by $\log(4\pi d_\mathrm{L}^2)$ with d_L being the luminosity distance of the object (see Eq. (8.11)). This way, the maximum energy output is given at the frequency where the SED has its peak. The integral of the SED over a certain frequency interval provides a direct measurement of the total energy output of the source in that energy range. In Figure 5.17 the spectral energy distribution of the radio-loud quasar 3C 273 is shown. At first glance it is readily apparent that a single component cannot be responsible for the entire emission of this AGN. In this section we will have an in-depth look at the SED of AGN in order to see which components can account for the various peaks and the overall shape. In general terms, beamed and nonbeamed sources show very different SEDs, thus we will start by discussing SED of blazars in Section 5.8.1 and then that of nonbeamed sources.

The main differences can be listed as follows:

- Blazars and radio-loud quasars: These sources are beamed, and characterized by nonthermal spectra and polarization. Nonthermal means in this respect that their spectral forms cannot be reproduced as a superposition of blackbody or bremsstrahlung spectra, but instead a power-law distribution of charged particles in a magnetic field is a source of synchrotron emission, or inverse Compton up-scattering of seed photons by a population of relativistic particles occurs. As we have seen in Section 2.2, this will lead to a power-law-shaped photon spectrum from the radio up to the optical domain, where the high-energy limit depends on the energy distribution of the charged particles. Then, through inverse Compton processes, either based on up-scattering of synchrotron photons

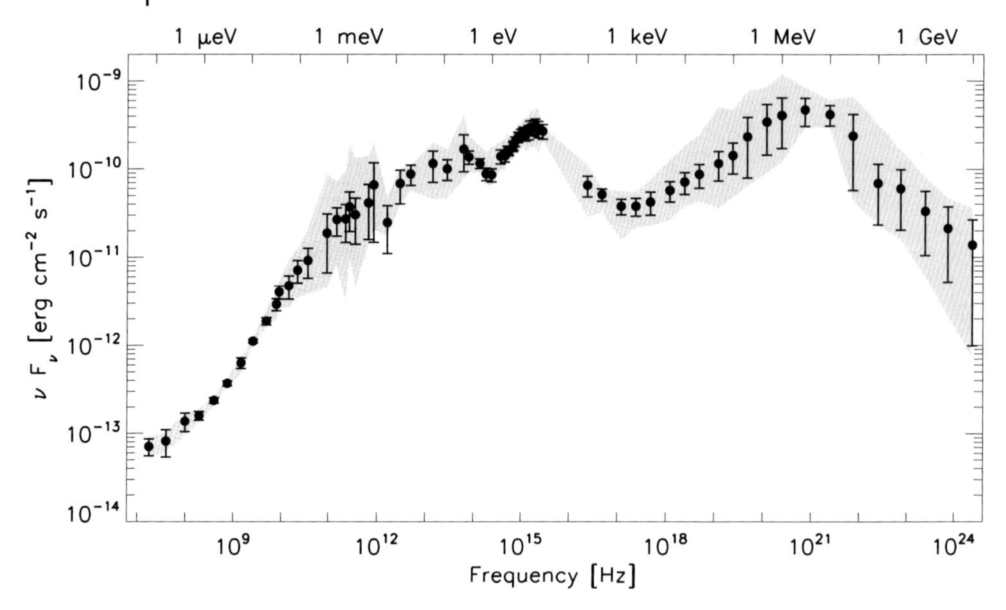

Figure 5.17 Average spectral energy distribution of the quasar 3C 273 from radio to gamma rays (points), spanning from 4 to 44 yr of observations depending on the wavelength. The error bars are calculated as the standard deviation from the mean values and the gray area represents the observed range of variations. See Soldi *et al.* (2008).

(synchrotron self-Compton, SSC), or of photons from the surrounding medium (external Compton, EC), these sources will have a second hump in their spectral energy distribution, typically peaking in the X-ray to gamma-ray domain. Again, the position of the peak depends on the energy distribution of the emitting plasma. Even when considering multiple emission regions or a mix of SSC and EC contributions, the double SED hump shape remains largely intact, with the possible addition of second-order structures superposed. Because these AGN are also embedded in a host galaxy, here one can expect the presence of components of a "standard" AGN, such as accretion flow, and broad- and narrow-line region. These components will also be observable unless outshined completely by the beamed emission.

- Nonbeamed sources, like Seyferts: Here the overall spectral energy distribution is more complex, as the dominance of the nonthermal components is missing and one has to consider a mix of emitting features when studying the SED. In the radio band there might still be some nonthermal emission visible. As we move into the infrared, the dust surrounding the AGN will be a strong source emitting thermal radiation. The SED is dominated, though, by the optical to UV band, which is probably driven by the accretion disk emission. Also in nonbeamed sources inverse Compton processes can create a significant X-ray component. The relativistic electrons are in this case close to the accretion disk which emits the seed photons for the inverse Compton up-scattering. In this case the high-energy limit of the electron energy distribution leads to an exponential cutoff of the inverse Compton component at hard X-rays, which makes Seyfert galaxies too faint to be detectable above several MeV. In summary, in Seyfert galaxies one sees a multipeak structure in the SED, with local maxima possible in the infrared, optical/UV band, and in the X-rays.

5.8.1 SED of Blazars: a Whole Different Story

Although the blazars are a special kind of AGN, we start with the detailed description of their SED. Because, as mentioned above, their SED often appears to be much simpler than that of a source whose spectrum is not dominated by beamed emission. In many cases, just the two-hump "camel's back" SED is detectable, with the two bumps being synchrotron and inverse Compton spectra, respectively. For a phenomenological description of the blazar SED, often two parabolic model fits are used in the logarithmic space, one accounting for the synchrotron branch and one for the inverse Compton component. Although this is obviously not a physical model, it often gives an acceptable representation of the data. Then, based on the parameters of the parabola, the peak frequencies ν_S and ν_{IC} of the synchrotron and inverse Compton branch are determined, and also the luminosity L_{sync} and L_{IC} of both branches. If these are the values one is interested in, a parabolic fit appears to be a valid approach.

Fossati *et al.* (1998) sorted the blazars in different classes according to their bolometric luminosity and derived average SEDs for each of those samples. It turned out that the blazars form a distribution in the ($\log \nu L_\nu$) versus ($\log \nu$) depiction, the so-called blazar sequence, as shown in Figure 5.18. The highest-luminosity objects, like FSRQ, show relatively low peak frequencies for both the synchrotron and the inverse Compton branch, while lower-power objects, for example BL Lac objects have their peaks in the X-ray and VHE domain, respectively.

Ghisellini *et al.* (1998) interpreted the sequence in terms of cooling power of the leptons depending on the power (i.e., density) of the jet. In this scenario, a stronger jet would provide a higher particle density and thus electrons are more efficiently cooled by the dense photon field produced in the synchrotron process. Though the synchrotron branch will therefore cutoff at lower frequencies, the inverse Compton branch will be strong and will dominate the total energy output of the blazar. In this picture, on one end of the blazar population one will find the lower-power, low jet density, high-peaked, synchrotron branch-dominated, less variable BL Lac objects. At the other end, the FSRQ show high-power, high particle density in the jet, lower-frequency peaks for the synchrotron and the inverse Compton branch, an SED dominated by inverse Compton, and stronger variability.

Although this scheme is in general very useful for our understanding of the blazar phenomenon, it was pointed out early on that there are several issues. In a brief review of the subject, Padovani *et al.* (2007) enumerated the main problems as evidenced from observations. First, the anticorrelation between radio power and peak frequency is observable only if one groups the blazars into large classes, as

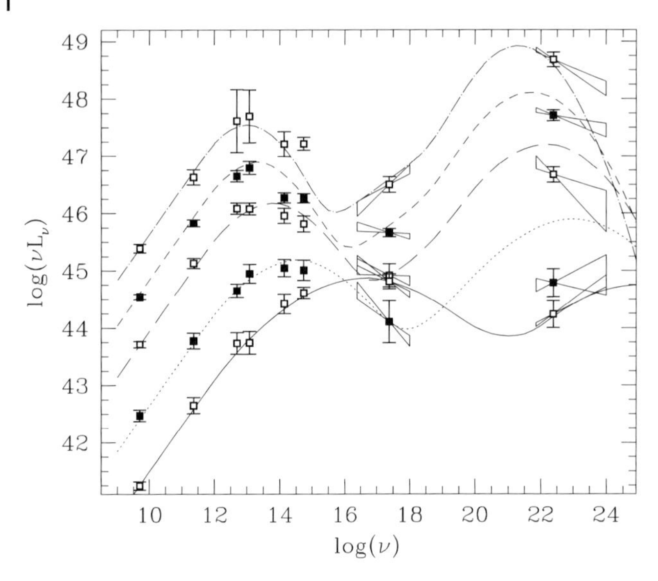

Figure 5.18 The blazar sequence as derived by Fossati *et al.* (1998). The more powerful a blazar appears to be, the lower the peak frequencies of the synchrotron and of the inverse Compton branch. Note that there are some issues with this simple scheme, as described in the text.

done in Figure 5.18. If instead for example, we use only the FSRQ, within this sample we do not see the same trend. In addition, the scatter in all blazar types is large, up to 5 orders of magnitude for a given peak frequency. Finally, there are outliers, like the high-frequency peaked BL Lac object H1517+656, which exhibits a bolometric luminosity typical for an FSRQ (Beckmann *et al.*, 1999).

Apparently, the blazar sequence scenario requires more parameters than particle density and jet power in order to explain the observations. Ghisellini and Tavecchio (2008) pointed out that early samples often included blazars in outburst, rather than the average state of the sources, which would further affect the blazar sequence. Over the years this problem became less severe, as we have established from long term monitoring what the typical properties of blazars are in a statistical context. Recent advances have led to the identification of a substantial number of TeV blazars. The black hole masses of the blazar host galaxies have become increasingly reliable allowing us to better determine the accretion power. This enables us to now consider both the mass of the central engine and its accretion rate when constructing the blazar sequence. Ghisellini and Tavecchio (2008) suggested in considering the blazar sequence, that the mass of the central engine and the accretion rate be accounted for. The jet power should scale with the accretion rate, and the larger the black hole mass, the further down the jet the dissipation will occur. In this extended scenario, the BL Lac objects would preferentially have lower mass black holes and accrete at a lower Eddington ratio, while FSRQ operate closer to the Eddington limit. By varying these parameters, one can indeed explain the range of observed parameters and their large scatter along the blazar sequence. A similar approach is applied by Meyer *et al.* (2011), who propose unification of radio-loud sources by considering the accretion efficiency. In their scheme, the sources with a weak jet exhibit radiatively inefficient accretion, and are distinguished from the population of strong-jet, nondecelerating, low-frequency peaking blazars like FSRQ and FR II radio galaxies which are thought to exhibit radiatively efficient accretion.

In order to clarify the physical parameters behind the observed emission, a number of different models have been proposed to explain the SED of blazars. In the most simplistic approach, the dominant physical processes are confined to one emitting region, which is responsible for both the synchrotron and the inverse Compton emission. Such a one-zone model using synchrotron self-Compton (SSC; see Section 2.2.5) processes is described in the next section.

5.8.1.1 The One-Zone Model

One of the defining aspects of quasars was, at least in the early years of AGN research, the radio emission of these sources. Radio emission in astrophysical sources is caused by synchrotron emission which is produced when charged particles move through a magnetic field. The electron changes the direction because the magnetic field exerts a force perpendicular to the direction of the motion. The energy of the emitted photons is a function of the electron energy, of the magnetic field strength B and of the angle between the electron's path and the magnetic field lines.

The main indicators for synchrotron radiation are its nonthermal spectrum and the presence of polarization. In a thin plasma where collisions among the particles can be ignored, relativistic electrons of any energy distribution will rapidly reach a state in which they are distributed in energy following a power law, due to synchrotron radiation losses. The emitted radiation as discussed in Section 2.2 will then also have the spectrum of a power law, suffering synchrotron self-absorption at low frequencies, as shown in Figure 2.4. The high-energy cutoff will be determined by the critical frequency ν_c of the highest-energy electrons of the synchrotron emitting plasma. In FSRQ and in most BL Lacs the Klein–Nishina formalism has to be applied (see Section 2.1.2). In this case, the highest electron energy is determined by the synchrotron cooling of the electrons.

In the case of blazars we assume that charged particles are traveling down a jet inside plasma blobs at relativistic speeds. Along the way, they interact with a magnetic field, which leads to synchrotron radiation. Ghisellini *et al.* (1996) and Tavecchio *et al.* (1998) have shown how the inverse Compton and synchrotron branches are correlated assuming synchrotron self-Compton processes and Doppler boosting. Instead of a distribution of the electron energies which follows a simple power law model, they apply a distribution which steepens with increasing energy. This can be approximated by a broken power law, with indices $p_1 < 3$ and $p_2 > 3$ below and above a break energy $\gamma_b m_e c^2$. The energy distribution of the electrons then has the form $N(\gamma) = k\gamma^{-p_1}$ if $\gamma < \gamma_b$ and $N(\gamma) = k\gamma_b^{p_2-p_1}\gamma^{-p_2}$ if $\gamma > \gamma_b$.

If we consider that the dominant synchrotron power is emitted close to the break in the electron spectrum ($\nu_S = \nu_b$), where electrons have a Lorentz factor γ_b and emitted photons have a frequency ν_b, then the peak frequency ν_{IC} of the inverse Compton branch is a function of ν_b :

$$\nu_{IC} = \frac{4}{3}\gamma_b^2 \nu_b \tag{5.12}$$

In other words, the ratio between the peak frequencies of the inverse Compton and synchrotron component directly gives the Lorentz factor γ_b of the dominant electrons in the plasma.

The Doppler factor is defined as

$$\delta \equiv \frac{1}{\gamma\left(1 - \frac{v}{c}\cos\theta\right)} \tag{5.13}$$

where γ is the Lorentz factor, and θ is the (small) angle between the plasma moving at speed v along the line of sight. At first glance Eq. (5.13) seems to imply that the Doppler factor δ is inversely proportional to the Lorentz factor γ. But the Lorentz factor itself is also a function of the velocity v

$$\gamma \equiv \frac{1}{\sqrt{1 - \frac{v^2}{c^2}}} \tag{5.14}$$

Therefore, we can write the Doppler factor as

$$\delta = \frac{\sqrt{1 - \frac{v^2}{c^2}}}{1 - \frac{v}{c}\cos\theta} \tag{5.15}$$

Because $0 < v/c < 1$, we also have $\sqrt{1 - v^2/c^2} > (1 - v/c) \geq (1 - v/c\cos\theta)$. Thus, as long as we observe the jet at a small angle with respect to the line of sight, for a larger velocity v, and thus for a larger Lorentz factor γ, one also derives a larger Doppler factor δ.

Considering the Doppler factor δ and with $\nu_b \simeq B \cdot 2.8 \times 10^6\,\mathrm{Hz\,G^{-1}}$ (Eq. (2.43)), the dominant synchrotron frequency then becomes

$$\nu_S = 3.7 \times 10^6 \gamma_b^2 B \frac{\delta}{1 + z} \tag{5.16}$$

with z being the redshift of the source and B the magnetic field. Because the Lorentz factor γ_b can be expressed as a function of the peak frequencies, we get:

$$B\delta = (1 + z)\frac{\nu_b^2}{2.8 \times 10^6 \nu_{IC}} \tag{5.17}$$

Thus, assuming that synchrotron self-Compton processes dominate, by measuring the SED peak frequencies we can determine the Lorentz factor γ and the product of Doppler factor and magnetic field $B\delta$ directly.

In addition we know from Section 2.2.5 that for synchrotron self-Compton (SSC) processes the ratio between synchrotron and inverse Compton luminosity is

$$\frac{L_{\rm IC}}{L_{\rm sync}} \simeq \frac{U_{\rm rad}}{U_B} = \frac{L_{\rm sync}}{4\pi R^2 c U_B \delta^4} \tag{5.18}$$

The δ^4 dependency arises from the Doppler boosting, that is, from the difference in speed between the observer's frame and that of the relativistic medium. Because the ratio of flux and ν^3 is Lorentz invariant (Rybicki and Lightman, 1986), we can write

$$\frac{f_\nu}{\nu^3} = \frac{f'_\nu}{\nu'^3} \tag{5.19}$$

For the observer, the frequency is shifted by the Doppler factor $\nu = \delta\nu'$ and we can substitute ν in Eq. (5.19)

$$\frac{f_\nu}{(\delta\nu')^3} = \frac{f'_\nu}{\nu'^3} \tag{5.20}$$

and thus we observe the flux $f_\nu = \delta^3 f'_\nu$. If we now integrate in frequency, we derive the δ^4 dependency

$$f = \int_0^\infty f_\nu \mathrm{d}\nu = \int_0^\infty \delta^3 f'_\nu \delta \mathrm{d}\nu' = \delta^4 f' \tag{5.21}$$

Because we treat the case as pure SSC, all the photons in the radiation field have their origin in the synchrotron process, thus $U_{\rm rad} = U_{\rm sync}$. In addition Ghisellini *et al.* (1996) assume a spheric "blob" as emitting region. The radius of this sphere can be estimated from the measured variability time scale: $R = ct_{\rm var}$. Because the emitting region is Doppler boosted towards us, we have to modify this to

$$R = ct_{\rm var}\frac{\delta}{1+z} \tag{5.22}$$

with δ being the Doppler factor. Thus, we can replace R in Eq. (5.18) and derive

$$B\delta^3 \simeq (1+z)\sqrt{\frac{2L^2_{\rm sync}}{L_{\rm IC}c^3 t^2_{\rm var}}} \tag{5.23}$$

Now we can express δ and B (measured in gauss) as functions of directly measurable quantities

$$\delta = 1.67 \times 10^4 \sqrt{\frac{\nu_{\rm IC}}{t_{\rm d}\nu^2_{\rm S}}}\left(\frac{L^2_{\rm sync,45}}{L_{\rm IC,45}}\right)^{1/4} \tag{5.24}$$

and

$$B = 2.14 \times 10^{-11}(1+z)\frac{\nu^3_{\rm S}\sqrt{t_{\rm d}}}{\nu^{3/2}_{\rm IC}}\left(\frac{L_{\rm IC,45}}{L^2_{\rm sync,45}}\right)^{1/4} \tag{5.25}$$

Here t_d is the variability time scale measured in days and the luminosities $L_{sync,45}$ and $L_{IC,45}$ are measured in units of 10^{45} erg s^{-1}. Note that these simple relations only apply if we consider one homogeneous spherical plasma to be the source of the synchrotron and inverse Compton emission. This model is also referred to as the *one-zone model*. Because of its simplicity, it can be used to derive physical parameters even from a poorly sampled SED. For example, following Eq. (5.12), if we can determine the approximate location of the peak of the inverse Compton branch ν_{IC} and of the synchrotron branch ν_S, we get an idea about the Lorentz factor of the electrons in the jet

$$\gamma_b = \sqrt{\frac{3\nu_{IC}}{4\nu_S}} \tag{5.26}$$

An example for an SSC fit to data of a blazar is shown in Figure 5.19. In order to make the SSC fit represent the radio data, it is necessary to constrain the Lorentz factor γ of the electrons in the jet to values with $\gamma > \gamma_{min}$. Thus it is assumed that there are no electrons with less energy in the jet. The radius R of the emitting region has been derived based on the observed variability time scale $t_{var} \simeq 1$ day following Eq. (5.22). With the blazar's redshift being $z = 0.03$ and a Doppler factor of $\delta \simeq 20$, the blob size is about $R \simeq 5 \times 10^{16}$ cm. Abdo *et al.* (2011b) use a double broken power law for the electron energy distribution, thus their model includes three power-law indices ($p_1 < p_2$, $p_2 < 3$, and $p_3 > 3$) and two break points, described by the electron Lorentz factors γ_{b1} and γ_{b2}. Figure 5.19 shows three different models with different power-law slopes, minimum Lorentz factors γ_{min}, and other physical parameters differing. What this shows is that rather different sets of parameters can give quite similar representations of the SED. This is demonstrated in Table 5.5, which gives the model parameters for two of the curves

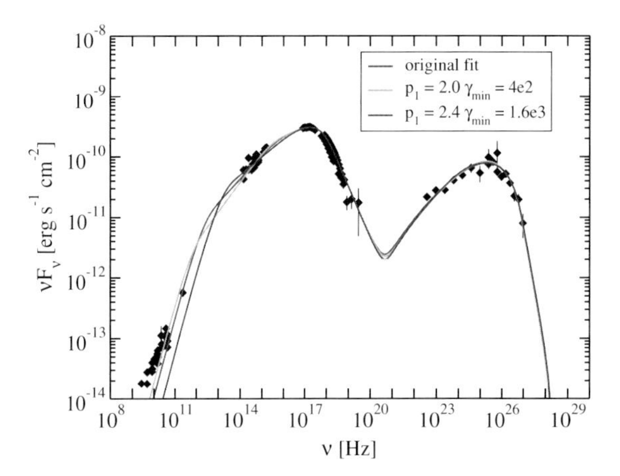

Figure 5.19 SSC model fit of the spectral energy distribution of Mrk 421 based on simultaneous data during a quiescent state of the source. Graphic from Abdo *et al.* (2011b), reproduced by permission of the AAS. For a color version of this figure, please see the color plates at the beginning of the book.

Table 5.5 Parameter values for a one-zone SSC model fit to the Mrk 421 SED (Abdo *et al.*, 2011b).

Parameter	Symbol	Model 1	Model 2
Variability time scale	t_{var}	1 day	1 h
Doppler factor	δ	21	50
Magnetic field (G)	B	0.038	0.082
Comoving blob radius (cm)	R	5.2×10^{16}	5.3×10^{15}
Low-energy electron spectral index	p_1	2.2	2.2
Medium-energy electron spectral index	p_2	2.7	2.7
High-energy electron spectral index	p_3	4.7	4.7
Minimum electron Lorentz factor	γ_{min}	800	400
Break1 electron Lorentz factor	γ_{b1}	50 000	22 000
Break2 electron Lorentz factor	γ_{b1}	390 000	170 000
Maximum electron Lorentz factor	γ_{max}	10^8	10^8
Jet power in B field ($erg\,s^{-1}$)	$P_{j,B}$	1.3×10^{43}	3.6×10^{42}
Jet power in electrons ($erg\,s^{-1}$)	P_{j,e^-}	1.3×10^{44}	1.0×10^{42}
Jet power in photons ($erg\,s^{-1}$)	$P_{j,ph}$	6.3×10^{42}	1.1×10^{42}

shown in Figure 5.19. Note here that the Lorentz factors are those of the electrons within the plasma traveling down the jet, while the Doppler factor δ is related to the bulk Lorentz factor of the plasma blob relative to the observer.

In the literature one will often find different models fitted to a given source, giving apparently contradicting results. It should be emphasized that none of these models represent a unique solution; there are a multitude (if not an infinite) number of functional forms that can be fitted to a given dataset. The task of the data analyst is to make his or her most educated guess as to which model is the best approximation to the physical reality of blazar emission and what parameters of that model can be constrained or fixed on the basis of sound physical arguments.

5.8.1.2 External Compton Scattering

It is not always possible to represent the emission of blazars by a pure synchrotron self-Compton model. In fact, the more data we gather on blazars, the more complex their spectra appear to be. In this context one always has to keep in mind that blazars are highly variable objects. Thus, strictly speaking, only truly simultaneous observations of the entire wavelength range will allow to constrain the emission model correctly. If we drop the SSC constraint that the up-scattering charged particles and the up-scattered photons have to originate in the same region (e.g., in the jet, where the photons are produced by synchrotron radiation by the very same charged particles), one gains more free parameters and thus more possibilities to model the spectral energy distributions of blazars. The inverse Compton process involving an additional source of seed photons is called external Compton scattering (EC). One can think of many sources for seed photons in the central region of an AGN. One natural source can be the accretion disk, which is also supposed to

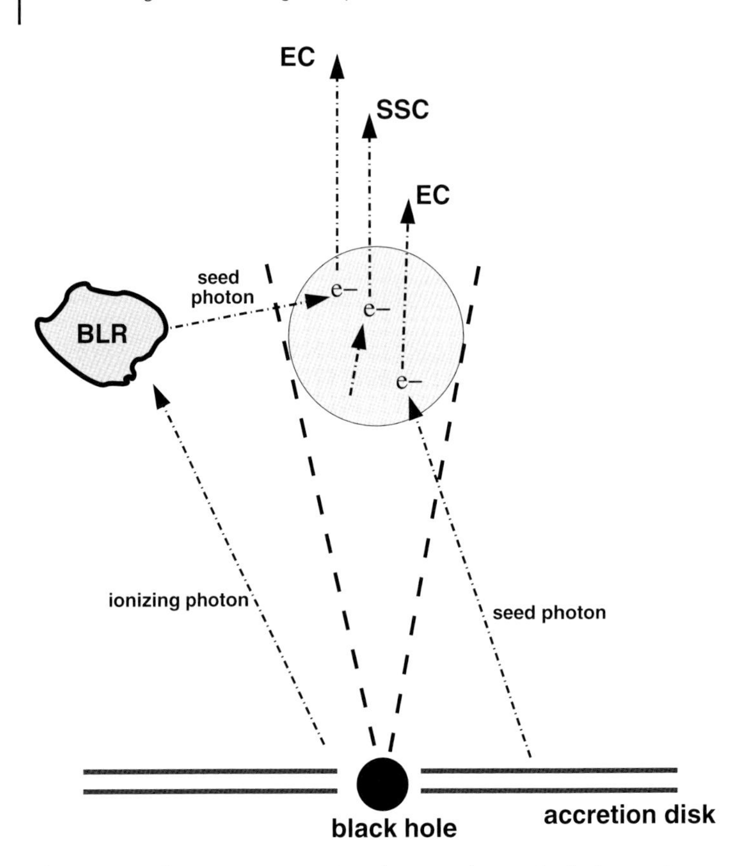

Figure 5.20 Schematic representation of the inverse Compton processes in the jet of a blazar. Here we assume a spherical emission region with radius R which is moving down the jet with a bulk Lorentz factor Γ. The electrons in the emission blob have Lorentz factors γ in the range $\gamma_{\text{min}} < \gamma < \gamma_{\text{max}}$. Note that usually one assumes $\gamma \gg \Gamma$. Seed photons for the inverse Compton process arise from synchrotron processes inside the jet (synchrotron self-Compton model, SSC), but can also be provided externally (external Compton, EC). In the latter case the photons might originate from the accretion disk, from the broad-line region, or another source in the vicinity of the AGN.

provide the seed photons in Compton processes in nonbeamed sources, as we will see in Section 5.8.2.4. Secondly, the broad-line region is a strong photon emitter close to the central engine and was used for example in the EC model of Sikora *et al.* (1994). A sketch of the possible inverse Compton processes in a single blob in the jet of an AGN is shown in Figure 5.20. Other possible sources are the dusty torus or the surrounding regions within the host galaxy. As all of those possible seed photon reservoirs have different characteristic black body temperatures, they will give rise to different inverse Compton components assuming the same up-scattering plasma of relativistic charged particles.

The principle model for external Compton scattering processes in blazars, and in particular for the case with the seed photons arising from the accretion disk, has been outlined by Dermer and Schlickeiser (1993). As in the SSC case, the electrons in the jet lose energy through synchrotron emission and synchrotron self-Compton processes. But, depending on the density of the additional photon field, inverse Compton losses of the electrons on these external photons can be dominating. This can push the high-energy branch of the SED up to higher flux values, giving rise to a strong dominance of the inverse Compton branch over the synchrotron branch in blazars. With the seed photon source no longer colocated with the charged particles, the overall geometry becomes crucial when computing the inverse Compton component: how far is the photon source from the jet, do we model an isotropic source or do the photons hit the charged particles from a certain direction, and so on, become important aspects. The model can be considered slightly less complicated than the SSC one, if we consider a constant photon seed source, as shown for example in Ghisellini *et al.* (1996).

As we have seen in Section 2.2, the peak of the synchrotron branch of a source at redshift z appears at

$$\nu_{\mathrm{S}} = \frac{4}{3}\nu_{\mathrm{g}}\gamma_{\mathrm{b}}^2\frac{\delta}{1+z} \tag{5.27}$$

with $\nu_{\mathrm{g}} = eB/(2\pi m_{\mathrm{e}}c) = B \cdot 2.8 \times 10^6\,\mathrm{Hz\,G^{-1}}$ (Eq. (2.43)) being the gyro- or Larmor frequency. The inverse Compton component will peak at ν_{IC} depending on the peak of the external photon field at frequency ν_{ext} with

$$\nu_{\mathrm{IC}} = \frac{4}{3}\nu_{\mathrm{ext}}\gamma_{\mathrm{b}}^2\Gamma\frac{\Delta\delta}{1+z} \tag{5.28}$$

and here Γ is the bulk Lorentz factor of the jet. Note also that now the difference in Doppler factor $\Delta\delta$ between the photon field and the charged particles is determining the Doppler shift. Recalling Eq. (5.18) for the SSC case, we can write for the external Compton scattering

$$\frac{L_{\mathrm{IC}}}{L_{\mathrm{sync}}} = \frac{U'_{\mathrm{rad}}}{U_B} \tag{5.29}$$

Note that the energy density of the radiation field U_{rad} is primed here while U_B is not. In the SSC case the photon field and the electrons were both moving with the jet, but in the case of external Compton scattering, there is likely a relative speed between the photon field and the relativistic electrons. Ghisellini and Madau (1996) have shown that

$$U'_{\mathrm{rad}} \simeq \frac{17}{12}\Gamma^2 U_{\mathrm{rad}} \tag{5.30}$$

Combining these formulas, we get as the ratio of the inverse Compton and synchrotron luminosities the expression

$$\frac{L_{\mathrm{IC}}}{L_{\mathrm{sync}}} \simeq \frac{17e^2}{6\pi m_{\mathrm{e}}^2c^2}\frac{U_{\mathrm{rad}}}{\nu_{\mathrm{ext}}^2}\frac{\nu_{\mathrm{IC}}^2}{\nu_{\mathrm{S}}^2} \tag{5.31}$$

For a constant seed photon field this gives a simple scaling between the ratios of luminosity and peak frequencies of the two branches

$$\frac{L_{\mathrm{IC}}}{L_{\mathrm{sync}}} \propto \frac{\nu_{\mathrm{IC}}^2}{\nu_{\mathrm{S}}^2} \tag{5.32}$$

Chen and Bai (2011) observe such a proportionality in a sample of blazars detected by Fermi and conclude that they are therefore dominated by external Compton rather than SSC processes. It should be noted however that other authors come to the opposite conclusion. For example, Gao *et al.* (2011) use Fermi data of blazars and do not seem to find the correlation described in Eq. (5.32). Gupta *et al.* (2012) also study Fermi/LAT detected blazars and argue that the scattering of external photons is energetically unimportant compared to the synchrotron self-Compton process. Other authors use the emission of several one-zone blobs with SSC emission in order to model the observed spectral energy distributions.

Other scenarios have also been considered. Up to now we have discussed inverse Compton scattering based on relativistic electrons, or more generally, on leptonic jets. Alternatively, the jet could also be dominated by hadrons. Hadronic models, where the γ-rays are produced by relativistic protons in the jet, have been considered. For a review, see for example Mücke *et al.* (2003). Here, the high-energy emission can either be produced by proton synchrotron radiation or through secondary emission from photo-pion and pair-production processes.

The difficulty in all these models is the large number of parameters. This, accompanied by often sparse coverage of the electromagnetic spectrum, leads to ambiguity and will most likely keep the discussion about what processes dominate the blazar SED alive for some time. Hybridizations of these different scenarios are of course also possible. It could be the case for example that either certain source classes preferentially exhibit SSC processes, or that the quiescent states of blazars require a different modeling than the outburst phases. Even in the case where EC is predominant, there should always be some contribution by SSC processes as its unavoidable that at least some photons produced in the jet will be up-scattered by the parent synchrotron electron population.

5.8.2 The Spectral Energy Distribution of Nonbeamed Sources

The SED of a Seyfert galaxy appears more complex than that of a bright blazar because there is no jet outshining all the other components. Thus, in nonbeamed sources we observe the resultant spectrum of an accretion disk, modified Compton emission upscattered by an ambient population of relativistic electrons, the broad- and narrow-emission line regions, dusty absorbing material which is potentially heated by the AGN itself, and perhaps star formation in the host galaxy. Often it is impossible to disentangle these various SED contributions and one has to rely on simulations and models. One of the most well-constrained Seyfert SEDs currently available is that of Mrk 509, a Seyfert 1 galaxy at redshift $z = 0.034$. Kaastra *et al.* (2011) obtained simultaneous measurements of the source flux from the radio up

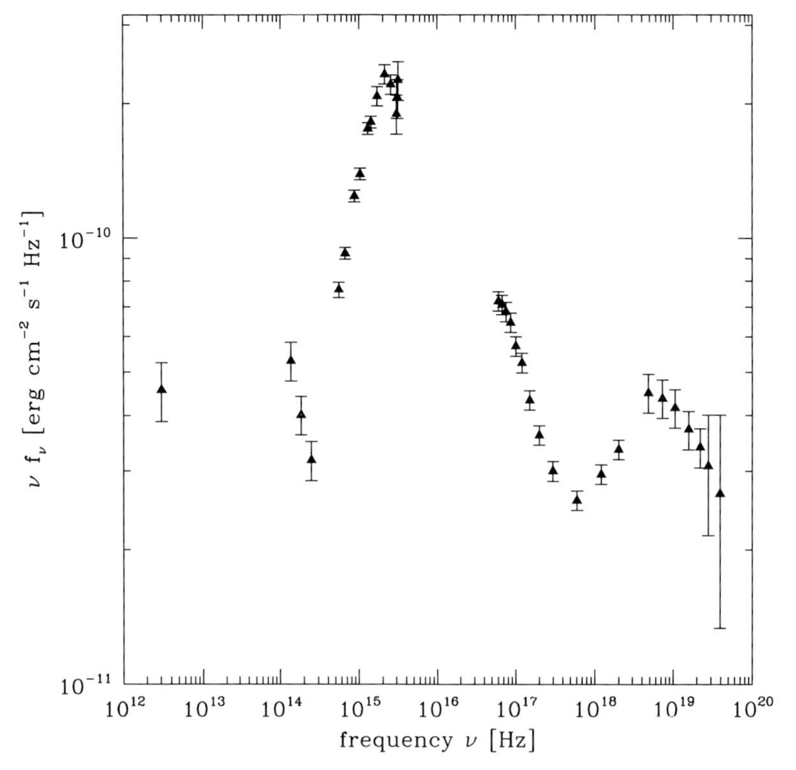

Figure 5.21 The spectral energy distribution of the Seyfert 1 galaxy Mrk 509, based on simultaneous data from the infrared up to the hard X-ray domain. Fluxes have been corrected for galactic and intrinsic absorption. The radio measurement ($\nu = 3 \times 10^{10}$ Hz) gives a flux level of $\nu f_\nu = 4.6 \times 10^{-18}$ erg cm^{-2} and is therefore not shown here. Data taken from Kaastra *et al.* (2011).

to the hard X-ray domain, involving a large number of ground- and space-based observatories. The resulting SED is shown in Figure 5.21. In the following we will go through the radio, infrared, optical, and X-ray to gamma-ray domain once more in order to see what the main contributors to the Seyfert spectrum are in the different bands.

5.8.2.1 **The Synchrotron Branch**

As we have seen earlier, the main indicators of synchrotron emission are a nonthermal spectrum and polarization. In most Seyfert galaxies the nonthermal component will be weak or absent, as are the jets in radio-quiet sources. While radio-loud AGN show flat radio spectra, the radio-quiet AGN have steep spectra with $\alpha \sim 0.7$ (for $f \propto \nu^{-\alpha}$) in this energy range and exhibit a high brightness temperature of $T_{\rm B} \geq 10^5$ K (Kukula *et al.*, 1998). In most Seyferts the radio emission seems to arise from the core region of the AGN, within less than a kiloparsec of the central engine. The radio emission might be associated with ongoing star formation near the core, but the high brightness temperature seems to be inconsistent with that: starbursts usually show temperatures of $T_B < 10^5$ K.

Synchrotron emission could alternatively be produced by the electron plasma of a putative hot corona surrounding the accretion disk, if we assume that a strong magnetic field is present (Ikhsanov, 1989; Di Matteo *et al.*, 1997). One also has to bear in mind that radio-quiet does not necessarily mean radio-silent: Seyfert galaxies often show nonthermal radio emission at some level, and their detection in the radio band is a question of sensitivity. In some sources it thus may be appropriate to treat the emission as that of a faint blazar with a brighter accretion disk and dust system overlaid. For the faint beamed component, the same considerations apply as for the SED of the blazar class which we discussed in Section 5.8.1.

5.8.2.2 **Dust in the SED**

The infrared part of the spectral energy distribution is difficult to disentangle in the case of nonbeamed AGN. Several components can contribute to it, and it depends on the single case which ones are dominant. First of all, star formation in the host galaxy can contribute significantly to the measurement one performs. This cannot easily be distinguished from the continuum emission from the AGN especially as AGN activity and starbursts in the host might go hand in hand as we will see in Section 7.2. In addition, dust in the vicinity of the AGN can absorb higher-energy photons and re-emit them in the infrared regime. From X-ray observations one can determine the column density of the absorbing matter, which for type 2 AGN is on the order of $N_H \sim 10^{22}–10^{25}\,\mathrm{cm}^{-2}$. But apart from the fact that the absorbed type 2 AGN have on average strong infrared emission, there is not a one-to-one correlation between the absorption and luminosity we observe in X-rays and the IR luminosity. In some cases one can clearly see thermal humps in the IR range with the dominant one at $\lambda \simeq 19\,\mu\mathrm{m}$. Assuming an ideal black body, one can apply Wien's displacement law, derived by Wilhelm Wien in 1893, in order to determine the temperature of the emitting dust

$$\lambda_{max} = \frac{b}{T} = \frac{2.898 \times 10^{-3}\,\mathrm{m\,K}}{T} \qquad (5.33)$$

with b being Wien's displacement constant and wavelength λ_{max} and temperature being given in SI units. Thus, the corresponding dust temperature of the infrared hump is of the order $T \sim 150$ K.

In Figure 5.22 the normalized SEDs of quasars are shown for strong and weak far-infrared emitters based on spectra taken by the Spitzer infrared space telescope. In this plot, Netzer *et al.* (2007) aim to disentangle the contribution from the starburst in the host galaxy and the AGN emission itself. The intrinsic quasar SED appears to be fairly similar in the case of bright and faint AGN. Both show three distinct near- and mid-infrared humps and consistent spectral slope at the low-energy end ($\lambda > 20\,\mu\mathrm{m}$). It is also apparent that most of the emission at wavelengths $\lambda > 30\,\mu\mathrm{m}$ is not due to the AGN itself, but rather to starburst processes in the surrounding. Considering that the last maximum is at $\lambda \simeq 19\,\mu\mathrm{m}$ corresponding

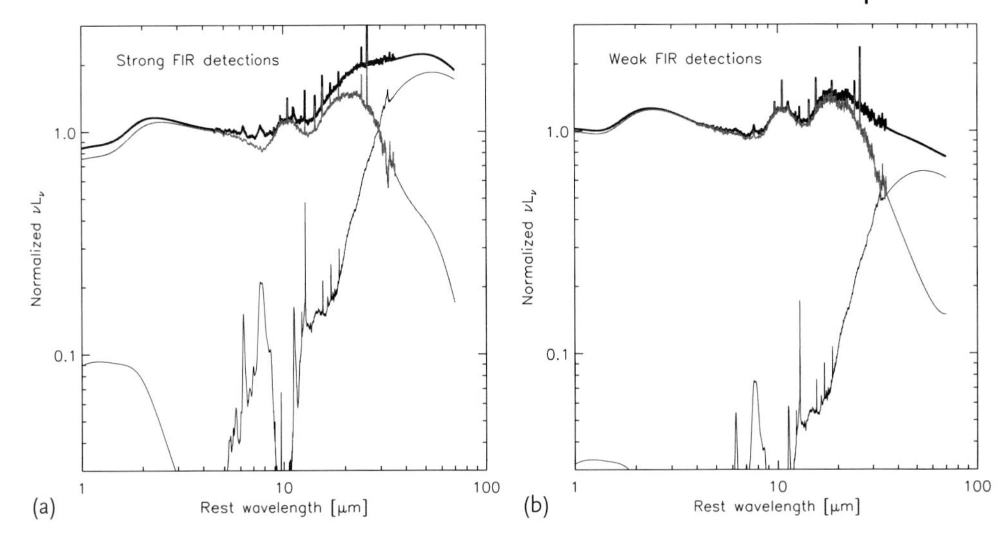

Figure 5.22 Normalized mean SEDs for strong far-infrared (FIR) emitting quasars (a, top curve) and weak FIR quasars (b, top curve). The adjacent red SED curves show "intrinsic" AGN SEDs obtained by the subtraction of the scaled mean starburst (ULIRG) spectrum (shown as the lowest curve in black) from the mean SEDs. From Netzer *et al.* (2007), reproduced by permission of the AAS. For a color version of this figure, please see the color plates at the beginning of the book.

to $T \sim 150$ K, the dust heated by the AGN has to have temperatures higher than that.

Building average SEDs, one can study the difference between a type 1 and a type 2 AGN in the IR range. Using Spitzer, optical, and XMM-Newton X-ray data of a sample of 132 AGN Polletta *et al.* (2007) showed that the average infrared SED of type 1 sources rises from 10 μm to the optical range, while the SED of type 2 sources and star-forming galaxies declines from ~ 1 μm to the optical, consistent with what we show in Figure 5.22. Although the average SEDs of type 2 and starburst galaxies appear to be quite similar, the mid-infrared band in AGN can be modeled by hot dust which has been heated by the central engine, while in starburst galaxies one detects the host galaxy.

5.8.2.3 **The Disk Component**

In the UV and optical band, radio-quiet AGN and also some radio-loud ones, show a peak in emission, the so-called big blue bump. The peak appears around $\lambda \simeq 1100$ Å (Shang *et al.*, 2005), as can be seen for example in the SED of 3C 273 in Figure 5.17. Because of its distinct shape, it was early on suggested that the bump might be caused by thermal emission. In order to peak at $\lambda \simeq 1100$ Å ($E \simeq 11$ eV), the emitting plasma has to have a temperature around $T \simeq 3 \times 10^4$ K. The big blue bump does not appear to be originating from a single temperature plasma though. The broad feature can instead be represented by a multitemperature plasma. As a production site, Shields (1978) suggested an optically thick accretion disk which is feeding the massive black hole in the center of the AGN. The disk would get hotter closer to the inner most stable orbit, giving rise to the temperature range observed in the big blue bump. As mentioned in Section 3.1.4, the temperature should decrease with increasing radius r as $T(r) \propto (r/R_S)^{-3/4}$, where R_S is the Schwarzschild radius (Eq. (3.1)). This temperature increase of matter toward the central engine has indeed been observed, although in the X-ray band: Wong *et al.* (2011) detect a hotter inner region in the temperature profile in Chandra data of NGC 3115.

Using the thin-disk model to explain the observed big blue bump faces several problems. Shang *et al.* (2005) pointed out that thin-disk models do not match the observed spectral shape in the far UV band, and that the measured mass and luminosity cannot fit into a thin-disk model which is intrinsically limited to an Eddington ratio of $L/L_{\mathrm{Edd}} < 0.3$. In addition, disk models would predict a break at Lyα ($\lambda \simeq 1216$ Å), because the opacity changes at the Lyman edge and therefore the depth down into which we can look into the emitting region. The fact that this break is only observed in some cases (e.g., in 3C 273), can be explained by Comptonization and reddening effects, which affect the observed UV spectrum and can smear out the Lyα edge. From a thin disk one would also expect to see a significant level of polarized light, but polarization in the optical and UV light of Seyfert galaxies is low, less than 1%. In the production of the big blue bump advection flows might also play a role here or magnetic torques across the innermost stable circular orbit (Blaes *et al.*, 2001). Both would enhance the radiative efficiency of the accretion process, thus the modeled inflow of matter could be lower. In this model the emission would arise from closer to the central engine which would push the peak of the emission to the hotter, that is bluer end of the spectrum. This could solve some of the problems of the thin-disk model with respect to the big blue bump.

Alternatives to the accretion disk as a source have been brought forward in order to explain the existence of the big blue bump. Lawrence (2012) gives a concise list of problems the accretion disk scenario has and proposes instead that a population of cold ($T \sim 10^4$ K), thick (density $n \sim 10^{12}\,\mathrm{cm}^{-3}$) clouds at a distance of 30 R_S is responsible for the observed hump in the SED. In this model, the clouds reprocess the continuum emission from the central engine. Most of the continuum emission would simply be reflected by such clouds, but the extreme UV (EUV) radiation would be absorbed and re-emitted in emission lines, such as He II Lyα ($\lambda = 304$ Å) and H I Lyβ (1025 Å). Because the clouds are in orbit around the black hole, the lines will be blurred by the rotation and the line emission would appear like a broad hump around 1100 Å – the big blue bump.

When comparing low- and high-luminosity quasar spectra in the UV, Scott *et al.* (2004) showed that the low-luminosity objects observed by FUSE had bluer spectra than AGN with large UV luminosity taken by HST (Telfer *et al.*, 2002). If the emission indeed arises from the accretion disk, this would indicate that the low-luminosity objects have disks with higher temperature than their high-luminosity counterparts. This is consistent with the Baldwin effect (see Section 3.5): Baldwin (1977) found that there is an anticorrelation between AGN luminosity and emission line equivalent width. The ionizing power is larger when the UV spectrum is bluer, as indicated by decreasing luminosities.

5.8.2.4 The Inverse Compton Branch

The X-ray domain of Seyfert galaxies is, as in the case of beamed sources, dominated by inverse Compton emission, as proposed by Shakura and Sunyaev (1973) for accretion onto galactic black holes and then by Katz (1976) for quasars. Different than in the case of blazars, though, the seed photons which are up-scattered and the relativistic particles are not located in a jet. Nevertheless, one observes an X-ray spectrum from Seyfert galaxies which resembles that of the blazars, as it shows an inverse Compton-like shape: Seyfert spectra have X-ray spectral slopes with a photon index close to $\Gamma = 2$ and show the characteristic high-energy cutoff at energies on the order of 100 keV, while blazar spectra extend up into the gamma-ray range. It also seems to be the case that the X-ray and UV emission have approximately the same power (per decade of frequency) in Seyfert galaxies. This suggests that perhaps the X-rays and UV emission regions lie at approximately the same distance from the central engine, receiving the same seed radiation power.

This led Liang and Price (1977) to introduce a model with a hot corona surrounding a cool accretion disk, first for the case of galactic black hole accretion and later also for AGN (Liang, 1979). Based on this, Haardt and Maraschi (1991, 1993) developed their two-phase accretion disk model for Seyfert galaxies (see Maraschi and Haardt, 1997, for a description of how the model evolved). Here the accretion disk itself provides the seed photons in the optical/UV range. Different than in the synchrotron self-Compton model for blazars, the seed photons have a multicolor black body spectrum rather than a power law-like one. The relativistic electrons on which these seed photons can be up-scattered, have to be in the vicinity of the disk, in order to allow a high efficiency of this process. This medium is in the form of a corona, a hot electron plasma cloud surrounding the accretion disk. The relativistic electrons are cooled mainly by the Comptonization processes. Some of the Comptonized high-energy photons will be scattered back onto cooler parts of the accretion disk, giving rise to the reflection hump observed around $\sim$ 30 keV, as shown in Section 5.5.

The main advance in the model by Haardt and Maraschi (1991) was the coupling of the emission of the optically thick disk to the corona, as the disk photons cool the electrons in the corona and the Comptonized photons in turn heat the accretion disk. If we assume an infinite parallel arrangement of the disk and its corona, half of the photons emitted by the corona will impinge on the disk. A small fraction (10–20%) will be reflected, but the majority will be absorbed. This model explains and couples the three main components: the observed multicolor black body optical/UV emission, the inverse Compton photons emitted by the corona and forming a power-law spectrum, and the Compton reflection component. The balance between the mechanisms depends on how much of the available accretion power is dissipated in the corona, and what fraction heats the disk. As shown by Haardt and Maraschi (1991), in order to produce the observed homogeneous

Seyfert spectra with $\Gamma \sim 2$, the model requires that most of the energy is dissipated in the corona and very little in the cool accretion disk. On the other hand, the actual seed photon temperature kT has little effect on the resulting spectrum, and a value of $kT \simeq 10\,\text{eV}$ as derived from the UV observations of the disk works fine. The model also predicted correctly that the Seyfert spectra will show an exponential cutoff. The energy of the turnover depends on the optical depth τ of the corona and is anticorrelated to it. For a value of $\tau \sim 0.1-1$ the cutoff appears at energies $E_{\text{cut}} \sim 250-70\,\text{keV}$ and the X-ray photon index of a simple power law would turn out to be $\Gamma = 1.4-2.4$ (Haardt and Maraschi, 1991), a range consistent with most of the X-ray spectra observed from Seyfert galaxies. As a result, spectral slope and corona temperature are anticorrelated: the more relativistic the electrons are in the corona, the higher energies can be reached by the seed photons through inverse Compton scattering. This leads to flatter spectra in the X-rays, and the cutoff energy will appear at higher frequencies.

A model which includes a two-phase disk and corona scenario in a self-consistent way has been brought forward by Poutanen and Svensson (1996) and is implemented in the X-ray fitting package XSPEC (Arnaud, 1996). Murphy and Yaqoob (2009) have presented a model which takes into account a toroidal reprocessor which can be in the Compton-thin or Compton-thick regime, thus the geometry in this case is more realistic than in previous model attempts.

A problem inherent to the disk-corona model is that it tightly couples the observed seed photon emission to the inverse Compton component. If we assume that the disk is seen in the UV range, and the Compton component is observed in the X-rays, this would mean a perfect correlation between these two energy ranges, including both, flux and spectral slope variations. But this is indeed not the case. In order to overcome these difficulties, Haardt *et al.* (1994) modified the model. Instead of a plane-parallel model in which the whole disk is covered by the corona, a patchy geometry was adopted. This way, the accretion disk is directly visible, at least partially. Below the corona clouds, the disk would have a higher temperature because of the inverse Compton photons impinging on the disk here, and the correlation of UV and X-ray emission would be blurred. This would also explain the wide range of temperatures of the disk as visible in the big blue bump. As a way to contain the coronal clouds, Haardt *et al.* (1994) suggested magnetic loops, which would be able to store energy and then to release it rapidly through magnetic reconnection.

Some more recent models locate the emitting electron plasma in a sphere with radius R_{plasma} surrounding the black hole, with the inner radius of the accretion disk $R_{\text{inner}} \simeq R_{\text{plasma}}$ (Poutanen *et al.*, 1997; Zdziarski *et al.*, 1999). The inflow then happens through a type of ADAF process (see Section 3.2.4). Similar as in the model by Haardt *et al.* (1994), the hot plasma in the sphere radiates via thermal Comptonization and the seed photons for up-scattering are the ones emitted by the disk. As has been shown in the case of NGC 4151, one of the brightest Seyfert galaxies in the X-rays and thus one providing high-quality spectra, this model seems to represent the intrinsic AGN spectrum well, especially when the spectrum extends to high energies, that is when the source is in a hard state (Lubiński *et al.*, 2010).

We point out that care must be taken in interpreting model fitting results. For example, it must be emphasized that there is no geometrical information inherent in the emergent AGN spectra alone. Rather, the spectra are the convolution of a radiation source and the line of sight optical depth. Specific geometrical parameters obtained from the models described above or the growing number of models in the literature are dependent on the starting assumptions of the model itself. Combined spectral and temporal analysis can, in principle, constrain the source geometry.

The high-energy exponential cutoff in the range $E_{\rm cut} \sim 50{-}500$ keV has as a consequence that there is no detectable gamma-ray emission from Seyfert galaxies. As we have seen earlier, this is not always the case. Some narrow-line Seyfert 1 galaxies have been detected by Fermi/LAT, indicating that in these sources some additional component produces a significant amount of gamma-ray emission. In radio galaxies like Cen A it is reasonable to assume that the gamma-ray emission is produced by the misaligned jet, which is not pointing in the direction of the observer but still gives rise to the observation of a faint BL Lac type core and the spectrum can extend all the way up into the TeV range. But except for these peculiar cases, the Seyfert core with its accretion disk and inverse Compton emission produced in a (patchy?) corona is not detectable above 1 MeV.

6
AGN Variability

Variable emission over the full range of the observable spectrum is among the defining characteristics of AGN and it has also been exploited as a powerful tool to probe their physical properties. For example, from characteristic time scales one can infer the sizes of the emitting regions and lags between the continuum and line emission have been used to estimate the scale of the broad-line regions (BLR) and the central black hole masses. Soon after the discovery of AGN, intensity variations in the optical and radio bands on time scales of months to years provided evidence for compact emitting regions. This quickly led to the first suggestions that accretion onto massive black holes were the source of the remarkable energetics these objects exhibit. Since then, the black-hole central-engine paradigm has been strengthened by a variety of observational evidence. Interpretation of the variability is still open to discussion at some level, but some basic ideas are widely accepted. In radio-loud objects variations are roughly understood in terms of models of jet propagation. In radio-quiet objects, the variability may originate in the accretion disk, although the time scales in the X-ray band in comparison to the slower and lower amplitude optical-UV variations add uncertainty to this picture. The mass accretion rate onto the black hole, $\dot{M}$, is likely the fundamental parameter underlying the AGN variability.

6.1
Variability in Radio-Quiet AGN

Early studies were conducted mainly in the optical band, but sampling was often sporadic. In the case of the radio-quiet objects, later efforts involving large teams of observers using ground-based and space-based telescopes were carried out. The main goal of most of these campaigns was to study the variations of certain broad emission lines in response to continuum variations, that is, to carry out the "reverberation mapping" campaigns discussed in Section 3.1.2. As an aside though, these campaigns produced well-sampled continuum light curves thus facilitating more comprehensive study of at least a small number of objects. The variable optical-UV emission is likely to be a combination of direct disk thermal emission plus scattered emission resulting from reprocessing of X-radiation in the inner disk and

Active Galactic Nuclei, First Edition. Volker Beckmann and Chris Shrader.

associated coronal plasma. The latter mechanism should lead to measurable time lags between the optical and X-ray light curves.

Generally, in radio-quiet AGN the continuum variability is color-dependent with blue bands having higher variability amplitude than red bands. This is roughly consistent with the hypothesis that the optical continuum is predominantly thermal emission from the accretion disk and that longer wavelength emission originates from larger radii where the disk is cooler. The larger area of the cool, outer disk is expected to respond more slowly than the more compact hot, inner regions. However, coordinated UV, optical and X-ray monitoring campaigns are not always consistent with this picture. Even the blue to UV bands exhibit variations that are markedly less dramatic, particularly on short time scales, compared to those seen in X-rays.

Examples of some of this phenomenology are illustrated in Figures 6.1 and 6.2. The relatively nearby ($z = 0.0097$, distance 44 Mpc) and extensively studied type 1 Seyfert galaxy NGC 3783 was monitored in X-rays, optical and ground-based near IR for a nearly three year baseline (Lira *et al.*, 2010, 2011). The 3–10 keV X-ray, B-, J- and H-band are plotted in the upper-to-lower panels. While a general correlation in the long-term trending is evident, the short term fluctuation in the X-rays is clearly much greater than the IR-optical, although the B-band is clearly more variable on those time scales than the IR bands. Also, the X-ray intensity exhibits surges from minimum to maximum about a factor of 5, over tens of days, while the optical-IR light curves are relatively smooth and slowly varying. It thus seems very unlikely that reprocessing of the X-rays is solely responsible for the optical variability. Another representation of these data presented by the same group is shown in Figure 6.2. Here the V-, J-, H- and K-band fluxes are plotted against the corresponding (i.e., simultaneously obtained) B-band flux. Clearly, there is a tight correlation between the V and B bands, which can be most easily interpreted as the fluctuations in those two bands being a response to the same input, presumably that being the incident X-radiation. However, the much weaker correlation between J and B, and the apparent absence of any correlations between B and H or K (although there may be transitions between "high" and "low" IR states as a function of B flux) are more difficult to interpret. The IR emission may emanate from multiple physical components of differing size scales; the cool, outer accretion disk and the putative molecular torus. The overall variability seen here can certainly be reproduced by models invoking both variable accretion rates and X-ray reprocessing. However, it is often the case that to reproduce the observed phenomenology one introduces ad hoc assumptions. Examples from the literature include irregularities in accretion rate and viscous time scales that can then be adjusted to propagate density enhancements from cooler to hotter parts of the disk thus accommodating the observed color variations. It also should be noted that the AGN shown here, NGC 3783, is just one arbitrary example. There is a wide range of phenomenology exhibited by the radio-quiet population as a whole to the extent that it has been sampled.

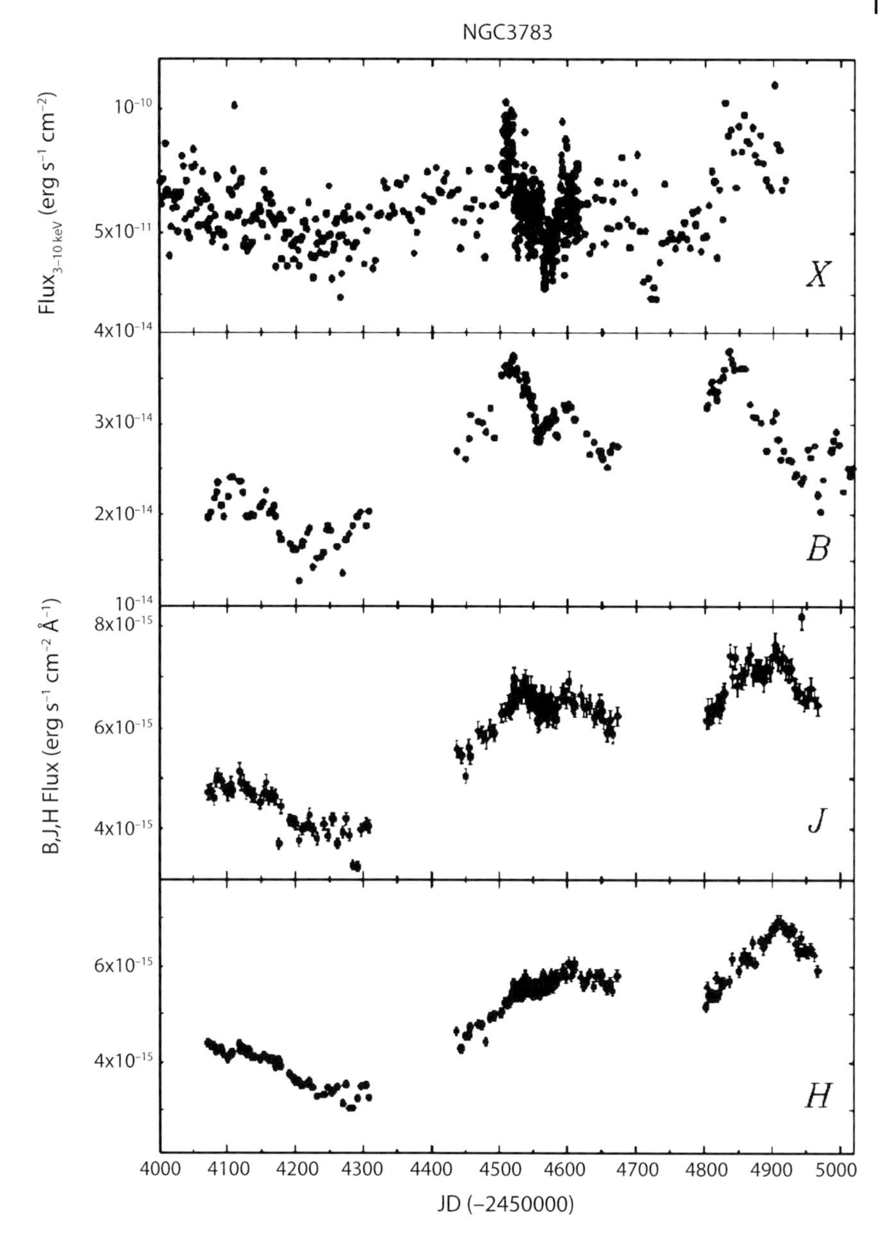

Figure 6.1 Well-sampled 3-year X-ray and optical-IR photometric light curves for the Seyfert 1 AGN NGC 3783 (Lira *et al.*, 2010). As noted in the text, the different light curves are generally well correlated on long time scales. The short term variability in the X-ray band is clearly much more significant than in the optical-IR bands while the latter exhibit greater variability in the bluer colors. The rapid, large amplitude surges seen in X-rays are absent in the photometric bands. (Reproduced with permission by Cambridge University Press.)

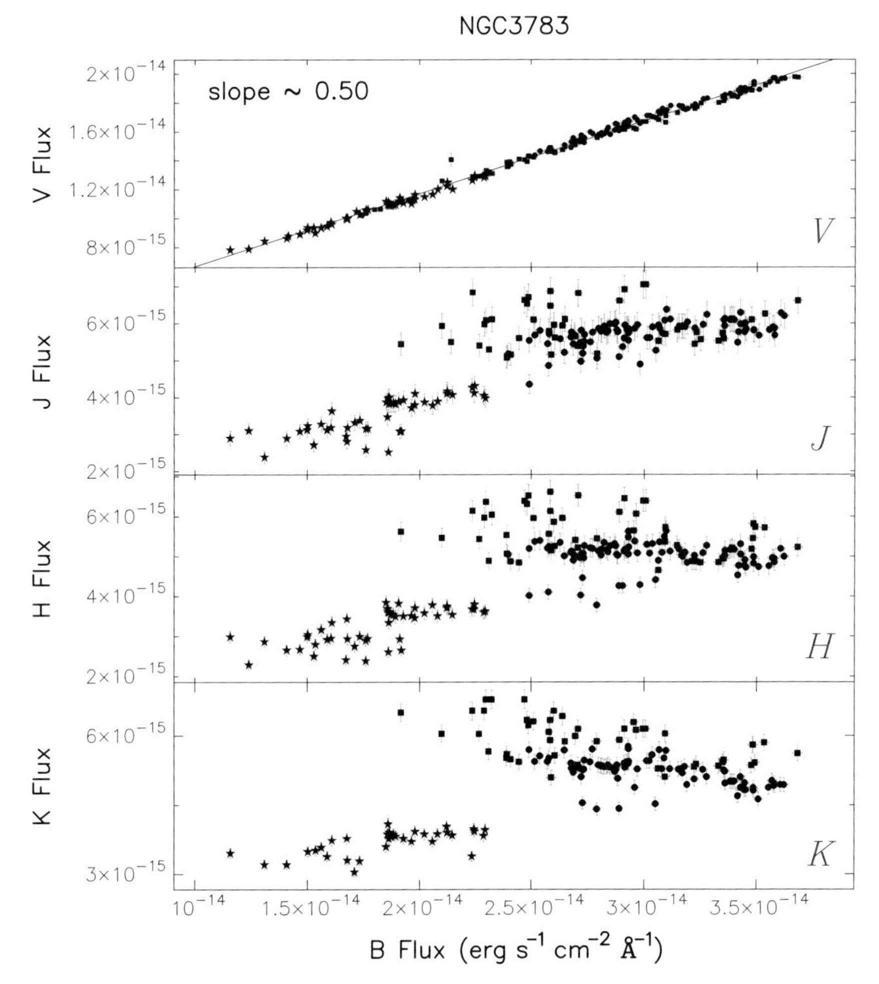

Figure 6.2 Alternate presentation of the Lira *et al.* (2011) photometric data for NGC 3783. Clearly, the B and V bands are tightly correlated, while J and B are only weakly correlated. It is not clear that H and K are correlated with B at all, and there may be a discrete "high-state"–"low-state" behavior in those two near IR bands.

6.2
Analysis Methods for Variability Studies

An important analysis tool available to astronomers in their assessment of AGN light curves is the cross-correlation function (CCF). This mathematical function of two time series, for example the X-ray and optical light curves, reveals the magnitude and significance of characteristic lag times between the two series. For example, in the reprocessing scenarios it would be natural to see a delay between the arrival of an X-ray light curve inversion or flare and the corresponding change in the optical light curve. The length of the delay, τ, is related to the light travel time between the X-ray source (e.g., inner accretion disk corona) and the reprocessing region (e.g., the outer accretion disk). We should note that in general, there is also a finite response time involved for the reprocessing media to absorb and re-emit radiation. At a detailed level, that depends on the microphysics of the reprocessing medium but for practical intent and purposes it can be adequately approximated by a simple analytical function. The form of the cross-correlation function is

$$\mathrm{CF}(\tau) = \frac{E\{[a(t) - \bar{a}][b(t + \tau) - \bar{b}]\}}{\sigma_a \sigma_b} \tag{6.1}$$

where a and b are the two time series, $\bar{a}$ and $\bar{b}$ their average values, and σ_a and σ_b are their standard deviations. $E(f)$ is the expectation value of the function $f = [a(t) - \bar{a}][b(t + \tau) - \bar{b}]$. The CCF formulation assumes that the data are evenly spaced. It is also necessary for the data to oversample the shortest time scales of interest and to span as large a baseline as possible, at least several times the reciprocal of the lowest frequency variations one hopes to assess.

Both of these requirements lead to practical issues in conducting observation campaigns. Weather, the lunar cycle and seasonal horizons are all potential issues as is scheduling pressure on telescopes, which are usually intended to support a broad range of astronomical investigations. Space-based observatories naturally circumvent some of these problems, however, they are fewer in number and thus generally much more oversubscribed and more expensive to operate than ground-based telescopes. Thus, obtaining the needed observation time with the rigid scheduling requirements to carry out a rigorous variability study is difficult. The need for uniformly sampled data can be relaxed to some degree by the use of an alternate formulation of the CCF known among the AGN community as the discrete cross-correlation function (DCF). First introduced by Edelson and Krolik (1988), the DCF incorporates measurement errors in addition to the standard deviation of each data series about its mean value as well as some binning to offset the effects of uneven sampling.

The discrete correlation function can be written as

$$\mathrm{DCF}(\tau) = \frac{1}{M} \sum \mathrm{UDCF}_{ij} \tag{6.2}$$

Here, $\mathrm{DCF}(\tau)$ is the average of all UDCF_{ij} for all couples which fall into the time interval $\Delta t_{ij} = t_j - t_i$, which is defined by the time lag τ

$$\tau - \frac{\Delta\tau}{2} \leq \Delta t_{ij} \leq \tau + \frac{\Delta\tau}{2} \tag{6.3}$$

UDCF_{ij} is computed on all pairs (a_i, b_j) and each pair is associated to the large time Δt_{ij}

$$\mathrm{UDCF}_{ij} = \frac{(a_i - \bar{a})(b_j - \bar{b})}{\sqrt{\left(\sigma_a^2 - e_a^2\right)\left(\sigma_b^2 - e_b^2\right)}} \tag{6.4}$$

Here, a, b and the σ are defined as before. The e are the measurement errors and M is the number of measurements within the symmetric interval of width $\Delta\tau$ about the lag time τ. The choice of $\Delta\tau$ is a tradeoff between statistical significance and the resolution of the result, that is, the achievable precision of a possible time-lag measurement, versus the statistical significance of that measurement. An example based on an exceptional AGN multiband monitoring dataset is illustrated in Figure 6.3, specifically from Arévalo *et al.* (2009) who collected about 2 years of X-ray and optical data for NGC 3783. In this case, the observations were sampled at intervals of 2–4 days over that time span. The authors then considered lags on time scales of 0–40 days. A peak in the DCF indicates that the optical emission is delayed relative to the X-rays by about 5 days.

Arguably the most powerful tool for studying variable light curves is the power density spectrum (PDS). The PDS is constructed by computing the Fourier transform of the time series $x(t)$ leading to $X(\nu)$, i.e., the frequency domain representation of the data. At a given frequency, the product of the Fourier transform of $x(t)$ and its complex conjugate, $X(\nu) * X(\nu)$, is the PDS amplitude or "power" at that frequency. Greater power is indicative of greater source variability. Here we are assuming some basic familiarity with Fourier analysis, as a detailed treatment is beyond the scope of this book. A useful reference is Howell (2001).

A function related to the PDS is the Fourier cross spectrum. This involves the simultaneously sampled time series, for example an X-ray light curve $x(t)$ and an optical light curve $o(t)$. The Fourier cross spectrum (herein cross spectrum) is then $X * O$, where $X(\nu)$ and $O(\nu)$ are the Fourier transforms of x and o and the asterisk indicates complex conjugation. Thus the cross spectrum is a complex quantity. It represents the phase lags, or equivalently the frequency-dependent time lags between the two concurrent light curves. This is a potentially powerful analysis tech-

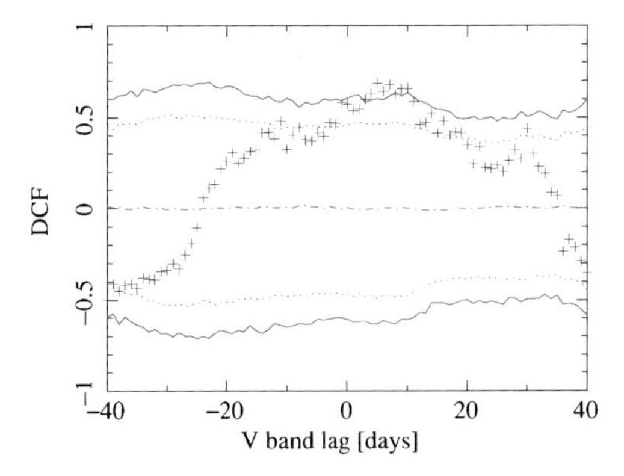

Figure 6.3 Discrete cross-correlation function for NGC 3783 computed using about a 2-year baseline of X-ray and optical monitoring observations with 2–4 day sampling (Arévalo *et al.*, 2009). The crosses are the points computed from the data. The solid and dotted lines represent 99 and 95% extremes based on simulations, that is, the points above the solid line in the $\tau \simeq 7$ day lag range represent a better than 99% confidence detection of X-ray-to-optical lag (X-rays lead optical).

nique in that it combines temporal and, at least crudely, spectral information. An example of its application to AGN is discussed in Section 3.6.

Other methods that are related to the Fourier power-spectrum technique are the structure function (see box below) and the autocorrelation function (ACF, i.e., Eq. (6.1) with $a(t) = b(t)$). Both are computed in the time domain. They similarly provide information on the distribution of rms variability or power with differing time scales, which like the PDS can in principle provide insight into the intrinsic nature of the source emission, and reveal any periodic or quasiperiodic variations. The ACF can also reveal a distinct signature of echo effects in light curves.

Structure function

Structure functions (SF) are similar to auto- and cross-correlation functions and have been introduced for analysis of radio light curves by Simonetti *et al.* (1985).

The structure function is a useful and simple tool to use in order to find characteristic variability time scales. The first-order structure function is defined as $D^1(\tau) = \langle[x(t) - x(t + \tau)]^2\rangle$. Here $x(t)$ is the measurement at time t, and τ is the time-lag, or variability time scale. For a stationary random process the structure function is related simply to the variance σ^2 of the process and its autocorrelation function $\rho(\tau)$ by $D^1(\tau) = 2\sigma^2[1 - \rho(\tau)]$. The function $D^1(\tau)$ can be characterized in terms of its slope: $b = \mathrm{d}\log D^1/\mathrm{d}\log\tau$. For lags longer than the longest correlation time scale, there is an upper plateau ($b = 0$) with an amplitude equal to twice the variance of the fluctuation ($2 \cdot (\sigma_Q^2 + \overline{\sigma_i}^2)$). For very short time lags, the structure function reaches a lower plateau which is at a level corresponding to the measurement noise ($2 \cdot \overline{\sigma_i}^2$). Between these two extremes, a positive slope of the SF indicates the presence of variability on the corresponding time scales. See however Emmanoulopoulos *et al.* (2010) for a critical view on the usefulness of structure functions.

The structure function (e.g., Hughes *et al.*, 1992), which is the mean variance of a signal $x(t)$ with the signal offset by τ, $x(t + \tau)$, avoids some of the problems of windowing and aliasing associated with Fourier analysis. It can thus be applied to sparser and less evenly sampled datasets (note however Emmanoulopoulos *et al.*, 2010, who argue that the effects of data gaps may have more significant effects than is commonly perceived), and it has been applied extensively in the AGN literature.

The quality of the data available, both in terms of its signal-to-noise and the density and uniformity of its temporal sampling, as well as its coverage baseline generally drive ones decision on which method(s) to pursue. The advantage of the PDS analysis is its enhanced information content: one now has information on the rms variability for the different frequency components present in the system. These may for example be due to different physical time scales characteristic of the AGN. The presence of any periodic or quasiperiodic phenomenon will also be revealed and other PDS structures, such as discontinuities or "breaks," may also be related to physical parameters of the AGN. The drawback with this approach however is that large amounts of data – specifically, measurements at numerous epochs over large time baselines – are necessary to construct PDS of sufficient quality. As could be seen in the light curve examples presented here for NGC 3783, the optical variability is less dramatic than in X-rays. This is readily evident when one examines the PDS of a typical object. In Figure 6.4, the optical (photometric B band) PDS for NGC 3783 is plotted as the dotted line. The overplotted dashed line is a power-law fit. The solid curve is the X-ray PDS, again with an overlying dashed curve representing a model fit. Clearly, the X-ray emission exhibits much greater variability at all frequencies above a few 10^{-8} Hz. The optical PDS, which is typical of many radio-quiet AGN, is basically a red-noise spectrum, i.e., a power law with index $\alpha \simeq 1.5–2$ ($P_\nu \propto \nu^{-\alpha}$).

Until relatively recently, the X-ray light curves were generally insufficiently sampled to construct useful power spectra, particularly ones extending to long time scales. The EXOSAT satellite which operated in the 1980s had an elongated elliptical orbit allowing for continuous viewing of a source for up to three days. The situation was improved even more dramatically with the launch of the RXTE satellite in 1995 and its successful operation for over 16 years. The flexible scheduling and rapid repointing capabilities of RXTE led to enough well-sampled AGN light curves to begin to assemble a number of reasonable quality PDS over years to days or even hour time scales. An important feature of these X-ray PDS was quickly pointed out. They bore a morphological resemblance to those of the more extensively studied galactic black hole X-ray binaries (BHXRBs) seen in their soft spectral state, that is,

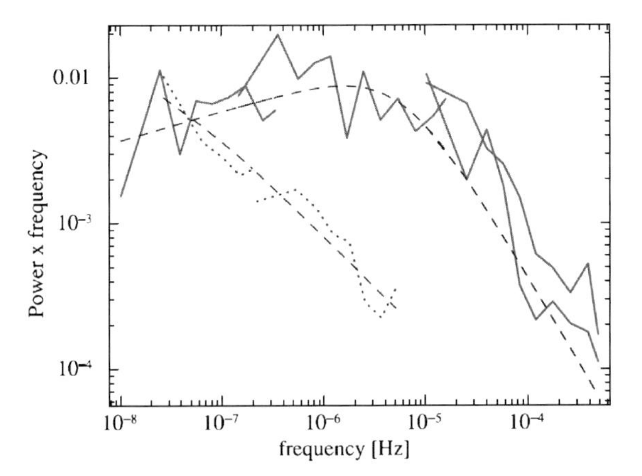

Figure 6.4 Optical (B-band) and X-ray power-density spectra (PDS) for NGC 3783 compiled by Arévalo *et al.* (2009). The optical and X-ray measurements are represented by the dotted and solid curves, respectively, with the overlaid dashed curves representing best fit phenomenological models. Visual inspection of the PDS can roughly quantify the variability amplitude as a function of the time scale or correspondingly, of Fourier frequency. The PDS is often normalized so that its integral over a frequency interval equals the normalized variance of the light curve at that frequency band. Note that PDS are often presented as νP_ν like in the example shown here. This representation is analogous to the commonly used νf_ν representation of AGN spectral energy distributions. It highlights deviations from a simple power-law form, facilitating ease of visual inspection and interpretation.

having one break only as seen in Figures 6.4 and 6.5, albeit over a frequency range shifted by 5 or 6 decades (Uttley and McHardy, 2004).

In addition to the shape of the PDS the AGN and BHXRBs were found to exhibit a strong linear correlation between the rms variability amplitude of the X-ray flux (e.g., Uttley and McHardy, 2004). Lag times between hard and soft band fluxes were also shown to be similar in terms of direction and their relative magnitude. Collectively, these similarities imply a commonality in the underlying variability process for the two object classes. Furthermore, the PDS exhibited by BHXRBs are distinct from that of galactic neutron star X-ray binaries. The latter do not exhibit the broken power-law, or smoothly connected two power-law, form. They are instead characterized by higher overall variability extending to higher frequencies. Thus, the similarity of the AGN and BHXRB PDS is likely rooted in the presence of accretion onto a black hole in both cases.

Uttley and McHardy (2004) have noted that the well known scaling relationship between the PDS break frequency and black hole mass in the BHXRBs may be applicable to AGN as well. As noted in Section 3.1, a number of methods for estimating AGN black hole masses have been developed and applied. So for objects

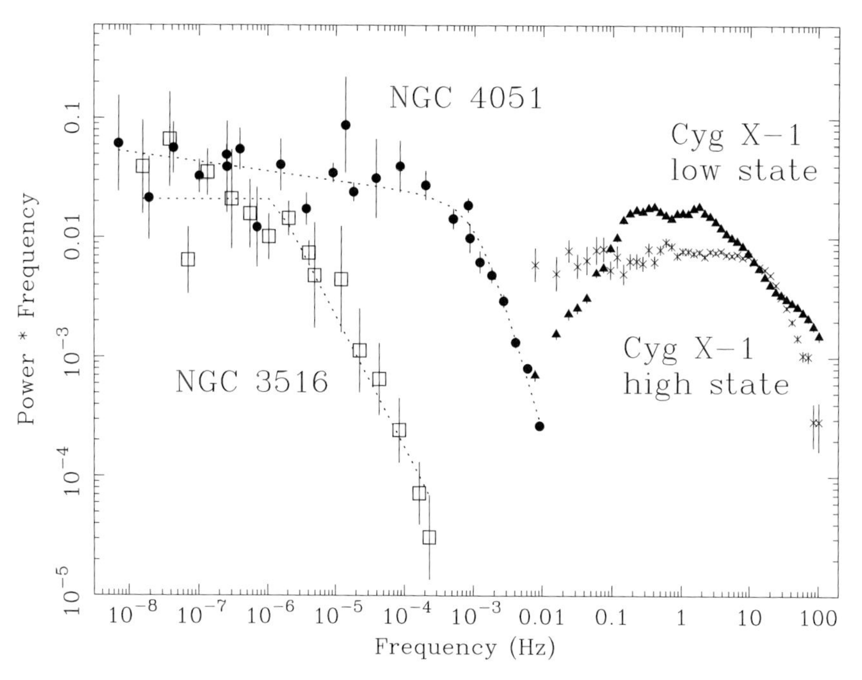

Figure 6.5 Examples of Seyfert 1 AGN (NGC 4051 and NGC 3516) power-density spectra (PDS) reproduced from McHardy *et al.* (2004). Also plotted in the higher frequency range are two PDS representative of the low-hard and high-soft spectral states of the BHXRB Cygnus X-1. Based on the apparent self-similarity between NGC 4051 and the Cyg X-1 high-soft state, the break frequency at approximately 10^{-3} Hz can be used to derive a black hole mass by scaling from Cyg X-1. McHardy *et al.* (2004) thus obtained a value of $3 \times 10^5\ M_\odot$, which is consistent with the estimates obtained from reverberation mapping.

with both a well-defined PDS and, for example, $M-\sigma$ or reverberation mapping black hole mass estimates, this idea can be tested. Some results are illustrated in Figure 6.6. This shows that indeed the correlation between time break in the PDS and mass of the black hole can be used to estimate the mass of an AGN core. The power of this method for black hole mass estimation may not reach the level of precision of the spectroscopic methods for some time. However, it seems clear that this line of analysis convincingly demonstrates a clear symbiosis between the BHXRBs and radio-quiet AGN, and establishes black-hole accretion as the underlying commonality. The correlation between time break and black hole mass can be improved still by considering the accretion rate of the central engine. As we have seen in Section 4.8.4, this can be expressed in a form combining the break time scale T_B, the mass of the central engine M_{BH}, and the bolometric luminosity (see Eq. (4.11)). As both, black hole mass and bolometric luminosity are in this equation, this implies a correlation with accretion rate.

It has been noted in the literature that the NLS1 subclass of Seyfert AGN exhibits greater variability in the X-ray band than their broad-line counterparts. This is evident in Figure 6.6, where the NLS1 like Akn 564 and Mrk 766 are outliers relative to the mass-PDS break relation. They are positioned such that their PDS breaks at higher frequencies than the broad-line objects, i.e., they exhibit greater power at frequencies exceeding the characteristic break frequency for broad-line objects of comparable mass. More comprehensively, for example, Leighly (1999) (but see also Turner *et al.*, 1999) demonstrates that the excess variance (see box below) as a function of luminosity computed for a large sample of Seyfert AGN reveals a clear separation between the narrow- and broad-line objects, with the former showing the greater excess.

Excess variance

Any light curve consisting of N flux measurements x_i varies due to measurement errors σ_i. In the case that the object is also intrinsically variable, an additional source variance σ_Q has to be considered. The challenge of any analysis of light curves of variable sources is to disentangle them in order to estimate the intrinsic variability. A common approach is to use the "excess variance" (Nandra *et al.*, 1997a) as such an estimator. The sample variance is given by

$$S^2 = \frac{1}{N-1} \sum_{i=1}^{N} (x_i - \overline{x})^2 \qquad (6.5)$$

where $\overline{x}$ is the average of the measurements, and the excess variance is given by

$$\sigma_{XS}^2 = S^2 - \overline{\sigma_i^2} \qquad (6.6)$$

with $\overline{\sigma_i^2}$ being the average uncertainty of the measurements. For the case of varying measurement errors ($\sigma_i \neq$ const.), Almaini *et al.* (2000) showed that σ_Q has to fulfill the following condition:

$$\sum_{i=1}^{N} \frac{(x_i - \overline{x})^2 - \left(\sigma_i^2 + \sigma_Q^2\right)}{\left(\sigma_i^2 + \sigma_Q^2\right)} = 0 \qquad (6.7)$$

It is also well known that the NLS1 objects have softer characteristic X-ray continuum spectra than the broad-line Seyfert 1 objects (Vaughan *et al.*, 1999). This has been interpreted as evidence for accretion rates that are a large fraction of the Eddington limit (e.g., Mathur, 2000) and the soft continuum has been interpreted as thermal emission from an accretion disk (e.g., Shrader and Titarchuk, 2003). This is in analogy with the BHXRBs, which are well known to transition between high and low states characterized by changes in mass accretion rate. The high ac-

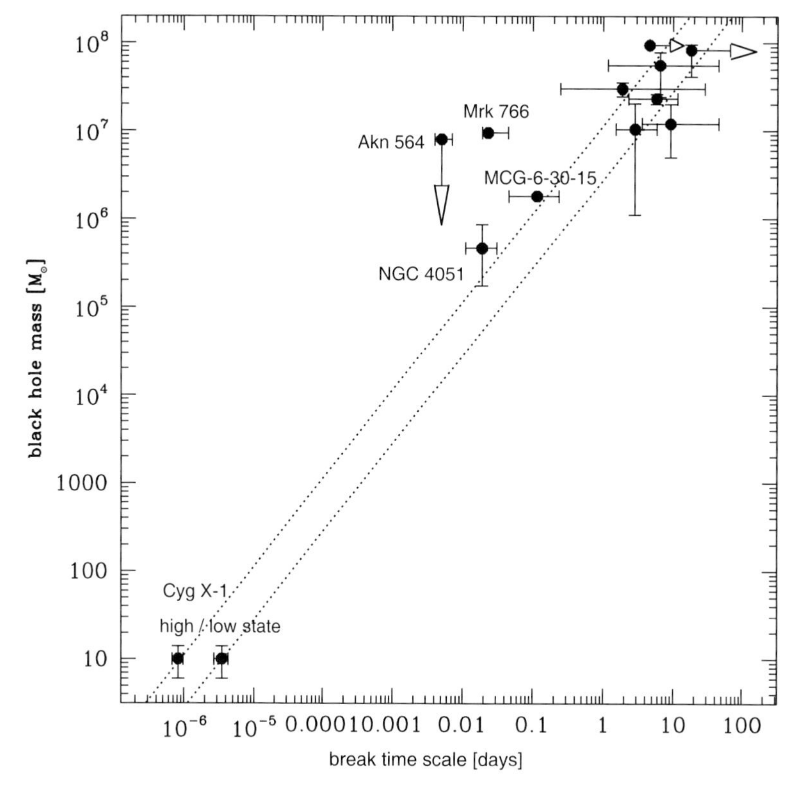

Figure 6.6 Extrapolation of the BHXRB black-hole mass–break-frequency scaling relationship to the ~ $10^8 M_\odot$ range of AGN. The data points in the upper right are reverberation mapping or $M-\sigma$ black hole mass estimates for a sample of AGN. The two lines indicate an extrapolation from the high/soft and low/hard state of the BHXRB Cyg X-1 assuming that the break time scale is proportional to the black hole mass. Data from McHardy *et al.* (2004); Summons *et al.* (2007); Marshall *et al.* (2009).

cretion state is characterized by a much softer X-ray spectral energy distribution than the low state. BHXRBs typically spend most of their time in the low state. The analogy then is that the NLS1s represent the high accretion rate, soft spectral state and other, more numerous, Seyferts represent the low-hard state.

Transitions from one state to another are not seen directly because of the much greater time scales involved with the AGN. Variability characteristics however pose a problem for this analogy. As noted, the NLS1s tend to be more variable in X-ray flux than the typical Seyfert 1, which is opposite to the well-established trend for the BHXRBs. For the latter, the greatest flux variability is exhibited by a given object when it is in its low-hard state. The NLS1 are assumed to accrete at higher Eddington ratios than the average Seyfert galaxy, which might explain also their high/soft state nature, again based on the BHXRB analogy.

6.3 Variability of Radio-Loud AGN

The radio-loud AGN have variability characteristics that are distinct from their radio-quiet counterparts. Variability in these objects is widely believed to be dominated by emission from a relativistic jet. The emission from the jet is relativistically beamed and as such is often significantly amplified and critically sensitive to our line of sight. In the case of blazars, which we believe are the most highly aligned objects, amplification can be an order of magnitude or more. The observed variability time scales documented by an observer are shorter than those in the rest frame of the jet plasma as a result of time dilation effects. The broad-band continuum radiation observed from radio to optical, and in some cases X-rays, is due to synchrotron radiation resulting from the accelerated charged particles in the jet interacting with the magnetic field ambient to the AGN central engine. The higher energy X-ray and gamma-ray radiation are believed to result from Compton scattering. The Comptonizing electrons are believed to be from the same population as those emitting the synchrotron radiation. The source of the seed photons is not precisely known. It could be the synchrotron photons generated within the jet scattering on the same electrons that produced them. In that case we observe synchrotron self-Compton (SSC) emission as described in Section 2.2.5. Alternatively, the seed photons could emanate from outside the jet, for example from the BLR or the accretion disk, both of which are copious sources of optical-UV photons. Also, there could be additional contribution from the putative torus surrounding the disk or possibly from stellar emission from the host galaxy. In these latter cases, the high-energy emission is referred to as external Compton (EC), as described in Section 5.8.1.2.

Another unresolved issue is the location of the production site within the jet of the individual spectral components. For example the TeV gamma-ray emission cannot be produced arbitrarily close to the AGN central engine. This is the case since the radiation field of the accretion disk and BLR lead to an effective photon–photon opacity which would be evident as a sharp spectral discontinuity between the GeV and TeV domains (microscopically, photon–photon scattering is the quantum mechanical inverse process of pair annihilation; in the literature, it is sometimes alternatively referred to as pair production opacity). This constraint also depends on the jet relativistic beaming factor. On the other hand, at larger distances, for example a few parsecs, the ambient photon density may be insufficient for the EC process to generate the observed high-energy gamma flux. More significantly, the variability time scales for high-energy gamma rays, certainly days and perhaps as short as 10 min (Aharonian *et al.*, 2007), render sources farther from the central engine and more spatially distributed less tenable due to light travel time arguments. Thus a balance satisfying these two basic constraints must be achieved by any viable jet model.

In any case, the best chance at addressing these issues is through the compilation and interpretation of multiwavelength monitoring data, including photometric, polarimetric and VLBI imaging data. Numerical models of jets can then be developed and adjusted to reproduce the basic features of the data presumably leading to new insights. Figure 6.7 shows the example of the spectral energy distributions of three well-studied blazars in low and high emission state.

Significant progress is currently being made on the observational front with the advent of new or improved ground- and space-based facilities coming on line. Notably, the Fermi Gamma-Ray Space Telescope, launched and commissioned in mid-

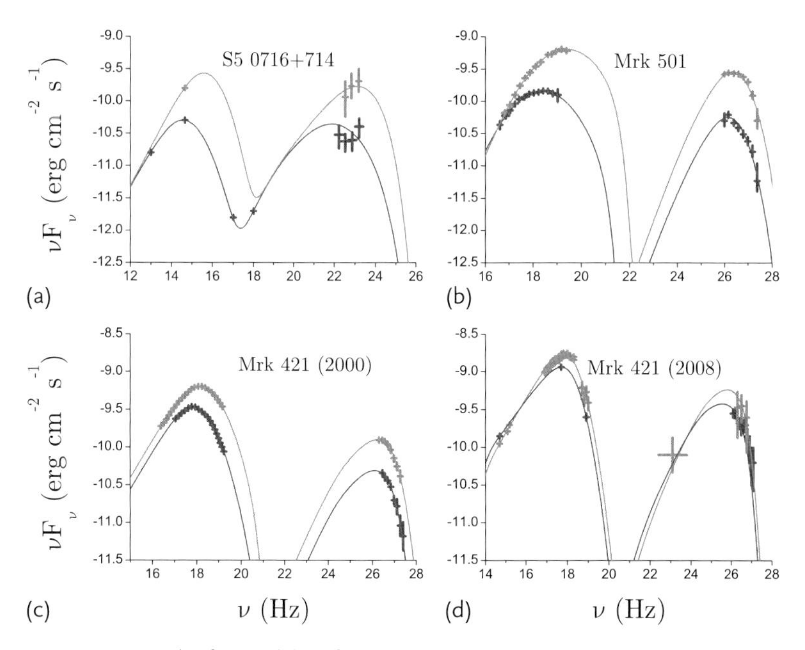

Figure 6.7 Examples for variability of the spectral energy distribution in blazars in low and high emission states (a–d). Blazars tend to shift the peak of the synchrotron and inverse Compton branch to higher frequencies, while when comparing different blazars, usually the ones with higher peak frequencies are the less luminous ones as discussed in Section 5.8.1. Graphic from Paggi *et al.* (2009), reproduced with permission © ESO. For a color version of this figure, please see the color plates at the beginning of the book.

2008, provides a nearly 30-fold improvement in sensitivity over its predecessors in the ∼ 0.1–100 GeV range, that is, where many blazar AGN emit the bulk of their radiative energy output. In addition to the improvement in sensitivity, perhaps more significant for AGN studies is the nearly continuous all-sky coverage offered by Fermi given its enormous (∼ 2 sr) field-of-view and its mode of operation. It continuously scans the sky with periodic 30–60° offset from the zenith, covering the full sky once every three spacecraft orbits (4.5 h), with an on-source duty cycle of about one part in six. In addition to, and in part in response to Fermi, radio to optical ground-based observatories as well as very-high energy (∼ TeV) Cherenkov telescopes (e.g., Cui, 2009) have enhanced their capabilities and focused on increased AGN programs.

Chatterjee *et al.* (2012) have compiled one of the most comprehensive sets of photometric monitoring data for a sample of Fermi-monitored blazars in five photometric bands in the optical to near IR. Their data are sufficiently sampled to construct optical PDS as well as discrete and autocorrelation functions (DCF and ACF) for a number of objects. They find that the ACF reveal no periodicity or characteristic time scales.

The optical PDS are consistent with a featureless red-noise power law, with slopes similar to that found for the gamma rays. No evidence for PDS breaks, as seen in many radio-quiet objects, is evident (Figure 6.8). The shortest time scales for which two-fold optical intensity variations are seen are ∼ 2–4 days.

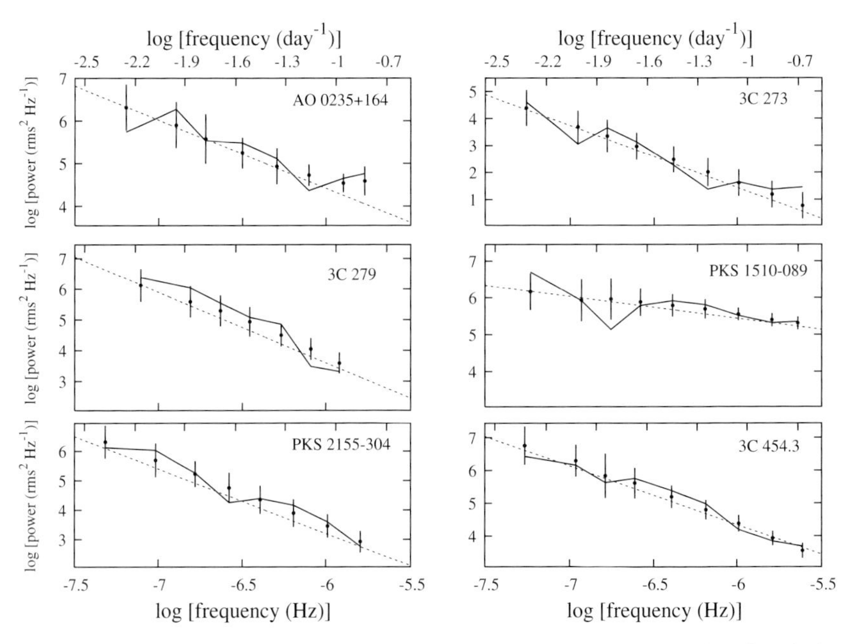

Figure 6.8 Optical PDS for six GeV-bright blazars based on data obtained by Chatterjee *et al.* (2012) since Fermi has been operating. The data are consistent with a red-noise power law (dotted curves illustrate best fit), which is nominally consistent with the gamma-ray PDS. (Reproduced by permission of the AAS.)

Another example of multiwavelength radio-loud AGN light curves is illustrated in Figure 6.9. The blazar BL Lac was monitored in radio, optical, X-ray and gamma-rays including radio interferometric imaging and polarization (Marscher *et al.*, 2011). Unprecedented datasets such as this allow astronomers to begin to establish phenomenological patterns, which will hopefully facilitate future theoretical development. Here for instance, the radio flares consistently lag the X-ray and gamma-ray flares by days to weeks. Furthermore, the radio flares seem to correspond to specific morphological patterns; specifically, the passage of an individual

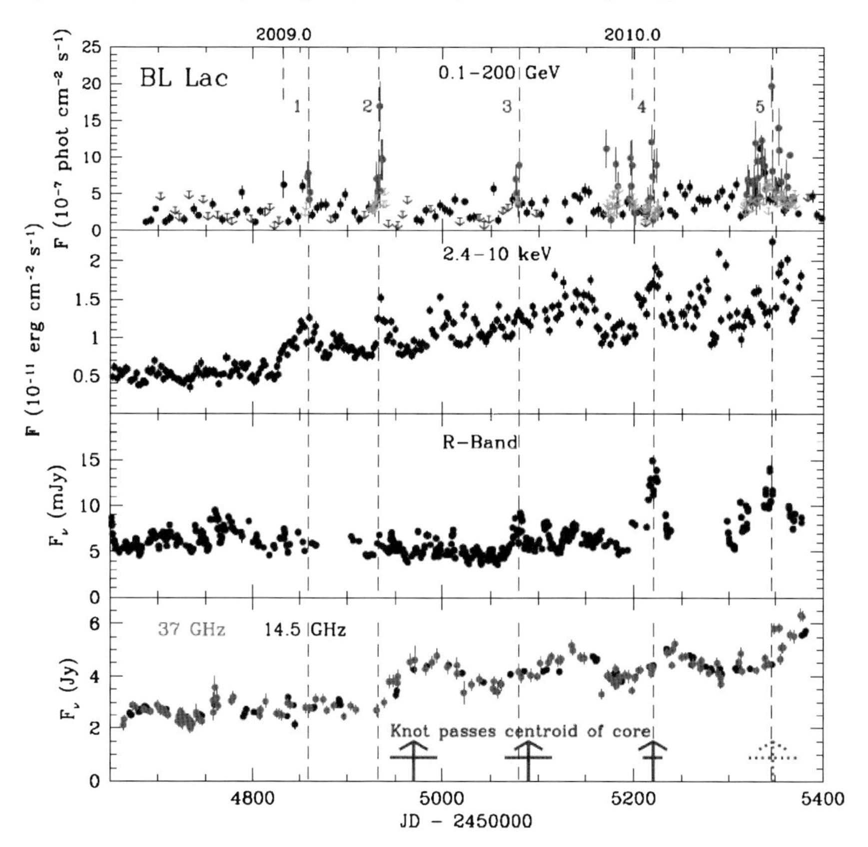

Figure 6.9 Multiwavelength light curve of BL Lac (Marscher *et al.*, 2011). As described in the text, patterns are beginning to emerge between the radio intensity, morphological evolution and flaring at X-ray and gamma-ray bands. Additional information is potentially forthcoming from comparing polarization measurements of individual jet components to optical polarization of flaring features; specifically, discrete increases in polarization are expected to coincide with gamma-ray flares. The gamma-ray peaks are sharp, and coincide with the rising phase of the radio flux. The radio peaks lag behind the gamma-ray peaks by days to weeks. The VLBA images, not shown here, indicate that the jet knots usually pass through the position corresponding to the radio core concurrently with the gamma-ray flares (vertical arrows). For a color version of this figure, please see the color plates at the beginning of the book.

jet knot through the jet core (note that these authors consider the radio core to be associated with a standing shock in the jet, at ~ parsec-scale distances downstream from the central engine). In some cases, the optical flares occur in tandem with the X-rays and gamma rays, but not in other cases. Correlating the polarization information obtainable for individual radio knots with the polarization in optical flares should allow one to begin to associate spatial and temporal components, leading to a much more complete observational picture of the jet than what presently exists. One should keep in mind however that this is just one particular AGN. Patterns must be established over a sample of objects before a comprehensive empirical picture emerges and from which reliable model constraints can be established. Notably, the VLBI monitoring is resource-intensive and the morphology of the jet structures evolves relatively slowly. It may take a ~ 10-year baseline of these types of observation campaigns to gain a true understanding.

Several other recent studies support the parsec-scale offset between the blazar core and gamma-ray emission site. León-Tavares *et al.* (2011) obtained densely sampled 37 GHz light curves for about 60 AGN that are also Fermi detected gamma-ray sources. They decompose their radio light curves into individual subflares with exponential rise and decay profiles. They then compile a distribution of the lag times between observed gamma-ray flares and the onset of the 37 GHz subflares. Although the latter peak after the gamma rays, they take time to grow and the decomposition technique allows one to predict the time of their origin. They find that the gamma-ray flares lag the onset of the 37 GHz flares by an average delay of about 70 days corresponding to about 30 days in the source frame. They estimate that this corresponds to a typical physical separation of 7 pc from the jet launching site to the gamma-ray producing jet knot.

Pushkarev *et al.* (2010) reached consistent conclusions based on their statistical analysis of VLBA monitoring results for a smaller sample of Fermi gamma-ray AGN. They found the average source frame lag time between the gamma-ray and radio light curve peaks to be about 30 days.

It should be noted though that the León-Tavares *et al.* (2011) light curve decomposition, upon which results are based, is not a unique process in a mathematical sense, and that the Pushkarev *et al.* (2010) results are preliminary in that VLBA monitoring over the ~ 10 year time scale of the Fermi mission will ultimately be needed for this type of study. In that vein, we further note that Poutanen and Stern (2011), following a different line of reasoning, interpret the GeV spectral breaks in terms of gamma–gamma opacity within the AGN broad-line region, which would necessitate a much smaller separation between the jet origin and the gamma-ray production region.

Another recent breakthrough comes from VLBI imaging at unprecedented resolution for the nearest radio-loud AGN, Cen A (see Section 3.7.2). These observations, from Müller *et al.* (2011), resolve individual jet knots at a physical scale of ~ 0.013 pc at two frequencies. They identify several of the jet knots as being consistent with flat-spectrum radio emitters, making those sites viable candidates for the gamma-ray emission in that source. The presence of more than one such structure argues for multiemission zone models for jets (as opposed to the simpler, often employed one-zone models as described in Section 5.8.1.1). Future VLBI facilities, incorporating space- plus ground-based antennas, can in principle reveal similar levels of detail in additional objects at greater distances.

6.4 Quasiperiodic Oscillations in AGN

As noted there is commonality between AGN, particularly radio-quiet objects and galactic black hole X-ray binaries (BHXRB). This is perhaps the most evident in the context of their X-ray variability characteristics, as accretion of matter onto the black hole leads to rapid light curve variations. The characteristic time scales involved are possibly related by mass scaling between the two classes of objects. And the general PDS morphologies are similar with the caveats noted earlier in this chapter, notably the low-frequency broad noise component. The BHXRBs however exhibit a variety of periods and quasiperiods presumably associated with the accretion disk and the orbital dynamics within the binary. While AGN are not expected to have binary central compact objects, other than in exceptional cases, one could reasonably suspect that the accretion dynamics associated with the disk accretion flow are analogous (it should be noted however that supermassive binary black holes at separations of less than 1 pc are predicted to be at an inevitable late stage in the evolution of galaxy mergers). This has motivated many searches for periods or quasiperiods in AGN, thus far leading mostly to null results.

In the case of one radio-loud object, the well known BL Lac object OJ 287, a 12 year periodicity in optical flaring activity has been reported (e.g., Nilsson *et al.*, 2006). However, from the available data a strict period cannot be claimed. Models invoking binary central black holes have been considered (e.g., Valtaoja *et al.*, 2000) and orbit "solutions" leading to primary and secondary black hole masses have been made. These models involve for example primary-disk – secondary black hole interactions leading to accretion driven events or jet precession. However, there is some documented phenomenology which is not readily explained; some outbursts occur in the optical but not in the radio. Some have double-peaked morphologies with $\sim$ year separations. Although optical monitoring data exists for over a century, only in the last 30–40 years – that is three putative orbital cycles – have the data been sufficiently well sampled over a long enough baseline for detailed analysis. So, patience is in order.

A perhaps more compelling case, albeit on a much shorter time scale, may be the radio-quiet object, the NLS1 object RE J1034+396. Gierlinski *et al.* (2008) obtained a 90 ks, nearly continuous X-ray observation of this AGN using the XMM-Newton satellite observatory. They identified a $\sim$ 1 h periodicity present in the PDS at significance level of about 5.6σ. Furthermore, the period is readily evident in a simple visual inspection of the light curve, although the phase shifts over the course of the data collection. Thus, the oscillations are appropriately designated as quasiperiodic in nature, thus called quasiperiodic oscillations (QPO).

The PDS derived by those authors is shown in Figure 6.10. Figure 6.10a shows the PDS normalized to the $(\text{rms/mean})^2$ in which the peak is at about 0.27 mHz, corresponding to a period of 62.2 min, is clearly seen. The overlaid curves show the best fit and uncertainties to a power-law continuum, the Poisson noise level and the confidence level they derive for the QPO feature. Figure 6.10b is a data to continuum-fit ratio plot.

Why this type of phenomenon is not seen more commonly in AGN is unclear. It could be that RE J1034+396 is unique in some manner. Its X-ray spectrum is exceptionally soft, extending into the extreme ultraviolet (Casebeer *et al.*, 2006), which could be indicative of an exceptionally high mass accretion rate which somehow drives the QPO. Again however, the BHXRBs mostly exhibit their various QPO phenomena while they are in their relatively low mass accretion rate state. They do

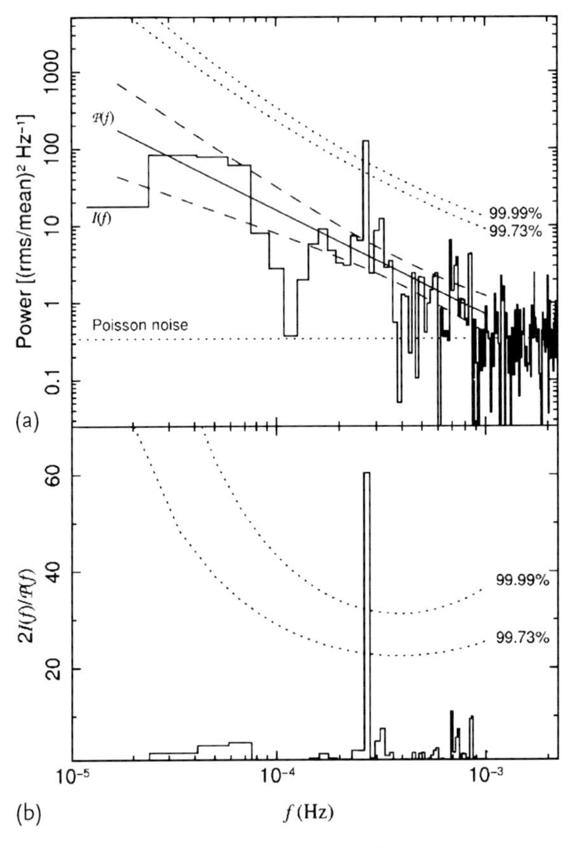

Figure 6.10 Power-density spectrum (PDS) obtained by Gierlinski *et al.* (2008) from a 90 ks observation of the NLS1 RE 1034+396. The quasiperiod $\nu = 2.7 \times 10^{-4}$ Hz is clearly detected, the formal confidence level being in excess of 99.9%. At present, this is a unique result for AGN. Speculatively, this QPO could correspond to the "high-frequency" feature seen occasionally in BHXRB, where pairs having 3 : 2 frequency ratios are documented. In that case, a mass estimate of about $10^7\ M_\odot$ is obtained.

exhibit two general classes of QPOs, referred to in the literature simply as low- or high-frequency.

The BHXRB high-frequency QPOs have peaks in the 10^2 Hz range, sometimes with multiple peaks at a 3 : 2 frequency ratio (Figure 6.11). These QPOs occur with a much smaller duty cycle than their low-frequency ($\sim$ 10 Hz) counterparts, so it may be that RE J1034+396 is a fortuitous example of catching an AGN in an analogous accretion-flow configuration. If that is the case, approximate mass scaling laws derived from the small sample of BHXRBs exhibiting the 3 : 2 high-frequency QPO pairs can be applied to RE J1034+396. Specifically, $\nu_0 = 931\,\text{Hz}\,(M/M_\odot)$, where the observed frequencies are $2\nu_0$ and $3\nu_0$ (Remillard and McClintock, 2006).

For RE J1034+396 it is unclear whether the observed feature is $2\nu_0$ or $3\nu_0$ (or neither), but assuming the latter a mass of $M_{\text{BH}} = 10^7\,M_\odot$ is obtained. However, this should be considered as highly uncertain. Detection of an actual 3 : 2 QPO ratio in this or some other object, and/or compilation of a sample of AGN QPO detections, among objects with alternative mass determinations from $M{-}\sigma$ or reverberation mapping methods would be needed before the QPOs become a viable tool for black-hole mass estimation in AGN.

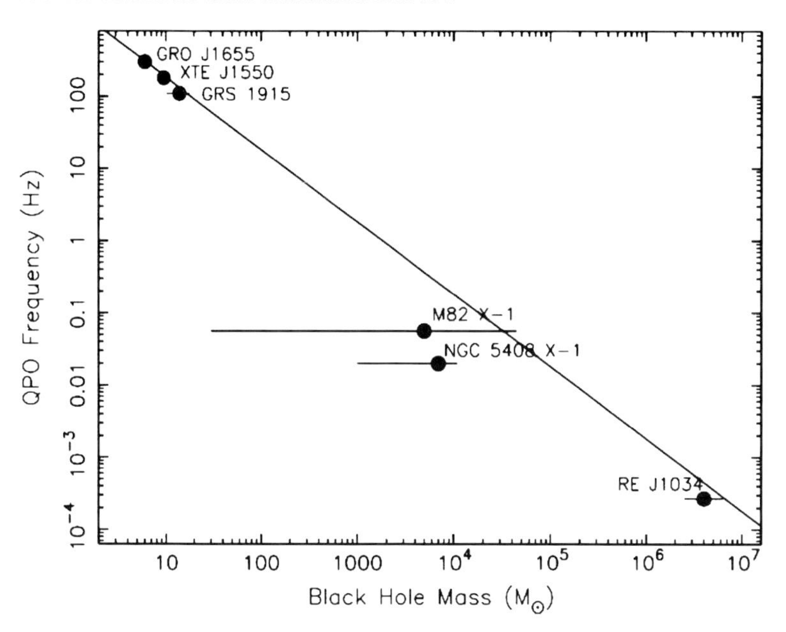

Figure 6.11 Relationship between the quasiperiodic oscillation (QPO) frequency and the mass of the black hole. Three X-ray binaries, XTE J1550-564, GRO J1655-40, and GRS 1915+105, display a pair of high-frequency QPOs with a 3 : 2 frequency ratio. The frequencies of X-ray binaries are plotted for the stronger QPO that represent $2 \times \nu_0$. The line denotes the relation, $2\nu_0$ (Hz) $= 1862\,(M_{\text{BH}}/M_\odot)^{-1}$ (Remillard and McClintock, 2006). The observed QPO frequency in the ultraluminous X-ray sources M82 X-1 and NGC 5408 X-1 is also plotted. From Zhou *et al.* (2010), reproduced by permission of the AAS.

6.5
Rapid Variability

AGN variability time scales are generally consistent with the physical or dynamical time scales of the system, such as the size of the broad-line region, viscous time scales of an accretion disk or structures within the jet. However, there have been examples of extremely rapid variations that are less easily interpreted. For example Carini and Miller (1992) or Carini (2006) have reported on optical band "microvariability" in blazar AGN on time scales as short as tens of minutes. At higher energies, Fermi observations have established hour time scale variations in

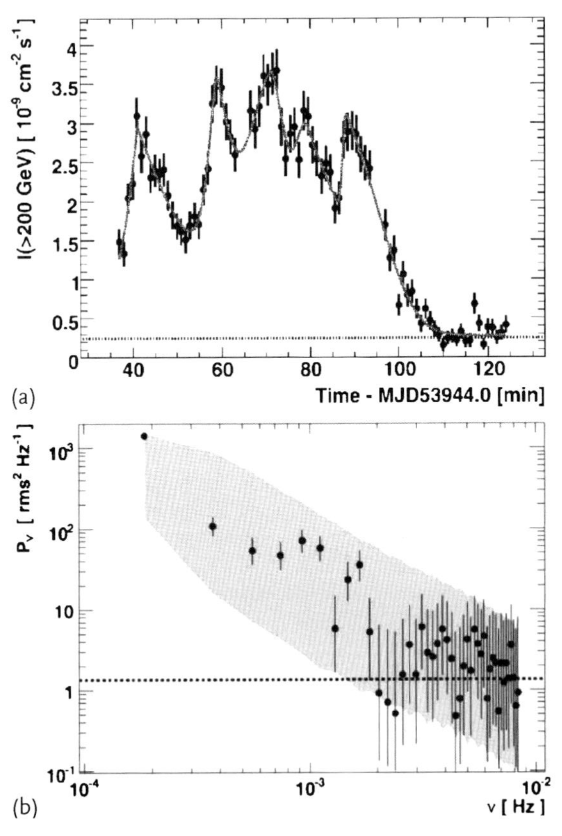

Figure 6.12 Gamma-ray flux history for photon energies above 200 GeV from PKS 2155-304 observed with HESS (a) (Aharonian *et al.*, 2007). The data are binned in 1 min intervals. The horizontal line represents the Crab Nebula flux and the curve is a model fitted to the data. PDS constructed from the same data and associated measurement error (b). The dotted line is the noise level and the gray shaded area corresponds to the 90% confidence interval for the PDS of a $P \propto \nu^{-2}$ fit to the data. Clearly, there is excess power at frequencies of about $\nu \simeq 1.6 \times 10^{-3}$ Hz corresponding to a $\sim$ 10 min variability time scale. (Reproduced by permission of the AAS.)

the GeV gamma-ray flux in 3C 454.3 (Abdo *et al.*, 2011a) and in particular a flux doubling in less than three hours.

Most remarkably though have been reports of high-energy (TeV) gamma-ray variations on minute time scales in several blazars, for example PKS 2155-304 (see Figure 6.12) and Mrk 501 (Albert *et al.*, 2007). Considering the likely $\sim 10^9\ M_\odot$ central black hole masses of such systems, minute time scales are an order of magnitude shorter than expected. The accretion processes presumed to underlie AGN variability should scale with the Schwarzschild radius R_S (Eq. (3.1)), since that is the smallest characteristic scale of the system. From this line of reasoning, an upper limit on the mass can be expressed in terms of the observed variability time scale t_{var} and a Doppler factor δ (Aharonian *et al.*, 2007),

$$M_{BH} \leq \frac{c^3 t_{var} \delta}{2G(1+z)} \frac{R_S}{R} \sim 1.6 \times 10^7\, M_\odot \delta \frac{R_S}{R} \tag{6.8}$$

Published black hole mass estimates for PKS 2155-304 (Bettoni *et al.*, 2003) suggest $M_{BH} \sim 10^9\ M_\odot$ thus requiring $\delta \geq 100 R/R_S$. Thus, emission from the most compact regions of a few R_S would require values of δ of 100 or greater, which is much greater than the values typically derived for blazars, $\delta \sim 10$ (note that δ is related to the bulk Lorentz factor γ by the expression $\delta = [\gamma(1 - \beta \cos\theta)]^{-1}$, see Eq. (5.13)).

This dilemma has led to suggestions for alternative jet emission models. For example, Begelman *et al.* (2008) have considered scenarios in which the variability involves enhanced emission in a small region within the jet. Enhancement emission could be triggered for example by dissipation in a portion of the black hole magnetosphere at the jet base within subregions smaller than R_S. Alternatively it could result from instabilities within the jet itself. Considering the energetics of the flares, along with the requirement that the photon–photon opacity allows the TeV photons to escape, those authors deduce that the bulk Lorentz factor in the jets must be $\gamma \sim 50$.

These ideas may be further tested in the near future. Ultimately, the shortest measurable variability time scales are limited by the telescope effective area. Future Cherenkov telescope arrays like CTA with significantly greater collecting area should reveal whether or not minute-time scale variations are ubiquitous properties of blazar gamma-ray flaring and if even more rapid variations are present.

7
Environment

Active galactic nuclei are not isolated in space. In fact, they interact extensively with surrounding matter, first through accretion of nearby matter onto the supermassive black hole, but additionally through radiation and particles emanating from the AGN and propagating into the surrounding medium. Concise reviews on the topic can be found in Veilleux (2008), Cattaneo *et al.* (2009), and Fabian (2010).

7.1
Host Galaxies of AGN

We first discuss briefly the question whether AGN reside in galaxies (they do!). Then we turn to the more difficult questions. Is there a preferred kind of galaxy type in which a supermassive black hole is accreting matter? A short answer seems to be following: the Hubble type does not matter as much as the mass of the host galaxy, i.e., a more massive galaxy is more likely to harbor an AGN, but whether that is a spiral or an elliptical galaxy matters only as far as ellipticals are in general more massive than spirals. Even more controversial is the possible connection between black hole mass and the host galaxy. At present time one may not be able to state more than the following: more massive black holes tend to sit in more massive galaxies. There seems to be some connection with the level of star formation in the host galaxy, but the question of how feedback (see Section 3.1.2) might work, if it is indeed present, is wide open.

Merging events involving AGN host galaxies as described in Section 7.3 were frequently invoked to explain AGN activity in the early 2000s, but presently there are lots of question marks around this theory. Instead minor mergers and secular growth seem likely be more important for the AGN than the rare merging of galaxies of similar size. Let us work our way through these questions, keeping in mind that the answers are not agreed on yet.

Active Galactic Nuclei, First Edition. Volker Beckmann and Chris Shrader.

7.1.1
Are There Naked Black Holes?

All AGN seem to be hosted in galaxies. Some claims have been raised of detection of "naked" AGN cores, which do not reside in a host galaxy. Early observations with the HST seemed to indicate that there was a significant fraction of quasars with no associated extended emission (Bahcall *et al.*, 1994). One prominent and persistent case was the quasar HE 0450-2958. Imaging with the HST failed to detect a host galaxy of this AGN at a moderate redshift of $z = 0.286$ (Magain *et al.*, 2005). At the same time, a nearby companion galaxy is visible and there is some evidence for an AGN also residing in this object. Deep infrared imaging has shown recently that there is indeed a very compact host galaxy, and that the AGN is consistent with a high-luminosity NLS1 (Jahnke *et al.*, 2009). This AGN is just one out of a large number of quasars, where the host galaxy had escaped detection, but with the deeper observations using improved instrumentation, the embedding environment was finally revealed. Today there seems to be no evidence that naked AGN exist: there is always a galaxy hosting the quasar core. In some cases it is difficult to determine the type of host galaxy, for example in the case of blazars where the beamed emission greatly outshines its surroundings, but even for bright nonbeamed AGN at high redshift this can be observationally challenging. Nevertheless, even in blazars we can see evidence of the surrounding matter through absorption lines which are produced when the radiation from the jet travels through the host before leaving the galaxy.

7.1.2
Morphological Classification of Galaxies

When discussing the connection of the AGN with its host, it is useful to have a look at the morphologies of galaxies in general and what is believed to be their evolutionary path. As one might expect, this is a vast topic for which we can only offer a brief overview.

The *Hubble sequence* (Hubble, 1926) is an observational classification of galaxies based on their morphological characteristics. Hubble divided the galaxies he observed into four categories: ellipticals, lenticulars, spirals, and irregulars. Ellipticals were sorted into subcategories according to their ellipticity, with E0 being spherical and E7 having the most flattened appearance. Lenticulars or S0 are at stages in between ellipticals and spirals, and can be thought of as E7 galaxies viewed edge-on. Spiral galaxies were divided into barred and nonbarred galaxies. A bar is visible in about two-thirds of all galaxies, and it connects the inner ends of the spiral arms and leads straight through the central bulge of the galaxy. Subgroups are defined according to the tightness of the spiral arms, with the nonbarred Sa and the barred SBa being tightly wound with a strong bulge component on one end, and the Sd (SBd) galaxies with loosely wound and fragmented spiral arms where most of the luminosity originates in the spiral arms and the bulge is comparably faint. Hubble subdivided the irregular galaxies into those with symmetric (Irr 1) and asymmetric

(Irr 2) appearance. The Hubble sequence aligned the classification along a "fork," with the round ellipticals on one end, the lenticulars at the dividing point, and with one prong of the fork for the nonbarred, and one prong for the barred spiral galaxies.

Hubble designated the ellipticals as "early-types" and the spirals as "late types," with E0 being the earliest, and Sd the latest galaxy types. He was aware and pointed out that this classification scheme should be interpreted as an evolutionary sequence, that is, "early" galaxies are not (necessarily) younger than the "late" ones. In fact, the presently favored scenario is that galaxies form as irregulars before they develop a spiral structure. Evolutionary events then can lead to an elliptical structure. This simple evolution, irregular–spiral–elliptical, is disturbed though by the merging events, as they can lead to intermediate steps in which the galaxy has again a more disturbed or distorted appearance due to the tidal forces during the interaction with another galaxy.

De Vaucouleurs (1959) and de Vaucouleurs *et al.* (1976) refined the classification scheme and also added a numerical Hubble stage T parameter, which aligns the Hubble sequence to values ranging from the compact ellipticals ($T = -6$) to the irregular galaxies ($T = 10$) on the other end of the scale. In Table 7.1 we give the Hubble stage in comparison with de Vaucouleurs class and the original Hubble classification. One still refers to the elliptical galaxies as "early-type," although the reader should keep in mind that those objects usually contain a rather old stellar population, while the "late type" spirals host larger amounts of gas and dust as well as young stars. The continued usefulness of the Hubble classification scheme and whether or not one should revise it based on our current understanding of galaxies has been a long-standing debate. We refer the reader to the review by Sandage (2005) for an in-depth discussion.

It was found early on that Seyfert cores mainly reside in spiral galaxies, primarily early-type spirals of Hubble type Sa and S0. The fact that most bright quasars are found in elliptical galaxies also supported the notion that Seyferts and quasars were assumed to be fundamentally different objects. If we believe on the other hand that they are intrinsically the same and differ from each other only through the black hole mass and/or their Eddington ratios, one would expect to find the brightest Seyferts and the fainter quasars to reside in both, spiral and elliptical galaxies. In other words, the host galaxy type should not influence the nature of the AGN at its center. But when looking at AGN in the local Universe, one finds that host galax-

Table 7.1 Galaxy classification following Hubble and de Vaucouleurs. The first line gives the Hubble stage T, the second line is de Vaucouleurs' classification, and the third line gives the Hubble class which corresponds to it.

−6	−5	−4	−3	−2	−1	0	1	2	3	4	5	6	7	8	9	10
cE	E	E^+	$S0^-$	$S0^0$	$S0^+$	S0/a	Sa	Sab	Sb	Sbc	Sc	Scd	Sd	Sdm	Sm	Im
E	E	E	S0	S0	S0	S0/a	Sa	Sa-b	Sb	Sb-c	Sc	Sc	Sc	Sc-Irr	Irr I	Irr I

ies are ellipticals, lenticulars, and bulge-dominated spirals, while LINER seem to be hosted by earlier type elliptical galaxies (Ho *et al.*, 1997). In fact, Cid Fernandes *et al.* (2010) argue that most LINER are not AGN but are instead driven by post-AGB stars and their ionizing emission. Most luminous Seyferts are found in spiral galaxies although these galaxies tend to be overluminous in comparison to typical spiral galaxies. In most cases one can find either a companion galaxy near an AGN, or finds the morphology of the host to be disturbed. This might point to the importance of merging events for the AGN activity, and we will come back to that point in Section 7.3. On the low-luminosity end of the Seyfert population (Section 4.2) one finds some dwarf systems as host galaxies, like the Small Magellanic Cloud (SMC). At the bright end of the AGN zoo most host galaxies are elliptical, as stated above. The separation seems to be rather sharp, with only one example of a bright FR-II known to be located inside a spiral galaxy. This AGN, known as 0313-192, is within the galaxy cluster Abell 428, which might explain this exception. Looking at radio-loud AGN, one finds that the majority are hosted by elliptical galaxies. The reason for the dearth of jets in spiral galaxies might be that the denser bulge interstellar medium absorbs the jet energy. Jets, as powerful as they are, appear to be sensitive to the environment in such a way that they can be easily switched off by an overly dense ambient environment. Moreover, it appears that there are no radio-loud AGN within galaxies with a stellar mass below $10^{10}\ M_{\odot}$. Best *et al.* (2005) studied host galaxies of a large sample of AGN at low redshifts ($0.03 < z < 0.3$) and found a steep rise in galaxies hosting AGN toward $M_{\rm gal} \simeq 10^{11}\ M_{\odot}$, where more than 50% of all galaxies harbor active cores. Figures 7.1 and 7.2 show the fraction of galaxies which are optically identified and radio-loud AGN, respectively, as a function of the host galaxy's mass.

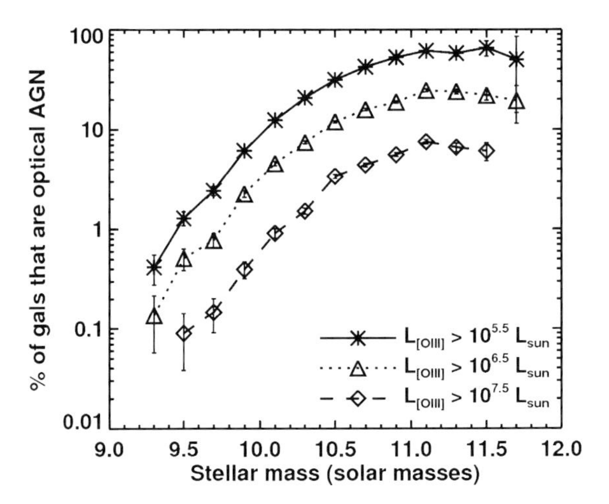

Figure 7.1 The fraction of galaxies which host an AGN showing [O III] line emission as a function of stellar mass in the galaxy. The AGN have been divided into three subsets according to their [O III] line luminosity. The fraction of [O III] luminous AGN does not depend on the mass of the host galaxy. Graphic from Best *et al.* (2005).

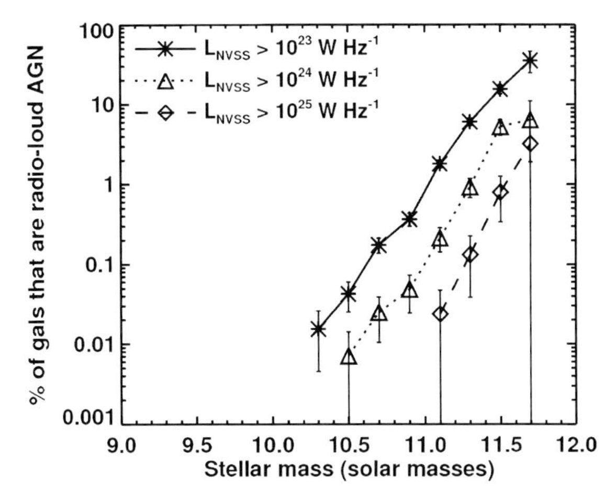

Figure 7.2 The fraction of galaxies which host a radio-loud AGN as a function of stellar mass in the galaxy. The radio-loud AGN have been divided into three subsets according to their radio luminosity at 1.4 GHz. Radio-loud AGN seem to reside in massive galaxies only. Graphic from Best *et al.* (2005).

As can be seen from these figures, the fraction of radio-loud AGN rises steeply, and at a host galaxy mass of $5 \times 10^{11}\ M_\odot$ more than half appear to be radio-loud. At the same time, the ratio between bright and faint AGN does not depend on the host galaxy's mass. This can be seen in Figure 7.2, where the ratio between the fraction of galaxies containing an AGN as a function of radio luminosity remains constant over the host mass range of $\sim 3 \times 10^{10}$–$5 \times 10^{11}\ M_\odot$. Similar results were derived earlier by Kauffmann *et al.* (2003a) when studying the dependence between host masses and AGN cores. Using [O III] as tracer of AGN activity it is found that all AGN reside in massive galaxies, but there is no indication for a correlation between AGN power and host galaxy mass. In addition, the high-luminosity AGN seem to reside in hosts with a young stellar population. Kauffmann *et al.* did not find a difference between hosts of Seyfert 2 galaxies and those of quasars when grouping the AGN into classes with the same [O III] luminosity, that is, with the same core activity.

Other recent studies also find an increase of the AGN fraction with the stellar mass of the host galaxies. Using Chandra and HST data, Gallo *et al.* (2010) find at a given luminosity limit an apparent increase of the AGN fraction with increasing host stellar mass. But they argue that this previously identified correlation is merely due to a selection effect and there is instead an anticorrelation between accretion efficiency and black hole mass. The correlation between Eddington ratio and M_{BH} has the form of $\lambda \propto L_X/L_{Edd} \propto M_{BH}^{-0.6}$. Lower-mass AGN seem on average to accrete matter closer to the Eddington limit than the large black holes, although the latter have on average higher bolometric luminosities. This anticorrelation is also referred to as *accretion downsizing*. Thus, for a sample comprised of objects above a given Eddington ratio λ, the AGN fraction indeed *decreases* with increasing black hole mass. The difference here is to look at accretion efficiency λ rather than at the luminosity, which is basically depending on the accretion rate $\dot{M}$.

As for the type of host galaxy, one hopes to find also other correlations between the AGN core and the host. This would enable us to determine the AGN type, or other AGN properties, from the appearance of the host. But there seems to be a lack in strong correlations, except for the fact that no bright radio galaxies are found in dwarf systems. Other than that, it seems that host galaxies with an AGN core have the same appearance as the parent population of galaxies without such a central engine. For example, when studying radio-loud AGN, O'Dowd *et al.* (2002) found that the elliptical host galaxies follow the well known *Kormendy* relation similar to nonactive elliptical galaxies; see Figure 7.3. The Kormendy relation compares the surface brightness of the galaxy with its effective radius and was derived by Kormendy (1977). The effective radius is defined as the radius which encloses half of the total luminosity of the galaxy. The relation thus basically tells us that more luminous objects are larger and more diffuse. Analysis of HST images of galaxies with and without an AGN core in the redshift range $0.3 < z < 1$ supports the claim that there is no difference in their host galaxies, and that there is no significant difference in the fraction of distorted galaxies (Cisternas *et al.*, 2011). Also when going toward higher redshifts this statement appears to be true. There is no difference in AGN host galaxies compared to the parent galaxy population without active

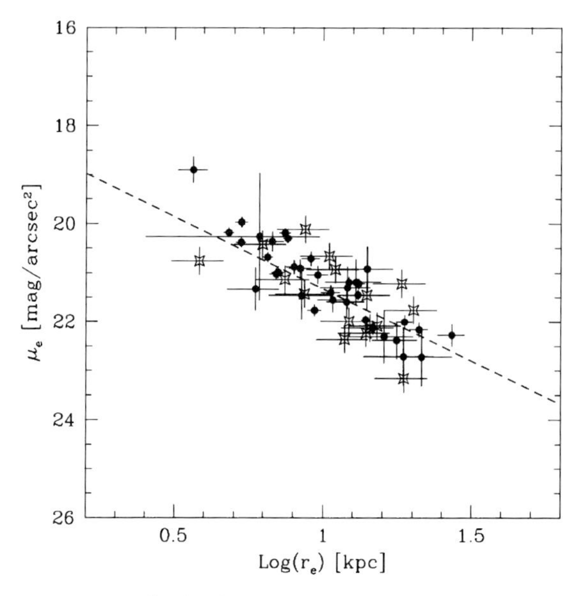

Figure 7.3 Surface brightness versus effective radius of host galaxies of radio-loud AGN. High-power AGN are marked by stars, low-power ones by circles. The dashed line indicates the *Kormendy relation* for elliptical galaxies as presented in Hamabe and Kormendy (1987). Apparently, the early-type host galaxies are not altered by low- or by high-luminosity AGN. Graphic from O'Dowd *et al.* (2002). (Reproduced by permission of the AAS.)

nuclei, and there is no evolution detectable in either effective radii or morphology distribution between $z \sim 2$ and the local Universe (Schawinski *et al.*, 2011a).

7.1.3
Host Galaxy and Black Hole Mass

One should bear in mind that even supermassive black holes of $10^9\ M_\odot$ are still only a small fraction of the galactic mass. The two masses, that of the host M_* and that of the central engine M_{BH}, seem to be correlated however. This goes hand in hand with the observation that the bulge velocity dispersion σ is correlated with M_{BH}, and that the velocity dispersion in elliptical galaxies is correlated with the total stellar mass. The ratio M_*/M_{BH} depends on M_{BH} at all redshifts. Studying systems at redshift $z \simeq 2$, Trakhtenbrot *et al.* (2011) found that for the lower black hole masses of $M_{BH} \simeq 10^8$ this ratio is higher ($M_*/M_{BH} \sim 280$) than for the high-mass systems with $M_{BH} = 10^9\ M_\odot$, where this ratio is about 40. This means that although high-mass AGN reside in larger host galaxies, there is no constant ratio between the host mass and the black hole mass. In the local Universe M_*/M_{BH} grows by a factor of $\sim$ 4–8, which seems to indicate that the host galaxies grow faster than the black hole cores. Thus, there seems to be a rather large scatter in the ratio between the black hole mass and the total mass of the galaxy. A tighter relation appears when one considers only the mass of the inner part of the galaxy, that is, of the bulge (Figure 7.4). In this context, one can see elliptical galaxies as "bulge-only" objects.

Silk and Rees (1998) found that there is a correlation between the velocity dispersion σ of stars in the bulge of galaxies, and the mass of the central black hole M_{BH}, the $M-\sigma$ relation mentioned earlier. As shown by Fabian (2010), this cor-

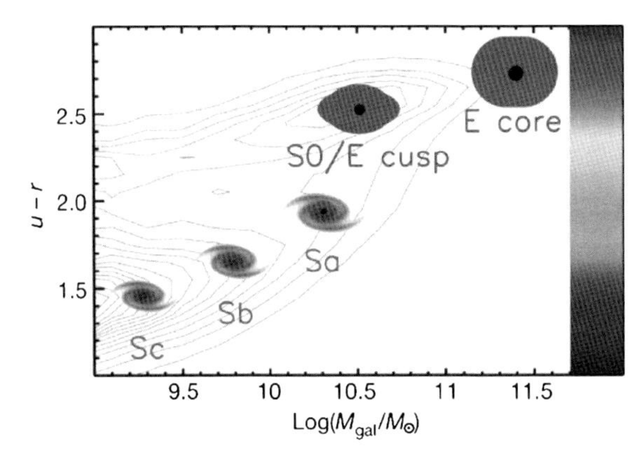

Figure 7.4 Color $u-r$ versus galaxy mass. The larger the $u-r$ value, the redder the galaxy. The central bulge of the spirals are similar to small elliptical galaxies. Ellipticals and spiral bulges have similar stellar mass to black hole mass ratio of the order of 0.1%. Ellipticals are located above a critical mass of $M_{crit} \simeq 10^{12}\ M_\odot$. Graphic from Cattaneo *et al.* (2009). For a color version of this figure, please see the color plates at the beginning of the book.

relation can be understood when assuming that the AGN core is operating close to the Eddington limit. The radiation pressure will drive out a fraction f of the gas mass M_{bulge}, which we assume to be spherical with radius R. In a steady mode, the force from the time-averaged luminosity L_{Edd}/c will be equal to the gravitational force, keeping the mass $f M_{\text{bulge}}$ just outside the bulge at radius R

$$\frac{L_{\text{Edd}}}{c} = \frac{G M_{\text{bulge}} f M_{\text{bulge}}}{R^2} \tag{7.1}$$

On the left side we can use Eq. (3.9) for the Eddington luminosity. Assuming that the bulge is an isothermal sphere, the mass inside R is a function of the velocity dispersion σ of the gas: $M(< R) = 2R\sigma^2/G$ and thus

$$\frac{M_{\text{bulge}}}{R} = \frac{2\sigma^2}{G} \tag{7.2}$$

Using this in Eq. (7.1) gives

$$\frac{4\pi G M_{\text{BH}} m_{\text{p}}}{\sigma_{\text{T}}} = G f \left(\frac{2\sigma^2}{G}\right)^2 \tag{7.3}$$

with σ_{T} being the Thomson cross-section. After rearranging the equation, we can use the velocity dispersion to measure the mass of the central engine

$$M_{\text{BH}} = \frac{f \sigma^4 \sigma_{\text{T}}}{\pi G^2 m_{\text{p}}} \tag{7.4}$$

It turns out that a fraction $f \simeq 0.1$ gives a good representation of the measured values. Fabian's arguments imply a spherically symmetric geometry with the AGN effecting the bulge omnidirectionally. Considering the unified model we described in Section 4.8, this seems questionable though: Seyfert galaxies have a plane in which obscuring matter should avoid direct outflows and radiation pressure on the surrounding medium. One therefore would not expect spherical symmetry regarding the AGN influence on its host. Jahnke and Macciò (2011) argue that the observed $M_{\text{bulge}}/M_{\text{BH}}$ relation does not have to imply a physically coupled growth of the central supermassive black hole and the bulge mass, but that it can be fully explained by the hierarchical assembly of black holes and of the bulge through galaxy merging. If this is indeed the case it means that there is no coupling between the bulge and the central engine, but that they simply grow in parallel when the host galaxy is undergoing a merging event with another galaxy. This would imply that the mechanism described by Fabian is not responsible for regulating the growth of the black hole. We will come back to these arguments when looking into the effects of galaxy mergers on the AGN in Section 7.3.

A relation similar to $M-\sigma$ was found by Faber and Jackson (1976) between the luminosities and stellar velocity dispersions of elliptical galaxies. Their luminosities are also seen to follow a σ^4 dependence, the so-called *Faber–Jackson relation*. As Murray *et al.* (2005) pointed out, it would be remarkable if this similarity in the

$M_{BH}-\sigma$ and in the $L_{bulge}-\sigma$ relation were a pure coincidence. They determine the limiting luminosity above which the central activity would drive away a significant amount of gas from the galaxy and thus stop the accretion process onto the bulge

$$L \simeq \frac{4fc\sigma^4}{G} \tag{7.5}$$

In the Faber–Jackson relation $f \simeq 0.1$ seems to apply. Although this formula had been derived considering the feedback between star-forming processes and interstellar gas in the galaxy, it should also apply to AGN cores in elliptical galaxies and in bulges of spiral galaxies. Again, this argument can be turned around if there is no physical coupling. If we assume that there is a direct near linear correlation between M_{BH} and M_{bulge} because they have been assembled simultaneously in the history of the Universe, this would also explain why the *Faber–Jackson relation* works in the case of AGN. If the bulge of an AGN is not different from the bulge in a nonactive galaxy, the mass scales with velocity dispersion (σ^4) and hence the black hole mass also scales with σ^4.

Correlating the bulge mass with the black hole mass, one finds a ratio of $M_{BH}/M_{bulge} = 0.1\%$ throughout the whole mass spectrum of AGN (e.g., Gebhardt *et al.*, 2000). As in the case where the more luminous AGN on average host larger central black holes, there is a trend for spheroidal hosts to become more prevalent with increasing AGN luminosity, as shown by Dunlop *et al.* (2003). They found out that for nuclear luminosities brighter than $M_V \sim -23.5$ mag practically all the host galaxies are ellipticals, regardless of whether the AGN itself is radio-loud or radio-quiet.

Mass outside the bulge does not seem to be influenced by the AGN at the center. There is neither a connection detectable between the black hole mass and the galactic disk (Kormendy *et al.*, 2011), nor with the dark matter halo of the galaxies (Kormendy and Bender, 2011). Also, the overall H I content of the galaxy has no influence on the central engine. It seems that bulge and central black hole grow in parallel, with little influence from the larger environment.

Some caution has to be applied in all these studies, though, to ensure that one samples the same AGN population at high redshift as in the local Universe: in a flux-limited study, obviously the high-redshift systems are the high-luminous objects, whereas in the local Universe one will sample many low-luminosity AGN. In addition, dynamical properties which are resolvable in the local Universe, will be blurred at high redshifts. As Lauer *et al.* (2007) point out, the low redshift relations between $\log M_{BH}$, $\log L$, $\log M_*$, and $\log \sigma$ are linear and high-redshift data are often interpreted by assuming that the same relations apply as in the local Universe. This can then result in an offset in the observed correlation, for example in $\Delta \log M_{BH}$ when studying the same host galaxy mass or velocity dispersion. Other selection effects have been discussed in Schulze and Wisotzki (2011). As one observes a down-sizing effect in the local Universe, galaxies with a high black hole mass are often much more quiescent than their lower black hole mass counterparts. The latter apparently have a higher probability to be found actively accreting, which results in a bias toward low-mass black holes when building the $M-\sigma$ relation.

7.1.4
AGN-Host Galaxy Feedback

At the same time, the AGN can also prevent further mass accretion. If the black hole accretes close to the Eddington limit, the radiation pressure can drive the gas outwards and thus prevent further growth. This leads to an interesting self-regulating mechanism: if gas is abundant in the vicinity of the AGN core, the enhanced emission that results will drive that gas out from its immediate surroundings. This will occur through winds and jets as well as radiation. In fact, photons can heat the gas by photoionization, that is, by ejecting one or more electrons from atoms. Simulations show that the interstellar medium can be heated up to 10^6 K by the AGN activity through photoionization and Compton heating (e.g., Kim *et al.*, 2011). This will cause the gas to expand and thus escape from the galaxy. Photons can also directly push material out through radiation pressure. The effect on ionized gas is low though because of the lower cross-section, thus this effect is limited to the vicinity of the AGN where the photon flux is strongest. Radiation pressure on dust particles is more efficient and the transfer of momentum from the photons to the dust can be close to 100% if photons can travel through the material in all directions, that is, if the dust occupies a large solid angle. In total, radiation pressure can expel about 10% of the stellar bulge mass.

Jets can also affect the matter in the galaxy in two ways. The jet can shock heat matter as it propagates outwards. This will lead, similarly to radiative heating effects, to an expansion of the enshrouding gas. In addition, ram pressure of jets can drive out mass directly. It has to be remembered however, that jets are by definition confined to a certain direction which therefore limits their effectiveness. In addition we observe many jets which reach far beyond the edges of the galaxies they start in. Thus, the transfer of momentum from the jet to the gas in the galaxy must be low in these cases. The effect of the jet on the surrounding medium will not set in immediately after the formation of the jet. As De Young (2010) points out, the heating will start only after the jet becomes fully turbulent sonic or subsonic due to the deceleration in the cold gas. Thus, it will take $\sim 10^8$ years for the interstellar medium to react to the onset of a radio-loud AGN.

As a consequence of winds, the gas in-flow will slow down, leaving the central engine starving. This will in turn reduce the driving pressure of the AGN on its environment. As discussed in the previous subsection, the AGN might not only influence the accretion flow onto itself, it can also steer the growth of the bulge in which it resides. This type of feedback might thus cause the firm link between the black hole mass and the velocity dispersion in the bulge and be responsible for the $M_{BH}-\sigma$ relation (Fabian, 2010). As long as the AGN remains active it will effectively regulate the host galaxy's ability to continue increasing its mass. Other authors called into question the physical link and interaction between AGN and host (Jahnke and Macciò, 2011).

AGN cores can remain active only as long as there is abundant cool gas available in the host galaxy. Thus, one can expect the AGN activity to slow down as time goes by. This effect is indeed seen. Whereas at redshift $z \simeq 2$–3 about 10% of the supermassive black holes are accreting and thus appear as AGN, this fraction is only $\simeq 0.1\%$ in the local Universe (Dunlop *et al.*, 2003). The question, whether this AGN evolution will in turn impact the host galaxy evolution, remains under debate. If the feedback scenario is correct, then the fraction of time the AGN is accreting, the so-called duty cycle, is critical to the growth of the host galaxy. As one now expects most galaxies to host a supermassive black hole, this implies that the feedback mechanism will directly impact the cosmological evolution of galaxies. On the other hand, if there is no physical link, and galaxies and black holes simply grow through the history of the Universe in parallel, but unrelated, AGN hosts will not behave any differently than normal inactive galaxies. Support for the latter hypothesis comes from Jiang *et al.* (2011) when studying low-mass black hole AGN with $M_{\mathrm{BH}} \sim 10^5$–$10^6\,M_\odot$. The majority of the AGN host galaxies with disks are likely to contain pseudobulges and very few of these low-mass black holes live in classical bulges. This suggests that AGN cores with mass $M_{\mathrm{BH}} < 10^6\,M_\odot$ live in galaxies that have evolved secularly over the majority of their history. A classical bulge is apparently not a prerequisite to host a black hole.

When studying jets and tori of AGN, one can measure their orientation with respect to the host galaxy. It was found early on that there seems to be little or no correlation between the AGN jet axis and the host symmetry axis (Keel, 1980; Lawrence and Elvis, 1982). This seems to indicate the lack of a large-scale physical coupling between the active nucleus and its surroundings. That is all the more surprising, given that undoubtedly the matter which flows onto the AGN originates from the larger galactic scale. If both rotations were co-aligned, one would expect that the black hole will be spun up by the accretion, as it would be reasonable to assume that at least some of the angular momentum of the matter rotating with the host galaxy and eventually falling onto the black hole will be conserved. Simulations trying to solve this problem face the problem that it is not yet possible to model the larger-scale galaxy evolution with the same spatial resolution necessary to understand the accretion process close to the AGN. One approach to overcome this problem is to simulate the larger scales in a more coarse grid and then to zoom into the central region, as done for example by Hopkins and Quataert (2010). Using these simulations, Hopkins *et al.* (2012) show that the instabilities in the in-flow can lead to the near random orientation of the nuclear disk with respect to the rotation axis of the host galaxy. Their work also shows that moderate values of black hole spin can be expected, on the order of $j \sim 0.3$–0.9 (Eq. (5.8)), although even lower values are possible if further processes increase the randomization of angular momentum of the in-falling matter.

7.2
The AGN–Starburst Connection

In the previous section we have seen that there might be a tight connection between the bulge of a galaxy and the AGN core. At the same time, the host galaxies seem to look similar to those without an active nucleus. An important property of a galaxy that one must consider is the number of stars created per unit time. Star-formation rates (SFR) can vary significantly between galaxies. Our Galaxy for example has a very low star-forming rate. Observations by the Spitzer infrared space telescope of young stellar objects in the Milky Way indicate a value as low as SFR $\simeq 1\ M_\odot\ \mathrm{yr}^{-1}$ (Robitaille and Whitney, 2010). Other methods use the total amount of ^{26}Al to estimate the SFR. Since this element is purely created in supernova explosions and has a half life time of $\simeq 7.2 \times 10^5$ years, it is an excellent tracer of the rate stars are created in the Milky Way. Based on INTEGRAL/SPI measurements of the ^{26}Al line at $E = 1808.65$ keV, Diehl *et al.* (2006) derive a star-formation rate of SFR $\simeq 4\ M_\odot\ \mathrm{yr}^{-1}$ for our Galaxy.

Spiral galaxies like the Milky Way have already used up most of the cold, star-forming gas. Simulations show that creation of stars occurred at a much higher rate in such galaxies back at redshifts $z \simeq 1$, as shown in Figure 7.5. In more extreme cases, the star-formation rate can reach values exceeding $10^3\ M_\odot\ \mathrm{yr}^{-1}$ (e.g., Smolčić *et al.*, 2011).

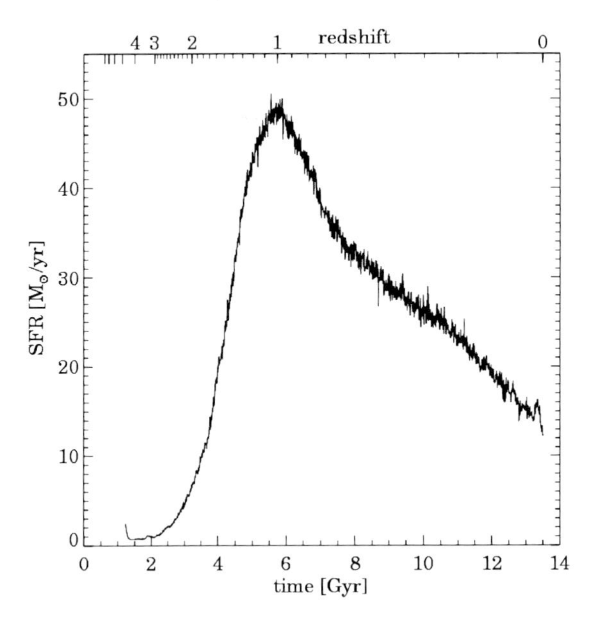

Figure 7.5 Simulated evolution of the star-formation rate within the 20 kpc halo of a galaxy similar to our Milky Way. The maximum of star formation is reached around redshift $z \simeq 1$ and then decreases steadily again thereafter. Graphic from Samland and Gerhard (2003), reproduced with permission © ESO.

7.2.1 Estimating the Star-Formation Rate

In order to study the connection between the host star-formation rate and the active nucleus, we first must clarify how to quantify the star-formation rate in terms of observables. There are several methods used to estimate the SFR, as described in a review by Kennicutt (1998). The basis of all these methods is to quantify the amount of hot, young, and short-lived stars in the galaxy. This can be done through measurement of those emission lines and continuum emission that are predominantly produced by these stars, or by measuring the energy that has been deposed in the dust of the interstellar medium by the star-formation process.

A close relation between young stars and the spectral energy distribution of galaxies can for example be derived in the UV region, because the short-lived but bright O and B stars are strong UV emitters. Taking the measurement in the 1500–2800 Å band, one can derive the star-formation rate directly, in case one assumes solar element abundances and an initial mass function (IMF) similar to that derived by Salpeter (1955)

$$\mathrm{SFR}(M_{\odot}\,\mathrm{yr}^{-1}) = 1.4 \times 10^{-28}\,L_{\nu}\,(\mathrm{erg\,s^{-1}\,Hz^{-1}}) \tag{7.6}$$

The Salpeter IMF happens to be flat in the UV region, thus Eq. (7.6) is frequency-independent as long as one measures the luminosity in the UV band.

Another approach to measuring SFR is based on the fact that the UV radiation field of young, massive stars will heat up and ionize the interstellar gas. Therefore, a starburst will lead to strong nebular emission lines. One can measure the strength of the Hα or Hβ line and deduce the star-formation rate using known scaling laws. Assuming again solar abundances and a Salpeter IMF, one derives for example for Hα

$$\mathrm{SFR}(M_{\odot}\,\mathrm{yr}^{-1}) = 7.9 \times 10^{-42}\,L_{\mathrm{H}\alpha}\,(\mathrm{erg\,s^{-1}}) \tag{7.7}$$

Here, $L_{\mathrm{H}\alpha}$ is the measured luminosity of the Hα line at $\lambda = 6562.8$ Å. Already at moderate cosmological distance of $z \sim 0.5$ the Hα line is redshifted out of the visible domain. The [O II] λ3727 forbidden line doublet can then be used instead. This line is sensitive though to the physical conditions in the gas, such as temperature and abundance, and also more sensitive than Hα to absorption by dust in the line of sight, therefore providing SFR estimates with larger uncertainties than those computed applying Eq. (7.7). An average calibration gives

$$\mathrm{SFR}(M_{\odot}\,\mathrm{yr}^{-1}) = (1.4 \pm 0.4) \times 10^{-41}\,L_{[\mathrm{OII}]}\,(\mathrm{erg\,s^{-1}}) \tag{7.8}$$

Finally, the interstellar dust also gets heated by the large number of hot stars, preferentially by the UV emission as the cross-section is highest for UV light. The dust will re-emit in the far-infrared, in the range $\lambda = 10{-}300$ μm. If one assumes that the dust is optically thick to the UV radiation of the starburst, then the star-formation rate can be directly determined from the far-infrared luminosity

$$\mathrm{SFR}(M_{\odot}\,\mathrm{yr}^{-1}) = 4.5 \times 10^{-44}\,L_{\mathrm{FIR}}\,(\mathrm{erg\,s^{-1}}) \tag{7.9}$$

A drawback of all these methods is the possible contamination of the measured quantities by the direct AGN emission. The accretion disk continuum provides peaks in the far UV flux and emission lines like Hα and [O II] are a key AGN spectral feature. There is also infrared reprocessed AGN emission, which can dominate that portion of the spectrum. Thus, these methods require spatially resolving the host from the AGN core and to derive spectra from each of the spatially separated components. In addition, star-formation rates as given above are averaged for entire galaxies, but the local SFR will depend where within the galaxy one observes. In case one measures for example the Hα line in the spiral arms, one will likely get a higher SFR than when measuring in the bulge, because star formation primarily can be found in the gas-rich arms. In addition, measurements of parts of the galaxy have to be extrapolated to the total amount in the host, adding further uncertainties.

An indicator for star formation, and not for AGN activity, are the polycyclic aromatic hydrocarbon (PAH) lines, already mentioned in Section 5.2. These lines are also easier to be detected in more distant galaxies. Thus, detection of strong PAH lines is a good indicator for star-forming activity. There are some issues though when determining the star-formation rate in galaxies, which host an AGN. In the vicinity of the central engine, PAH gets destroyed and thus the equivalent width of those lines is reduced. Therefore, in the case of a bright AGN, the star-formation rate derived from the PAH line strength might be underestimated.

In view of the large sky surveys with more than 10^5 objects such as the SDSS, new classification criteria to estimate the amount of star formation have been developed. As described by Kauffmann *et al.* (2003b) a combination of the calcium break strength at 4000 Å and the Balmer absorption line Hδ can be used to measure the SFR. The main contributor to the calcium break are ionized metals, which are rare in young stellar populations and abundant in old, metal-rich galaxies. Hδ is produced mainly in late-B to early-F stars and thus is strongest some time after the end of the starburst. Assuming a star-forming rate development of the form $\mathrm{SFR}(t) \propto \exp(-\gamma t\,[\mathrm{Gyr}])$ with starting time t_{form} and time scale parameter γ, the strength of the calcium break in the galaxy will increase monotonically over the galaxy's lifetime, while the Hδ absorption line will increase in strength until $t \simeq t_{\mathrm{form}} + 4 \times 10^8$ years and decrease afterwards. In the case of type 1 AGN the Hδ absorption line is blended with the emission line from the broad-line region, which has to be considered when estimating the SFR.

7.2.2
AGN–Starburst Feedback

As previously discussed, there have been claims that a feedback mechanism is occurring in which star-formation processes in the host are caused by AGN activity. If true, the question arises as to how strong these effects are, and how they manifest among the various classes, sizes, and luminosity ranges. If such dependencies would indeed be observable, it lends support to the AGN host galaxy feedback scenario.

In reality, it is often difficult to distinguish between the star formation and the AGN activity, especially in the case of low-luminosity AGN. Seyfert 2 show Ca II triplets like normal galaxies but the Mg Ib line at 5175 Å is often weaker. This can be explained by a superposition in the spectrum of young red supergiants plus AGN continuum emission (Cid Fernandes *et al.*, 2004). The bolometric luminosity of the starburst in these cases is of the same order as the bolometric luminosity of the obscured AGN core ($10^{10}\ L_{\odot}$).

When studying a large sample of 2×10^4 local ($z < 0.3$) narrow-line AGN from the SDSS (see box of Section 4.6), Kauffmann *et al.* (2003a) found that low-luminosity AGN rather live in galaxies with older stellar populations, while the high-luminosity ones are hosted by galaxies with young populations. The study shows that the low-luminosity AGN reside mainly in early type galaxies, in ellipticals, in lenticulars and in bulge-dominated spirals. Concerning the connection to star formation, it seems that the more luminous AGN are connected to young stellar populations in their hosts. This statement has been altered somewhat by Jahnke *et al.* (2004a) and by Sánchez *et al.* (2004) who find that the disk-dominated AGN have quite normal stellar populations for their Hubble type but that the early-type ones have much younger stellar populations than normal early-types. However, the disks and spheroids were nearly indistinguishable from each other regarding their optical colors hence at least the young part of their stellar populations. Therefore, host galaxies appear to have colors and thus star-forming activity at least as blue and active as the comparison nonactive galaxy.

In fact, one finds strong Hδ absorption lines in the high-luminosity sources, which, as pointed out earlier, indicates that these galaxies have experienced a starburst some 0.1–1 Gyr ago. The starburst is found not to primarily happen in the central region of the galaxy, but is rather distributed over scales of several kiloparsecs or more. Another result from the SDSS is that there is apparently no difference between the SFR of type 1 and type 2 AGN of similar luminosity. One has to keep in mind that because the SDSS contains such a large number of objects, the galaxies are no longer classified by their Hubble or de Vaucouleurs type, which would require inspection by eye, but instead are sorted on the basis of parameters such as the concentration index C and the stellar surface mass density μ. One then finds that the early-type galaxies (Hubble type E but also Sa and S0) have a μ in the range $3–30 \times 10^9\ M_{\odot}\ \mathrm{kpc}^{-2}$ and a higher concentration of $C > 2.6$. The late-type (spiral) galaxies have $\mu < 3 \times 10^9\ M_{\odot}\ \mathrm{kpc}^{-2}$, and $C < 2.6$.

The AGN–starburst connection can also be seen in the infrared spectra of quasars. Using Spitzer spectra of radio-loud local quasars, Schweitzer *et al.* (2006) detected PAH lines in 11 out of 26 AGN and fine-structure line emission in all of the objects. Thus, tracers of starbursts also correlate with the AGN activity indicators. The link between starburst and AGN can also be found in high-redshift quasars, for example in samples of AGN up to redshifts $z \sim 2.8$ (Sánchez *et al.*, 2004; Jahnke *et al.*, 2004b), and in single case studies beyond $z = 6$ (e.g., Goto *et al.*, 2009). Among radio-loud quasars, about 20–30% show an excess in infrared compared to the optical emission with luminosity ratios of $L_{\mathrm{IR}}/L_{\mathrm{B}} > 0.4$. The infrared spectra of high-redshift quasars however seem to lack the PAH signatures present in local quasar samples. The strong AGN activity may destroy the PAH molecules in these objects, although they still exhibit strong infrared excess indicative of dust.

An alternative possibility is that radio jets might expel the gas necessary for the star formation. In fact, Shabala *et al.* (2011) found that luminous FR-II radio galaxies tend to inhibit star formation in the host by expelling the cold interstellar gas. The same effect is not seen in the less powerful FR-I galaxies. More evidence for the connection comes from UV observations of $\sim 5 \times 10^4$ local galaxies ($z \sim 0.1$) using the GALEX satellite (Salim *et al.*, 2007). They find that the star-formation history depends strongly on whether the galaxy hosts an AGN, or when none is apparent, whether the Hα line is visible in emission. Also when looking into the X-ray-selected AGN, one finds a tight connection to younger host galaxies. Silverman *et al.* (2009) studied the hosts of 147 Chandra detected AGN, which appear to be preferentially inside environments equivalent to massive galaxies with substantial levels of star formation. The same conclusion was reached by Rumbaugh *et al.* (2012), studying host galaxies of X-ray-selected AGN detected at redshifts $z \sim 0.7$–0.9. Using the optical [O II] and Hδ lines as indicators for star formation, they find that on average the AGN hosts either are still active starburst galaxies or experienced one within the previous ~ 1 Gyr.

Apparently, the galaxies start out having active star formation, followed by a period of simultaneous star formation plus AGN activity. Both subsequently decline leaving a nonactive galaxy with no strong Hα emission, indicative of a halt in star-formation activity. A more prudent statement might be to say that at a certain point in time, sufficient cool gas is available in the host galaxy, and this gas is feeding the star-formation areas as well as the nucleus. Simulations by Hopkins and Quataert (2010) estimate that the black hole's mass accretion rate $\dot{M}_{\rm BH}$ is correlated with the star-formation rate $\dot{M}_*$ with $\dot{M}_{\rm BH} \simeq 0.003\,\dot{M}_*$. Although there is a large scatter, this correlation seems to hold for low star-formation rates of $\sim 1\,M_\odot\,{\rm yr}^{-1}$. The study also shows that within the central 10 pc around the black hole, the star-formation rate is equal to the mass accretion rate. These results from simulations are consistent with the observations by Silverman *et al.* (2009) who observe black hole accretion rates of $\dot{M}_{\rm BH} \sim 0.001$–$0.01\,\dot{M}_*$ with high star-formation rates of $\dot{M}_* > 10\,M_\odot\,{\rm yr}^{-1}$.

AGN and star-formation activity do not necessarily end at the same time. While the star-formation process requires a large amount of cold gas, only a small fraction of this is needed to keep a supermassive black hole active. The mass involved in an active star-formation phase is some hundreds to thousand of times larger than the mass accreted onto the black hole. Thus, when the starburst dies out, the AGN activity can continue for another ~ 0.1 Gyr, as shown in feedback simulations (Hopkins, 2012). In addition, the heating up to 10^6 K of the interstellar medium by the AGN will cause the starburst to shut down, at least within the galactic inner core. This has been derived from simulations also earlier by Springel *et al.* (2005a,b): once an accreting supermassive black hole has grown to a critical mass, feedback terminates its further growth and expels gas from the central region in a powerful quasar-driven wind. The remnant is then gas-poor. A similar conclusion has been reached by Bell (2008): a bulge, and by implication a supermassive black hole, is an absolute requirement for full quenching of star formation in galaxy centers. Schawinski *et al.* (2009) derive a time delay of 100 Myr between the shutdown of star formation and the onset of AGN activity, with a time delay of 0.5 Gyr between the peak of star formation and the peak of the accretion onto the black hole. This kind of feedback naturally explains the population of extremely red and thus relatively inactive massive early-type galaxies in the local Universe. Other studies deny a large impact of the AGN on the SFR, simply because its influence will be limited to the central region of the bulge only.

AGN outflows may influence not only the star formation in their hosts, but the ambient intergalactic (IGM) medium as well. Germain *et al.* (2009) estimate that from 65 to 100% of the entire Universe at the present epoch received matter from AGN. Thus, just as we assume that matter surrounding the solar neighborhood has been processed through supernova explosions, it is possible that all matter might have been affected by AGN over the course of cosmic evolution. The affected volume fraction increases with increasing kinetic fraction and decreasing AGN lifetime, and level of clustering.

Finally, one also finds an AGN and star-formation activity connection in dwarf galaxies. In the case of the galaxy Henize 2-10, a highly irregular companion of our Galaxy, extreme star-forming activity and an actively accreting supermassive black hole of approximately $10^6\ M_\odot$ are observed (Reines *et al.*, 2011). This might mark the beginning of all three: onset of galaxy formation, the starburst, and the creation of an AGN core. Bulge and AGN core would grow simultaneously, at least at the beginning, before feedback processes set in. At the other end of the evolution, weak AGN seem to have massive host galaxies and their starburst activity is low at the same time. Here, cold gas has been used up by the star-formation processes or has been accreted onto the central black hole, or it has been pushed out of the galaxy by the AGN feedback. These seem to be the final stages in galaxy evolution of AGN hosted in starburst galaxies, which, at the end, will remain silent, red, and massive.

7.3 Merging

A central question regarding the evolution of galaxies, and thus of AGN, is how they grew from the early Universe until today. There are two basic scenarios that have been purported. Either the galaxies form out of a large number of stars and then accrete more mass and form stars over their lifetime, or the seed galaxies merge with nearby galaxies, creating larger systems which then again undergo further merging. This hierarchy of galaxy growth and evolution through merging events is the commonly favored one. It is not only supported by the observation of large-scale structures which become increasingly pronounced toward lower redshifts. Furthermore, cosmological simulations indicate that growth through merging is the prime mechanism.

When discussing merging events, one often distinguishes between *major merger*, where the interacting galaxies have approximately the same mass, and *minor merger* with mass ratios in the range $m_1/m_2 = 0.1-0.5$.

The next question to ask, is, how do the merging events of the host galaxy affect the black hole in the center? Is a correlation between merging and AGN activity observable? Morphological studies of AGN host galaxies indicated early on that these interacting galaxies show more signs of disturbance and irregularity than the parent population of galaxies (Barnes and Hernquist, 1992). In more than 70% of the cases, AGN host galaxies have nearby companions, some showing tidal tails which are strong evidence for an interaction. This supported the idea that tidal interactions trigger activity in AGN. A more complex picture emerged over the recent years. If we assume that, on average, a large bulge implies a large black hole mass (either simply because more mass is available to grow both, or because there is a real feedback), one can look into merger-induced bulge formation studies. Hopkins *et al.* (2010) examined through simulations the effect of mergers on the mass of the bulge in galaxies. Although major mergers seem to dominate the formation of the bulges, minor mergers add to that a significant fraction of about 30%. In addition, the situation does not seem to be that simple, as the result is dependent on the mass. While in high-mass systems the bulges are indeed mainly formed through major mergers, the minor mergers seem to dominate in low-mass systems.

Finally, Hopkins *et al.* (2010) point out that today's uncertainty in the average major merger rate of a factor of ~ 2 greatly affects the simulation results: the errors still include the possibility that half of the bulge mass is accumulated through non-merging activity. Other recent studies do not find a significant correlation between disturbed morphology, that is, indication for recent merger processes, and AGN activity. Kocevski *et al.* (2012) compared Chandra-detected AGN hosts at $1.5 < z < 2.5$ with a control sample of nonactive galaxies at the same redshift. Although this study cannot rule out that minor mergers might play a role, major merger do not seem to trigger AGN at $z \sim 2$.

Barnes and Hernquist (1991) showed through simulations that the merging of two spiral galaxies of equal mass leads to the formation of an elliptical galaxy. The merging additionally causes the formation of a bar in the inner part of the remnant, which allows gas to lose angular momentum. This gas can be funneled in to form an AGN or to feed an existing supermassive black hole. A major merger would transform nearly all of the gas in a way that most of the cold gas will be within the galaxy remnant in the core ($r < 0.5$ kpc). The hot gas ($T \simeq 10^4$ K) will instead form an "atmosphere." The transformation time is thought to be quite short, i.e., in the range of 0.2–0.9 Gyr (Lotz *et al.*, 2008b).

The idea that bars in galaxies may be responsible for efficient fueling of a central engine suggests that barred galaxies should preferentially host AGN. Recent investigations seem to call this into question. Cardamone *et al.* (2011) investigated a sample of 10^4 face-on barred galaxies at low redshifts ($0.02 < z < 0.05$). Although 31% of these barred galaxies seem to host an AGN, this fraction is consistent with the fraction of all galaxies which harbor an active core, which is in the range 24–34% (Gallo *et al.*, 2010). It appears that if there is indeed a correlation between AGN and bars in their host galaxies, this relation is weak and possibly diluted. Again, this would suggest that merging and AGN activity may not be related.

As indicated in the previous section, the interplay between starburst and AGN activity is complex. Following a merger event with a large in-flow of gas into the central region, a feedback process between starburst and AGN could be triggered. It has been suggested that this might occur after some $\sim$ 0.1 Gyr (Schawinski *et al.*, 2009). Alternatively, the in-falling gas could simply allow simultaneous star formation and black hole growth, without physical connection between the two. One must distinguish in this context between *dry* and *wet* mergers. Dry mergers merely rearrange matter in the galaxies, while wet mergers, carrying a significant amount of cold gas, will indeed cause a significant amount of star formation. Early type galaxies might lead to dry merging events, while the tidal interaction of spiral or irregular galaxies would be considered a wet merger as plenty of cool gas is involved. Wet mergers of these gas-rich systems are likely to cause a starburst and AGN activity. Dry mergers might be responsible for the large number of massive elliptical galaxies.

Figure 7.6 shows results of a galaxy merger simulation taking into account radiative cooling, star formation, black hole growth through accretion, energetic feedback from supernovae explosions, and gravitational dynamics of gas, stars, and dark matter; see Di Matteo *et al.* (2005). The merger leads to a fueling of the AGN because the tidal forces lead to a disturbance which allows matter to flow into the center toward the black hole. It has been pointed out that there might be a "final parsec problem." If we assume that the two merging galaxies both host a supermassive black hole, their angular momentum might prevent a coalescence and the merged system will host a binary supermassive black hole pair. Only if an exchange of angular momentum with the surrounding stars and gas takes place can this SMBH binary finally merge (Khan *et al.*, 2011). In other simulations the in-fall of matter arriving at $\sim$ 10 kpc-scale down to the accretion disk, which is on the order of $\sim$ 0.1 pc, appears possible because of a cascade of gravitational instabilities on smaller scales (Hopkins and Quataert, 2010). This would allow to transport as much as $10\,M_\odot\,\mathrm{yr}^{-1}$ down to $\lesssim 0.1$ pc, more than sufficient to feed the central engine.

If an efficient mass in-flow takes place, the SFR and black hole growth accelerate rapidly after a merging event, within a few 10^8 years. Then, the feedback of the AGN leads to heating of the surrounding gas, causing reduced accretion and a declining SFR. Some studies argue though that the AGN can influence the host galaxy only in its very core, on a scale of $\lesssim 300$ pc (Debuhr *et al.*, 2010). Thus, from the point of view of the whole galaxy, this would have very little impact on the over all star-formation rate. In cases of major mergers without the presence of an AGN (or if feedback processes can be ignored), the star-formation rate remains high for a long period after the merger. This kind of simulation is also able to reproduce the observed $M_{\mathrm{BH}}-\sigma$ relation, and thus seems to also describe the growth of the bulge correctly.

Based on simulations of the activation of AGN through major merging events, Hopkins *et al.* (2005) derive a black hole growth time scale of about $\sim$ 100 Myr. The

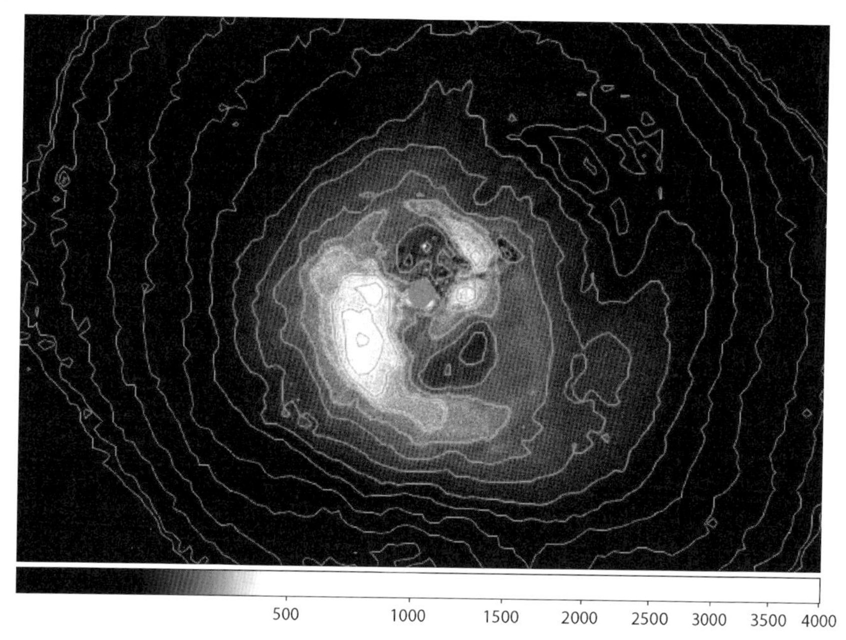

Figure 7.7 The Perseus Cluster as seen by the Chandra X-ray telescope. The center of the cluster is dominated by the emission of the Seyfert and radio galaxy NGC 1275. North and south of the AGN, cavities are visible as well as a larger dark bubble in the northwest. Chandra data obtained from the High Energy Astrophysics Science Archive Research Center (HEASARC), provided by NASA's Goddard Space Flight Center. For a color version of this figure, please see the color plates at the beginning of the book.

and the work required to inflate it. The latter depends on the pressure p of the intracluster medium, which will in turn depend on the richness of the cluster. Then, the energy needed to excavate a volume V in the ICM (assuming a uniform gas pressure) is $E = pV$. This can range from $E = 10^{55}$ erg in relatively poor clusters to $E = 10^{61}$ erg in rich ones. The energy needed to heat the cavity is several times greater. The age of a bubble can be estimated assuming the buoyant forces carry it away from the center of the cluster with approximately the free fall velocity.

Thus, through observation of cavities in galaxy clusters, we possess a calorimeter for AGN activity. Although the measurement is a rough approximation, it allows us to determine whether or not the energy deposited by the AGN in the cluster is enough to offset cooling. Fabian *et al.* (2006) studied the Perseus Cluster using the Chandra X-ray telescope. The structure of the cluster turned out to be highly complex, with concentric ripples around the central object. These are interpreted as sound waves which are created by the central pressure peaks caused by the activity of the Seyfert and radio galaxy NGC 1275 at its center. Apparently, the energy flux from the AGN is indeed sufficient to stop the cluster from cooling.

An AGN in the core of a galaxy cluster will have further impact on its surroundings in addition to the heating of the ICM. AGN jets will transport heavy elements from the center into the ICM. As cool, metal-poor gas can flow to the cluster center, the AGN outflows will therefore lead to an efficient mixing. In simulations of bubbles created by AGN in galaxy clusters, Sijacki and Springel (2006) showed that the cavities can alter the cluster properties. This effect is more important for cases involving relaxed, massive clusters in which the gas morphology is otherwise undisturbed. Because of the AGN heating, the cold baryon content is significantly reduced in the central galaxy and thus inhibiting its star formation.

In analogy to the case of AGN–host galaxy feedback, this AGN–cluster feedback can affect the entire inner core of a massive clusters. The heating and cooling at the cluster core are kept in balance by the AGN. When the gas cools down in the central ICM, more matter gets accreted onto the supermassive black hole, leading to enhanced activity causing reheating of the ICM. This in turn slows down the accretion processes, causing the AGN to fall dormant again. And thus a new feedback cycle can begin.

8
Quasars and Cosmology

AGN have been observed up to extreme redshifts, indicating that they were present in the early Universe and throughout much of the subsequent cosmic evolution. Thus, AGN provide a window into the structure and evolution of the Universe we live in. Used as background light sources for example, the line of sight absorption allows us to study the matter of intergalactic space which in turn has major cosmological implications. Conversely, to best interpret our observations, and to gain a deeper understanding of the AGN phenomenon, we need to understand the geometry and dynamics of the Universe itself. In this chapter we will look at models of the Universe in that context. We will then consider how distant quasars may in turn facilitate cosmological studies.

8.1
The Universe We Live in

The current status of the standard cosmological model with its benchmark parameters is presented in the box of Section 8.1.1 and in Table 8.1. A proper discussion of cosmology is beyond the scope of this book. Thorough reviews have been provided by Bennett (2006), Springel *et al.* (2006), Hu and Cowie (2006), and Cowan and Sneden (2006) in a special review in *Nature*, addressing the topics of the general cosmological model, large-scale structure, high-redshift galaxies, and nucleosynthesis, respectively. An introduction to cosmology can be found for example in the book by Ryden (2003). An excellent starting point in order to learn about the earliest times in the Universe is still the book "The First Three Minutes" by Weinberg (1993). Here we offer only a very rapid tour through the basics of cosmology, mainly to provide the reader a general picture rather than a comprehensive discussion. We include the formulas which we deem essential for working with AGN samples and their observable quantities affected by the expansion of the Universe. Because we focus on the application of cosmology to AGN-related problems, we consider only the model of our Universe which is at present considered the most realistic.

Active Galactic Nuclei, First Edition. Volker Beckmann and Chris Shrader.

8.1.1
Geometry and Distances

In Newton's theory the gravitational force scales with the mass of the attracting body, in our case the mass of the Earth. Mass is an intrinsic property of matter and a measure of its inertia, governed by Newton's famous equation $F = ma$. There was however nothing inherent in his theory relating these separate concepts of gravitational and inertial mass. Based on thought experiments (like the light beam emitted in a gravitational field and in an accelerated lift) Albert Einstein used the equivalence principle – the assertion that gravitational and internal mass were equivalent – to define the interaction of matter and space differently. In his theory, the energy–momentum tensor (from which the famous $E = mc^2$ relation can be derived) tells space–time how to curve, and curved space–time tells mass-energy how to move. Based on these principles Einstein formulated the *Einstein field equations*. Their solutions are called *metrics*. The Minkowski metric describes a flat space-time

$$ds^2 = -c^2 dt^2 + dr^2 + r^2 d\Omega^2 \tag{8.1}$$

Here we used $d\Omega^2 \equiv d\theta^2 + \sin^2\theta \, d\phi^2$, the speed of light c, and ds is the differential space-time separation between two points. Light will travel along a *null geodesic* with $ds = 0$. Considering a radial light path with $d\Omega = 0$, we see that $ds^2 = 0 = -c^2 dt^2 + dr^2$, or $c dt = dr$, which gives us the expected speed of light: $dr/dt = \pm c$. The Minkowski metric is applied in special relativity. Noncurved, Euclidean (or Minkowski) space is a valid approximation for the local Universe (up to $z \simeq 0.5$) but it is not a valid assumption for the Universe on large scales. We can constrain the multitude of possible models by applying two basic assumptions which have been confirmed by observations:

a) The Universe is isotropic: in whatever direction we point our telescopes, the Universe at large scales has the same appearance.
b) The Universe is homogeneous: although that assumption is not true on small (e.g., galactic) scales, matter seems to be equally distributed once one turns to distance scales of $\gtrsim 150$ Mpc.

Under these two assumptions the *Friedmann–Lemâitre–Robertson–Walker metric* describes the Universe correctly:

$$ds^2 = -c^2 dt^2 + a(t)^2 \left[dr^2 + S_\kappa(r)^2 d\Omega^2\right] \tag{8.2}$$

In this notation we use $S_\kappa(r)$ which includes the information about the spatial curvature of the Universe we assume:

$$S_\kappa(r) = \begin{cases} R \sin\left(\frac{r}{R}\right) & (\kappa = +1) \\ r & (\kappa = 0) \\ R \sinh\left(\frac{r}{R}\right) & (\kappa = -1) \end{cases} \tag{8.3}$$

κ describes the curvature (flat Universe $\kappa = 0$, positively curved $\kappa = +1$, negatively curved $\kappa = -1$). R is the curvature radius, also called the world radius. In addition we introduced the scale factor $a(t)$, which describes how distances change in time for an isotropic and homogeneous Universe. For the present we set $a(t = t_0) = 1$. Distances in an expanding Universe depend on how we define the term "distance." Because the Universe is expanding, the distance between two points in the Universe is also increasing and we have to specify the time t of our measurement. Assuming that we freeze the Universe at a given time t with a fixed scale factor $a(t)$ we get the proper distance $d_{\text{p}}(t)$ between the two points. The proper distance is the length of the geodesic connecting these two points and Eq. (8.2) applies here. Because our measurements were made at a fixed time, we have $-c^2 \mathrm{d}t^2 = 0$ in this case and the metric is given by

$$\mathrm{d}s^2 = a(t)^2 \left[\mathrm{d}r^2 + S_\kappa(r)^2 \mathrm{d}\Omega^2\right] \tag{8.4}$$

Along the radial geodesic, the angles θ and ϕ do not change, so in this case we have $\mathrm{d}s = a(t)\mathrm{d}r$ and the proper distance is the integral over the comoving coordinate r: $d_{\text{p}}(t) = a(t)r$. Because the scale factor a is a function of time, the proper distance changes with time and we can write the proper speed as:

$$\frac{\mathrm{d}d_{\text{p}}}{\mathrm{d}t} = v_{\text{p}} = \frac{\mathrm{d}a}{\mathrm{d}t} r = \frac{\dot{a}}{a} d_{\text{p}} \tag{8.5}$$

The Standard Model and Inflation

The standard model describes our current understanding of the Universe. We seem to live in a spatially flat Universe (i.e., $\kappa = 0$, $\Omega_{\text{total}} = 1$; see Section 8.1.3), which has been dominated by dark energy for about 4 Gyr and which will continue expanding forever (a scenario sometimes referred to as the Big Chill). Important epochs in the Universe and properties at those times are given in Table 8.1. The density parameters for the components in the Universe today are listed in Table 8.2. One has to keep in mind that all these values are only true in the case that our standard model describes the Universe correctly. The standard model alone, without inflation, has three major problems: the horizon problem (why is the temperature of the cosmic microwave background the same even in places which had no contact throughout the life of the Universe?), the flatness problem (why is $\Omega = 1$ exactly?), and the apparent absence of magnetic monopoles. Inflation solves these problems through a period of exponential expansion at the very beginning, starting at $t = 10^{-36}$ s and lasting until $t = 10^{-33}$–10^{-32} s after the Big Bang, which inflated submicroscopic sizes to parsec scale. Through this, the Universe as we see it in the CMB, has been connected before inflation. Inflation also flattens the Universe and it explains why there are no magnetic monopoles: the starting density of monopoles was lowered by inflation down to a number density which is too small to be detectable in the local Universe.

Table 8.1 Important epochs in the Universe (standard model).

Epoch	Redshift	T (K)	Time	Scale factor $a(t)$
Neutron freeze-out	5×10^9	9×10^9	1 s	—
Nucleosynthesis finished	2×10^8	4×10^8	10 min	—
Radiation = matter	3570	9730	47 000 yr	0.00028
Recombination	1370	3740	240 000 yr	0.00073
Photon decoupling	1100	3000	350 000 yr	0.0009
Last scattering	1100	3000	350 000 yr	0.0009
Reionization	6–20	—	$1.5{-}11 \times 10^8$ yr	—
Matter = Λ	0.33	3.7	9.8×10^9 yr	0.75
Today	0	2.75	1.35×10^{10} yr	1

Table 8.2 Density parameters in the present Universe (*standard model*).

Component	Density parameter
Photons	$\Omega_{\gamma,0} = 5.0 \times 10^{-5}$
Neutrinos	$\Omega_{\nu,0} = 3.4 \times 10^{-5}$
Total radiation	$\Omega_{\mathrm{r},0} = 8.4 \times 10^{-5}$
Baryonic matter	$\Omega_{\mathrm{bary},0} = 0.04$
Dark matter	$\Omega_{\mathrm{dm},0} = 0.26$
Total matter	$\Omega_{\mathrm{m},0} = 0.30$
Dark energy	$\Omega_{\Lambda,0} \simeq 0.70$

In the 1920s Edwin Hubble made a pioneering study of galaxies and their redshifts

$$z = \frac{\Delta\lambda}{\lambda_0} \tag{8.6}$$

Here $\Delta\lambda$ is the difference between the Doppler shifted wavelength of a line in the spectrum compared to its laboratory wavelength λ_0. Using Cepheid variables within individual galaxies as distance indicators Hubble discovered that there is a linear correlation between the speed of receding galaxies and their distance. This can be written in the form

$$v_\mathrm{r} = H_0 \cdot d \tag{8.7}$$

where v_r is the radial velocity (in $[\mathrm{km\,s^{-1}}]$), d is the distance in Mpc and $H_0 = (73 \pm 3)\ \mathrm{km\,s^{-1}\,Mpc^{-1}}$ is the *Hubble constant*. This relation is only valid in the local Universe because the v_r is not properly defined over large distances. In addition, the Hubble parameter is a function of time ($H(t)$), so the Hubble constant is indeed not at all constant. The observations are consistent with an expanding Universe in which the scale factor $a(t_\mathrm{em})$ at the time the light was emitted is related to the redshift z by

$$a(t_{\mathrm{em}}) = \frac{1}{1+z} \tag{8.8}$$

The inverse of the Hubble parameter is referred to as the Hubble time: $t_0 = H_0^{-1} \simeq 13\,\mathrm{Gyr}$. This time is equal to the time passed since the Big Bang in the case that the expansion of the Universe had not accelerated or decelerated, that is, if $H(t) \equiv H_0 = \text{constant}$. The first attempt by Hubble (1929) to determine the expansion of the Universe resulted in $H_0 \sim 500\,\mathrm{km\,s^{-1}\,Mpc^{-1}}$, which causes immediate problems because the age of the Universe would be only $t_0 \sim 2\,\mathrm{Gyr}$. The precise determination of distances in the Universe underwent extensive and arduous debate until ultimately converging upon today's cosmological model.

The identification of a Hubble flow – the region of space sufficiently distant for the recession velocity to be larger than local peculiar velocities – is direct evidence for an expanding Universe: Eqs. (8.7) and (8.5) are equivalent when we set

$$H_0 = \left(\frac{\dot{a}}{a}\right)_{t=t_0} \tag{8.9}$$

The second derivation with respect to time of the scale parameter $\ddot{a}(t)$, that is, the change of the relative velocity between two points in the Universe, can be used to define the deceleration parameter q_0

$$q_0 \equiv -\left(\frac{\ddot{a}\,a}{\dot{a}^2}\right)_{t=t_0} = -\left(\frac{\ddot{a}}{a\,H^2}\right)_{t=t_0} \tag{8.10}$$

If $q_0 > 0$ the expansion is decelerating, and for $q_0 < 0$ we get $\ddot{a} > 0$ and the Universe is in accelerated expansion. The current measurements indicate that $q_0 \simeq -0.55$, which indicates a Universe with accelerated expansion. We will come back to the calculation of q_0 later.

Through observations we are not able to measure the proper distance directly. Instead we measure the effect the expanding Universe has on the emitted light. One therefore defines the luminosity distance as

$$d_L \equiv \sqrt{\frac{L}{4\pi f}} \tag{8.11}$$

The luminosity distance is equivalent to the proper distance when we assume that the space is static, that is, $\dot{a} = 0$, and that the spatial curvature is small. In a Universe in which the Robertson–Walker metric applies, we can write the luminosity distance in the following way

$$d_L = S_\kappa(r)(1+z) \tag{8.12}$$

with $S_\kappa(r)$ as defined in Eq. (8.3). The Universe we live in appears to be flat ($\kappa = 0$), or at least very close to being flat, in which case the world radius today

($R_0 \equiv R(t = t_0)$) is very large compared to r. We can therefore assume that $S_\kappa(r) \simeq r$ and we get a simple relation between proper and luminosity distance

$$d_L = r(1+z) = d_\mathrm{p}(t_0)(1+z) \tag{8.13}$$

In the very local Universe, where the redshift is small, the proper and luminosity distance are similar

$$d_L \simeq d_\mathrm{p}(t_0) \simeq \frac{c}{H_0} z \tag{8.14}$$

Another important measurement is the angular size α of a distant object of size x, for example the angular size of a galaxy. The distance related to this measurement is called the angular size distance d_A. The angular size distance is equivalent to the proper distance if the Universe is static and Euclidean. The angle α will be small compared to the distance of the object, so we can write $d_\mathrm{A} = x/\alpha$. It can be shown that

$$d_\mathrm{A} = \frac{x}{\alpha} = \frac{S_\kappa(r)}{1+z} \tag{8.15}$$

In a Universe, which is flat or close to being flat with a large world radius R, we can write therefore

$$d_\mathrm{A} = \frac{d_\mathrm{p}(t_0)}{1+z} = \frac{d_L}{(1+z)^2} \tag{8.16}$$

Thus, in the local Universe, proper, angular-size and luminosity distance are equivalent. With this relation at hand and assuming that we live indeed in a homogeneous, isotropic, expanding and (nearly) flat Universe, it is sufficient to determine the redshift z and one of the three distance measures (d_A, d_p, or d_L) to derive the others. Without deriving it here, we present the luminosity distance as a function of Hubble parameter H_0, redshift z, and deceleration parameter q_0 in the form derived by Terrell (1977) for cases of $q_0 > 0$

$$d_L = \frac{cz}{H_0}\left[1 + \frac{z(1-q_0)}{1 + q_0 z + \sqrt{1 + 2q_0 z}}\right] \tag{8.17}$$

In the case of a nearly flat Universe we can use the approximation

$$d_L = \frac{cz}{H_0}\left(1 + z\frac{1-q_0}{2}\right) \tag{8.18}$$

For small redshifts this becomes

$$d_L \simeq d_\mathrm{p}(t_0) \simeq d_\mathrm{A} \simeq \frac{cz}{H_0} \tag{8.19}$$

We would like to stress that it is important to establish first which metric applies to a given problem and what approximations can be made before computing distances. For a rigorous approach to calculating cosmological distances we refer the

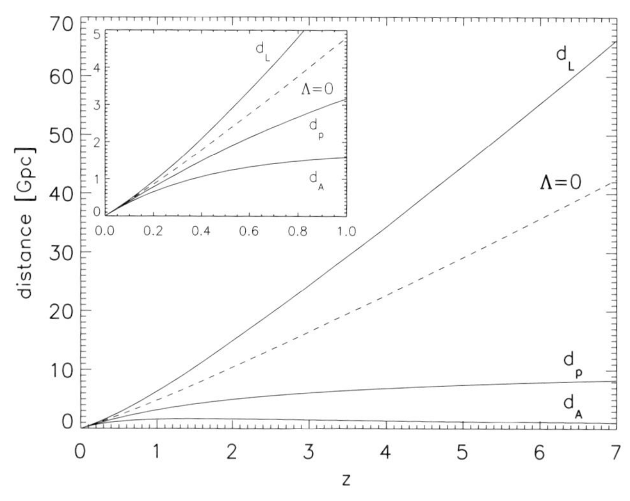

Figure 8.1 Luminosity distance d_L, proper distance d_p and angular size distance d_A as a function of redshift z for a flat Universe with $\Omega_{\mathrm{m},0} = 0.3$, $\Omega_{\Lambda,0} = 0.7$, $q_0 = -0.55$, and $H_0 = 73\,\mathrm{km\,s^{-1}\,Mpc^{-1}}$. The dashed line shows the luminosity distance for a flat Universe without dark energy ($\Lambda = 0$, $q_0 = 0.5$). The inset shows the redshift range up to $z = 1$. Computation using the numerical solution presented in Wright (2006). Graphic courtesy of Simona Soldi.

reader to Wright (2006). In Figure 8.1 we show the luminosity, angular size, and proper distance for the standard model, and d_L for a flat Universe with $\Omega_\Lambda = 0$ as comparison.[1] The latter presents the standard model before 1998, by which time it was known that the Universe was approximately flat from CMB observations by COBE, but the accelerated expansion had not yet been discovered.

8.1.2 Measuring Fluxes

The expanding Universe leads to a difference between the wavelength of an emitted versus an observed photon. Let us first consider monochromatic fluxes f_ν, which are measured in ($\mathrm{erg\,cm^{-2}\,s^{-1}\,Hz^{-1}}$). We observe a flux $f_{\nu,\mathrm{obs}}$ at a frequency ν_obs which has been emitted as flux $f_{\nu,\mathrm{em}}$ at frequency ν_em

$$f_{\nu,\mathrm{obs}}(\nu_\mathrm{obs}) = \frac{f_{\nu,\mathrm{em}}(\nu_\mathrm{em})}{1+z} \tag{8.20}$$

with a frequency shift of

$$\nu_\mathrm{obs} = \frac{\nu_\mathrm{em}}{1+z} \tag{8.21}$$

1) For a quick estimate of distances of a given redshift and cosmological model, see also Ned Wright's online calculator under http://www.astro.ucla.edu/~wright/CosmoCalc.html, last accessed: 15 May 2012.

In case we measure the monochromatic flux f_λ in (erg cm^{-2} s^{-1} Å^{-1}) we have to consider that

$$\lambda_{\rm obs} = \lambda_{\rm em}(1+z) \tag{8.22}$$

and therefore we get

$$f_{\lambda,\rm obs}(\lambda_{\rm obs}) = \frac{f_{\lambda,\rm em}(\lambda_{\rm em})}{(1+z)^3} \tag{8.23}$$

These formulas are dependent on the cosmological model we apply. Often we measure fluxes over some energy band, for example the optical V-band or in the X-rays in the 2–10 keV range, then the total energy flux $f_{\rm obs}$ is derived by integrating the monochromatic flux $f_{\nu,\rm obs}(\nu_{\rm obs})$ over frequency

$$f_{\rm obs} = \int f_{\nu,\rm obs}(\nu_{\rm obs}) {\rm d}\nu_{\rm obs} = \int \frac{f_{\nu,\rm em}(\nu_{\rm em})}{1+z} \frac{{\rm d}\nu_{\rm em}}{1+z} = \frac{f_{\rm em}}{(1+z)^2} \tag{8.24}$$

We can then substitute the emitted flux by the luminosity L of the object in the source frame, applying its luminosity distance d_L

$$f_{\rm obs} = \frac{f_{\rm em}}{(1+z)^2} = \frac{L}{4\pi d_L^2} \tag{8.25}$$

Note that by including the luminosity distance here, we have to specify which cosmology applies when deriving source luminosities out of observed fluxes.

Other quantities we are interested in are also affected by the redshift. Because of Eq. (8.22), the equivalent width, defined in the box of Section 8.1.4, is a function of redshift

$$E\,W_{\rm obs}(\lambda_{\rm obs}) = E\,W_{\rm em}(\lambda_{\rm em})(1+z) \tag{8.26}$$

A strong dependence can be seen in the surface brightness Σ, measured in (erg cm^{-2} s^{-1} arcsec^{-2})

$$\Sigma_{\rm obs} = \frac{\Sigma_{\rm em}}{(1+z)^4} \tag{8.27}$$

and thus also in intensity I, measured in (erg cm^{-2} s^{-1} sr^{-2})

$$I_{\rm obs} = \frac{I_{\rm em}}{(1+z)^4} \tag{8.28}$$

The specific intensity, or monochromatic intensity, I_ν (erg cm^{-2} s^{-1} Hz^{-1} sr^{-1}) evolves like

$$I_{\nu,\rm obs}(\nu_{\rm obs}) = \frac{I_{\nu,\rm em}(\nu_{\rm em})}{(1+z)^3} \tag{8.29}$$

A black body with characteristic temperature $T_{\rm em}$ will be observed to have

$$T_{\rm obs} = \frac{T_{\rm em}}{1+z} \tag{8.30}$$

8.1.3
The Three-Component Universe

The Friedmann equations are a special case of Einstein's field equations for the case of a homogeneous and isotropic Universe. There are several ways to write the Friedmann equations, for example

$$H^2 = \left(\frac{\dot{a}}{a}\right)^2 = \frac{8\pi G}{3c^2}\epsilon - \frac{\kappa c^2}{R_0^2 a^2} + \frac{\Lambda}{3} \tag{8.31}$$

H is again the Hubble parameter, a is the scale factor and $\dot{a}$ its derivative in time, ϵ is the energy density, R_0 is the world radius, and Λ is the cosmological constant. Λ was introduced by Einstein in order to accomodate a static Universe, but it also can represent the effect of a dark energy component. In this type of Universe the conservation of mass-energy is described by

$$\dot{\epsilon} = -3H(\epsilon + P) = -3\frac{\dot{a}}{a}(\epsilon + P) \tag{8.32}$$

where ϵ is the total mass-energy density, and $\dot{\epsilon}$ is its derivative in time. P is the total pressure of all the components in the Universe. Thus, if we apply the first law of thermodynamics to the Universe at large, we can write $\dot{E} + P\dot{V} = 0$: the change in energy is balanced by the pressure times the change of volume of the Universe.

In order to take into account the different components we want to consider, we can write the Friedmann equation also in the following form

$$\dot{a}^2 = \frac{8\pi G}{3c^2}\sum_{\omega} \epsilon_{\omega,0} a^{-1-3\omega} - \frac{\kappa c^2}{R_0^2} \tag{8.33}$$

Here ω is the equation of state parameter. This dimensionless parameter is $\omega = 0$ for nonrelativistic matter, $\omega = 1/3$ for radiation, and $\omega = -1$ for the dark energy. The change of the scale factor with time ($\dot{a}$) depends on the components (matter, radiation, dark energy/cosmological constant) we consider in the model of the Universe.

The density parameter Ω is defined as $\Omega = \epsilon/\epsilon_c$, where ϵ is again the mass-energy density and ϵ_c is the critical energy density. In the case that the energy density is equal to the critical density, we get $\Omega = 1$ and the Universe is flat. If the energy density is smaller than the critical value, we have an open Universe; with an energy density larger than critical energy density we have a closed Universe. One should keep in mind that the Universe does not change its curvature parameter κ, for example a Universe which is flat will always be and always has been flat. Looking at Eq. (8.31), we see that adding the Λ term is equivalent to the addition of another component with an energy density

$$\epsilon_\Lambda \equiv \frac{c^2}{8\pi G}\Lambda \tag{8.34}$$

This is the so-called dark energy with density parameter $\Omega_\Lambda = \epsilon_\Lambda/\epsilon_c$. We can write the Friedmann equation as a function of the density parameters of the single components Ω_r, Ω_m, and Ω_Λ, for radiation, matter, and dark energy, respectively

$$\frac{H^2}{H_0^2} = \frac{\Omega_{r,0}}{a^4} + \frac{\Omega_{m,0}}{a^3} + \Omega_{\Lambda,0} + \frac{1-\Omega_0}{a^2} \tag{8.35}$$

Here $1 - \Omega_0$ is the spatial curvature density.

Thus we consider in our models three components for the Universe. Nonrelativistic matter has no pressure ($P = \omega\epsilon = 0$), radiation has positive pressure, and dark energy has negative pressure. The three components today are believed based on measurements to have density parameters of $\Omega_{m,0} = 0.3$, $\Omega_{rad,0} = 8 \times 10^{-5}$, and $\Omega_{\Lambda,0} = 0.7$, respectively. This implies that the Universe is flat and at present dominated by dark energy. Because of the negative pressure of dark energy, the Universe will continue expanding. This can be seen also in the deceleration parameter q_0, which can be written as a function of the equation of state and the density parameters of the components we consider in the Universe

$$q_0 = \frac{1}{2}\sum_\omega \Omega_{\omega,0}(1 + 3\omega) \tag{8.36}$$

Using the three-component Universe and our current best estimates of density parameters, gives

$$q_0 = \frac{1}{2}\Omega_{m,0} + \Omega_{r,0} - \Omega_{\Lambda,0} \simeq 0.15 + 8.4 \times 10^{-5} - 0.7 \simeq -0.55 \tag{8.37}$$

When studying the importance of the different components throughout the history of the Universe, one has to keep in mind that $\epsilon_m \propto a^{-3}$, $\epsilon_{rad} \propto a^{-4}$ (as can be seen from Eq. (8.35)), and that the dark energy has a constant energy density. The scale factor is related to the redshift $a = 1/(1 + z)$ in *all* universes described by the *Friedmann equation*. The early Universe was radiation-dominated until 47 000 years after the Big Bang, then matter dominated, and now, since $a \simeq 0.75$ and for about 4×10^9 years, dominated by dark energy. The matter in the Universe consists mainly of *dark matter* ($\Omega_{dm,0} = 0.26$). This matter has a mass, but no charge and is nonbaryonic in nature. It can be detected through the rotation curves of galaxies and through the velocity dispersion of galaxies in clusters of galaxies. Baryonic matter can be found predominantly in the gas which is trapped in galaxy clusters ($\Omega_{gas,0} = 0.04$) and which can be detected in the X-rays, while the visible matter in stars and in the interstellar medium accounts for only $\Omega_{*,0} = 0.004$. The amount of baryonic matter we can detect is consistent with the value we get from the nucleosynthesis theory.

Measuring distances is crucial for determining the true nature of our Universe. For an in-depth discussion of this topic see the book of de Grijs and Cartwright (2011). Distances can be measured directly through gravitational lenses or masers in the accretion disk, as we will see later. In order to measure the large distances of more common objects, one first has to measure the local distances and then hierarchically work one's way up the *distance ladder*. Distances to stars in our Galaxy (up to a few kpc) can be determined using the parallax. Brunthaler *et al.* (2005) have demonstrated that through VLBI observations of proper motions and angular rotation of masers in galaxies of the Local Group their distance can be determined. In the example of M33 this resulted in a measurement of $d = (730 \pm 168)$ kpc. Some variable stars, the Cepheids, show a characteristic correlation between luminosity and period, thus can be used to determine their distance out to the nearest galaxies. Cosmological distances can be determined by using supernovae type Ia as standard candles. Supernovae type Ia progenitors are believed to be binary systems of a white dwarf and a "normal" star although there is evidence that some may be double-degenerate binaries. A white dwarf is a star at the end of its life time. It is supported against gravitational collapse by electron degeneracy pressure. When a white dwarf is located in a binary system and accretes matter from the companion, it can reach the critical mass of $1.4\,M_\odot$. Because this always happens at the same mass, the light curves of those explosions have a characteristic peak and shape and thus can be used as standard candles. The observation of supernovae led in the late 1990s to the discovery that our Universe is undergoing an accelerated expansion. In 2011 this discovery was awarded the Nobel Prize in Physics, given to two groups which performed the observations in parallel, and thus one half was awarded to Saul Perlmutter, the other half jointly to Brian Schmidt and Adam Riess.

8.1.4
From the Big Bang to the Cosmic Microwave Background

The cosmic microwave background (CMB) is a nearly isotropic background radiation which has an accurately determined temperature of $T_0 = 2.725$ K. The CMB radiation has a nearly perfect black body shape. It was created at the moment the Universe became transparent, some 350 000 years after the Big Bang at $z = 1100$ when the Universe had a temperature of $T_{\mathrm{ion}} \simeq 3400$ K. This moment also defines the so-called surface or "last scattering". At this point in time photons decoupled from matter, because the free electrons got bound by protons to form hydrogen. Before this, photons interacted efficiently with electrons through Thomson scattering and the Universe was opaque. Once we consider the bipolar structure of the CMB because of the solar system's relative movement, the COBE, WMAP and Planck satellite-experiments show that the CMB is nearly isotropic, with deviations of only 30 μK. Beside the bipolar structure, the CMB shows anisotropies mainly on two different angles. Both anisotropy effects cause temperature fluctuations in the CMB, as measured by COBE, WMAP, and Planck. The 1° angular-scale temperature anisotropy corresponds to the event horizon at the time of the CMB emission at $z = 1100$. At angular scales of about 0.6° we see acoustic oscillations, caused by the baryon–photon fluid being affected by density fluctuations. It is believed that these baryon acoustic oscillations (BAO) are the seed for the large-scale structure we observe today with a size on the order of ~ 150 Mpc. The COBE results were so important for our understanding of the Universe that two of the leaders of that project, John Mather and George Smoot, were awarded the Nobel Prize in Physics for their work in 2006. The precise measurement of BAO in the Universe is one of the main objectives of today's dark energy projects, like ESA's Euclid satellite and the Baryon Oscillation Spectroscopic Survey (BOSS), which is going to be part of the SDSS-III survey. Measuring the BAO precisely constrains the Hubble parameter ($H(z)$) and the angular size distance $d_A(z)$ as a function of redshift and thus helps to determine the type of Universe we live in.

The basic elements of our Universe were created within the first 10 min after the Big Bang. The fact that apparently there is mainly matter and nearly no antimatter in the Universe points to a small overabundance of matter over antimatter in the very early Universe. While most matter and antimatter particles annihilated in the first second into radiation, the slightly higher amount of matter remained intact and formed the cosmos we live in. The heavier an elementary particle is, the earlier it was formed in the Universe, because the higher mass of the particle requires a higher energy density for its creation and the energy density drops with the expansion of the cosmos. Once the energy density is too low for the production of the particular particle, it will "freeze out," and it will not be produced in significant quantities subsequently, but might annihilate with its antiparticle or decay into lighter particles. Atomic nuclei cannot form until after the fundamental particles, in particular the hadrons have frozen out. Assuming equilibrium between protons and neutrons one can estimate how that ratio evolved by considering the process $n \rightarrow p + e^- + \nu$ which is mediated by the *weak force*. This requires extremely high temperatures, $T \gg 1.5 \times 10^{10}$ K, a condition fulfilled only until the Universe was 1 s old. Since neutrons ($m_n = 939.6$ MeV) are slightly more massive than protons ($m_p = 938.3$ MeV), their precise freeze out times are slightly different. Calculations show that at the time of the freeze out ($t \simeq 1$ s) there is an overabundance of protons over neutrons, with $n_n/n_p \simeq 0.2$. Subsequently, free neutrons continue to decay into protons with a half life of about 10 min unless they become bound in atomic nuclei. The first of these to form were the deuterium nuclei starting at $t \simeq 200$ s, and thus the neutron-to-proton ratio is observed to be $n_n/n_p \simeq 0.15$. Deuterium (1 neutron and 1 proton in the core) was formed starting at about $t \simeq 2$ s, followed by tritium (2 neutrons, 1 proton) and helium. Because there is a sharp peak in binding energy per nucleon in helium, this element effectively represented a dead end in nucleosynthesis. After about 10 min the temperature in the Universe was too low for nuclear fusion to proceed. Thus the element mix after 10 min stayed more or less constant until the first stars started to produce heavier elements (but this is after the CMB creation starting at times $t \gg 10^7$ years). The Universe consisted mostly of protons (which will later catch electrons and form hydrogen), ^{4}He, deuterium (^{2}H), ^{3}He, and tritium (^{3}H) (in decreasing abundance). Only traces of beryllium (^{7}Be) and lithium (^{7}Li) were created.

8.1.5
The Dark Matter Universe

Table 8.2 gives some information concerning what we know (perhaps more appropriately: what we do not know!) about the contents of the local Universe. Through direct observation, we are able to detect less than 15% of the matter in the local Universe. Most of this visible matter we see in the X-rays in the medium trapped in the gravitational potential of galaxy clusters. The overwhelming part of the matter we can infer for example from rotation curves of galaxies which show that invisible ("dark") matter reaches out far beyond the optically detected galaxy mass, as first suggested by Zwicky (1937), with a hundred times more mass in the dark matter than in the visible baryonic one. It had been assumed that the sources behind dark matter are compact objects in the halos of galaxies, the so-called MACHOS (massive compact halo objects). These objects should show up in observations through microlensing events. When a MACHO would pass through the line of sight of an extragalactic object, the light of the latter should be amplified and peak due to gravitational lensing. Extensive programs in order to search for such microlensing events have been initiated, for example monitoring observations of the Large Magellanic Cloud (LMC) or analyzing light curves of objects near the galactic center. Many microlensing events have been indeed detected in the direction of the bulge of the Milky Way, but those are most probably caused by normal stars in the line of sight. Only a few MACHO candidates have been found and their masses appear to be low, i.e., less than a solar mass. Primordial black holes (see second half of Section 9.1) could be the sources behind the MACHOs, should they indeed be detected.

Alternatively, exotic particles could be responsible for the dark matter content of the Universe. It has been proposed that dark matter is made of particles which interact only through the weak force and through gravity, therefore escaping observation throughout the electromagnetic spectrum.

WIMPs could possibly be produced and characterized in high-energy particle accelerator experiments, they could be observed via WIMP-nucleon scattering in sensitive laboratory experiments, or perhaps most promisingly due to their self-annihilation in the astrophysical environments. Among the different decay products likely produced in WIMP annihilation, gamma-rays are of particular importance, as they propagate freely through space and can thus reveal their sources. This information is currently being exploited for efficient signal/background discrimination in searches for gamma-ray signatures from, for example, dwarf galaxies or towards the galactic center. The two-body annihilation of WIMP dark matter into photons producing monochromatic gamma-ray signal, that is, line emission from one of these sources would represent clear smoking gun signature. Such signals, however, are expected to be extremely faint. A number of searches have been made, notably using the Fermi observatory which offers unprecedented coverage and sensitivity over a spectral range of interest. For example, in an upcoming paper Christoph Weniger examines the galactic center region using 43 months of Fermi data, applying customized background models and employing data-driven spatial-region cuts to search for possible WIMP annihilations. He reports an excess consistent with a line at about 130 GeV, however, the significance of their result is only significant at about the 3.3-σ level. If Fermi continues to operate its anticipated ten-year lifetime it should be able to unambiguously confirm or refute this result.

Estimates on the mass of the WIMP particle come for example also from the modeling of AGN spectra. Saxena *et al.* (2011) did a synchrotron self-Compton (SSC) model fit to the data of the radio galaxy M87. At the highest energies, they noticed a deficiency between model and data. In order to account for the difference, the authors add a component based on dark matter annihilation of WIMP particles. The resulting mass of the WIMP particle turns out to be in the TeV range. These kinds of studies can give a hint on what the WIMP mass might be, but obviously suffer from simplified models and from nonsimultaneous data used in the fit – the missing component might just be caused by variability of the source between the measurements in the various energy bands. A review on our current understanding of dark matter and the experiments searching for it has been presented by Zacek (2008).

It is likely that dark matter plays an important role in the creation and evolution of supermassive black holes in the Universe. Identifying what the nature of the dark matter is, might have profound impact on our understanding and could possibly change the current models substantially.

8.2 AGN and the Distribution of Matter on Large Scales

The study of AGN has contributed significantly to our understanding of the large-scale structure and evolution of the Universe. In particular, since quasars have high luminosity and a broad nonthermal continuum extending into the far ultraviolet portion of the spectrum, they are valuable probes of the Universe out to a redshift of $z \simeq 6$. This represents a view of the Universe over a substantial fraction of a Hubble time.

The large Lyα atomic cross-section provides astronomers with a powerful tool for probing the baryon content and matter distribution of the Universe. As previously noted (see Section 9.1), in an expanding Universe characterized by a uniform intergalactic medium (IGM) containing neutral hydrogen, the continuously redshifted Lyα resonance will lead to an absorption trough blueward of the source frame Lyα line. In their seminal 1965 paper, Gunn and Peterson (1965) used this type of measurement to establish upper limits on the neutral hydrogen content of putative uniform IGM. What became evident in high-resolution spectroscopy as the technology led to improved capabilities was the presence in the nearby Universe, out to redshifts $z \sim 5$, of the so-called Lyα forest (e.g., Weymann *et al.*, 1981).

The Lyα forest is manifest as a set of densely and unevenly spaced absorption lines observed in high-resolution spectra of background quasars. An example is shown in Figure 8.2. It has been extensively studied in the UV and optical wavelength range, from the local Universe up to the highest known quasar redshifts.

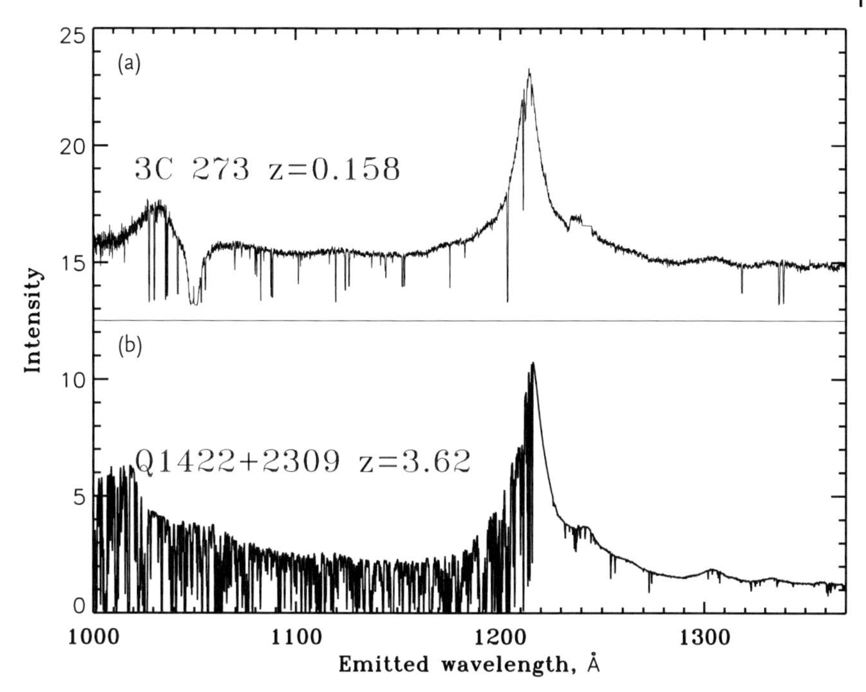

Figure 8.2 Example of the Lyα forest observed in low- and high-redshift quasar spectra (obtained with permission from the website of Professor William Keel, Univ. Alabama, http://www.astr.ua.edu/keel/agn/forest.html, last accessed: 15 May 2012). While the dense distribution of Lyα absorption lines is readily evident in the high-redshift object, Q 1422+2309 (a), there are also absorptions, albeit much less densely spaced, attributable to Lyα at intervening redshifts in 3C 273 (b). This is indicative of the strong cosmic evolution of the physical gas structures, e.g., filaments or clouds, responsible for the absorption lines.

Neutral hydrogen along the quasar line of sight leads to line absorption of the source continuum emission by the redshifted Lyα (λ1215 Å) resonance line. The number of lines detected for a given object, along with the line strengths and profiles to the extent that they are measurable, probe the intervening intergalactic medium.

Originally, it was suggested that the lines might be due to matter ejected from and ambient to the individual quasars rather than being cosmologically distributed. However, that idea was convincingly refuted on the basis of several lines of argument. An excessive amount of momentum would be required for ejection of gas from the quasar to reproduce the observations (Goldreich and Sargent, 1976). It is not at all clear how such a mechanism could be viable. Furthermore, the distribution of absorption systems over multiple objects reveals a cosmological signature, rather than a randomness that would be expected for the case of ejection scenarios (e.g., Bergeron, 1986).

Numerical simulations of plausible IGM configurations have been performed and the results have been compared to the Lyα forest data. These analyses suggest a hierarchical sequence of gaseous structures, typically ranging from sheets to filaments to spherical clouds consistent with galactic halos as column density increases. Metallic absorption lines, associated with the same galactic halos, have additionally been searched for and identified in some cases, although this is observationally challenging. The intergalactic gas clouds that are distinct from galaxies exhibit higher-redshift density distributions and weaker clustering compared to galaxies. The cloud redshift distributions can also be used as a tracer of gravitationally induced density fluctuations in the cosmological distribution of matter. Roughly speaking, the number density of Lyα clouds inferred from quasar spectra is flat out to about $z \sim 1$, then steepens as a power law $(1+z)^{\gamma}$ where γ is in the range of 2–3 out to redshift $z \sim 2$ and steeper yet out to redshifts 4–5. The fraction of matter currently in the IGM relative to that incorporated into galaxies is a parameter to which structure formation models are sensitive.

The predominance of the Lyα forest and the apparent lack of a Gunn–Peterson trough out to redshifts of $z \sim 5$ significantly limits the neutral hydrogen fraction of the IGM. On the other hand, the early Universe consisted almost completely of ionized hydrogen and helium up to the recombination epoch characterized by $z \simeq 1100$. Thus, gaseous matter in the Universe was "reionized" at some subsequent epoch and the most likely mechanism for this reionization involves the intense UV radiation field of quasars and the first generation of (putative Population III) stars. As these first sources of UV radiation turned on, they began to reionize the Universe, ending the so-called "dark ages" that persisted since the recombination epoch. In this scenario, which likely coincides with the start of galaxy formation, the Universe undergoes a phase transition from being neutral to being mostly ionized. Observationally, this transition should be discernible as anisotropic structure in the cosmic microwave background radiation (Bennett *et al.*, 2003) but also in high-resolution spectra of $z \sim 5$–6 quasars. The latter would be expected to exhibit a transition from a Lyα forest absorption signature to a Gunn–Peterson trough with decreasing wavelength. The onset of the trough would establish the threshold of the reionization epoch.

The advent of 10-m class telescopes and technology driven improvements in instrumentation have made such observations plausible in the last decade or so. For example, Djorgovski *et al.* (2001) obtained a high signal-to-noise spectrum of the quasar SDSS 1044-0125 which is at a redshift of $z = 5.73$. They reported a dramatic increase in the density of the Lyα forest at λ7550 Å corresponding to a redshift of $z \geq 5.2$, which they interpret as the onset of the reionization era at $z \sim 6$. This is illustrated in Figures 8.3 and 8.4, which show two different representations of a high-resolution spectrum of SDSS 1044-0125. In the first plot the quasar spectrum above the source frame Lyα is clearly seen to be dramatically absorbed below that wavelength. Note, however, the change in the Lyα forest spectrum above about λ7550 Å: the optical depth seems to change discretely there in a manner consistent with the appearance of a Gunn–Peterson trough. This can be seen perhaps more clearly in Figure 8.4, where the spectrum has been normalized to the continuum level and plotted in redshift space. The dramatically greater opacity above redshift 5.2 is readily evident (in this case, the effective redshift coverage has been extended

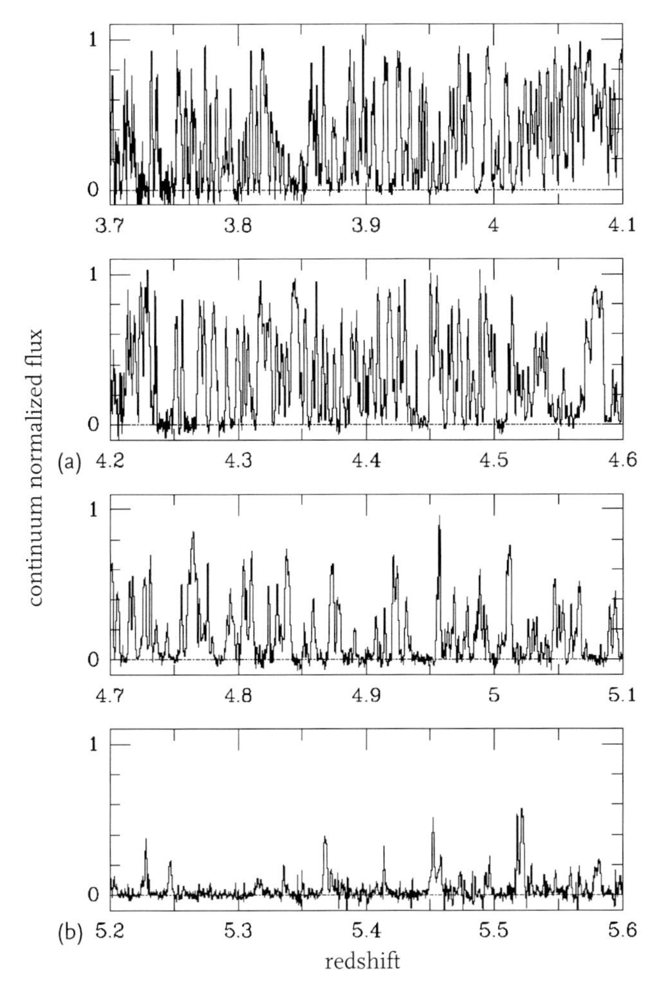

Figure 8.3 The two panels are from the quasar SDSS 1737+5828 spectrum extending down to $z = 3.7$, normalized to the continuum level and plotted in redshift space (a). The others represent the spectrum of SDSS 1044-0125 spectrum (b) (from Djorgovski *et al.*, 2001, reproduced by permission of the AAS). The dramatic increase in opacity and transition to a trough-like rather than line-like absorption is evident above redshift $z = 5.2$.

down to $z = 3.7$ by the inclusion of another high-luminosity, $z = 4.94$ quasar; refer to Figure 8.4).

Additional luminous, high-redshift quasars have been subsequently observed at comparable resolution and signal-to-noise. The general conclusion is that the onset of reionization occurs at $z \sim 6$ but that it is "patchy," that is, at slightly differing redshifts for different lines of sight. The cosmic microwave background studies reached conclusions on the reionization era which were initially highly discrepant with the quasar results; $11 < z < 30$. However, these numbers were later revised to $7 < z < 11$, which is nearly consistent with the quasar-based results.

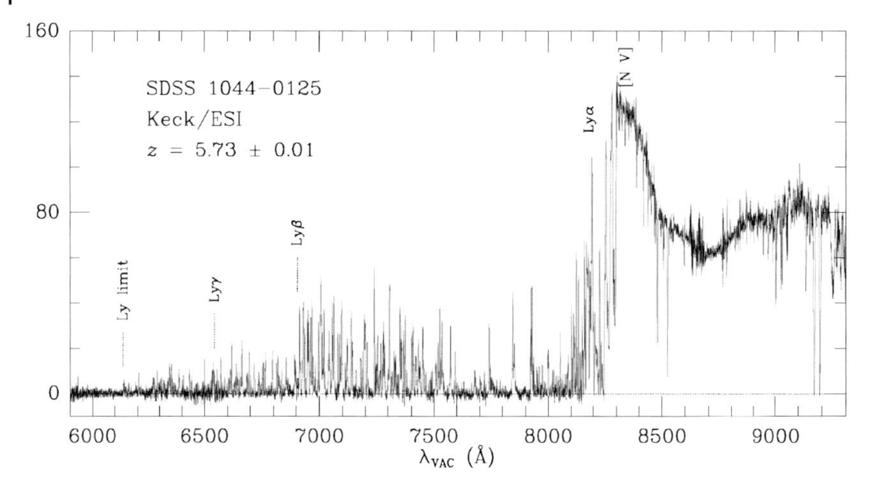

Figure 8.4 This high-resolution spectrum of the luminous, $z = 5.73$ quasar SDSS 1044-0125 obtained by Djorgovski *et al.* (2001) with the Keck 10 m telescope suggest that the reionization epoch may have extended to about $z \sim 6$. The quasar continuum is seen relatively unaltered by absorption at wavelengths above source frame Lyα. The blue-wards onset of substantial opacity is clearly evident, but in addition, there is a transition from a discrete Lyα forest spectrum to an apparent Gunn–Peterson trough above about λ7550 Å. The authors interpret this as strong evidence for the trailing edge of the reionization epoch at about $z \sim 6$. (Reproduced by permission of the AAS.)

Even accounting for the observed cosmological distributions of galaxies and the intergalactic H I clouds, there is still a discrepancy of a factor of 2 or more between the cosmic baryon density thus inferred and that inferred to exist in the early Universe from cosmic microwave background measurements (e.g., Bennett *et al.*, 2003). An apparent resolution to this dilemma comes from numerical hydrodynamic simulations of the formation of structures in the Universe (Cen and Ostriker, 1999). These calculations suggest that the "missing" baryons reside in a tenuous filamentary web of ionized gas at temperatures of $\log(T/\mathrm{K}) \sim 5.5$–$7$.

This so-called warm-hot intergalactic medium (WHIM) is in principle detectable, in analogy to the intergalactic HI clouds, using background quasars to search for absorption line systems. In this case, the observations need to be made in the far UV or X-ray portion of the spectrum, as the plausibly detectable ions reside in those wavelength ranges. Several claims of WHIM detections have been reported, for example, Nicastro *et al.* (2005) obtained high-resolution X-ray spectroscopy of the BL Lac object Mrk 421 during a high-amplitude flaring event. They report the detection of O VII and N VI at discrete redshifts lower than that of Mrk 421. Such measurements are currently difficult to make and are thus rare, but future space-based observatories offer the possibility of using quasar observations to obtain a much more accurate characterization of this illusive baryonic component, which is in turn critical to our understanding of the hierarchical, large-scale structure of the Universe.

Another area where quasar studies have contributed to cosmology involves the phenomena of *gravitational lensing*. A gravitational lens involves a foreground distribution of matter, typically a galaxy or a cluster of galaxies on the line of sight between a distant source such as a quasar and an observer. The foreground object is capable of bending, or lensing, the light from the source on its path towards the observer. This effect is known as gravitational lensing. Sir Isaac Newton speculated whether gravitational force would be able to bend light, and the German astronomer Johann von Soldner presented a calculation of this effect, based on Newtonian mechanics. The effect was then correctly predicted by Albert Einstein in the context of his General Theory of Relativity. Fritz Zwicky first proposed in 1937 that the lensing effect could be observable using fortuitously oriented galaxy clusters which should lead to altered images of distant background galaxies. In a theoretical work, Refsdal (1964, 1966) pointed out that the measurement of lensed images of a quasar can be used directly to determine the Hubble constant H_0. If we assume a simple configuration in which two images appear from the same quasar through gravitational lensing by a massive object (e.g., a galaxy), then the time delay between the two images can be used to determine H_0, without independent knowledge of the distance to either the lens or the quasar. It took about four decades after Zwicky's claim of gravitational lensing until this effect was observationally confirmed. The so-called twin quasar SBS 0957+561 was reported in 1979 by Walsh *et al.* (1979). The twin quasar's two images are separated by 6 arcsec and both have a magnitude of approximately 17 mag; specifically the A component has magnitude 16.7 mag and the B component 16.5 mag. Monitoring observations also revealed a 417 ± 3 days time delay between the two images due to light travel times along the separate paths to the two images (Kundic *et al.*, 1997). With some knowledge of the mass distribution of the lensing object, such lag time measurements can be used to constrain cosmological parameters. For example, those authors inferred a value for the Hubble constant of $H_0 = 63 \pm 13\,\mathrm{km\,s^{-1}\,Mpc^{-1}}$, which is consistent with more recent determinations from cosmic microwave background studies. More precise light curves and the usage of several gravitationally lensed quasars allows nowadays to derive an independent and consistent determination of the Hubble constant with $H_0 = 76 \pm 3\,\mathrm{km\,s^{-1}\,Mpc^{-1}}$ (Paraficz and Hjorth, 2010).

Quasars have long been considered as potential cosmological indicators, given their high luminosities and large distances. However, given the diversity of the population and their steep luminosity function identifying the required "standard candle" to place them on a Hubble diagram has proven illusive. The use of type-Ia supernovae (SNe Ia) light curves to establish those objects as standard candles has led to the extraordinary findings that the expansion of the Universe is actually accelerating leading to "dark energy" cosmologies. However, the SNe Ia studies can only be carried out to redshifts less than 2. Recently, a method that can potentially utilize observable AGN properties as a cosmological distance indicator out to redshifts beyond that currently accessible to the SNe Ia studies, in principle up to $z \sim 4$, has been proposed by Watson *et al.* (2011).

The basis for the method is that the flux emitted by an AGN BLR gas is due to reprocessed continuum photons and that it thus varies in response to variations in the luminosity of the central source. The time delay, or "lag," τ is determined essentially by the light travel time, $\tau = r/c$. Measuring the time lag thus allows a determination of the BLR radius r for a given ion, which comprises the technique of reverberation mapping and which has been used to gain major advances in our understanding of AGN intrinsic properties; see Section 3.6 and references therein. The size of the BLR is determined by the incident ionizing flux, which drops inversely with the square of the distance from the central engine. The time lag should thus be proportional to the square root of the luminosity of the central

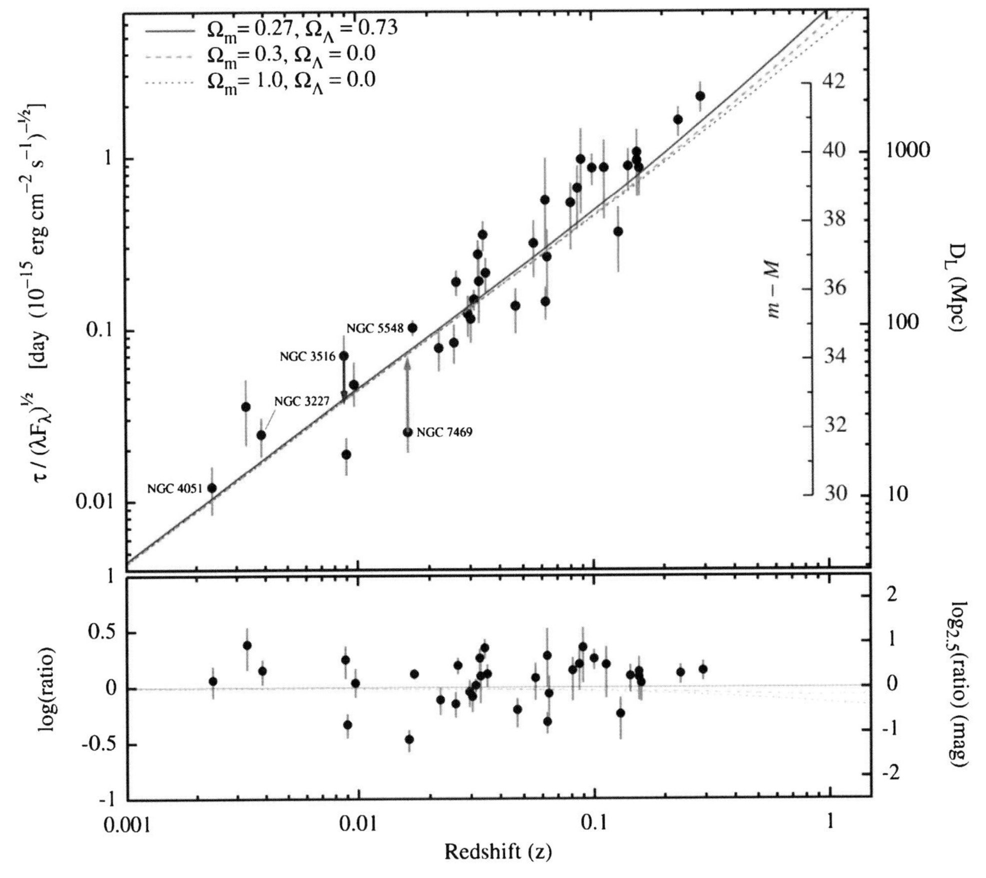

Figure 8.5 The AGN Hubble diagram based on a sample of 38 reverberation mapped objects reproduced from Watson *et al.* (2011) by permission of the AAS. The luminosity distance indicator $\tau/\sqrt{F}$, which is directly related to the luminosity distance (right-hand vertical axis) is plotted as a function of redshift. The solid curve represents a best-fit cosmological model from the current literature. Cosmologies with no dark energy components are plotted as dashed and dotted lines. Extending this type of diagram to $z \sim 3$–4, which is arguably plausible, offers the potential to meaningfully constrain dark energy models.

source $\tau \propto \sqrt{L}$. Thus, the observable quantity $\tau/\sqrt{F}$, τ being the lag time for a given emission line and F the continuum flux, is directly related to the luminosity distance.

At present, there is a large sample of AGN (nearly 40) for which reverberation mapping measurements have been obtained, for example Bentz *et al.* (2009b) and also Section 3.6, however, they are primarily low-redshift objects. Figure 8.5 is an example of a Hubble diagram constructed by Watson *et al.* (2011) using the current reverberation mapped AGN sample comprised of 38 AGN with redshifts out to about $z = 0.3$.

In principle, these types of measurements can be carried out for more luminous AGN at higher redshifts. However, there are practical problems that need to be overcome. The typical lag times involved, due to the fact that higher luminosity objects have larger BLR sizes but also due to time dilation effects, can become on the order of years. This renders the needed observational campaigns difficult to organize and maintain since they span multiple observatory time-allocation cycles. Also, the inherent scatter in the AGN Hubble plots needs to be minimized. Multiple measurements can help to reduce statistical error bars, but again this adds to the resource-intensive nature of these campaigns. Dust extinction in the host galaxies introduces another potential source of scatter. For higher-redshift objects UV resonant lines such as C IV λ1549 are accessible from the ground. Since C IV in particular has a smaller BLR ionization radius than for example common Balmer lines and since it is free of contamination by stellar light from the host it offers advantages. However, it is also strongly affected by dust extinction so that must be accounted for to the extent possible. Some of the methods employed to estimate dust extinction in the low-redshift objects such as using the Balmer decrement and the Na I D or K lines are more difficult for higher-redshift objects. These problems notwithstanding, this methodology for using AGN as cosmological indicators holds future promise.

9
Formation, Evolution and the Ultimate Fate of AGN

In Chapter 7 we studied the environment of AGN, considering the nature of the galaxies and clusters of galaxies in which they reside and their relationship with their surrounding media. In this chapter we will step back further and ask the question how might AGN have come into existence in the first place. We will then study their evolution in numbers and luminosity through redshift space following several approaches. This will in turn provide further insight into the question whether AGN types are evolving, and what the nature of this evolution might be. It will also tell us when in cosmic history the AGN were most active, and finally where AGN evolution is heading in the future.

9.1
The First AGN: How Did They Form?

AGN have been observed up to redshifts greater than 7 with ULAS J112001.48+064124.3 at $z = 7.085$ currently being the record holder (Mortlock *et al.*, 2011). It is unlikely that this particular object is the first quasar in the Universe, and by the time you are reading this, higher-redshift AGN have likely been discovered. Already the highest-redshift objects require the formation of supermassive black holes at a time some 800 million years after the Big Bang. If we consider today's widely accepted standard cosmological model, the Universe became transparent and neutral some 3.5×10^5 years after the Big Bang. It then took a considerable amount of time for the first stars to form at around $z \sim 50$ when the Universe was about 50 million years old. This phase of the Universe is believed to have lasted for some 1.5×10^8 years until redshift $z \simeq 20$ when the reionization of the matter begins, caused by the radiation of hot and massive stars. If we assume that massive black hole formation commenced roughly in tandem with reionization, we are left with $\Delta t \sim 6 \times 10^8$ years from the first significant wave of star formation to grow the supermassive black holes with masses $M_{\mathrm{BH}} \geq 10^8\ M_{\odot}$.

Active Galactic Nuclei, First Edition. Volker Beckmann and Chris Shrader.

The Gunn–Peterson effect and the end of the reionization phase

The effect first noted by and named after Gunn and Peterson (1965) describes the absorption troughs observed in the continua of high-redshift quasars due to neutral hydrogen in the line of sight. Because the matter is distributed in redshift space, rather than a single, narrow absorption line the absorption is spread out, forming a wide valley or trough in wavelength space. Through observations of high-redshift quasars one can determine the amount of neutral hydrogen at different cosmological distances by measuring the optical depths of the relevant absorption lines, such as Lyα. This can be used to identify the end of the reionization era by looking for the highest-redshift quasars which still show the Gunn–Peterson trough. It is even possible to track the transition from a partially to the fully reionized Universe around $z \simeq 6$, as shown in Figure 9.1.

A central question of AGN research is: how can supermassive black holes grow fast enough in order to accumulate a mass of $M_{\rm BH} \simeq 10^9\,M_\odot$ within 600 million years? A concise review addressing this question has been provided for example by Volonteri (2010b). One can do a very simple Gedankenexperiment in order to estimate the final mass. If we assume abundant matter in the vicinity of the seed black hole with starting mass M_0, so that the growth can continue at the maximum possible rate, the only restriction for accumulation of mass is that of the Eddington limit as given in Eq. (3.9). Therefore we have the Eddington ratio

$$\lambda = \frac{L_{\rm bol}}{L_{\rm Edd}} = L_{\rm bol} \cdot 8 \times 10^{-39} \frac{M_\odot}{M_{\rm BH}}\,{\rm s\,erg}^{-1} \tag{9.1}$$

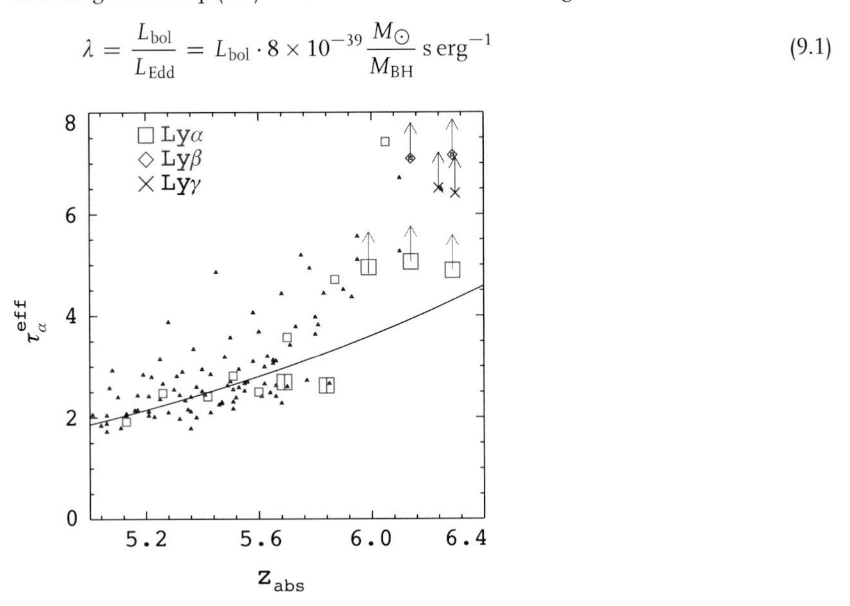

Figure 9.1 Measured optical depth τ_α of neutral hydrogen at redshifts in the range $5.0 < z < 6.4$. Note that the values above $z = 6$ are mainly lower limits, indicating that a large fraction of the matter at these redshifts is neutral. Graphic from Goto *et al.* (2011b).

The bolometric luminosity resulting from the accretion process is

$$L_{\text{bol}} = \eta \dot{M} c^2 \tag{9.2}$$

Here η is the efficiency with which the accretion is converted into emission, and $\dot{M}$ is the mass accretion rate. The efficiency depends on the details of the accretion process we assume. For example, a spherically symmetric accretion will have a different efficiency than an accretion disk. Through studying local AGN for which the accretion processes can be best constrained, it seems that an efficiency of about $\eta = 10\%$ is a good approximation (e.g., Ho, 2008). The mass M_{BH} after n years of accretion will then be

$$M_{\text{BH}} = M_0(1+\delta)^n \simeq M_0 \exp(\delta n) \tag{9.3}$$

with δ being the fractional accretion rate per year, which is

$$\delta = \frac{\lambda}{\eta} \frac{4.1 \times 10^{45}\,\text{g}\,\text{cm}^2\,\text{s}^{-2}}{M_\odot c^2} = 2.28 \times 10^{-9} \frac{\lambda}{\eta} \tag{9.4}$$

For example, if we assume a starting mass of $M_0 = 10\,M_\odot$, an efficiency $\eta = 0.1$ and accretion at the Eddington limit ($\lambda = 1$), we derive after $n = 6 \times 10^8$ years a final mass of $M_{\text{BH}} = 8.5 \times 10^6\,M_\odot$, which falls short of the required mass for a bright quasar by approximately 2 orders of magnitude. In the early Universe, when the metal abundance was low, stars would have been able to grow to larger sizes. If we assume zero metal abundance, Population III stars may have reached masses in excess of 200 $M_\odot$. The very massive stars with $M_* > 260\,M_\odot$ are especially interesting candidates for the supermassive black hole seeds, as they are likely to form black holes with masses of at least half the original stellar mass (Fryer *et al.*, 2001). In addition, evolution of these heavy stars would be rapid, with typical lifetimes of only 2×10^6 years. We are then left to assume that either sustained super-Eddington accretion occurred and/or that mass accretion proceeded with a lower than expected radiative efficiency factor η. Already lowering the efficiency to $\eta = 0.09$ and assuming $M_0 = 150\,M_\odot$, $n = 6 \times 10^8$ years gives us a supermassive black hole of $M_{\text{BH}} = 6 \times 10^8\,M_\odot$. Thus, theoretically it appears to be possible to achieve the required AGN masses at redshift $z = 7$ just assuming secular growth. Whether these high accretion rates over significant time scales are realistic or not is a different question. In addition, the initial mass function (IMF), that is the distribution of stellar masses of the Population III stars is not known. Recently for example Stacy *et al.* (2012) have presented new calculations suggesting typical Population III objects result from primordial mass sinks that asymptotically approach $\sim 30\,M_\odot$, which would further compound the problem. If the IMF is indeed as top heavy as indicated by some simulations, these stars seem to be the most promising seed candidates for the currently observed massive black holes, but this issue remains unresolved at present.

It might also be possible to achieve black holes directly without previously forming a "real" massive star. The gas in the early Universe was very metal-poor. If the angular momentum of protogalaxies was sufficiently low, dynamical instabilities in the gas can then result in the creation of local overdensities. The low metallicity leads to low opacity and therefore the gas cloud, or quasistar, can collapse directly into a black hole with a mass in the range $10^3 M_\odot < M_{BH} < 10^6 M_\odot$ (e.g., Begelman *et al.*, 2008). Gas mass halos of (proto)galaxies have to be larger in this case than for the creation of Population III (Pop III) stars, and thus these seeds would form later. Nevertheless, their large initial mass would allow them to evolve directly into supermassive black holes until redshift $z \sim 7$ or even earlier. As for the Pop III stars, this channel for forming black hole seeds comes to a halt as soon as the metallicity in the Universe is too high due to supernovae production of heavy elements. Thus, by the time the Universe is about 1.2 billion years old ($z \sim 5$), we would no longer expect the formation of these massive black hole seeds. This may also explain, if Pop III stars and massive "quasistars" existed, how they have escaped observation. From the colors of the high-redshift galaxies we are beginning to see observational evidence for the low-metallicity era finishing earlier, at redshift $z > 5$. For example, McLure *et al.* (2011) studied a sample of galaxies at redshift $6.0 < z < 8.7$. Their stellar content does not appear to be bluer or more actively evolving than that of star-forming galaxies in the local Universe, as one would expect for a low metal abundance environment. In addition, the star-formation rate is similar to that of starburst galaxies at $z \sim 2$. We also note that evidence for dust in the $z \simeq 5$ host galaxies of gamma-ray bursts has been reported by Perley *et al.* (2010). Collectively, these constraints indicate that the circumstances for star formation as we see them today were already in place at $z > 6$.

Galaxies have been detected already out to redshift $z \sim 8.7$. Observations of the first stars, as might be achieved by the James Webb Space Telescope (JWST), will certainly be of importance in identifying the seeds of supermassive black holes. If the favored evolutionary path involving the "quasistars" exists, JWST should be able to detect several of these massive stars within each field it observes in the 2–10 μm band (Volonteri and Begelman, 2010).

A more exotic possibility for seeding massive black hole growth involves the weakly interacting massive particles (WIMPs) mentioned earlier in Section 8.1.5. Under certain circumstances the density of WIMPs inside the first generation of stars can be high enough that WIMP annihilation causes heating of the protostar. That could stop its cooling process and further collapse is prevented. Therefore, gas cloud collapse could create large "dark stars" with predicted masses up to $M_* = 10^3 M_\odot$ (Freese *et al.*, 2008). The massive black hole resulting from their final collapse would provide large seed black holes. Needless to say that until the existence of WIMPs is actually established this possibility is rather speculative.

Primordial black holes offer another way to create supermassive black holes. These black holes would have formed shortly after the Big Bang out of density fluctuations, resulting in local regions sufficiently dense in order to create a black hole (e.g., Carr *et al.*, 2010, and references therein). These black holes can start within a large range of masses, because their size simply depends on the time of their creation and thus on the particle horizon at this time

$$M \simeq \frac{c^3 t}{G} \simeq 10^{38} t \, \mathrm{g\,s^{-1}} \tag{9.5}$$

with the possible time for this event after the Big Bang being $10^{-43}\,\mathrm{s} < t < 1\,\mathrm{s}$, resulting in black hole masses from $M_{\mathrm{BH}} \sim 10^{-5}\,\mathrm{g}$ to $M_{\mathrm{BH}} \sim 10^5\,M_\odot$. The lightest of these primordial black holes would evaporate quickly because of Hawking radiation (Hawking, 1974) within

$$t_{\mathrm{ev}} = \frac{1620\pi^2 G^2 M_{\mathrm{BH}}^3}{hc^4} = 1.33 \times 10^{-26}\, M_{\mathrm{BH}}^3\, \mathrm{s\,g^{-3}} \tag{9.6}$$

with M_{BH} in grams. But the larger primordial black holes $M_{\mathrm{BH}} \gg 5 \times 10^{14}\,\mathrm{g}$ would have survived until today, and possibly have grown further through accretion. Current upper limits on Hawking radiation observed in the Milky Way restrict the possible number of primordial mass black holes. This in turn constrains the spectrum of density fluctuations in the early Universe. Nevertheless, at least some AGN cores might be primordial black holes about which a galaxy subsequently formed. Hawkins (2011) argues that primordial black holes could also be candidate sources for the dark matter content of the Universe. He finds that the light curves and spectral changes observed in quasars are consistent with the predictions of microlensing by a putative population of stellar mass compact bodies.

Alternative to scenarios which start with a relatively massive black hole seed, one can imagine the formation of massive black holes through the merging of smaller ones (Volonteri, 2010a). But the dynamics of such merging events are subtle. A black hole pair might experience strong recoil with velocities $v \gtrsim 100\,\mathrm{km\,s^{-1}}$. Campanelli *et al.* (2007) determine possible recoil velocities as high as $v \simeq 4000\,\mathrm{km\,s^{-1}}$, depending on the configuration of the premerger black hole pairs including their spins and relative orientation. In the early stages of the Universe, the protogalaxy halo masses can be assumed to be low. Correspondingly, the escape velocity of such environments can be lower than $v_{\mathrm{esc}} \sim 10\,\mathrm{km\,s^{-1}}$. Thus, the merged black hole could be kicked out of the dense environment into the intergalactic space, leaving a naked black hole which will undergo no further growth. Since we have limited knowledge of the actual conditions in the early Universe the problem is poorly constrained and results must be considered speculative. Other studies come to the contrary conclusion that supermassive BHs should rarely be able to escape from massive galaxies at high redshifts (e.g., Sijacki *et al.*, 2011).

Once a massive black hole with $M_{\mathrm{BH}} \gtrsim 10^3\,M_\odot$ is formed in the early Universe ($z \gtrsim 15$) it is fairly "easy" to concoct scenarios to grow supermassive black holes until redshift $z \simeq 6$. Simulations show that once these relatively massive seeds are available within dark matter halos of some 10^9–$10^{10}\,M_\odot$, the required mass assembly can be achieved through periods of Eddington-limited accretion (Sijacki *et al.*, 2009). Then once the supermassive state is reached, feedback processes associated with the resulting AGN will prevent further rapid growth. The result is a down-sizing of AGN activity in the sense that the highest accretion rates are observed in the lower-mass objects. Alternatively, if we consider mass seeds which are produced by Pop III stars, these require super-Eddington accretion over significant time scales ($\Delta t \sim 10^8$ years), in order to achieve quasars at $z > 6$. In addition, dense and large-mass supply regions with $M > 10^{10}\ M_\odot$ have to surround the black hole, e.g., as shown in simulations by Kawakatu and Wada (2009).

The resulting AGN at redshift $z \sim 6$ tend to be highly obscured with column densities of $N_H \gtrsim 10^{24}\ \mathrm{cm}^{-2}$. Treister *et al.* (2011) derived this value from stacked Chandra X-ray data. Although individual high-redshift galaxies, as observed by HST are usually not detectable individually by today's X-ray telescopes, their high masses of 10^9–$10^{11}\ M_\odot$ suggest that they might nevertheless host supermassive black holes, which are possibly highly absorbed. Stacking of the X-ray data of 197 galaxies at $z \sim 6$ shows a very hard (i.e., flat) X-ray flux ratio. Apparently, strong absorption in the line of sight, probably close to the central engines, suppresses the soft X-ray flux. This being said, other studies do not confirm their results and instead find that a more conservative extraction technique leads to only upper limits for the $z \gtrsim 6$ sample (Fiore *et al.*, 2012a,b). If indeed the majority of AGN at this high redshift is considered to be Compton-thick they would not have contributed significantly to the reionization of the Universe. Instead most of the UV and X-ray radiation from those objects would have been absorbed and re-emitted as infrared radiation.

9.2 Tools to Study AGN Evolution

The early stages of formation and subsequent growth of supermassive black holes are currently impossible to constrain observationally and can be investigated only by means of numerical simulations. The evolution of AGN on the other hand can be observed once sufficiently large and statistically complete samples are available over a given redshift range. In the following three subsections we will first briefly describe some tools that can be used in order to get a handle on AGN evolution. Afterwards, we will take a look at the application of these tools to actual data.

9.2.1 The Number-Flux Relation

The most simple study of a population distributed in space is probably the *number counts* relation, often also called the "$\log N - \log S$ test." Here the number of objects above a certain flux is counted and plotted versus the flux in log–log space. Figure 9.2 shows the number counts for the second INTEGRAL AGN catalog (Beckmann *et al.*, 2009, 2010). The number counts give some insight as to the distribution in space. If a complete sample is considered, that is, only objects down to a given flux limit over a specified region of sky have been included, the number counts should have a slope of $-3/2$ in the double logarithmic representation, if the objects are evenly distributed in Euclidean space and they are drawn from a symmetrical luminosity distribution. The number counts are independent of the

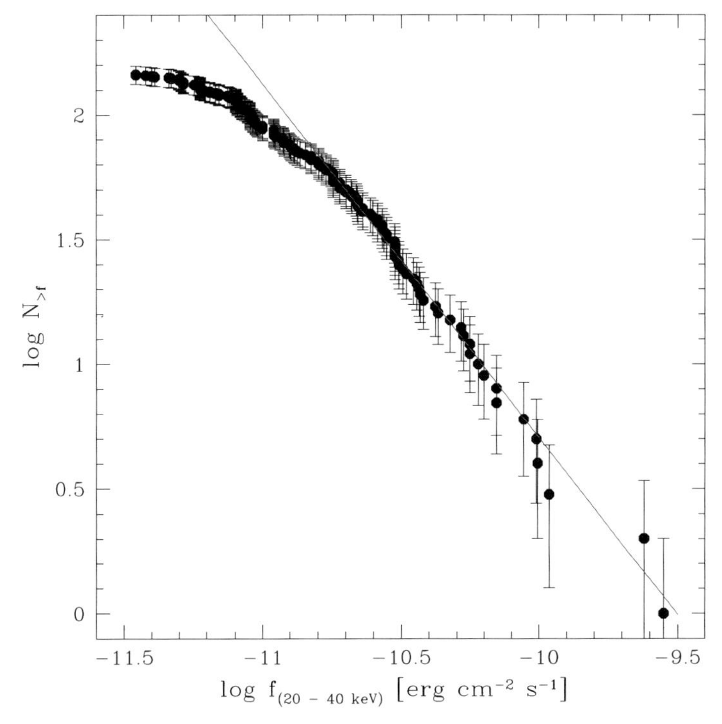

Figure 9.2 Number counts of Seyfert galaxies detected by INTEGRAL in the hard X-rays above 20 keV. The line indicates a fit to the objects with $f_{(20-40\,\text{keV})} >$ 1.7×10^{-11} erg cm^{-2} s^{-1} and has a gradient of -1.42 ± 0.07, consistent with the value of -1.5 expected for normal distribution in Euclidean space.

luminosity function of the objects studied. This means that we cannot infer the luminosity function of the sources from the number counts. One then expects the number N of objects being correlated with the flux limit f_{limit} in a sense that $N \propto f_{\text{limit}}^{-3/2}$. This is just based on the fact that the number N of sources increases with the radius r of the sphere in which we search by $N \propto r^3$ and that at the same time the measured flux f of an object is correlated with the distance r by $f \propto r^{-2}$. Although in the example of Figure 9.2 one can infer that the objects occupy a Euclidean space (low redshifts) and have normally distributed luminosities, the number counts begin to flatten towards the lowest fluxes represented, around $f \simeq 1.5 \times 10^{-11}$ erg cm^{-2} s^{-1}. This is simply because the sample starts to become incomplete at this flux level and thus not all sources which should contribute to the $N \propto f_{\text{limit}}^{-3/2}$ law are included. Therefore the number counts are often of limited use to study evolution near the flux limit of a sample. If on the other hand, the flattening occurs at a flux level significantly above the sample limit, it is indicative of a breakdown of the Euclidean space approximation or evolution of the sample population.

From cosmology we learn that the Universe is isotropic. We see for example the same cosmic microwave background in all directions. But cosmology also tells us that space is non-Euclidean on large scales. The Euclidean space as we have used it thus far is a valid assumption in the local Universe, as long as $z \ll 1$. Number counts of any sample covering a larger redshift range will be affected by cosmology.

If the sample of objects studied covers a large range in redshift, one has to account for the fact that the flux we measure in a given band has been emitted at a shorter wavelength. Assuming that the spectrum has a power-law form with an energy index $\alpha \neq 0$ (or photon index $\Gamma \neq 1$), we will also get a different flux level depending on the redshift, as we have seen in Section 8.1.2. The observed flux is a function of the emitted flux and the redshift

$$f_{\rm obs} = \frac{f_{\rm em}}{(1+z)^2} = \frac{L}{4\pi d_L^2} \tag{9.7}$$

See the discussion of the "K-correction" in Section 9.2.3. Therefore, strictly speaking, a correct treatment of the number counts, requires not only a statistically complete sample, but also redshifts have to be available for all sources. This can be ignored though and becomes negligible in low-redshift samples in which the redshift corrections can be neglected. Other applications where a correction is not necessary are small redshift ranges, like the number counts for highest-redshift quasars (e.g., $z > 5$), where one observes a large shift from $\nu_{\rm em}$ to $\nu_{\rm obs}$, but the shift is similar for all sources in the sample. Therefore here the redshift correction does not present a major problem, also because in these kinds of studies other aspects such as sample completeness and measurement precision (e.g., for the spectral slope) are more severe. Often, the number counts relation is given in number of objects per unit sky area. These curves allow one to estimate the number of expected sources in a sky survey of given flux limit and sky coverage. The steepness of the number counts relation is often a striking argument when thinking about deeper surveys or new telescopes and instruments. Assuming Euclidean space, doubling the sensitivity will lead to a factor of $2^{1.5} \simeq 2.8$ more sources detected in the same observation time, and by increasing the sensitivity by an order of magnitude, one derives an increase in source numbers by a factor of 32.

9.2.2
The $V/V_{\rm max}$ Test

Another method to establish evolution in a complete sample of objects for which the redshifts are known, is the application of a $V/V_{\rm max}$ test (Schmidt, 1968). This test is based on the ratio of the redshift of the objects in relation to the maximal allowed redshift $z_{\rm max}$ within the survey. If we have a sample of n objects, of which every object encloses a volume V_i, determined by its redshift z_i, and the object with the highest redshift $z_{\rm max}$ in the survey encloses a volume $V_{\rm max}$, then the mean

$$\left\langle \frac{V}{V_{\rm max}} \right\rangle = \frac{1}{n} \cdot \sum_{i=1}^{n} \frac{V_i}{V_{\rm max}} \tag{9.8}$$

will have a value in the interval $[0 \ldots 1]$. A value of $\langle V/V_{\rm max} \rangle = 0.5$ would refer to an equally distributed sample in space, $\langle V/V_{\rm max} \rangle > 0.5$ indicates positive evolution, and a smaller value negative evolution. The area of the survey is not important for this value, because $V_i/V_{\max} = d_p^3/d_{p,\max}^3$, with d_p being the proper distance of the object with redshift z and $d_{p,\max}$ the value for $z_{\max}$. This test is very sensitive to the maximum detected redshift $z_{\max}$. Therefore, Avni and Bahcall (1980) improved the test by using V_e/V_a. Here V_e stands for the volume that is enclosed by the object, and V_a is the accessible volume in which this individual object could have been found (e.g., due to a flux limit of a survey). Thus different surveys with different flux limits in various energy bands can be combined by the V_e/V_a test. In addition, multiple flux limits for each object can be taken into account. If we assume for example a sample of radio-selected quasars, we not only have to consider the maximum redshift $z_{\max,1}$ which results from the radio flux limit applied, but must also consider the maximum possible redshift $z_{\max,2}$ up to which a candidate could have been identified as a quasar and up to which limit $z_{\max,3}$ the redshift itself could have been determined. As such, several limits can apply for each source, and one carefully has to identify and determine each one. The final $z_{\max}$ to be applied is then the smallest of all such redshift limits for that source.

The error of $\langle V_e/V_a\rangle$ can be determined as follows. For an equally distributed sample the mean value $\langle m\rangle = \langle V_e/V_a\rangle$ is:

$$\langle m\rangle = \frac{\int_0^1 m \mathrm{d}m}{\int_0^1 \mathrm{d}m} = 0.5 \tag{9.9}$$

The mean square divergence of the mean value is

$$\sigma_m^2 = \frac{\int_0^1 (m-0.5)^2 \mathrm{d}m}{\int_0^1 \mathrm{d}m} = \left(\frac{1}{3}m^3 - \frac{1}{2}m^2 + \frac{1}{4}m\right)_0^1 = \frac{1}{12} \tag{9.10}$$

Therefore for n objects we get an error of

$$\sigma_m(n) = \frac{1}{\sqrt{12n}} \tag{9.11}$$

For an arbitrary mean value $\langle m\rangle$ we get an error of

$$\sigma_m(n) = \sqrt{\frac{1/3 - \langle m\rangle + \langle m\rangle^2}{n}} \tag{9.12}$$

A first application of the $V/V_{\max}$ test by Schmidt (1968) gave a result of $\langle V/V_{\max}\rangle \simeq 0.7$. This indicated that AGN must have been more luminous and/or more numerous at cosmological distances than they are today. Since then the test has been applied to an increasing number of new observational samples, confirming the earlier results. In addition, it has been shown that not all AGN evolve in the same way. Blazars for example have a type-dependent evolution, as mentioned in Section 4.8.3.

The $\langle V_e/V_a\rangle$ value can also be used in order to verify the completeness threshold of a given sample. For a complete sample the test should give the true value, for example $\langle V_e/V_a\rangle_{\text{true}}$. One can then verify at what flux or significance level this value

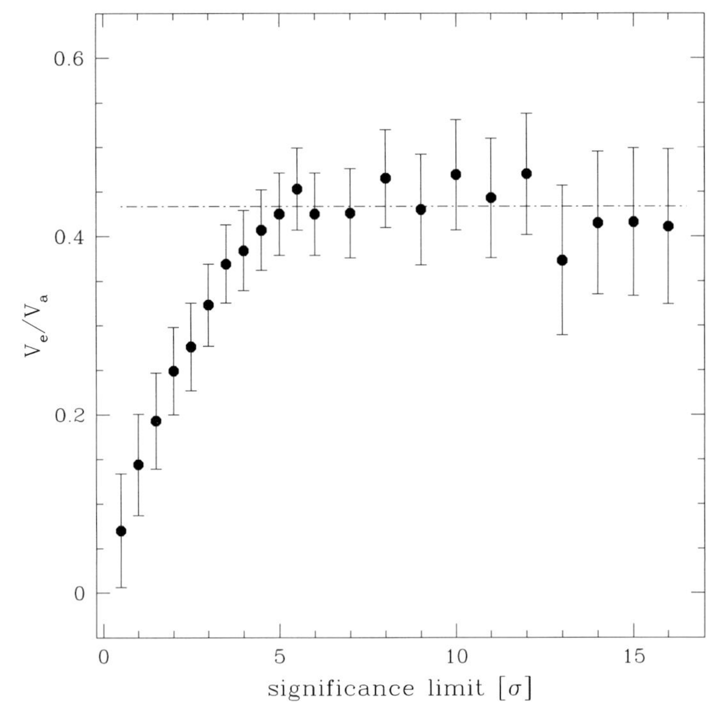

Figure 9.3 Using the $\langle V_e/V_a \rangle$ test in order to verify the completeness level of a given sample. Here the $\langle V_e/V_a \rangle$ is computed for a sample including sources above a given significance σ. For the complete sample the $\langle V_e/V_a \rangle$ value is distributed around the true value $\langle V_e/V_a \rangle_{true}$. In this example, completeness is reached for detection significance $\geq 5\sigma$. From Beckmann *et al.* (2006c). (Reproduced by permission of the AAS.)

is reached. If the sample is not complete, it is likely to miss the fainter, higher-redshift sources, resulting in $\langle V_e/V_a \rangle < \langle V_e/V_a \rangle_{true}$. Above the completeness limit the $\langle V_e/V_a \rangle$ values should be distributed around $\langle V_e/V_a \rangle_{true}$ within the statistical uncertainties. An example for such a completeness test based on the $\langle V_e/V_a \rangle$ value is shown in Figure 9.3. This test can be applied only if there is no gradient in the $\langle V_e/V_a \rangle$ value within the sample, that is the redshift range of the sample should be small compared to cosmological time scales.

9.2.3
Luminosity Function

Ultimately, one would like to be able to answer the questions: how many objects of a given luminosity do we find in a given volume? And how did this distribution change with time? Do AGN luminosities change on cosmological time scales, or did they become more or less numerous? A more refined technique to study the evolution of objects with redshift is the *luminosity function* (LF) analysis. To determine the cumulative LF, one has to count all objects within a complete sample above a given luminosity L and divide this number by the volume V_a, which has been surveyed for these objects. For each object the maximum redshift z_{max}, for which this object would have been found above the survey limit, is computed by using the individual flux limit of the object, as shown in the previous section, and its redshift. Here, the same rules apply as for the V_e/V_a test concerning the consideration of multiple survey limits in order to determine z_{max}. The space density ϕ above a given luminosity L is then described by

$$\phi(> L) = \sum_{i=1}^{n} \frac{1}{V_{a,i}} \tag{9.13}$$

The corresponding 68% error σ on $\phi(> L)$ is then determined using the formula

$$\sigma = \left(\sum_{i=1}^{n} V_{a,i}^{-2} \right)^{1/2} \tag{9.14}$$

which weighs each object by its contribution to the sum (see, e.g., Marshall, 1985, for further details). An example of a cumulative luminosity function is shown in Figure 9.4a. The cumulative LF is useful in order to determine how many objects one can expect from a survey. Figure 9.4a tells us for example that one can expect to find a BL Lac brighter than $L_X = 10^{45}\ \mathrm{erg\,s^{-1}}$ in a volume of $10^{8.3}\ \mathrm{Mpc^3} = 0.2\ \mathrm{Gpc^3}$.

A more common way of presenting the distribution of luminosities within a sample is the differential luminosity function as shown in Figure 9.4b. Here the number of objects within a luminosity bin is divided by the accessible volume V_a. One has to keep in mind that in this representation the density values depend on the luminosity binning one chooses.

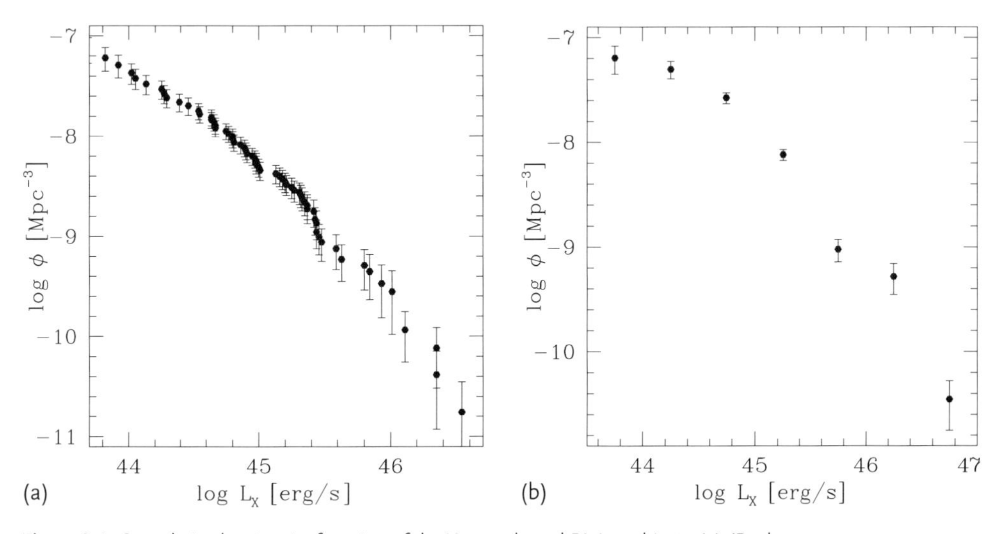

Figure 9.4 Cumulative luminosity function of the X-ray-selected BL Lac objects (a) (Beckmann *et al.*, 2003). Differential luminosity function of the same sample (b). The *x*-axis is binned to $\Delta \log L_X = 0.5$. The density refers to object density per $\Delta \log L_X = 1.0$.

One problem of any luminosity function including a significant redshift range is that the observed spectral band λ_{obs} is not the energy band in which the radiation has been emitted at the source. If we assume that the spectrum is continuous between the two bands, we extrapolate from the observed band to the emitting band, an effect more severe for large z_{max} values. This correction is the so-called *K-correction* (Schmidt and Green, 1986). For a given spectral slope with photon index Γ for $f_\nu \propto \nu^{-\Gamma}$ the transformation from the observed flux f_{obs} to the emitted flux f_{em} at a redshift z is given by

$$f_{em} = f_{obs} \cdot (1 + z)^{\Gamma - 1} \tag{9.15}$$

or, when optical magnitudes are used

$$m_{em} = m_{obs} - K(z) = m_{obs} - 2.5(\Gamma - 1)\log(1 + z) \tag{9.16}$$

where the term $K(z)$ is called the K-correction. This means that the observed flux is lower than the emitted flux if $\Gamma > 1$, because the frequency region with the lower flux is shifted into the observed wavelength region by the redshift z. This effect can be ignored in the case where one studies objects at low redshifts, or, in general, in a narrow redshift range.

If the sample of AGN covers a large redshift range and is sufficiently large, one can derive the luminosity function per redshift interval. If there were no evolution of density or luminosity, then the different LF curves extracted in each of the eras of the Universe should overlay and show the same slope. As soon as sufficient samples were accumulated in the 1970s, it became apparent that most of them turn over at a certain luminosity. Thus, luminosity functions appear as two smoothly connected power laws, with a flat slope η at low luminosity, and a steeper slope above a certain break luminosity L^*. Schechter (1976) successfully modeled this shape of the differential luminosity function the following way:

$$\phi(L)\mathrm{d}L = n_* \left(\frac{L}{L^*}\right)^{-\eta} \mathrm{e}^{-\frac{L}{L^*}} \left(\frac{\mathrm{d}L}{L^*}\right) \tag{9.17}$$

This depiction became known as the *Schechter function*. Here n_* is the normalization at the turnover luminosity L^*. At low luminosities, this gives the shape $\phi \propto L^{-\eta}$, while at high luminosities the curve approaches $\phi \propto \mathrm{e}^{-L}$. The curvature indicates that low-luminosity objects are more common than the highly luminous ones. The more commonly used form today was introduced by Marshall (1987), where instead of the exponential cutoff at high luminosities, a power law of the form $\phi \propto L^{-\eta_2}$ is applied

$$\phi(L)\mathrm{d}L = \phi^* \left[\left(\frac{L}{L^*}\right)^{-\eta} + \left(\frac{L}{L^*}\right)^{-\eta_2}\right]^{-1} \frac{\mathrm{d}L}{L^*} \tag{9.18}$$

It must be pointed out that the question of what underlies luminosity functions of this specific form has been a matter of many investigations and is far from being settled. The fact is that the luminosity function break comes in very handy when comparing ones derived for different redshift intervals. The break indicates how much the function shifted in luminosity and in density when moving from one redshift interval to the next. The shift of the break luminosity with redshift can be parameterized like the following:

$$L^*(z) = L_0^*(1+z)^k \tag{9.19}$$

where L_0^* refers to the break luminosity in the local Universe. This parameterization reflects a pure luminosity evolution (PLE), that is the number of AGN in a comoving volume does not change, just the luminosities are altered. Since measurements show that $k \sim 3.5$ up to $z \sim 2$, this indicates that the AGN population becomes more luminous when we look back in the Universe to 1/3 of its current age. With larger samples extending to higher redshifts, this simple model for the luminosity evolution is often replaced by a slightly more complex formula:

$$L^*(z) = L_0^* 10^{k_1 z + k_2 z^2 + k_3 z^3} \tag{9.20}$$

Here recent values are $k_1 = 1.3$, $k_2 = -0.3$, and $k_3 = 0.02$ (Assef *et al.*, 2011). For example, at redshift $z = 1$, when the Universe was half the age as it is now, we derive $L^*(z=1) = 11.3 L_0^*$ (Eq. (9.19)) and $L^*(z=1) = 10.5 L_0^*$ (Eq. (9.20)). AGN appear to have been 11 times more luminous at $z = 1$ compared to the population in the local Universe. In more complex models, not only an evolution of the luminosity is considered, but also a density evolution which depends on the bolometric luminosity of the objects studied. This gives the luminosity-dependent density evolution models (LDDE), as suggested first by Schmidt and Green (1983), but see also Ueda *et al.* (2003) and Hasinger *et al.* (2005) for applications to more recent data.

Currently, we can already see the evolution towards $z \sim 2$ in the bare numbers of AGN we observe at different redshifts. As seen in Figure 4.12, there are more quasars known at redshift $z \sim 1.8$ than at $z \sim 1$, although objects of similar luminosity appear fainter and thus are more difficult to detect at the higher redshift.

The Soltan argument

Soltan (1982) estimated the black hole mass density in the Universe ρ_{BH}, which he postulated should exist in the local Universe as "dead quasars," based on a set of simple assumptions. He first assumed that all accretion onto black holes leads to radiation following the formula $L_{bol} = \eta \dot{M} c^2$, where η is the radiation efficiency and $\dot{M}$ is the mass accretion rate. The latter one is a function of the black hole mass if we consider accretion which is Eddington-limited with $\lambda = 1$ (Eqs. (3.9) and (3.10)). Thus, the bolometric luminosity of AGN together with their luminosity function gives us a means to estimate the black hole mass density. Following Yu and Tremaine (2002), the black hole mass density accreted during bright AGN phases is then

$$\rho_{BH}(z) = \int_z^\infty \frac{dt}{dz} dz \int_0^\infty \frac{(1-\eta) L_{bol}}{\eta c^2} \Phi(L, z) dL \tag{9.21}$$

At high redshifts, one can further simplify this expression as the luminosity function will not be a strong function of z: $\Phi(L, z) \simeq \Phi(L)$. The bolometric luminosity can be approximated for example by the X-ray or optical luminosity, applying a correction factor, as discussed earlier in the second box of Section 4.8.1. Note that this expression is independent on the cosmological model, that is the formula for $\rho_{BH}(z)$ does not include q_0 or H_0.

9.3 Luminosity Functions of AGN

If we want to trace the density of AGN and its dependence on luminosity all the way back to the early Universe, we face several challenges. We must first consider the consequences of which is the energy band we choose to construct the LF from. If we use the radio band, we are more likely to pick up the radio-loud quasars and especially blazars, but we will miss many of the Seyfert galaxies, especially those among the type 2 class. In the infrared, active star-forming galaxies will likely contaminate our sample. The optical band, for which we have the largest samples, gives a good representation of unabsorbed objects but again type 2 AGN will be underrepresented. X-rays seem to be a good tracer of AGN activity, especially when approaching higher redshifts, because they are less affected by absorption than for example optical radiation. However as the X-rays do not usually tell us the redshift of an object, we still rely on optical and infrared follow-up for that information. In order to construct the LF, those follow-up observations have to be complete, that is, all the sources in our sample need redshifts, and, if we want to study subpopulations, the identification of each AGN type.

The early results of Schmidt (1968) regarding evolution of quasars still hold: when looking back in time, we observe a higher density of luminous AGN than in the local Universe. Where the peak of the AGN activity in the Universe lies, has been a matter of numerous studies, and the answer might depend on source class as well as on luminosity. There appears to be a peak in the density of optically detected type 1 quasars in the range $2.2 < z < 2.8$. This can be seen for example in an analysis of more than 15 000 quasars from the SDSS presented by Richards *et al.* (2006). A consistent result was achieved using an infrared-selected sample from Spitzer data of type 1 quasars by Brown *et al.* (2006). In order to reduce contamination by bright starburst galaxies, in their study only optically compact sources were included. This resulted in a sample of 183 quasars at $z > 1$ and 292 at lower redshift. The maximum density of quasars in this study is reached at a redshift of $z = 2.6 \pm 0.3$, which seems to indicate that the infrared and optically selected type 1 quasar samples follow a consistent cosmological evolution. One does however detect a dependence on the luminosity.

Croom *et al.* (2009) studied an AGN sample collected in a deep field of $\sim 192\,\text{deg}^2$, selected from the SDSS. They constructed the luminosity function for

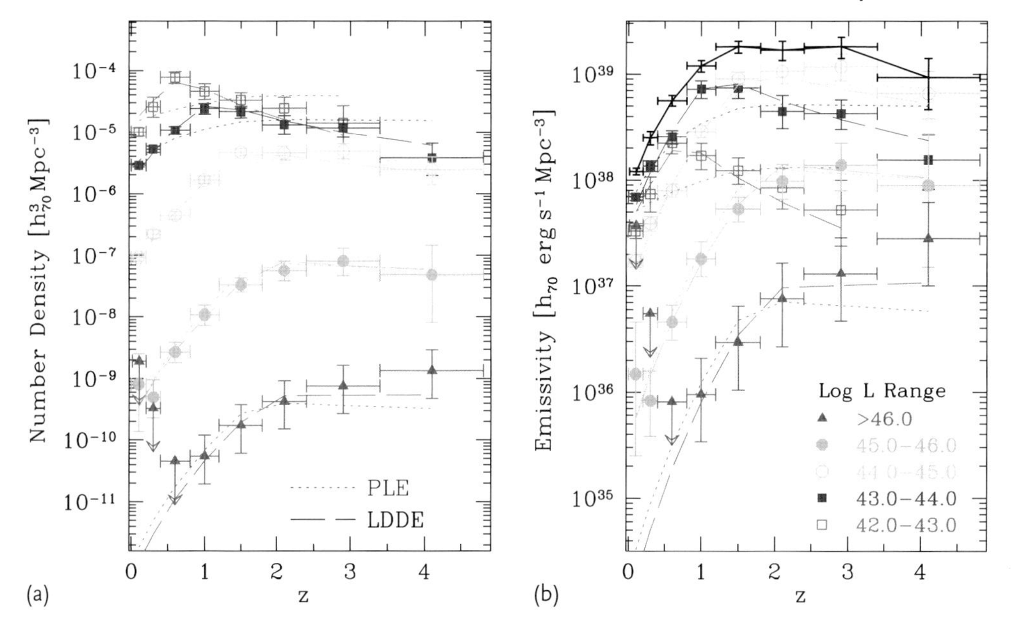

Figure 9.5 X-ray luminosity function as presented by Hasinger *et al.* (2005); reproduced with permission © ESO. The space density of AGN as a function of redshift for several X-ray luminosity bins (a). The lines shown are for a pure-luminosity evolution (PLE) and for a luminosity-dependent density evolution (LDDE). Emissivity of AGN as a function of redshift (b). The uppermost curve shows the summed emissivity of all luminosity classes considered here. For a color version of this figure, please see the color plates at the beginning of the book.

faint and for bright quasars separately and found that while the density of bright quasars peaks indeed at $z > 2$, the faint AGN seemed predominate in the local Universe, with a peak in density at $z < 1$. This is also called *cosmic downsizing* (Barger *et al.*, 2005) and indicates that the most massive and active AGN were accreting in the earlier Universe, while the less massive ones are more active at the present time. In Figure 9.5a we show the X-ray luminosity function for different brightness classes. As stated earlier, the X-ray emission can be used as a proxy for the bolometric luminosity. The higher the luminosity of the AGNs investigated, the earlier in the Universe the luminosity function seems to peak. In the emissivity plot (Figure 9.5b) we see that although the bright AGN with $10^{44}\,\mathrm{erg\,s^{-1}} < L_X < 10^{45}\,\mathrm{erg\,s^{-1}}$ were dominating the total energy output at $z \gtrsim 1.5$, the emissivity in the local Universe is produced by those objects with $L_X < 10^{44}\,\mathrm{erg\,s^{-1}}$.

This can also be seen in the optical luminosity function. We show as an example the luminosity functions in various redshift bins in Figure 9.6. The higher the redshift bin, the more bright objects per unit volume. This trend seems to slow down above $z \simeq 1.5$. All luminosity functions derived to date show a steep decline beyond $z \sim 3$, some 2.2 Gyr after the Big Bang. In terms of density, for example for

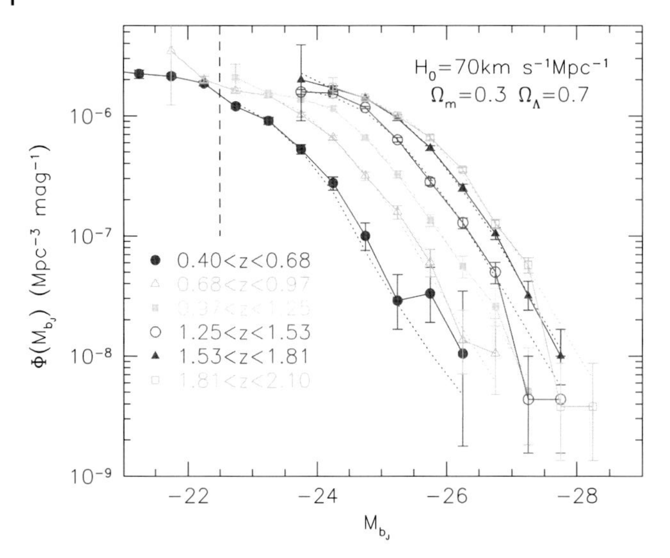

Figure 9.6 Optical luminosity function in several redshift bins derived from SDSS data. The dotted lines denote the predictions of the best-fitting double power law exponential evolution mode. From Croom *et al.* (2004). For a color version of this figure, please see the color plates at the beginning of the book.

bright X-ray-selected sources ($L_X > 10^{44}\,\mathrm{erg\,s^{-1}}$), this means that while at $z \sim 3$ one finds an AGN density of $\sim 7 \times 10^{-6}\,\mathrm{Mpc^{-3}}$, it is 10 times lower at redshift $z \sim 5$, as shown in Figure 9.7 (Civano *et al.*, 2011). Thus, it is in this redshift range that we are most likely to observe rapid evolution and supermassive black hole growth.

What kind of AGN population can we expect at high redshifts? Today, reliable studies can be done out to a redshift of $z \sim 5$. For example, Trakhtenbrot *et al.* (2011) studied 40 type 1 AGN around $z \simeq 4.8$. These sources are characterized by high accretion rates: some seem to exceed the Eddington limit by a factor of 3.9, while the average Eddington ratio is $\lambda \sim 0.6$. Their luminosities are in the range $3 \times 10^{46}\,\mathrm{erg\,s^{-1}} < L_{\mathrm{bol}} < 2.4 \times 10^{47}\,\mathrm{erg\,s^{-1}}$. This seems consistent with luminosities and accretion rates one infers from other surveys up to redshift $z \sim 6$, although toward the highest redshifts, samples are not complete. For example, the highest-redshift quasar, ULAS J112001.48+064124.3 at $z = 7.085$, has a bolometric luminosity of $L_{\mathrm{bol}} = 2.4 \times 10^{47}\,\mathrm{erg\,s^{-1}}$, black hole mass $M_{\mathrm{BH}} \simeq 2 \times 10^9\,M_\odot$, and is accreting close to the Eddington limit (Mortlock *et al.*, 2011). In general, the central black holes of the $z \gtrsim 5$ AGN appear to have fully evolved, with a median mass of $8.4 \times 10^8\,M_\odot$. Therefore, these objects can be regarded as likely progenitors for the most massive AGN with $M_{\mathrm{BH}} \gtrsim 10^{10}\,M_\odot$ at $z \sim 3$. Following Eq. (9.3) this evolution within 1 Gyr can be achieved by growing at an average Eddington ratio of $\lambda = L_{\mathrm{bol}}/L_{\mathrm{Edd}} \sim 0.1$, when assuming an efficiency of $\eta = 10\%$. At the same time this is another indication of the cosmic downsizing: the large supermassive black holes underwent rapid growth early in cosmic time, within the first

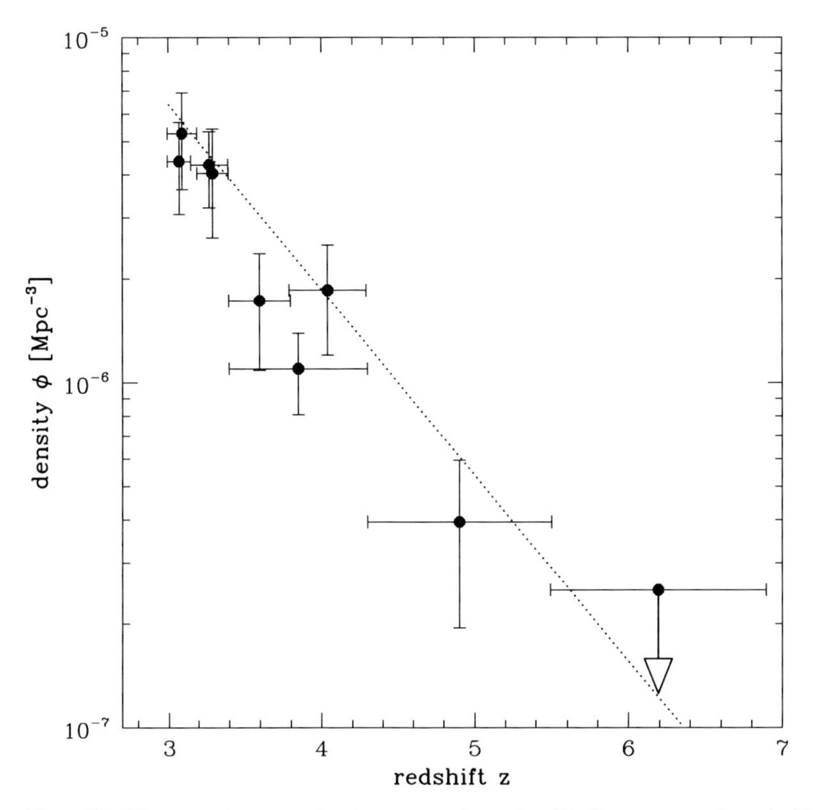

Figure 9.7 The comoving space density in different redshift bins above $z = 3$ for bright X-ray detected AGN with $L_{2-10\,\mathrm{keV}} > 10^{44}\,\mathrm{erg\,s^{-1}}$ from the XMM-Newton COSMOS survey. Values have been corrected for absorption. The line corresponds to the X-ray-selected AGN space density computed for the same luminosity limit from a model presented by Gilli *et al.* (2007). Data from Civano *et al.* (2011).

5 Gyr, and then ceased being active in the local Universe. From X-ray observations we know that the sources at $z > 4$ seem to be absorbed, but not Compton-thick, with a hydrogen column density N_H less than a few $10^{22}\,\mathrm{cm}^{-2}$ and X-ray spectral slopes consistent with what is observed in the local Universe (Saez *et al.*, 2011). Thus, the cosmic downsizing goes hand in hand with a reduction of the obscuring gas and dust surrounding the AGN central engine. The intrinsic X-ray spectrum, on the other hand, has a consistent shape, indicating that the central engine works the same way at high and low redshift.

Absorption is an important consideration when studying luminosity functions. So far, we looked primarily at type 1 objects, as they are the predominant component of high-redshift optically selected samples. With X-ray observations for example by Chandra and XMM-Newton we also pick up absorbed sources, except for extreme cases, with around $N_\mathrm{H} \gtrsim 10^{24}\,\mathrm{cm}^{-2}$. At those column densities it becomes practically impossible for most photons below 10 keV to pass through the absorbing material. Another way to look for absorbed sources is to use the submillimeter range. Using observations at 850 μm from the SCUBA array, Almaini *et al.* (1999) pointed out that in considering the observed X-ray background intensity, an increasing fraction of the higher-redshift sources must be obscured. We will further explore arguments concerning the cosmic background radiation in the next section.

For the time being, it is important to note that any assessment of the net accretion power in the Universe must carefully account for the effects of absorption. A large fraction of the cosmic infrared background has to come from obscured AGN. Almaini *et al.* (1999) estimated that there might be three times as many obscured AGN, that is, with $N_H > 10^{22}\,\mathrm{cm}^{-2}$ than unobscured ones at larger redshifts. The same ratio should also apply for the comparison between Seyfert 2 and Seyfert 1 galaxies. In the local Universe this ratio appears to be different. Seyfert 2 galaxies are more difficult to assess, as their weaker optical emission lines can be hard to detect. One way to tackle the problem is to study the luminosity function of a specific emission line, for example of the [O III] line at $\lambda = 5007$ Å. The ratio between type 1 and type 2 objects in the local Universe ($z < 0.15$) seems to be a function of the luminosity of the objects, with the more luminous objects being preferentially of type 1. The numbers of Seyfert 1 and Seyfert 2 galaxies are comparable at low luminosity, while at high luminosity, Seyfert 1 galaxies outnumber Seyfert 2 galaxies by a factor of 2–4 (Hao *et al.*, 2005). Thus, it appears that we observe an evolution of AGN absorption, with a correspondingly higher fraction of absorbed objects at redshifts $z \sim 2$.

Again, let's turn to the X-rays which give us a better view of the central engine, regardless of the extent to which it may be absorbed. Barger *et al.* (2005) studied the LF of AGN using Chandra data in the 2–10 keV band. For a pure luminosity evolution they find that the turnover L^* of the LF evolves like $L^*(z) = L_0^*(1+z)^k$ with $k = 3.2$ up to redshift $z \sim 1.2$ for all AGN, and $k = 3.0$ when considering only type 1 sources. Apparently, all objects dropped in luminosity by a factor of 8–9 from redshift $z = 1$ until today, that is within ~ 7.7 Gyr. Only 25–50% of the total mass density residing in SMBH is contained in broad-line AGN. The similar slope of the LF evolution of broad-line and absorbed sources seems to indicate that at least since $z \sim 1$ there has been no significant change in the luminosity ratio between these classes. At the same time, one observes accretion rates which on average are the same as in the high-redshift Universe, with an Eddington ratio of $\lambda \sim 0.15$. If we look at all X-ray detected AGN, for example in the Chandra deep fields, the peak activity appears instead to be at $z \sim 1.2$, which coincides with the peak at around $z \sim 1$ of the low-luminosity sources with $L_{\mathrm{bol}} < 10^{44}\,\mathrm{erg\,s}^{-1}$. Apparently the absorbed and faint AGN dominate here, and their evolution peaks some 2.6 Gyr after that of their bright and broad-line counterparts. This was pointed out by Steffen *et al.* (2003), and the fact that broad-line AGN dominate the number densities at high X-ray luminosities and narrow-line AGN at low X-ray luminosities is sometimes referred to as the *Steffen effect*. This indicates a break in the AGN unification paradigm in the sense that in addition to orientation and beaming effects, the luminosity or accretion rate are also a determining factor in the type of AGN we observe.

The growth of supermassive black holes in the Universe was rather constant between redshift $z = 5$ and $z = 2$ in terms of their Eddington ratios, but it seems to have subsequently slowed significantly. The black hole mass density at $z = 5$ was about 200 $M_\odot$ Mpc^{-3}, increasing within 2 Gyr to about 30 000 $M_\odot$ Mpc^{-3} at $z = 2$. After that the accretion, again assessed in terms of Eddington rate, has slowed down. The mass density of SMBHs has grown until the present day to $\sim$ 200 000 $M_\odot$ Mpc^{-3} during a time period of 10 Gyr (Silverman *et al.*, 2008). The X-ray data show us that while the absolute mass accretion rate for AGN peaked around $z \sim 1.5$, the star-formation rate had already peaked at $z \sim 2$–3. This is consistent with what we discussed in terms of AGN versus starburst connection in Section 7.2: AGN activity can follow star formation in the host galaxy with some delay, and the supermassive black hole can stay active some time after the starburst has died out.

Lately, Fanidakis *et al.* (2012) have put these observations into a theoretical framework. Their model predicts that the comoving space density of AGN peaks at $z \sim 3$, similar to cosmic star-formation history. However, when taking into account obscuration, the space density of faint AGN peaks at lower redshift ($z < 2$) than that of bright AGN ($z \sim 2$–3). This implies that the cosmic evolution of AGN is shaped in part by obscuration. Indeed, it appears that absorption plays a major role and has to be taken into account when studying the evolution of AGN. The evolution of absorption can be observed directly in X-ray data. La Franca *et al.* (2005) found more absorbed AGN at low luminosities and at high redshifts. But bright quasars, which are significantly absorbed, have also been found. Polletta *et al.* (2006, 2008) found heavily absorbed bright quasars at $z = 1.3$–3 using infrared observations. Despite the risk of sample contamination by inclusion of active star-forming galaxies, the authors claim that $< 20\%$ of the observed flux comes from star-formation processes rather than from AGN.

So far, we have drawn a reasonably consistent picture of AGN evolution. First the heavy supermassive black holes are generated, and they dominate the high-redshift Universe. The fainter AGN and the type 2 sources are the main class of objects accreting since $z \sim 1.5$. The fraction of absorbed sources increases with redshift, at least to redshift $z \sim 2$ but potentially even further back in time. While the type 1 sources dominate optically selected samples, X-ray surveys pick up both, type 1 and type 2. It seems apparent that the absorbed sources outnumber the unabsorbed ones. But what if we are still missing a significant fraction of AGN in these observations? What if there is a significant population of sources with $N_H > 10^{24}$ cm^{-2}, which are Compton-thick and therefore do not show up in X-ray data below 10 keV, and will therefore be hard to find? These sources can be studied at hard X-rays, above the energy range accessible to Chandra or XMM-Newton.

The first hard X-ray local ($z < 0.1$) luminosity function at $E > 20$ keV using INTEGRAL data has been presented by Beckmann *et al.* (2006b). The result was a LF which is fairly consistent with the one below 10 keV, which means that the source population is not significantly different than that seen at 5–10 keV. The values are also consistent with the luminosity function for AGN in the 3–20 keV band as derived by Sazonov and Revnivtsev (2004) from the RXTE all-sky survey. A large number of Compton-thick AGN has not been found, and this result was confirmed later by other studies using larger INTEGRAL-selected samples (e.g., Sazonov *et al.*, 2007) and Swift/BAT data (Tueller *et al.*, 2008). Concerning the ratio between unabsorbed and absorbed sources, these surveys detect a ratio of 1 : 1 for the very local Universe. The same is true for the comparison of Seyfert 1 and Seyfert 2 objects. Evolution with redshift cannot currently be assessed in the hard X-ray surveys. The Swift and INTEGRAL surveys cover the whole sky, but compared to the "pencil beam" surveys of Chandra and XMM-Newton, which have only a few square degrees at most, the hard X-ray surveys are very shallow, probing the Universe at redshift $z \leq 0.1$. Progress will hopefully come from future hard X-ray experiments using focusing optics and having deployable optical benches to accommodate the required large focal lengths. Forthcoming examples include NASA's NuSTAR satellite and the Japanese Astro-H mission.

9.4 AGN and the Cosmic X-ray Background

Throughout the electromagnetic spectrum, if one looks deep enough one can observe a diffuse background radiation from the sky. The best-studied background is that in the microwave range, which has its origin from the time when the Universe became transparent. The cosmic microwave background (CMB) therefore shows the temperature $T = 3000$ K of the Universe some 350 000 years after the Big Bang (Table 8.1). As this corresponds to a redshift of $z \simeq 1100$, the peak of the emission at $\nu_{\max}(z = 1100) \simeq 3.1 \times 10^{14}$ Hz is redshifted down to $\nu_{\max}(z = 0) = 2.8 \times 10^{11}$ Hz, which corresponds to a black body temperature of $T = 2.725$ K.

Apart from the CMB, all other background radiation has its origin after the Universe became transparent, the so-called recombination epoch. Thus, in wavelengths other than the microwave range, unresolved sources and putative diffuse emission processes are responsible for the observed background. The cosmic infrared background (CIB) and the cosmic optical background (COB), as shown in Figure 5.16 are dominated by galaxies and stars. The IR and optical backgrounds are very weak with an intensity of $I = 9.4 \times 10^{-5}$ erg cm^{-2} s^{-1} sr, which is only 9% of the CMB. Observations of an UV extragalactic background is difficult as the Universe begins to become opaque at redshifts of a few as a result of cosmologically distributed Lyman-limit absorbers (Section 5.4).

A domain more relevant for AGN-related research is the extragalactic background in the X-rays and at higher energies. X-ray luminosity functions of AGN are therefore also important to test whether the emission from AGN at hard X-rays can explain the cosmic X-ray background (CXB), which was first detected during the early rocket flights at the dawn of X-ray astronomy (Giacconi *et al.*, 1962). A high-quality measurement of the shape of the CXB was achieved by the Cosmic X-ray Experiment on NASA's HEAO-1 satellite in the 3–250 keV band. Marshall *et al.* (1980) showed that the spectral energy distribution of the CXB peaks around 30 keV with a total intensity about 10% of that of the CMB (Figure 9.8). Apart from

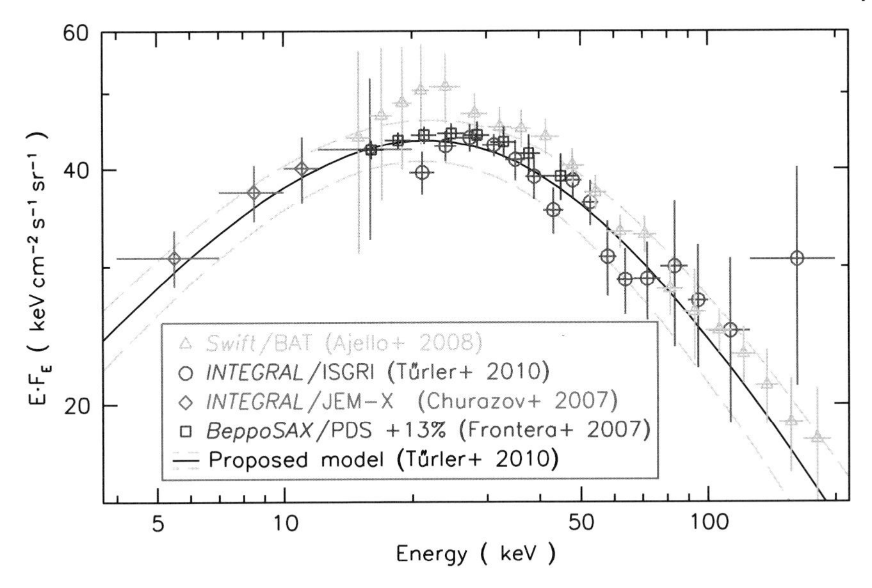

Figure 9.8 Recent measurements of the cosmic X-ray background by BeppoSAX, Swift and INTEGRAL. The continuous line shows the best-fitting model derived by Türler *et al.* (2010). Figure courtesy of Marc Türler. For a color version of this figure, please see the color plates at the beginning of the book.

the microwave background, which does not originate from individual sources, this makes the CXB the background which contributes most to the electromagnetic output in the Universe. It is thus vital to understand which sources are the main emitters. A peak around 30 keV indicates that the sources contributing to the CXB have to have an averaged spectrum with $\Gamma < 2$ before the peak and $\Gamma > 2$ at energies $E \gg 30\,\text{keV}$. Therefore, galaxy clusters with a temperature of a few keV cannot be the main contributors at the peak, nor can starburst galaxies. Thus, already early on it was speculated that most, if not all, of the CXB emission around its peak is coming from AGN.

In order to resolve the background, it is necessary to perform deep observations in a given energy band. Ideally, one would then take all the sources in the field, sum up their fluxes, and derive the total contribution by point sources to the background radiation. Thus, number counts are needed reaching far enough down to low flux limits in order to unambiguously solve the problem. Alternatively, if not all sources of the background can be resolved in an observation, one can use the local luminosity function of AGN and estimate the total contribution to the background by extrapolating to the whole population in the Universe. Here, evolution has to be taken into account, which introduces a large uncertainty in the procedure.

The resolution of the CXB into sources was tackled from its lower-energy end. Complete and deep surveys are required, and the first sufficient survey in the 0.5–2 keV range was provided by the German satellite ROSAT in the 1990s. In this soft X-ray band absorption is critical, and even the absorption of our Galaxy can suppress many of the X-ray sources. Thus, ROSAT measured the source distribution in the *Lockman hole*, an area in the sky with low galactic hydrogen column density as described in the first box of Section 4.8.1. The summed flux of the sources detected down to a flux limit of $f_{0.5-2\,\mathrm{keV}} = 10^{-15}\,\mathrm{erg\,cm^{-2}\,s^{-1}}$ was sufficient to account for 70–80% of the CXB (Hasinger *et al.*, 1998). Combined with later studies using XMM-Newton and Chandra data we reach now a flux limit of $f_{0.5-2\,\mathrm{keV}} \simeq 2.5 \times 10^{-17}\,\mathrm{erg\,cm^{-2}\,s^{-1}}$. At this flux level one is able to resolve 94 ± 7% of the CXB (Moretti *et al.*, 2003). Therefore, we can safely consider that all the X-ray emission in this band comes from single sources and not from a diffuse background. The source population dominating in this energy band is formed by type 1 AGN (Seyfert 1 and type 1 quasars). About 5% is contributed by bremsstrahlung emission from clusters of galaxies. Seyfert 2 galaxies do not contribute significantly at fluxes higher than a few $10^{-16}\,\mathrm{erg\,cm^{-2}\,s^{-1}}$, as their X-ray emission is absorbed and re-emitted in the infrared band. Only at the faint end with $f_{0.5-2\,\mathrm{keV}} < 10^{-16}\,\mathrm{erg\,cm^{-2}\,s^{-1}}$ do type 2 AGN dominate the background (Gilli *et al.*, 2007). This picture changes as we move to the harder X-rays. At 2–10 keV we can reach down to a flux limit of $f_{2-10\,\mathrm{keV}} \simeq 2 \times 10^{-16}\,\mathrm{erg\,cm^{-2}\,s^{-1}}$ using Chandra data of the southern deep field (see first box of Section 4.8.1). At this level 93% of the background can be resolved into sources. Again, about 5% are contributed by galaxy clusters, and unabsorbed AGN dominate the bright flux end, with $f_{2-10\,\mathrm{keV}} \gtrsim 10^{-14}\,\mathrm{erg\,cm^{-2}\,s^{-1}}$. This trend, that faint sources are rather obscured, and bright ones are predominantly of type 1, is not only seen in flux, but also in luminosity. As Gilli *et al.* (2007) pointed out, the fraction of absorbed sources declines as we go to higher and higher luminosities. While at $L_{2-10\,\mathrm{keV}} = 10^{42}\,\mathrm{erg\,s^{-1}}$ about 70% of the sources are absorbed, one finds only 40% of them at $L_{2-10\,\mathrm{keV}} = 10^{45}\,\mathrm{erg\,s^{-1}}$. One finds the same trend going to higher photon energies, as seen by RXTE, INTEGRAL, and Swift. Thus, in the 2–10 keV band the CXB is also dominated by AGN, but here the type 2 sources are more numerous.

As we move to higher energies, the X-ray surveys are less sensitive, but cover again the whole sky or a significant fraction of it, like the ROSAT All-Sky Survey (RASS) did below 2 keV. The Rossi X-ray Timing Explorer (RXTE) sky survey in the 3–20 keV energy band revealed about 100 AGN, showing an even higher fraction of absorbed sources of about 60% (Sazonov and Revnivtsev, 2004). Using these shallow surveys, one is no longer able to directly resolve the X-ray background into sources. The total flux of the RXTE detected AGN accounts for only a few percent of the background radiation. In order to estimate the contribution to the CXB of the type of sources found in the survey, the total AGN emissivity W, computed from the luminosity function, is compared to the emissivity of the background in the same energy band. The emissivity is simply computed by multiplying the differential luminosity function by the luminosity represented in each data point. The total emissivity is then the integral over this curve, that is for the discrete differential emissivity function, the sum of the densities over the luminosities probed by the survey. In this manner, one extrapolates the local luminosity function, as measured in the survey, to higher redshifts and sums the expected flux from the source population as it should arrive at the observer, ignoring the possible effects of evolution.

Sazonov *et al.* (2008) showed that the integrated emissivity of AGN in the RXTE energy range is about $W_{3-20\,\mathrm{keV}} \sim 5 \times 10^{38}\,\mathrm{erg\,s^{-1}\,Mpc^{-3}}$, which does not account for the total X-ray volume emissivity in the local Universe as measured in the CXB. This indicates that some sources in the local Universe might be missing from the survey, for example low-luminosity objects like galaxy clusters or low-luminosity AGN. In addition, a strong evolution with redshift can account for the difference. If, as seen in the luminosity function studies derived in other energy bands, the emissivity of all AGN was higher at higher redshifts, this might fill in the missing fraction of the CXB. Above 20 keV, INTEGRAL and Swift have determined that the local luminosity function and the associated emissivity can be compared to the CXB as well. Because the CXB peaks at these energies, this is indeed the most interesting band to resolve the background into point sources. Similar to the RXTE measurements, the hard X-ray telescopes are only able to resolve about 3% of the CXB within deep field observations. The total emissivity is $W_{20-40\,\mathrm{keV}} \sim 3 \times 10^{38}\,\mathrm{erg\,s^{-1}\,Mpc^{-3}}$ (Beckmann *et al.*, 2006c). Assuming a simple power-law spectrum with photon index $\Gamma = 2$ for the sources responsible for the background, the total emissivity of AGN measured above and below 20 keV appears to be consistent. We observe the same trend between luminosity and intrinsic absorption as at lower energies, also at the peak of the CXB emission: $\sim 65\%$ of the sources with $L_{20-100\,\mathrm{keV}} \sim 10^{42}\,\mathrm{erg\,s^{-1}}$ are absorbed, but this fraction drops to $\sim 35\%$ when bright AGN with $L > 3 \times 10^{44}\,\mathrm{erg\,s^{-1}}$ are considered (Beckmann *et al.*, 2009).

A large fraction of Compton-thick AGN, which could explain the 30 keV peak because the Compton hump in their spectra would fall within the same energy range, is also not observed. The detected fraction of Compton-thick sources above 20 keV is somewhere between 5 and 10%, significantly below the value required in order to close the gap to the CXB. One has to keep in mind that the total number of Compton-thick (CT) AGN detected above 20 keV is still somewhere around a dozen or so. Some attempts have been made to define complete subsamples of hard X-ray-selected Seyfert galaxies in which the fraction of highly absorbed sources is higher. Especially in view of the small numbers of CT AGN, the validity of the thus derived higher fraction is difficult to verify. The fact remains that very few CT AGN have been detected. Thus, a safe statement today is that we do not detect a large fraction of those sources in the hard X-rays, and that the sources necessary to explain the peak of the CXB at 30 keV are outside the parameter space observable for example by Swift and INTEGRAL. This might be because these sources are of low luminosity, or they are predominantly at higher redshift and have thus escaped the shallow hard X-ray surveys. The latter scenario seems to be supported by the observation at softer X-rays, that number density and intrinsic absorption of the AGN population increase with z. In the deep fields taken at energies below 10 keV there seems to indeed be some indication for highly absorbed sources at higher redshifts. Using data of the Chandra Deep Field South (CDF-S), Fiore *et al.* (2012a) find a fraction of $18^{+17}_{-10}\%$ Compton-thick sources among 40 AGN at redshift $z > 3$. Other, more indirect methods use the nondetection in the X-rays of optically or infrared-selected AGN as a signature of Compton thickness. Obviously, this requires assumptions about the intrinsic spectral energy distribution of the sources in question. In another approach, Luo *et al.* (2011) stacked the X-ray data of 23 AGN candidates in the CDF-S, deriving a hard X-ray signal with a flat photon index $\Gamma = 0.6^{+0.3}_{-0.4}$. Under the assumption that the intrinsic spectrum should have a photon index of $\Gamma \simeq 2$, this indicates strong absorption, which affects the soft X-ray band stronger than the hard X-rays, thus leading to a flattening of the spectrum. Together with simulations in order to derive the AGN fraction within the sample, this results in a fraction of Compton-thick AGN in the range 60–80% for $z \simeq 0.5-1$. If this measurement turns out to be correct, these rather low-luminous CT sources could indeed provide the missing flux in order to explain the CXB around its peak.

Clearly, there is some discrepancy between all these methods. While the directly observed local CT AGN fraction is $\ll 10\%$, the fraction based on CXB modeling is supposed to be $\sim 10-20\%$, and the direct measurements at high redshifts give $\sim 20\%$. Further identification of sources in the Chandra and XMM-Newton deep fields will help in this respect, but the problem might be solved only through direct hard X-ray observations by focusing hard X-ray telescopes, like NuSTAR, Astro-H, or ATHENA.

9.5 The Late Stages of an AGN's Life and Reignition SMBH

In this chapter we have seen that supermassive black hole activity peaked at $3 < z < 1$, between ~ 11.5 and 7.5 Gyr ago. The largest supermassive black holes are rather dormant in the local Universe, smaller sized AGN seem to dominate now, but the peak of the accretion history of our cosmos took place long before the solar system formed. Where does the AGN evolution go from here? The answer to this question is closely linked to the general evolution of the cosmos. Assuming a cosmology with a flat Universe ($\Omega_{r,0} + \Omega_{m,0} + \Omega_{\Lambda,0} \simeq 1$, see Table 8.2) with a cosmological constant $\Omega_{\Lambda,0} \simeq 0.7$, we are heading toward a Big Chill, with a constantly decreasing matter and radiation density. In such an environment, galaxy collisions will become less frequent. Merging, which can be a trigger of star formation and black hole growth, will therefore not be a dominant mechanism anymore. This indicates, that the large supermassive black holes with $M_{BH} = 10^9-10^{10}\ M_\odot$ will remain in red and dead galaxies. The smaller AGN will continue to grow, but it is questionable whether they have enough material around them in order to grow to the size of the giant black hole cores, formed at $z \sim 2$ and earlier. An example is the black hole in our own galaxy, which is comparably small with $M_{BH} \simeq 4 \times 10^6\ M_\odot$ and accreting at very low Eddington ratio $\lambda \simeq 10^{-4}$. Eating like this it will never grow up to be a full-grown quasar core before the end of times.

But at the present time, black hole growth has not come to a halt. We see AGN activity all around in the local Universe even in our own Galaxy. There are indications that the Milky Way's central black hole was accreting matter at a much larger rate than today just a few hundred years ago. The past activity may be seen as the afterglow of the molecular cloud Sgr B2, which is located several hundred light years away from our line of sight to the galactic center. The cloud, which on its own would not produce significant high-energy radiation, is visible at the hardest X-rays above 20 keV as observed by INTEGRAL (Revnivtsev *et al.*, 2004). It is bright in the iron K_α line at $E = 6.4$ keV in ASCA data, consistent with a Compton reflection spectrum, and has a spectral slope in the X-rays with photon index $\Gamma \simeq 1.8$, similar to what one would expect from an AGN. Thus, the interpretation of this cloud with an approximate mass of $M \sim 2 \times 10^6\ M_\odot$ as a "Compton mirror" of the activity of our very own supermassive black hole seems possible. It should be pointed out however that competing models, in which Sgr B2 is illuminated by cosmic rays are considered to be a plausible alternative explanation. Monitoring of the source on decadal time scales could resolve the issue, as the mini-AGN hypothesis would entail a gradual dimming.

Our Galaxy is not the only one which shows sporadic outbursts of activity. At a distance of 14–21 kpc from the nonactive galaxy IC 2497, an ionized cloud of some $10^9\ M_\odot$ was found.[1] Because this cloud does not host a star-forming region, nor an active core, it seems that in this case the ionizing source has to be outside; the nearest candidate as illuminating source is IC 2497, a galaxy which does not host an AGN, as deduced from the absence of X-ray emission (Lintott *et al.*, 2009). The observed [O III] luminosity of the ionized cloud requires a source with $L_X \sim 8 \times 10^{44}$ erg s^{-1} at the distance of IC 2497, more than 4 orders of magnitude brighter than today's X-ray emission from the galaxy. If the connection between the ionized cloud and the galaxy indeed exists, this would mean a rapid shut down of AGN activity over a time span of less than 70 000 years (Schawinski *et al.*, 2010).

The onset of accretion onto a supermassive black hole might also be directly observable. The Swift X-ray telescope detected an outburst in March 2011, something which looked at first glance like a gamma-ray burst (GRB 110328A). But where the X-ray emission of gamma-ray bursts fades away quickly, on time scales of seconds to hours, this source showed rebrightening and activity lasting over months and is apparently still detectable in the X-rays. Follow-up observations show that the source, now called SWIFT J164449.3+573451, is located at the core of a small host galaxy at redshift $z = 0.354$ with half-light radius of only 1 kpc. As a comparison, the half-light radius of our Galaxy is about 6 kpc. Compared to the X-ray flux before the outburst, the object increased its luminosity by at least four orders of magnitude. Assuming spherical emission of the observed radiation, the source appears to have had a luminosity of 10^{48} erg s^{-1} within the first 1 ks of the outburst and a total energy output of 2×10^{53} erg within the first two weeks. This combined evidence led Burrows *et al.* (2011) to the conclusion that one observes in this case the tidal disruption of a star with subsequent accretion onto a super massive black hole, which had sat dormant at the center of the host galaxy. The mass of the central engine is assumed to be in the range $M_{\rm BH} \simeq 10^6 - 10^7\ M_\odot$, or

1) The cloud was named "Hanny's Voorwerp" after the citizen scientist Hanny van Arkel who discovered the object in the course of the Galaxy Zoo project.

even $M_{BH} < 10^5\ M_\odot$ (Krolik and Piran, 2011). Considering the brightness of the outburst and the relatively small mass of the black hole, this indicates Eddington ratios of $\lambda \simeq L_X/L_{Edd} \gtrsim 10^3$. Bloom *et al.* (2011) argue that the emission is most likely beamed, which would reduce the effective luminosity of the outburst and thus the Eddington ratio needed to explain it. It raises the question though, how the system can create a jet so rapidly, which requires the presence of a stable magnetic field, usually considered being anchored in the accretion disk. Considering that we observe the onset of a jet in galactic black hole binaries ($M_{BH} \sim 10\ M_\odot$) on $\sim$ hour time scales, even a black hole mass as low as $M_{BH} = 10^4\ M_\odot$ should require around month-long time scales for the creation of a jet.

In summary we can conclude that although the growth of the largest supermassive black holes appears to be in the past, we still observe a rich variety of AGN-related phenomena in the local Universe. Notable in this regard is the role of low-luminosity AGN and the onset of AGN activity, as observed in Henize 2-10 (see last paragraph of Section 7.2.2) and in SWIFT J164449.3+573451, which is something best-studied in our cosmic environment.

10
What We Don't Know (Yet)

We want to end this book by considering the open questions which arise from our current knowledge regarding the AGN phenomenon. For some of these open issues we have a reasonably good idea of what is needed to solve them. A frequent response to a lack of understanding is clarifying the situation through observations. This often requires better instruments with improved sensitivity as well as higher spectral, imaging, and/or timing resolution. In other cases we lack sufficiently robust numerical simulations to assess observations in the context of our theoretical models. Then again, our models might not be sufficient and additional theoretical effort will be necessary to solve the problem. In the following we have divided the open issues which are apparent today into three categories. First we consider the central engine itself including absorption and particle acceleration in its immediate vicinity. Second, we discuss questions arising from the interaction of an AGN with its environment. Finally, in Section 10.3 we discuss the questions which concern the evolution of AGN, including where they originate and what is their ultimate fate.

Each of these topics are parts of the larger puzzle which one must ultimately put together in order to truly understand the AGN phenomenon. We are certain that current and future research will ask new questions as we find new ways to solve the known issues. In the end, we might yet draw a drastically different picture than the one we envision today.

10.1
The Central Engine

Our questions about the AGN phenomenon begin with the most central of issues. It is widely believed that the primary mechanism for powering AGN is accretion onto a supermassive black hole (Section 3.1). But whether this black hole is slowly rotating or nonrotating and can therefore be treated as a Schwarzschild black hole, or it is rapidly spinning close to the theoretical limit and has to be understood as a Kerr black hole (Section 3.1.4), is an unresolved issue. One suspects that at least some AGN harbor Kerr black holes, but how can we prove this beyond doubt? At present the Fe K_α line at $E = 6.4\,\mathrm{keV}$ seems to be the best tool for assessing black

Active Galactic Nuclei, First Edition. Volker Beckmann and Chris Shrader.

hole rotation as we discuss in Section 5.5.3 – if it is indeed created close to the event horizon (X-ray polarization experiments may offer a powerful and independent tool in the not too distant future). In order to accurately measure the iron line profile, one must also model the underlying continuum very precisely, considering all possible effects such as Comptonization and absorption. Here observations which cover the range from significantly below to significantly above 6.4 keV are required. Future X-ray spectrometers may be able to exploit the iron L-shell lines, in the photon-rich $\sim$ 1 keV spectral region for this purpose as well. Ultimately, the answer as to whether or not and how fast the black hole is spinning depends on the model applied. That being the case, alternative interpretations must all be given careful consideration. Hopefully, future missions such as Astro-H and GEMS will help to settle the question.

We believe that accretion of matter onto the supermassive black hole is the process that drives AGN. But how is this accretion flow taking place? Is the standard model of an axisymmetric accretion disk, an α-disk the right way to describe the process? At least in some cases we seem to observe an emissivity inconsistent with standard accretion disk scenarios. In such cases alternative mechanisms such as the advection-dominated accretion flow (ADAF; Section 3.2.4) may be occurring. Are there other forms of spherically symmetric accretion possible which can somehow be accommodated within our observational picture of AGN? This is a field that is likely to benefit from bigger and better numerical simulations that will inevitably be facilitated by faster computers and increasingly clever algorithms. These will have to include magnetohydrodynamic effects, and will ultimately need to be done three-dimensionally if we are to achieve a realistic picture. Although at some level accretion disk structure may be directly observable, for example as inferred from temperature gradients in the innermost region of nearby Seyferts or in HST direct imaging of M87, this will help us only for a few of the nearest AGN.

Our best window on the accretion of matter occurring in AGN is the X-ray emission believed to emanate from an area close to the central engine. Commonly this is modeled by an accretion disk modified by Compton scattering from electrons in an ambient hot cloud or disk corona (Section 5.5.3). The geometry of this electron plasma cloud is unknown. A variety of ideas have been considered such as an arrangement of clouds, forming a "patchy corona." This scenario allows for transmission of the soft excess components sometimes seen, but is problematic for modeling observed variability. How close to the black hole is the corona, how far out does it extend? Spectroscopy alone cannot tell us, instead time-resolved spectroscopy is our most promising tool. In recent years, advances in this field have been driven by new observations using current state of the art facilities such as Chandra, XMM-Newton and INTEGRAL. To continue to make gains though, parallel advances in modeling are needed that can be applied to these data, as is happening already today for the photon-rich galactic X-ray binaries.

Understanding the launching, collimation and propagation of astrophysical jets remains one of the holy grails of not only AGN research but of modern astrophysics in general. From current observations we stand at an unprecedented level in terms of the empirical picture. From VLBI radio imaging, one is able to pin-point the starting point of the jet as well as begin to show us how the jet evolves. For example we can directly observe spatial components, or "blobs," as they are being emitted and then as they travel downstream. Shocks and knots in the jet have been identified and sensitive observations in other wavebands allow us to correlate morphological evolution in the radio to the multiwavelength light curves. But the main question still persists: how are the jets formed in the first place? The answer hinges on the gaps in our knowledge regarding the accretion process powering the AGN. If the jet originates from matter being expelled from the accretion disk as a wind, knowledge of the detailed disk structure is essential to answering this question. While we are fairly certain that magnetic fields play an important role in raising and collimating the jet, we can only guess as to what is the exact geometry? Another major question remains: why do some AGN develop jets while some, apparently most, do not? The galactic black hole binaries seem to be most luminous when the jet is absent, whereas the most luminous quasars seem to sustain jet production. Once the matter is launched and collimated into a jet, what is the acceleration mechanism leading to the relativistic particles we observe, or more accurately whose existence we infer from the observed spectral energy distribution involving the synchrotron and inverse Compton processes? It is surprising that after many years of blazar research, we are unclear as to what particles dominate the outflow. Is this a hadronic or a leptonic jet? Do we have to consider protons, which need more energy than electrons to be accelerated up to the required speeds but which can more naturally explain the remarkably collimated and expansive structures extending to ~ 100 kpc distances? This begs the next open question: what collimates the jets so effectively over such vast distances that can exceed for example the diameter of our Galaxy? Here polarization mapping might give us more hints on the constraining magnetic field, and clearly higher-resolution, higher-cadence radio observations along with higher-fidelity numerical modeling are our pathways towards understanding jet structures.

Although we now possess an unprecedented collection of well-sampled – in time and energy – spectral energy distributions (SEDs) of blazars, there is still a lack of agreement on what is the emission mechanism at highest observed energies. Whereas there is a consensus that we observe synchrotron radiation from blazars all the way up to the optical/UV/X-ray regime (depending on where this component peaks), it is unclear where the seed-photon population leading to the inverse Compton component originates. Are blazars predominantly one-zone synchrotron self-Compton emitters (SSC; Section 5.8.1), where the synchrotron branch provides seed photons for the IC process? Do we see instead a multicomponent SSC? Or is there an alternative seed photon source such as the accretion disk or broad-line region, giving rise to external Compton (EC) processes as described in Section 5.8.1.2? Apparently, the SEDs of blazars still need to be better sampled in time and frequency, from the radio to the gamma-ray domain. This is a costly and time-consuming effort. Furthermore the handful of well-studied examples (Mrk 421, Mrk 501, 3C 454.3, 3C 279, ...), might represent special, caveated cases among a diverse zoo of objects. The situation may be further complicated, in that the emission mechanism depends on the state of the object when it is observed – in particular flaring objects, which are the most frequent targets of observers, might be dominated by a different type of inverse Compton scattering than the nonflaring persistent emission.

There seems to be a growing number of extragalactic gamma-ray sources which are not bona-fide blazars. This raises the question as to what drives them (Section 5.6). Can we classify all such sources which are dominated by a misaligned jet or by cosmic rays originating from star-forming processes, a situation we find in NGC 253 and M82? How do the gamma-ray bright (and radio-loud) narrow-line Seyfert 1 (NLS1; Section 4.1.4) fit into this picture? The ongoing Fermi Gamma-Ray Space Telescope is compiling a database of cosmic gamma-ray sources unprecedented in quality and quantity, and offers our best opportunity to assess these classes.

Although it would be of great interest to connect blazar or other gamma-ray AGN SEDs in the X-ray band with sensitive observations in the MeV range, telescopes to accomplish this do not currently exist and are not currently foreseen for the coming decade. Thus, simultaneous observations and reliance on model-based interpolation between the X-ray and gamma-ray bands are our only approach to the problem. Most of the nonblazar AGN which are now known to be gamma-ray sources, are however too close to the detection limit of Fermi/LAT, which constrains the possibilities for time-resolved studies.

There is now a general acceptance of the unification scheme for AGN. Two of its main elements were identified early on. One involves the system orientation with respect to the observer's line of sight, that is, whether or not the emission from the central engine and broad-line region is obscured or not. The second one involves radio-loudness, and whether or not the AGN produces a significant jet. Still, this simple unification scheme as it is summarized in Figure 4.16 fails to explain all the different varieties of AGN phenomena that we observe. Other important considerations may be the mass of the central black hole, the accretion rate, or the specific geometry of the absorber. Expanded samples of Seyfert galaxies should help bolster statistics needed to clearly identify the underlying factors. The SDSS has been invaluable in this respect and future progress can be expected from SDSS-III, LSST, Euclid, and other surveys, which will catalog $\sim 10^7$ AGN including their redshifts. In the low-luminosity regime, work remains in order to bridge the gap between the galactic black hole binaries and the supermassive black holes. The Ultraluminous X-ray sources (ULX; Section 4.3) are a candidate for intermediate mass black holes (IMBH), which could help bridge the gap, but further study and in particular improved classification of ULX counterparts will be necessary to settle this question. Other LLAGN classes need to be separated beyond ambiguity from the nonactive galaxies. In particular H II galaxies and LINER tend to become indistinguishable below some signal-to-noise threshold. The forthcoming large survey telescopes should bring clarification. Finally, the illusive link between AGN and nonactive supermassive black holes, like Sgr A* in our very own Galaxy, needs to be understood.

10.2 Environment, Interaction, and Feedback

One of the largest controversies in AGN physics is the putative interplay between the central engine and the host galaxy. One finds a variety of modeling approaches in the current literature. This begins with one extreme claiming little to no interaction: there is no connection other than big black holes reside in big galaxies simply because they grew simultaneously. At the other extreme, some purport a strong link: there is a close correlation between the central engine and the environment, with a feedback process regulating AGN growth and starburst activity in the host. One problem faced in resolving this issue is the need for unbiased samples, both for the AGN group but also for a control sample of nonactive galaxies. Another issue is that starburst and AGN activity cannot always be cleanly separated. Spatial resolution is one aspect of the problem: whereas we can distinguish the star-forming region from the AGN core in the local Universe, this becomes difficult even at moderate redshifts. Thus, larger samples with high-quality spectroscopic and imaging data are necessary to progress on this aspect. Simultaneously, it is important to work on self-consistent numerical simulations which take into account the black hole, the host galaxy, the star formation, all down to very small scales. This is certainly challenging, but recent progress in this field gives hope that we will see this type of investigation advancing rapidly.

We are facing a similar dilemma regarding the importance of merging for the feeding of the AGN. Although merging is undoubtedly an important mechanism for forming large galaxies, its significance for the AGN is unclear. In the early years of this century the idea that AGN are fed through gas which is transported toward the host galaxy's center by a merging event with another galaxy had been very popular. There is currently less evidence that this mechanism is central to SMBH growth. A scenario whereby the host galaxy undergoes a major merger that triggers a rapid increase in star formation, followed by enhanced AGN activity has yet to be verified. In fact, minor mergers may be of larger importance, although they are less easily identified than major merging events which leave their imprint on the galaxies for far longer. Again, large and well-defined samples with both spectroscopic and imaging information are necessary to address this issue. In addition, one seeks a clearer picture from cosmologists regarding the frequency of major and minor mergers. This goes hand in hand with understanding the distribution and clustering of dark matter in the Universe. Thus, projects to detect and map dark energy and dark matter such as the ESA Euclid mission or the ground-based attempts to track dark matter, like DES, PanSTARRS, LSST, PAU, and others are of great importance here.

10.3 Origin, Evolution, and Fate

We know that quasars existed in the Universe as early as 800 million years after the Big Bang at redshift $z > 7$. One task will be to identify the first quasar generation. JWST is hopefully going to reach out to these most distant AGN. ESO's planned 40 m class telescope E-ELT should be not only able to detect them but to obtain spectra. Already with quasars at $z = 7$ we are facing the question: how did AGN evolve so fast? Tightly connected with this question is the discussion about the black hole seeds of the SMBH. This can be heavy Population III stars, quasi-stars, or even primordial black holes and dark stars made of WIMPs. The different possibilities will have to be evaluated further both from a purely theoretical perspective as well as through numerical simulations. We then still face the problem of how black hole seeds can build up enough matter within a couple of hundred million years to power a mature quasar. In addition to establishing the starting conditions of this problem – how massive is the seed to start with – one has to determine how fast one can feed the black hole. Is it possible to maintain accretion at the Eddington limit for a substantial period of time? Is merging of individual black holes a plausible way to form supermassive ones, and how often did such merging events occur? Here also, theory and numerical modeling have to progress in tandem to provide reasonable answers. Further regarding this topic, one may pose the chicken and the egg question: did black holes grow first and then accumulate galaxies around them or is a host galaxy needed at the beginning in order to provide sufficient fuel for the AGN in the making? If we assume the latter, we obviously need to understand how early galaxies were formed. Here deep observations by JWST or E-ELT for example will be essential to establish the first stars and galaxies.

A remaining unsettled topic is the evolution of absorption close to AGN central engines. Whether or not there is a large fraction of Compton-thick AGN at high redshifts has important consequences for the total energy output, and thus also for the total accretion history of AGN. Establishing the fraction of Compton-thick AGN throughout the history of the Universe, tells us how many have been missed so far in surveys. Connected to this history of absorption is the quest for the sources which dominate the cosmic X-ray background (CXB) at 30 keV. Deep hard X-ray observations, as will be provided by NuSTAR and later by Astro-H, ATHENA, and other proposed space observatories, should provide a better handle on what the true fraction of Compton-thick AGN in the local Universe is, and how it has evolved since redshift $z \sim 5$.

We observe AGN throughout a large fraction of the Universe, but the time spans covered by our observations are too short to see AGN switching on and off. Thus the question remains as to what is the duty cycle of AGN? How much time do they spend dormant? An answer to this question could come from a larger census of dormant supermassive black holes in the local Universe. If we are confident that each and every massive galaxy harbors a supermassive black hole at its center, we could establish the fraction of actively accreting galaxies today. This is certainly a question which can be answered by the previously mentioned forthcoming survey telescopes.

10.4 Continuing the Quest

The selection of issues we raise here and more importantly, our approach to seeking answers, is very subjective. Our colleagues might feel that some of these questions have already been settled or that we have overlooked important aspects of AGN research here. In any case, half a century after their discovery, quasars are still enigmatic. How they are fed, their geometry, what the driving parameters are, in what kind of galaxies we find them and how they interact with their hosts, how they were first formed and how they evolve, are all still very much open issues. In order to gain a better understanding the combined talents and unrelenting efforts of a broad community are necessary:

Theoretical models This includes accretion, large scale structure and galaxy formation, radiation transport close to the black hole, and jet physics.

Numerical simulations Considering the black hole and its environment, including star formation and mass accretion. In addition, galaxy formation, merging events, and the impact on the black holes have to be simulated self-consistently.

Observations Although it appears as a standard answer to astrophysical questions, we need higher resolution and more sensitive imaging and spectral observations of AGN. In the radio and submillimeter domain, LOFAR and ALMA are going to open new windows, in the optical the JWST is needed together with efforts on the ground to increase the collecting size by an order of magnitude or more (E-ELT). At the same time, in order to achieve a breakthrough in statistics of AGN and to achieve deep surveys, Euclid, LSST, SDSS-III, and others are going to open up the field for really large and high-redshift sample studies. Similar advances should be made in the X-ray domain with the survey capability of eROSITA, accompanied by the deep observations of NuSTAR, Astro-H, and hopefully ATHENA. The MeV range remains a rather white spot on existing SED plots. Above 100 MeV, Fermi will continue to improve the situation, and in the TeV range CTA will be the next breakthrough. Finally eLISA is going to revolutionize the way we see the cosmos by showing the gravitational wave signature of black holes in the Universe.

Processing and data mining Data sets are becoming large, and, with the next generation of experiments, so complex that new ways of data processing have to be developed. That is true for all the large survey experiments, especially LSST and

Euclid, but also for CTA and eLISA. The choice of utilizing the local computer, the small computing cluster, the supercomputing facility, distributed processing on the grid, or cloud processing, has to be evaluated carefully. Data mining will be more and more essential for astrophysical research, and the virtual observatory initiative has to be developed further in order to meet the needs of tomorrow's science projects.

AGN research remains a rich field, worthy of our investments of time, energies and talents that will continue to provide unexpected future insights into the nature of the Universe we live in.

"Même pour ceux qui ne jugent des découvertes que par leur utilité directe, l'astronomie sera toujours la science plus digne des nobles efforts et de l'application soutenue de l'intelligence de l'homme." François Arago

References

Abdo, A.A., Ackermann, M., Ajello, M. *et al.* (2010) The first catalog of active galactic nuclei detected by the Fermi large area telescope. *Astrophys. J.*, **715**, 429–457, doi:10.1088/0004-637X/715/1/429.

Abdo, A.A., Ackermann, M., Ajello, M. *et al.* (2011a) Fermi gamma-ray space telescope observations of the gamma-ray outburst from 3C454.3 in November 2010. *Astrophys. J. Lett.*, **733**, L26, doi:10.1088/2041-8205/733/2/L26.

Abdo, A.A., Ackermann, M., Ajello, M. *et al.* (2009a) Fermi observations of TeV-selected active galactic nuclei. *Astrophys. J.*, **707**, 1310–1333, doi:10.1088/0004-637X/707/2/1310.

Abdo, A.A., Ackermann, M., Ajello, M. *et al.* and the Fermi/LAT Collaboration, Ghisellini, G., Maraschi, L., and Tavecchio, F. (2009b) Radio-loud narrow-line Seyfert 1 as a new class of gamma-ray active galactic nuclei. *Astrophys. J. Lett.*, **707**, L142–L147, doi:10.1088/0004-637X/707/2/L142.

Abdo, A.A., Ackermann, M., Ajello, M. *et al.* (2011b) Fermi large area telescope observations of Markarian 421: The missing piece of its spectral energy distribution. *Astrophys. J.*, **736**, 131, doi:10.1088/0004-637X/736/2/131.

Abramowicz, M.A. (1988) Thick accretion disks ten years later. *Adv. Space Res.*, **8**, 151–156, doi:10.1016/0273-1177(88)90232-3.

Ackermann, M., Ajello, M., Allafort, A. *et al.* (2012) Search for gamma-ray emission from X-ray selected Seyfert galaxies with Fermi-LAT. *Astrophys. J.*, **747**, 104, doi:10.1088/0004-637X/747/2/104.

Ackermann, M., Ajello, M., Allafort, A. *et al.* (2011) The second catalog of active galactic nuclei detected by the Fermi large area telescope. *Astrophys. J.*, **743**, 171, doi:10.1088/0004-637X/743/2/171.

Aharonian, F., Akhperjanian, A.G., Bazer-Bachi, A.R. *et al.* (2007) An exceptional very high energy gamma-ray flare of PKS 2155-304. *Astrophys. J. Lett.*, **664**, L71–L74, doi:10.1086/520635.

Aharonian, F., Akhperjanian, A.G., Bazer-Bachi, A.R. *et al.* (2006) A low level of extragalactic background light as revealed by γ-rays from blazars. *Nature*, **440**, 1018–1021, doi:10.1038/nature04680.

Aharonian, F.A. (2000) TeV gamma rays from BL Lac objects due to synchrotron radiation of extremely high energy protons. *New Astron.*, **5**, 377–395, doi:10.1016/S1384-1076(00)00039-7.

Albert, J., Aliu, E., Anderhub, H. *et al.* (2007) Variable very high energy γ-ray emission from Markarian 501. *Astrophys. J.*, **669**, 862–883, doi:10.1086/521382.

Alfvén, H. and Herlofson, N. (1950) Cosmic radiation and radio stars. *Phys. Rev.*, **78**, 616–616, doi:10.1103/PhysRev.78.616.

Allevato, V., Finoguenov, A., Cappelluti, N. *et al.* (2011) The XMM-Newton wide field survey in the COSMOS field: Redshift evolution of AGN Bias and subdominant role of mergers in triggering moderate-luminosity AGNs at redshifts up to 2.2. *Astrophys. J.*, **736**, 99, doi:10.1088/0004-637X/736/2/99.

Almaini, O., Boyle, B.J., Griffiths, R.E. *et al.* (1995) A deep ROSAT survey – IX. The discovery of a high-redshift type 2 QSO. *Mon. Not. R. Astron. Soc.*, **277**, L31–L34.

Almaini, O., Lawrence, A., and Boyle, B.J. (1999) The AGN contribution to deep submillimetre surveys and the far-infrared

Active Galactic Nuclei, First Edition. Volker Beckmann and Chris Shrader.

background. *Mon. Not. R. Astron. Soc.*, **305**, L59–L63.

Almaini, O., Lawrence, A., Shanks, T. *et al.* (2000) X-ray variability in a deep, flux-limited sample of QSOs. *Mon. Not. R. Astron. Soc.*, **315**, 325–336, doi:10.1046/j.1365-8711.2000.03385.x.

Antonucci, R. (1993) Unified models for active galactic nuclei and quasars. *Annu. Rev. Astron. Astrophys.*, **31**, 473–521, doi:10.1146/annurev.aa.31.090193.002353.

Antonucci, R.R.J. and Miller, J.S. (1985) Spectropolarimetry and the nature of NGC 1068. *Astrophys. J.*, **297**, 621–632, doi:10.1086/163559.

Arago, F. and Barral, J.A. (1854) *Astronomie populaire*, Gide et J. Baudry, Paris and T.O. Weigel, Leipzig, 1854–1857.

Arévalo, P., Uttley, P., Lira, P. *et al.* (2009) Correlation and time delays of the X-ray and optical emission of the Seyfert galaxy NGC 3783. *Mon. Not. R. Astron. Soc.*, **397**, 2004–2014, doi:10.1111/j.1365-2966.2009.15110.x.

Arnaud, K.A. (1996) XSPEC: The first ten years, in *Astronomical Data Analysis Software and Systems V*, Astronomical Society of the Pacific Conference Series, Vol. 101 (eds G. H. Jacoby and J. Barnes), Astronomical Society of the Pacific, San Francisco, p. 17.

Assef, R.J., Kochanek, C.S., Ashby, M.L.N. *et al.* (2011) The mid-IR- and X-ray-selected QSO luminosity function. *Astrophys. J.*, **728**, 56, doi:10.1088/0004-637X/728/1/56.

Atlee, D.W. and Mathur, S. (2009) GALEX measurements of the big blue bump in soft X-ray-selected active galactic nucleus. *Astrophys. J.*, **703**, 1597–1611, doi:10.1088/0004-637X/703/2/1597.

Atoyan, A.M. and Dermer, C.D. (2004) Neutrinos and γ-rays of hadronic origin from AGN jets. *New Astron. Rev.*, **48**, 381–386, doi:10.1016/j.newar.2003.12.046.

Atwood, W.B., Abdo, A.A., Ackermann, M. *et al.* (2009) The large area telescope on the Fermi gamma-ray space telescope mission. *Astrophys. J.*, **697**, 1071–1102, doi:10.1088/0004-637X/697/2/1071.

Avni, Y. and Bahcall, J.N. (1980) On the simultaneous analysis of several complete samples – The $V/V_{\rm max}$ and $V_{\rm e}/V_{\rm a}$ variables, with applications to quasars. *Astrophys. J.*, **235**, 694–716, doi:10.1086/157673.

Awaki, H., Kunieda, H., Tawara, Y., and Koyama, K. (1991) Discovery of strong absorption in the X-ray spectrum of Seyfert 2 galaxy NGC 4507. *Publ. Astron. Soc. Japan*, **43**, L37–L42.

Badnell, N.R. (2006) Dielectronic recombination of Fe $3p^q$ ions: A key ingredient for describing X-ray absorption in active galactic nuclei. *Astrophys. J. Lett.*, **651**, L73–L76, doi:10.1086/509739.

Bahcall, J.N., Kirhakos, S., and Schneider, D.P. (1994) HST images of nearby luminous quasars. *Astrophys. J. Lett.*, **435**, L11–L14, doi:10.1086/187582.

Balbus, S.A. and Hawley, J.F. (1992) A powerful local shear instability in weakly magnetized disks. IV. Nonaxisymmetric perturbations. *Astrophys. J.*, **400**, 610–621, doi:10.1086/172022.

Baldwin, J., Ferland, G., Korista, K., and Verner, D. (1995) Locally optimally emitting clouds and the origin of quasar emission lines. *Astrophys. J. Lett.*, **455**, L119, doi:10.1086/309827.

Baldwin, J.A. (1977) Luminosity indicators in the spectra of quasi-stellar objects. *Astrophys. J.*, **214**, 679–684, doi:10.1086/155294.

Baldwin, J.A. and Netzer, H. (1978) The emission-line regions of high-redshift QSOs. *Astrophys. J.*, **226**, 1–20, doi:10.1086/156578.

Barger, A.J., Cowie, L.L., Mushotzky, R.F. *et al.* (2005) The cosmic evolution of hard X-ray-selected active galactic nuclei. *Astron. J.*, **129**, 578–609, doi:10.1086/426915.

Barnes, J.E. and Hernquist, L. (1992) Dynamics of interacting galaxies. *Annu. Rev. Astron. Astrophys.*, **30**, 705–742, doi:10.1146/annurev.aa.30.090192.003421.

Barnes, J.E. and Hernquist, L.E. (1991) Fueling starburst galaxies with gas-rich mergers. *Astrophys. J. Lett.*, **370**, L65–L68, doi:10.1086/185978.

Barr, P. and Mushotzky, R.F. (1986) Limits of X-ray variability in active galactic nuclei. *Nature*, **320**, 421–423, doi:10.1038/320421a0.

Barthel, P.D. (1989) Is every quasar beamed? *Astrophys. J.*, **336**, 606–611, doi:10.1086/167038.

Battaglia, G., Helmi, A., Morrison, H. *et al.* (2005) The radial velocity dispersion profile of the galactic halo: constraining the density profile of the dark halo of the Milky Way. *Mon. Not. R. Astron. Soc.*, **364**, 433–442, doi:10.1111/j.1365-2966.2005.09367.x.

Beckmann, V., Bade, N., and Wucknitz, O. (1999) The extreme high frequency peaked BL Lac 1517+656. *Astron. Astrophys.*, **352**, 395–398.

Beckmann, V., Barthelmy, S.D., Courvoisier, T. *et al.* (2007) Hard X-ray variability of active galactic nuclei. *Astron. Astrophys.*, **475**, 827–835, doi:10.1051/0004-6361:20078355.

Beckmann, V., Engels, D., Bade, N., and Wucknitz, O. (2003) The HRX-BL Lac sample – Evolution of BL Lac objects. *Astron. Astrophys.*, **401**, 927–938, doi:10.1051/0004-6361:20030184.

Beckmann, V., Gehrels, N., Shrader, C.R., and Soldi, S. (2006a) The First INTEGRAL AGN Catalog. *Astrophys. J.*, **638**, 642–652, doi:10.1086/499034.

Beckmann, V., Soldi, S., Ricci, C. *et al.* (2009) The second INTEGRAL AGN catalogue. *Astron. Astrophys.*, **505**, 417–439, doi:10.1051/0004-6361/200912111.

Beckmann, V., Soldi, S., Ricci, C. *et al.* (2010) The unified scheme seen with INTEGRAL detected AGN, in *Proc. "X-ray Astronomy 2009"*, Bologna, September 2009.

Beckmann, V., Soldi, S., Shrader, C.R., and Gehrels, N. (2006b) Results from the first INTEGRAL AGN catalogue, in *The X-ray Universe 2005*, ESA Special Publication, Vol. 604 (eds A. Wilson), ESA Publications Division, Noordwijk, p. 777.

Beckmann, V., Soldi, S., Shrader, C.R. *et al.* (2006c) The hard X-ray 20–40 keV AGN luminosity function. *Astrophys. J.*, **652**, 126–135, doi:10.1086/507510.

Begelman, M.C., Blandford, R.D., and Rees, M.J. (1980) Massive black hole binaries in active galactic nuclei. *Nature*, **287**, 307–309, doi:10.1038/287307a0.

Begelman, M.C., Rossi, E.M., and Armitage, P.J. (2008) Quasi-stars: accreting black holes inside massive envelopes. *Mon. Not. R. Astron. Soc.*, **387**, 1649–1659, doi:10.1111/j.1365-2966.2008.13344.x.

Behar, E., Rasmussen, A.P., Blustin, A.J. *et al.* (2003) A long look at NGC 3783 with the XMM-Newton reflection grating spectrometer. *Astrophys. J.*, **598**, 232–241, doi:10.1086/378853.

Bell, E.F. (2008) Galaxy bulges and their black holes: a requirement for the quenching of star formation. *Astrophys. J.*, **682**, 355–360, doi:10.1086/589551.

Bennert, N., Falcke, H., Shchekinov, Y., and Wilson, A.S. (2004) Comparing AGN broad and narrow line regions, in *The Interplay Among Black Holes, Stars and ISM in Galactic Nuclei*, IAU Symposium, Vol. 222 (eds T. Storchi-Bergmann, L.C. Ho, and H.R. Schmitt), Cambridge University Press, Cambridge, pp. 307–308, doi:10.1017/S1743921304002340.

Bennett, A.S. (1962) The revised 3C catalogue of radio sources. *Mem. R. Astron. Soc.*, **68**, 163.

Bennett, C.L. (2006) Cosmology from start to finish. *Nature*, **440**, 1126–1131, doi:10.1038/nature04803.

Bennett, C.L., Halpern, M., Hinshaw, G. *et al.* (2003) First-year Wilkinson Microwave Anisotropy Probe (WMAP) observations: Preliminary maps and basic results. *Astrophys. J. Suppl.*, **148**, 1–27, doi:10.1086/377253.

Bentz, M.C., Peterson, B.M., Netzer, H. *et al.* (2009a) The radius-luminosity relationship for active galactic nuclei: The effect of host-galaxy starlight on luminosity measurements. II. The full sample of reverberation-mapped AGNs. *Astrophys. J.*, **697**, 160–181, doi:10.1088/0004-637X/697/1/160.

Bentz, M.C., Walsh, J.L., Barth, A.J. *et al.* (2009b) The lick AGN monitoring project: broad-line region radii and black hole masses from reverberation mapping of Hβ. *Astrophys. J.*, **705**, 199–217, doi:10.1088/0004-637X/705/1/199.

Bergeron, J. (1986) Properties of the sharp metal-rich absorption lines observed in QSO spectra, in *Structure and Evolution of Active Galactic Nuclei*, Astrophysics and Space Science Library, Vol. 121 (eds G. Giuricin, M. Mezzetti, M. Ramella, and F. Mardirossian), D. Reidel Publishing Co., Dordrecht, pp. 421–434.

Best, P.N., Kauffmann, G., Heckman, T.M. *et al.* (2005) The host galaxies of radio-loud active galactic nuclei: mass dependences, gas cooling and active galactic nuclei feedback. *Mon. Not. R. Astron. Soc.*, **362**, 25–40, doi:10.1111/j.1365-2966.2005.09192.x.

Bettoni, D., Falomo, R., Fasano, G., and Govoni, F. (2003) The black hole mass of low redshift radiogalaxies. *Astron. Astrophys.*, **399**, 869–878, doi:10.1051/0004-6361:20021869.

Bicknell, G.V., Jones, D.L., and Lister, M. (2004) Relativistic jets. *New Astron. Rev.*, **48**, 1151–1155, doi:10.1016/j.newar.2004.09.005.

Blaes, O., Hubeny, I., Agol, E., and Krolik, J.H. (2001) Non-LTE, relativistic accretion disk fits to 3C 273 and the origin of the Lyman limit spectral break. *Astrophys. J.*, **563**, 560–568, doi:10.1086/324045.

Blandford, R.D. and Konigl, A. (1979) Relativistic jets as compact radio sources. *Astrophys. J.*, **232**, 34–48, doi:10.1086/157262.

Blandford, R.D., McKee, C.F., and Rees, M.J. (1977) Super-luminal expansion in extragalactic radio sources. *Nature*, **267**, 211–216, doi:10.1038/267211a0.

Blandford, R.D. and Payne, D.G. (1982) Hydromagnetic flows from accretion discs and the production of radio jets. *Mon. Not. R. Astron. Soc.*, **199**, 883–903.

Blandford, R.D. and Rees, M.J. (1978) Some comments on radiation mechanisms in Lacertids, in *BL Lac Objects* (ed. A.M. Wolfe), University of Pittsburgh, Pittsburgh, pp. 328–341.

Blandford, R.D. and Znajek, R.L. (1977) Electromagnetic extraction of energy from Kerr black holes. *Mon. Not. R. Astron. Soc.*, **179**, 433–456.

Bloom, J.S., Giannios, D., Metzger, B.D. *et al.* (2011) A possible relativistic jetted outburst from a massive black hole fed by a tidally disrupted star. *Science*, **333**, 203–206, doi:10.1126/science.1207150.

Blumenthal, G.R. and Gould, R.J. (1970) Bremsstrahlung, synchrotron radiation, and Compton scattering of high-energy electrons traversing dilute gases. *Rev. Mod. Phys.*, **42**, 237–271, doi:10.1103/RevModPhys.42.237.

Boller, T., Brandt, W.N., and Fink, H. (1996) Soft X-ray properties of narrow-line Seyfert 1 galaxies. *Astron. Astrophys.*, **305**, 53.

Bondi, H. (1952) On spherically symmetrical accretion. *Mon. Not. R. Astron. Soc.*, **112**, 195.

Bordoloi, R., Lilly, S.J., and Amara, A. (2010) Photo-z performance for precision cosmology. *Mon. Not. R. Astron. Soc.*, **406**, 881–895, doi:10.1111/j.1365-2966.2010.16765.x.

Borys, C., Blain, A.W., Dey, A. *et al.* (2006) MIPS J142824.0+352619: A hyperluminous starburst galaxy at $z = 1.325$. *Astrophys. J.*, **636**, 134–139, doi:10.1086/497983.

Böttcher, M. and Dermer, C.D. (2002) An evolutionary scenario for Blazar unification. *Astrophys. J.*, **564**, 86–91, doi:10.1086/324134.

Brenneman, L.W. and Reynolds, C.S. (2006) Constraining black hole spin via X-ray spectroscopy. *Astrophys. J.*, **652**, 1028–1043, doi:10.1086/508146.

Brenneman, L.W. and Reynolds, C.S. (2009) Relativistic broadening of iron emission lines in a sample of active galactic nuclei. *Astrophys. J.*, **702**, 1367–1386, doi:10.1088/0004-637X/702/2/1367.

Brenneman, L.W., Reynolds, C.S., Nowak, M.A. *et al.* (2011) The spin of the supermassive black hole in NGC 3783. *Astrophys. J.*, **736**, 103, doi:10.1088/0004-637X/736/2/103.

Bridle, A.H., Hough, D.H., Lonsdale, C.J. *et al.* (1994) Deep VLA imaging of twelve extended 3CR quasars. *Astron. J.*, **108**, 766–820, doi:10.1086/117112.

Broderick, J.W. and Fender, R.P. (2011) Is there really a dichotomy in active galactic nucleus jet power? *Mon. Not. R. Astron. Soc.*, **417**, 184–197, doi:10.1111/j.1365-2966.2011.19060.x.

Brown, M.J.I., Brand, K., Dey, A. *et al.* (2006) The $1 < z < 5$ infrared luminosity function of type I quasars. *Astrophys. J.*, **638**, 88–99, doi:10.1086/498843.

Brownstein, J.R. and Moffat, J.W. (2006) Galaxy cluster masses without non-baryonic dark matter. *Mon. Not. R. Astron. Soc.*, **367**, 527–540, doi:10.1111/j.1365-2966.2006.09996.x.

Brunthaler, A., Reid, M.J., Falcke, H. *et al.* (2005) The geometric distance and proper motion of the triangulum galaxy (M33). *Science*, **307**, 1440–1443, doi:10.1126/science.1108342.

Burbidge, G.R. (1956) On synchrotron radiation from Messier 87. *Astrophys. J.*, **124**, 416, doi:10.1086/146237.

Burrows, D.N., Kennea, J.A., Ghisellini, G. *et al.* (2011) Relativistic jet activity from the tidal disruption of a star by a massive black hole. *Nature*, **476**, 421–424, doi:10.1038/nature10374.

Byram, E.T., Chubb, T.A., and Friedman, H. (1966) Cosmic X-ray sources, galactic and extragalactic. *Science*, **152**, 66–71, doi:10.1126/science.152.3718.66.

Caccianiga, A., Severgnini, P., Della Ceca, R. *et al.* (2008) The XMM-Newton bright serendipitous survey. Identification and optical spectral properties. *Astron. Astrophys.*, **477**, 735–746, doi:10.1051/0004-6361:20078568.

Campanelli, M., Lousto, C.O., Zlochower, Y., and Merritt, D. (2007) Maximum gravitational recoil. *Phys. Rev. Lett.*, **98**(23), 231102, doi:10.1103/PhysRevLett.98.231102.

Capetti, A. (2011) Revisiting the census of low-luminosity AGN. *Astron. Astrophys.*, **535**, A28, doi:10.1051/0004-6361/201117937.

Cardamone, C.N., Schawinski, K., Masters, K. *et al.* (2011) Galaxy Zoo: The connection between AGN activity and bars in late type galaxies, in *American Astronomical Society Meeting Abstracts #218*, American Astronomical Society, Washington DC, p. #206.03.

Carilli, C.L. and Barthel, P.D. (1996) Cygnus A. *Astron. Astrophys. Rev.*, **7**, 1–54, doi:10.1007/s001590050001.

Carini, M. (2006) Multicolor Microvariability in S5 0716+714 in 2003, in *Blazar Variability Workshop II: Entering the GLAST Era*, Astronomical Society of the Pacific Conference Series, Vol. 350 (eds H.R. Miller, K. Marshall, J.R. Webb, and M.F. Aller), Astronomical Society of the Pacific, San Francisco, p. 55.

Carini, M.T. and Miller, H.R. (1992) The optical variability of PKS 2155-304. *Astrophys. J.*, **385**, 146–150, doi:10.1086/170923.

Carr, B.J., Kohri, K., Sendouda, Y., and Yokoyama, J. (2010) New cosmological constraints on primordial black holes. *Phys. Rev. D*, **81**(10), 104 019, doi:10.1103/PhysRevD.81.104019.

Casebeer, D.A., Leighly, K.M., and Baron, E. (2006) FUSE observation of the narrow-line Seyfert 1 galaxy RE 1034+39: Dependence of broad emission line strengths on the shape of the photoionizing spectrum. *Astrophys. J.*, **637**, 157–182, doi:10.1086/498125.

Cattaneo, A., Faber, S.M., Binney, J. *et al.* (2009) The role of black holes in galaxy formation and evolution. *Nature*, **460**, 213–219, doi:10.1038/nature08135.

Cen, R. and Ostriker, J.P. (1999) Where are the baryons? *Astrophys. J.*, **514**, 1–6, doi:10.1086/306949.

Chatterjee, R., Bailyn, C., Bonning, E.W. *et al.* (2012) Similarity of the optical-infrared and γ-ray time variability of Fermi blazars. *Astrophys. J.*, **749**, 191, doi:10.1088/0004-637X/749/2/191.

Chatterjee, R., Marscher, A.P., Jorstad, S.G. *et al.* (2009) Disk-jet connection in the radio galaxy 3C 120. *Astrophys. J.*, **704**, 1689–1703, doi:10.1088/0004-637X/704/2/1689.

Chen, L. and Bai, J.M. (2011) Implications for the blazar sequence and inverse Compton models from Fermi bright blazars. *Astrophys. J.*, **735**, 108, doi:10.1088/0004-637X/735/2/108.

Cid Fernandes, R., Gu, Q., Melnick, J. *et al.* (2004) The star formation history of Seyfert 2 nuclei. *Mon. Not. R. Astron. Soc.*, **355**, 273–296, doi:10.1111/j.1365-2966.2004.08321.x.

Cid Fernandes, R., Stasińska, G., Schlickmann, M.S. *et al.* (2010) Alternative diagnostic diagrams and the "forgotten" population of weak line galaxies in the SDSS. *Mon. Not. R. Astron. Soc.*, **403**, 1036–1053, doi:10.1111/j.1365-2966.2009.16185.x.

Cisternas, M., Jahnke, K., Inskip, K.J. *et al.* (2011) The bulk of the black hole growth since $z \sim 1$ occurs in a secular universe: No major merger-AGN connection. *Astrophys. J.*, **726**, 57, doi:10.1088/0004-637X/726/2/57.

Civano, F., Brusa, M., Comastri, A. *et al.* (2011) The population of high-redshift active galactic nuclei in the Chandra-COSMOS survey. *Astrophys. J.*, **741**, 91, doi:10.1088/0004-637X/741/2/91.

Clavel, J., Reichert, G.A., Alloin, D. *et al.* (1991) Steps toward determination of the size and structure of the broad-line region in active galactic nuclei. I – an 8 month campaign of monitoring NGC 5548 with IUE. *Astrophys. J.*, **366**, 64–81, doi:10.1086/169540.

Colbert, E.J.M. and Mushotzky, R.F. (1999) The nature of accreting black holes in nearby galaxy nuclei. *Astrophys. J.*, **519**, 89–107, doi:10.1086/307356.

Colless, M., Dalton, G., Maddox, S. *et al.* (2001) The 2dF galaxy redshift survey:

spectra and redshifts. *Mon. Not. R. Astron. Soc.*, **328**, 1039–1063, doi:10.1046/j.1365-8711.2001.04902.x.

Collin, S. and Kawaguchi, T. (2004) Super-Eddington accretion rates in narrow line Seyfert 1 galaxies. *Astron. Astrophys.*, **426**, 797–808, doi:10.1051/0004-6361:20040528.

Cowan, J.J. and Sneden, C. (2006) Heavy element synthesis in the oldest stars and the early Universe. *Nature*, **440**, 1151–1156, doi:10.1038/nature04807.

Crenshaw, D.M., Kraemer, S.B., Boggess, A. *et al.* (1999) Intrinsic absorption lines in Seyfert 1 galaxies. I. Ultraviolet spectra from the Hubble space telescope. *Astrophys. J.*, **516**, 750–768, doi:10.1086/307144.

Crenshaw, D.M., Kraemer, S.B., and George, I.M. (2003) Mass loss from the nuclei of active galaxies. *Annu. Rev. Astron. Astrophys.*, **41**, 117–167, doi:10.1146/annurev.astro.41.082801.100328.

Croom, S.M., Richards, G.T., Shanks, T. *et al.* (2009) The 2dF-SDSS LRG and QSO survey: the QSO luminosity function at $0.4 < z < 2.6$. *Mon. Not. R. Astron. Soc.*, **399**, 1755–1772, doi:10.1111/j.1365-2966.2009.15398.x.

Croom, S.M., Smith, R.J., Boyle, B.J. *et al.* (2004) The 2dF QSO Redshift Survey – XII. The spectroscopic catalogue and luminosity function. *Mon. Not. R. Astron. Soc.*, **349**, 1397–1418, doi:10.1111/j.1365-2966.2004.07619.x.

Crummy, J., Fabian, A.C., Gallo, L., and Ross, R.R. (2006) An explanation for the soft X-ray excess in active galactic nuclei. *Mon. Not. R. Astron. Soc.*, **365**, 1067–1081, doi:10.1111/j.1365-2966.2005.09844.x.

CTA Consortium (2011) Contributions from the Cherenkov Telescope Array (CTA) Consortium to the ICRC 2011. *Exp. Astron.*, **32**, 191–316, doi:10.1007/s10686-011-9247-0.

Cui, W. (2009) TeV gamma-ray astronomy. *Res. Astron. Astrophys.*, **9**, 841–860, doi:10.1088/1674-4527/9/8/001.

Curtis, H.D. (1918) Descriptions of 762 nebulae and clusters photographed with the crossley reflector. *Publ. Lick Obs.*, **13**, 9–42.

Curtis, H.D. (1920) Modern theories of the spiral nebulae. *J. R. Astron. Soc. Can.*, **14**, 317.

Dadina, M. (2008) Seyfert galaxies in the local Universe ($z < 0.1$): the average X-ray spectrum as seen by BeppoSAX. *Astron. Astrophys.*, **485**, 417–424, doi:10.1051/0004-6361:20077569.

Daly, R.A. (2009) Black hole spins of radio sources. *Astrophys. J. Lett.*, **691**, L72–L76, doi:10.1088/0004-637X/691/2/L72.

Daly, R.A. (2011) Estimates of black hole spin properties of 55 sources. *Mon. Not. R. Astron. Soc.*, **414**, 1253–1262, doi:10.1111/j.1365-2966.2011.18452.x.

Davidsen, A.F., Hartig, G.F., and Fastie, W.G. (1977) Ultraviolet spectrum of quasi-stellar object 3C 273. *Nature*, **269**, 203–206, doi:10.1038/269203a0.

Davies, R.I., Thomas, J., Genzel, R. *et al.* (2006) The Star-forming torus and stellar dynamical black hole mass in the Seyfert 1 nucleus of NGC 3227. *Astrophys. J.*, **646**, 754–773, doi:10.1086/504963.

de Grijs, R. and Cartwright, S. (2011) *An Introduction to Distance Measurement in Astronomy*, John Wiley & Sons, Chichester, ISBN 0-470-51180-X.

de Vaucouleurs, G. (1959) Classification and morphology of external galaxies, in *Handbuch der Physik*, vol. 53, Springer Verlag, Berlin, Göttingen, Heidelberg, pp. 275–310.

de Vaucouleurs, G., de Vaucouleurs, A., and Corwin, J.R. (1976) Second reference catalogue of bright galaxies, in *Second Reference Catalogue of Bright Galaxies*, University of Texas Press, Austin, p. 1.

de Vries, W.H., Becker, R.H., White, R.L., and Loomis, C. (2005) Structure function analysis of long-term quasar variability. *Astron. J.*, **129**, 615–629, doi:10.1086/427393.

De Young, D.S. (2002) *The Physics of Extragalactic Radio Sources*, University of Chicago Press, Chicago.

De Young, D.S. (2010) How does radio AGN feedback feed back? *Astrophys. J.*, **710**, 743–754, doi:10.1088/0004-637X/710/1/743.

Debuhr, J., Quataert, E., Ma, C.P., and Hopkins, P. (2010) Self-regulated black hole growth via momentum deposition in galaxy merger simulations. *Mon. Not. R. Astron. Soc.*, **406**, L55–L59, doi:10.1111/j.1745-3933.2010.00881.x.

Della Ceca, R., Lamorani, G., Maccacaro, T. *et al.* (1994) The properties of X-ray selected active galactic nuclei. 3: The radio-quiet versus radio-loud samples. *Astrophys. J.*, **430**, 533–544, doi:10.1086/174428.

Deluit, S. and Courvoisier, T.J.L. (2003) The intrinsic emission of Seyfert galaxies observed with BeppoSAX/PDS. I. Comparison of the average spectra of the three classes of Seyfert galaxies. *Astron. Astrophys.*, **399**, 77–90, doi:10.1051/0004-6361:20021794.

Dermer, C.D. and Schlickeiser, R. (1993) Model for the high-energy emission from blazars. *Astrophys. J.*, **416**, 458, doi:10.1086/173251.

Dermer, C.D., Schlickeiser, R., and Mastichiadis, A. (1992) High-energy gamma radiation from extragalactic radio sources. *Astron. Astrophys.*, **256**, L27–L30.

Dewangan, G.C., Griffiths, R.E., Dasgupta, S., and Rao, A.R. (2007) An investigation of the origin of soft X-ray excess emission from Ark 564 and Mrk 1044. *Astrophys. J.*, **671**, 1284–1296, doi:10.1086/523683.

Di Matteo, T., Blackman, E.G., and Fabian, A.C. (1997) Two-temperature coronae in active galactic nuclei. *Mon. Not. R. Astron. Soc.*, **291**, L23–L27.

Di Matteo, T., Springel, V., and Hernquist, L. (2005) Energy input from quasars regulates the growth and activity of black holes and their host galaxies. *Nature*, **433**, 604–607, doi:10.1038/nature03335.

Dicken, D., Tadhunter, C., Axon, D. *et al.* (2009) The origin of the infrared emission in radio galaxies. II. Analysis of mid- to far-infrared Spitzer observations of the 2 Jy sample. *Astrophys. J.*, **694**, 268–285, doi:10.1088/0004-637X/694/1/268.

Dickinson, M., Giavalisco, M., and GOODS Team (2003) The great observatories origins deep survey, in *The Mass of Galaxies at Low and High Redshift* (eds R. Bender and A. Renzini), Springer Verlag, Berlin, Heidelberg, New York, p. 324, doi:10.1007/10899892_78.

Diehl, R., Halloin, H., Kretschmer, K. *et al.* (2006) Radioactive ^{26}Al from massive stars in the Galaxy. *Nature*, **439**, 45–47, doi:10.1038/nature04364.

Djorgovski, S.G., Castro, S., Stern, D., and Mahabal, A.A. (2001) On the threshold of the reionization epoch. *Astrophys. J. Lett.*, **560**, L5–L8, doi:10.1086/324175.

Dole, H., Lagache, G., Puget, J.L. *et al.* (2006) The cosmic infrared background resolved by Spitzer. Contributions of mid-infrared galaxies to the far-infrared background. *Astron. Astrophys.*, **451**, 417–429, doi:10.1051/0004-6361:20054446.

Dondi, L. and Ghisellini, G. (1995) Gamma-ray-loud blazars and beaming. *Mon. Not. R. Astron. Soc.*, **273**, 583–595.

Done, C. (2007) The origin of the soft excess in high $L/L_{\rm Edd}$ AGN. *Progr. Theor. Phys. Suppl.*, **169**, 248–251.

Donea, A.C. and Protheroe, R.J. (2003) Radiation fields of disk, BLR and torus in quasars and blazars: implications for γ-ray absorption. *Astropart. Phys.*, **18**, 377–393, doi:10.1016/S0927-6505(02)00155-X.

Donnarumma, I., De Rosa, A., Vittorini, V. *et al.* (2011) The remarkable γ-ray activity in the gravitationally lensed blazar PKS 1830-211. *Astrophys. J. Lett.*, **736**, L30, doi:10.1088/2041-8205/736/2/L30.

Dorodnitsyn, A. and Kallman, T. (2011) X-ray polarization signature of warm absorber winds in AGN. *Astrophys. Space Sci.*, **336**, 245–250, doi:10.1007/s10509-011-0656-3.

Draine, B.T. and Li, A. (2007) Infrared emission from interstellar dust. IV. The silicate-graphite-PAH model in the post-Spitzer era. *Astrophys. J.*, **657**, 810–837, doi:10.1086/511055.

Dunlop, J.S., McLure, R.J., Kukula, M.J. *et al.* (2003) Quasars, their host galaxies and their central black holes. *Mon. Not. R. Astron. Soc.*, **340**, 1095–1135, doi:10.1046/j.1365-8711.2003.06333.x.

Dunn, R.J.H. and Fabian, A.C. (2006) Investigating AGN heating in a sample of nearby clusters. *Mon. Not. R. Astron. Soc.*, **373**, 959–971, doi:10.1111/j.1365-2966.2006.11080.x.

Edelson, R.A. and Krolik, J.H. (1988) The discrete correlation function – A new method for analyzing unevenly sampled variability data. *Astrophys. J.*, **333**, 646–659, doi:10.1086/166773.

Edelson, R.A. and Malkan, M.A. (1986) Spectral energy distributions of active galactic nuclei between 0.1 and 100 microns. *Astrophys. J.*, **308**, 59–77, doi:10.1086/164479.

Einstein, A. (1916) Die Grundlage der allgemeinen Relativitätstheorie. *Ann. Phys.*, **354**, 769–822, doi:10.1002/andp. 19163540702.

Elder, F.R., Gurewitsch, A.M., Langmuir, R.V., and Pollock, H.C. (1947) Radiation from electrons in a synchrotron. *Phys. Rev.*, **71**, 829–830, doi:10.1103/PhysRev.71.829.5.
Elitzur, M. (1992) Astronomical masers. *Annu. Rev. Astron. Astrophys.*, **30**, 75–112, doi:10.1146/annurev.aa.30.090192.000451.
Elitzur, M. (2007) Unification issues and the AGN torus, in *The Central Engine of Active Galactic Nuclei*, Astronomical Society of the Pacific Conference Series, Vol. 373 (eds L.C. Ho and J.-W. Wang), Astronomical Society of the Pacific, San Francisco, p. 415.
Elitzur, M. and Shlosman, I. (2006) The AGN-obscuring torus: The end of the "Doughnut" paradigm?, *Astrophys. J. Lett.*, **648**, L101–L104, doi:10.1086/508158.
Elvis, M., Maccacaro, T., Wilson, A.S. *et al.* (1978) Seyfert galaxies as X-ray sources. *Mon. Not. R. Astron. Soc.*, **183**, 129–157.
Elvis, M., Schreier, E.J., Tonry, J. *et al.* (1981) Two optically dull galaxies with strong nuclear X-ray sources. *Astrophys. J.*, **246**, 20–27, doi:10.1086/158894.
Emmanoulopoulos, D., McHardy, I.M., and Uttley, P. (2010) On the use of structure functions to study blazar variability: caveats and problems. *Mon. Not. R. Astron. Soc.*, **404**, 931–946, doi:10.1111/j.1365-2966.2010.16328.x.
Eracleous, M. and Halpern, J.P. (1994) Doubled-peaked emission lines in active galactic nuclei. *Astrophys. J. Suppl.*, **90**, 1–30, doi:10.1086/191856.
Faber, S.M. and Jackson, R.E. (1976) Velocity dispersions and mass-to-light ratios for elliptical galaxies. *Astrophys. J.*, **204**, 668–683, doi:10.1086/154215.
Fabian, A. and Miniutti, G. (2009) The X-ray spectra of accreting Kerr black holes, in *Kerr Spacetime: Rotating Black Holes in General Relativity* (eds D.J. Wiltshire, M. Visser, and S.M. Scott), Cambridge Univ. Press, Cambridge, pp. 236–280.
Fabian, A.C. (1994) Cooling flows in clusters of galaxies. *Annu. Rev. Astron. Astrophys.*, **32**, 277–318, doi:10.1146/annurev.aa.32.090194.001425.
Fabian, A.C. (2008) XMM-Newton and broad iron lines. *Astron. Nachr.*, **329**, 155, doi:10.1002/asna.200710902.
Fabian, A.C. (2010) Cosmic feedback from AGN, in *IAU Symposium*, Vol. 267, Cambridge Univ. Press, Cambridge, pp. 341–349, doi:10.1017/S1743921310006691.
Fabian, A.C. and Canizares, C.R. (1988) Do massive black holes reside in elliptical galaxies? *Nature*, **333**, 829–831, doi:10.1038/333829a0.
Fabian, A.C., Iwasawa, K., Reynolds, C.S., and Young, A.J. (2000) Broad iron lines in active galactic nuclei. *Publ. Astron. Soc. Pac.*, **112**, 1145–1161, doi:10.1086/316610.
Fabian, A.C., Sanders, J.S., Taylor, G.B. *et al.* (2006) A very deep Chandra observation of the Perseus cluster: shocks, ripples and conduction. *Mon. Not. R. Astron. Soc.*, **366**, 417–428, doi:10.1111/j.1365-2966.2005.09896.x.
Fabian, A.C., Zoghbi, A., Wilkins, D. *et al.* (2012) 1H 0707-495 in 2011: an X-ray source within a gravitational radius of the event horizon. *Mon. Not. R. Astron. Soc.*, **419**, 116–123, doi:10.1111/j.1365-2966.2011.19676.x.
Falocco, S., Carrera, F.J., Corral, A. *et al.* (2012) Averaging the AGN X-ray spectra from deep Chandra fields. *Astron. Astrophys. A*, **538**, 83, doi:10.1051/0004-6361/201117965.
Fanaroff, B.L. and Riley, J.M. (1974) The morphology of extragalactic radio sources of high and low luminosity. *Mon. Not. R. Astron. Soc.*, **167**, 31P–36P.
Fanidakis, N., Baugh, C.M., Benson, A.J. *et al.* (2012) The evolution of active galactic nuclei across cosmic time: what is downsizing?, *Mon. Not. R. Astron. Soc.*, **419**, 2797–2820, doi:10.1111/j.1365-2966.2011.19931.x.
Fargion, D. (2008) Light nuclei solving the AUGER puzzles: the Cen-A imprint. *Phys. Scr.*, **78**(4), 045 901, doi:10.1088/0031-8949/78/04/045901.
Farnier, C., Walter, R., and Leyder, J. (2011) η Carinae: a very large hadron collider. *Astron. Astrophys.*, **526**, A57, doi:10.1051/0004-6361/201015590.
Farrell, S.A., Webb, N.A., Barret, D. *et al.* (2009) An intermediate-mass black hole of over 500 solar masses in the galaxy ESO 243-49. *Nature*, **460**, 73–75, doi:10.1038/nature08083.
Faucher-Giguère, C.A., Quataert, E., and Murray, N. (2012) A physical model of FeLoBALs: implications for quasar feedback.

Mon. Not. R. Astron. Soc., **420**, 1347–1354, doi:10.1111/j.1365-2966.2011.20120.x.

Fender, R.P., Belloni, T.M., and Gallo, E. (2004) Towards a unified model for black hole X-ray binary jets. *Mon. Not. R. Astron. Soc.*, **355**, 1105–1118, doi:10.1111/j.1365-2966.2004.08384.x.

Feng, H. and Soria, R. (2011) Ultraluminous X-ray sources in the Chandra and XMM-Newton era. *New Astron. Rev.*, **55**, 166–183, doi:10.1016/j.newar.2011.08.002.

Ferland, G.J., Korista, K.T., Verner, D.A. *et al.* (1998) CLOUDY 90: Numerical simulation of plasmas and their spectra. *Publ. Astron. Soc. Pac.*, **110**, 761–778, doi:10.1086/316190.

Ferland, G.J. and Mushotzky, R.F. (1982) Broad line region clouds and the absorbing material in NGC 4151. *Astrophys. J.*, **262**, 564–577, doi:10.1086/160448.

Ferland, G.J., Peterson, B.M., Horne, K. *et al.* (1992) Anisotropic line emission and the geometry of the broad-line region in active galactic nuclei. *Astrophys. J.*, **387**, 95–108, doi:10.1086/171063.

Ferrarese, L. and Merritt, D. (2000) A fundamental relation between supermassive black holes and their host galaxies. *Astrophys. J. Lett.*, **539**, L9–L12, doi:10.1086/312838.

Fine, S., Jarvis, M.J., and Mauch, T. (2011) Orientation effects in quasar spectra: the broad- and narrow-line regions. *Mon. Not. R. Astron. Soc.*, **412**, 213–222, doi:10.1111/j.1365-2966.2010.17898.x.

Fiore, F., Puccetti, S., Grazian, A. *et al.* (2012a) Faint high-redshift AGN in the Chandra deep field south: the evolution of the AGN luminosity function and black hole demography. *Astron. Astrophys. A*, **537**, 16, doi:10.1051/0004-6361/201117581.

Fiore, F., Puccetti, S., and Mathur, S. (2012b) Demography of high redshift AGN. *Adv. Astron.*, doi:10.1155/2012/271502.

Fischer, T.C., Crenshaw, D.M., Kraemer, S.B. *et al.* (2011) Hubble space telescope observations of the double-peaked emission lines in the Seyfert galaxy Markarian 78: Mass outflows from a single active galactic nucleus. *Astrophys. J.*, **727**, 71, doi:10.1088/0004-637X/727/2/71.

Forman, W., Jones, C., Cominsky, L. *et al.* (1978) The fourth Uhuru catalog of X-ray sources. *Astrophys. J. Suppl.*, **38**, 357–412, doi:10.1086/190561.

Foschini, L. (2011) Evidence of powerful relativistic jets in narrow-line Seyfert 1 galaxies, in *Narrow-Line Seyfert 1 Galaxies and their Place in the Universe*, Proceedings of Science, Trieste, Italy, p. id 24.

Foschini, L., Ghisellini, G., Kovalev, Y.Y. *et al.* (2011) The first gamma-ray outburst of a narrow-line Seyfert 1 galaxy: the case of PMN J0948+0022 in 2010 July. *Mon. Not. R. Astron. Soc.*, **413**, 1671–1677, doi:10.1111/j.1365-2966.2011.18240.x.

Fossati, G., Maraschi, L., Celotti, A. *et al.* (1998) A unifying view of the spectral energy distributions of blazars. *Mon. Not. R. Astron. Soc.*, **299**, 433–448, doi:10.1046/j.1365-8711.1998.01828.x.

Franceschini, A., Vercellone, S., and Fabian, A.C. (1998) Supermassive black holes in early-type galaxies: Relationship with radio emission and constraints on the black hole mass function. *Mon. Not. R. Astron. Soc.*, **297**, 817–824, doi:10.1046/j.1365-8711.1998.01534.x.

Frank, J., King, A., and Raine, D. (1992) *Accretion Power in Astrophysics*, Cambridge Astrophys. Ser., Vol. 21, Cambridge Univ. Press, Cambridge.

Freese, K., Bodenheimer, P., Spolyar, D., and Gondolo, P. (2008) Stellar structure of dark stars: A first phase of stellar evolution resulting from dark matter annihilation. *Astrophys. J. Lett.*, **685**, L101–L104, doi:10.1086/592685.

Fryer, C.L., Woosley, S.E., and Heger, A. (2001) Pair-instability supernovae, gravity waves, and gamma-ray transients. *Astrophys. J.*, **550**, 372–382, doi:10.1086/319719.

Fu, H., Zhang, Z.Y., Assef, R.J. *et al.* (2011) A kiloparsec-scale binary active galactic nucleus confirmed by the expanded very large array. *Astrophys. J. Lett.*, **740**, L44, doi:10.1088/2041-8205/740/2/L44.

Fukumura, K., Kazanas, D., Contopoulos, I., and Behar, E. (2010) Magnetohydrodynamic accretion disk winds as X-ray absorbers in active galactic nuclei. *Astrophys. J.*, **715**, 636–650, doi:10.1088/0004-637X/715/1/636.

Gabuzda, D.C., Vitrishchak, V.M., Mahmud, M., and O'Sullivan, S.P. (2008) Radio circular polarization produced in helical magnetic fields in eight active galactic nuclei. *Mon. Not. R. Astron. Soc.*, **384**, 1003–1014, doi:10.1111/j.1365-2966.2007.12773.x.

Gallo, E., Treu, T., Marshall, P.J. *et al.* (2010) AMUSE-Virgo. II. Down-sizing in black hole accretion. *Astrophys. J.*, **714**, 25–36, doi:10.1088/0004-637X/714/1/25.

Gallo, L.C., Miniutti, G., Miller, J.M. *et al.* (2011) Multi-epoch X-ray observations of the Seyfert 1.2 galaxy Mrk 79: bulk motion of the illuminating X-ray source. *Mon. Not. R. Astron. Soc.*, **411**, 607–619, doi:10.1111/j.1365-2966.2010.17705.x.

Gao, X.Y., Wang, J.C., and Zhou, M. (2011) External photon fields in Fermi bright blazars. *Res. Astron. Astrophys.*, **11**, 902–908, doi:10.1088/1674-4527/11/8/004.

Garofalo, D., Evans, D.A., and Sambruna, R.M. (2010) The evolution of radio-loud active galactic nuclei as a function of black hole spin. *Mon. Not. R. Astron. Soc.*, **406**, 975–986, doi:10.1111/j.1365-2966.2010.16797.x.

Garrington, S.T., Leahy, J.P., Conway, R.G., and Laing, R.A. (1988) A systematic asymmetry in the polarization properties of double radio sources with one jet. *Nature*, **331**, 147–149, doi:10.1038/331147a0.

Gaskell, C.M. (2009) What broad emission lines tell us about how active galactic nuclei work. *New Astron. Rev.*, **53**, 140–148, doi:10.1016/j.newar.2009.09.006.

Gebhardt, K., Adams, J., Richstone, D. *et al.* (2011) The black hole mass in M87 from Gemini/NIFS adaptive optics observations. *Astrophys. J.*, **729**, 119, doi:10.1088/0004-637X/729/2/119.

Gebhardt, K., Bender, R., Bower, G. *et al.* (2000) A relationship between nuclear black hole mass and galaxy velocity dispersion. *Astrophys. J. Lett.*, **539**, L13–L16, doi:10.1086/312840.

Genzel, R., Lutz, D., Sturm, E. *et al.* (1998) What powers ultraluminous IR. Astron. Soc. galaxies?, *Astrophys. J.*, **498**, 579, doi:10.1086/305576.

Genzel, R., Tacconi, L.J., Rigopoulou, D. *et al.* (2001) Ultraluminous infrared mergers: Elliptical galaxies in formation?, *Astrophys. J.*, **563**, 527–545, doi:10.1086/323772.

George, I.M. and Fabian, A.C. (1991) X-ray reflection from cold matter in active galactic nuclei and X-ray binaries. *Mon. Not. R. Astron. Soc.*, **249**, 352–367.

George, I.M., Turner, T.J., Netzer, H. *et al.* (1998) ASCA observations of Seyfert 1 galaxies. III. The evidence for absorption and emission due to photoionized gas. *Astrophys. J. Suppl.*, **114**, 73, doi:10.1086/313067.

Germain, J., Barai, P., and Martel, H. (2009) Anisotropic active galactic nucleus outflows and enrichment of the intergalactic medium. I. Metal distribution. *Astrophys. J.*, **704**, 1002–1020, doi:10.1088/0004-637X/704/2/1002.

Ghisellini, G. (2011) Jetted active galactic nuclei. *Int. J. Mod. Phys.*, in press, arXiv:1109.0015.

Ghisellini, G., Celotti, A., Fossati, G. *et al.* (1998) A theoretical unifying scheme for gamma-ray bright blazars. *Mon. Not. R. Astron. Soc.*, **301**, 451–468, doi:10.1046/j.1365-8711.1998.02032.x.

Ghisellini, G. and Madau, P. (1996) On the origin of the gamma-ray emission in blazars. *Mon. Not. R. Astron. Soc.*, **280**, 67–76.

Ghisellini, G., Maraschi, L., and Dondi, L. (1996) Diagnostics of inverse-Compton models for the γ-ray emission of 3C 279 and MKN 421. *Astron. Astrophys. Suppl.*, **120**, C503.

Ghisellini, G., Padovani, P., Celotti, A., and Maraschi, L. (1993) Relativistic bulk motion in active galactic nuclei. *Astrophys. J.*, **407**, 65–82, doi:10.1086/172493.

Ghisellini, G. and Tavecchio, F. (2008) The blazar sequence: a new perspective. *Mon. Not. R. Astron. Soc.*, **387**, 1669–1680, doi:10.1111/j.1365-2966.2008.13360.x.

Ghisellini, G., Tavecchio, F., Foschini, L. *et al.* (2010) General physical properties of bright Fermi blazars. *Mon. Not. R. Astron. Soc.*, **402**, 497–518, doi:10.1111/j.1365-2966.2009.15898.x.

Giacconi, R., Gursky, H., Paolini, F.R., and Rossi, B.B. (1962) Evidence for X-rays from sources outside the solar system. *Phys. Rev. Lett.*, **9**, 439–443, doi:10.1103/PhysRevLett.9.439.

Giacconi, R. and Rossi, B. (1960) A "Telescope" for soft X-ray astronomy. *J. Geophys. Res.*, **65**, 773, doi:10.1029/JZ065i002p00773.

Giavalisco, M., Ferguson, H.C., Koekemoer, A.M. *et al.* (2004) The great observatories origins deep survey: Initial results from optical and near-infrared imaging. *Astrophys. J. Lett.*, **600**, L93–L98, doi:10.1086/379232.

Gierliński, M. and Done, C. (2004) Is the soft excess in active galactic nuclei real? *Mon. Not. R. Astron. Soc.*, **349**, L7–L11, doi:10.1111/j.1365-2966.2004.07687.x.

Gierliński, M., Middleton, M., Ward, M., and Done, C. (2008) A periodicity of $\sim$ 1 hour in X-ray emission from the active galaxy RE J1034+396. *Nature*, **455**, 369–371, doi:10.1038/nature07277.

Gilli, R., Comastri, A., and Hasinger, G. (2007) The synthesis of the cosmic X-ray background in the Chandra and XMM-Newton era. *Astron. Astrophys.*, **463**, 79–96, doi:10.1051/0004-6361:20066334.

Ginzburg, V.L. (1989) *Applications of Electrodynamics in Theoretical Physics and Astrophysics*, Gordon and Breach, New York, 482 p., ISBN 2-88124-719-9.

Gioia, I.M., Maccacaro, T., Schild, R.E. *et al.* (1990) The Einstein observatory extended medium-sensitivity survey. I – X-ray data and analysis. *Astrophys. J. Suppl.*, **72**, 567–619, doi:10.1086/191426.

Gitti, M., Brighenti, F., and McNamara, B.R. (2012) Evidence for AGN feedback in galaxy clusters and groups. *Adv. Astron.*, **2012**, doi:10.1155/2012/950641.

Giustini, M., Cappi, M., Chartas, G. *et al.* (2011) Variable X-ray absorption in the mini-BAL QSO PG 1126-041. *Astron. Astrophys.*, **536**, A49, doi:10.1051/0004-6361/201117732.

Goldreich, P. and Sargent, W. (1976) Quasar absorption lines. *Comments Astrophys.*, **6**, 133–137.

Gondek, D., Zdziarski, A.A., Johnson, W.N. *et al.* (1996) The average X-ray/gamma-ray spectrum of radio-quiet Seyfert 1s. *Mon. Not. R. Astron. Soc.*, **282**, 646–652.

González-Martín, O., Masegosa, J., Márquez, I., and Guainazzi, M. (2009) Fitting liner nuclei within the active galactic nucleus family: a matter of obscuration? *Astrophys. J.*, **704**, 1570–1585, doi:10.1088/0004-637X/704/2/1570.

Goto, T., Arnouts, S., Inami, H. *et al.* (2011a) Luminosity functions of local infrared galaxies with AKARI: implications for the cosmic star formation history and AGN evolution. *Mon. Not. R. Astron. Soc.*, **410**, 573–584, doi:10.1111/j.1365-2966.2010.17466.x.

Goto, T., Utsumi, Y., Furusawa, H. *et al.* (2009) A QSO host galaxy and its Lyα emission at $z = 6.43$. *Mon. Not. R. Astron. Soc.*, **400**, 843–850, doi:10.1111/j.1365-2966.2009.15486.x.

Goto, T., Utsumi, Y., Hattori, T. *et al.* (2011b) A Gunn–Peterson test with a QSO at $z = 6.4$. *Mon. Not. R. Astron. Soc.*, 415, L1–L5, doi:10.1111/j.1745-3933.2011.01063.x.

Goulding, A.D., Alexander, D.M., Lehmer, B.D., and Mullaney, J.R. (2010) Towards a complete census of active galactic nuclei in nearby galaxies: the incidence of growing black holes. *Mon. Not. R. Astron. Soc.*, **406**, 597–611, doi:10.1111/j.1365-2966.2010.16700.x.

Graves, G.J., Faber, S.M., Schiavon, R.P., and Yan, R. (2007) Ages and abundances of red sequence galaxies as a function of LINER emission-line strength. *Astrophys. J.*, **671**, 243–271, doi:10.1086/522325.

Green, A.R., McHardy, I.M., and Lehto, H.J. (1993) On the nature of rapid X-ray variability in active galactic nuclei. *Mon. Not. R. Astron. Soc.*, **265**, 664.

Green, R.F., Schmidt, M., and Liebert, J. (1986) The Palomar–Green catalog of ultraviolet-excess stellar objects. *Astrophys. J. Suppl.*, **61**, 305–352, doi:10.1086/191115.

Greene, J.E. and Ho, L.C. (2006) The $M_{BH}-\sigma$ relation in local active galaxies. *Astrophys. J. Lett.*, **641**, L21–L24, doi:10.1086/500507.

Greene, J.E. and Ho, L.C. (2007) The mass function of active black holes in the local universe. *Astrophys. J.*, **667**, 131–148, doi:10.1086/520497.

Grimm, H.J., Gilfanov, M., and Sunyaev, R. (2003) High-mass X-ray binaries as a star formation rate indicator in distant galaxies. *Mon. Not. R. Astron. Soc.*, **339**, 793–809, doi:10.1046/j.1365-8711.2003.06224.x.

Grogin, N.A., Conselice, C.J., Chatzichristou, E. *et al.* (2005) AGN host galaxies at $z \sim 0.4-1.3$: Bulge-dominated and lacking merger-AGN connection. *Astrophys. J. Lett.*, **627**, L97–L100, doi:10.1086/432256.

Grupe, D., Komossa, S., Leighly, K.M., and Page, K.L. (2010) The simultaneous optical-to-X-ray spectral energy distribution of soft X-ray selected active galactic nuclei observed by Swift. *Astrophys. J. Suppl.*, **187**, 64–106, doi:10.1088/0067-0049/187/1/64.

Gu, Q. and Huang, J. (2002) Seyfert 2 galaxies with spectropolarimetric observations. *Astrophys. J.*, **579**, 205–213, doi:10.1086/342703.

Guainazzi, M., Bianchi, S., de La Calle Pérez, I. *et al.* (2011) On the driver of relativistic effect strength in Seyfert galaxies. *Astron. Astrophys.*, **531**, A131, doi:10.1051/0004-6361/201016245.

Gültekin, K., Richstone, D.O., Gebhardt, K. *et al.* (2009) The $M-\sigma$ and $M-L$ relations in galactic bulges, and determinations of their intrinsic scatter. *Astrophys. J.*, **698**, 198–221, doi:10.1088/0004-637X/698/1/198.

Gunn, J.E. and Peterson, B.A. (1965) On the density of neutral hydrogen in intergalactic space. *Astrophys. J.*, **142**, 1633–1641, doi:10.1086/148444.

Gupta, J.A., Browne, I.W.A., and Peel, M.W. (2012) Blazar Compton efficiencies: Fermi, external photons and the sequence. arXiv:1106.5172, submitted to MNRAS.

Gurvits, L.I. (2004) Surveys of compact extragalactic radio sources. *New Astron. Rev.*, **48**, 1211–1220, doi:10.1016/j.newar.2004.09.043.

Haardt, F. and Maraschi, L. (1991) A two-phase model for the X-ray emission from Seyfert galaxies. *Astrophys. J. Lett.*, **380**, L51–L54, doi:10.1086/186171.

Haardt, F. and Maraschi, L. (1993) X-ray spectra from two-phase accretion disks. *Astrophys. J.*, **413**, 507–517, doi:10.1086/173020.

Haardt, F., Maraschi, L., and Ghisellini, G. (1994) A model for the X-ray and ultraviolet emission from Seyfert galaxies and galactic black holes. *Astrophys. J. Lett.*, **432**, L95–L99, doi:10.1086/187520.

Hada, K., Doi, A., Kino, M. *et al.* (2011) An origin of the radio jet in M87 at the location of the central black hole. *Nature*, **477**, 185–187, doi:10.1038/nature10387.

Hagen, H., Groote, D., Engels, D., and Reimers, D. (1995) The Hamburg quasar survey. I. Schmidt observations and plate digitization. *Astron. Astrophys. Suppl.*, **111**, 195.

Halpern, J.P. (1984) Variable X-ray absorption in the QSO MR 2251-178. *Astrophys. J.*, **281**, 90–94, doi:10.1086/162077.

Hamabe, M. and Kormendy, J. (1987) Correlations between $R/1/4$ – law parameters for bulges and elliptical galaxies, in *Structure and Dynamics of Elliptical Galaxies*, IAU Symposium, Vol. 127 (eds P.T. de Zeeuw), D. Reidel, Dordrecht, p. 379.

Hao, L., Strauss, M.A., Fan, X. *et al.* (2005) Active galactic nuclei in the sloan digital sky survey. II. Emission-line luminosity function. *Astron. J.*, **129**, 1795–1808, doi:10.1086/428486.

Hardcastle, M.J., Kraft, R.P., Sivakoff, G.R. *et al.* (2007) New results on particle acceleration in the Centaurus a jet and counterjet from a deep Chandra observation. *Astrophys. J. Lett.*, **670**, L81–L84, doi:10.1086/524197.

Harris, D.E. and Krawczynski, H. (2006) X-ray emission from extragalactic jets. *Annu. Rev. Astron. Astrophys.*, **44**, 463–506, doi:10.1146/annurev.astro.44.051905.092446.

Harrison, F.A., Boggs, S., Christensen, F. *et al.* (2010) The Nuclear Spectroscopic Telescope Array (NuSTAR), in *Society of Photo-Optical Instrumentation Engineers (SPIE) Conference Series*, SPIE, Bellingham, Washington, doi:10.1117/12.858065.

Hartigan, P. (1989) The visibility of the Mach disk and the bow shock of a stellar jet. *Astrophys. J.*, **339**, 987–999, doi:10.1086/167353.

Hartman, R.C., Bertsch, D.L., Bloom, S.D. *et al.* (1999) The third EGRET catalog of high-energy gamma-ray sources. *Astrophys. J. Suppl.*, **123**, 79–202, doi:10.1086/313231.

Hartman, R.C., Bertsch, D.L., Fichtel, C.E. *et al.* (1992) Detection of high-energy gamma radiation from quasar 3C 279 by the EGRET telescope on the Compton gamma ray observatory. *Astrophys. J. Lett.*, **385**, L1–L4, doi:10.1086/186263.

Hasinger, G., Burg, R., Giacconi, R. *et al.* (1998) The ROSAT deep survey. I. X-ray sources in the Lockman field. *Astron. Astrophys.*, **329**, 482–494.

Hasinger, G., Miyaji, T., and Schmidt, M. (2005) Luminosity-dependent evolution of soft X-ray selected AGN. New Chandra and XMM-Newton surveys. *Astron. Astrophys.*, **441**, 417–434, doi:10.1051/0004-6361:20042134.

Hawking, S.W. (1974) Black hole explosions?, *Nature*, **248**, 30–31, doi:10.1038/248030a0.

Hawkins, M.R.S. (2011) The case for primordial black holes as dark matter. *Mon. Not. R. Astron. Soc.*, **415**, 2744–2757, doi:10.1111/j.1365-2966.2011.18890.x.

Heidt, J. and Nilsson, K. (2011) Polarimetry of optically selected BL Lacertae candidates from the SDSS. *Astron. Astrophys.*, **529**, A162, doi:10.1051/0004-6361/201116541.

Heinz, S. and Sunyaev, R.A. (2003) The nonlinear dependence of flux on black hole mass and accretion rate in core-dominated jets. *Mon. Not. R. Astron. Soc.*, **343**, L59–L64, doi:10.1046/j.1365-8711.2003.06918.x.

Hennawi, J.F., Myers, A.D., Shen, Y. *et al.* (2010) Binary quasars at high redshift. I. 24 New quasar pairs at $z \sim 3-4$. *Astrophys. J.*, **719**, 1672–1692, doi:10.1088/0004-637X/719/2/1672.

Hewett, P.C., Foltz, C.B., and Chaffee, F.H. (1995) The large bright quasar survey. 6: Quasar catalog and survey parameters. *Astron. J.*, **109**, 1498–1521, doi:10.1086/117380.

Hjellming, R.M. and Rupen, M.P. (1995) Episodic ejection of relativistic jets by the X-ray transient GRO J1655-40. *Nature*, **375**, 464–468, doi:10.1038/375464a0.

Ho, L.C. (2002) On the relationship between radio emission and black hole mass in galactic nuclei. *Astrophys. J.*, **564**, 120–132, doi:10.1086/324399.

Ho, L.C. (2008) Nuclear activity in nearby galaxies. *Annu. Rev. Astron. Astrophys.*, **46**, 475–539, doi:10.1146/annurev.astro.45.051806.110546.

Ho, L.C., Filippenko, A.V., and Sargent, W.L.W. (1997) A search for "Dwarf" Seyfert nuclei. III. Spectroscopic parameters and properties of the host galaxies. *Astrophys. J. Suppl.*, **112**, 315, doi:10.1086/313041.

Holt, S.S., Boldt, E.A., and Serlemitsos, P.J. (1969) Search for line structure in the X-ray spectrum of SCO X-1. *Astrophys. J. Lett.*, **158**, L155, doi:10.1086/180454.

Hopkins, P.F. (2012) Dynamical delays between starburst and AGN activity in galaxy nuclei. *Mon. Not. R. Astron. Soc.*, **420**, L8–L12, doi:10.1111/j.1745-3933.2011.01179.x.

Hopkins, P.F., Bundy, K., Croton, D. *et al.* (2010) Mergers and bulge formation in Λ CDM: Which mergers matter?, *Astrophys. J.*, **715**, 202–229, doi:10.1088/0004-637X/715/1/202.

Hopkins, P.F., Hernquist, L., Cox, T.J. *et al.* (2005) Black holes in galaxy mergers: Evolution of quasars. *Astrophys. J.*, **630**, 705–715, doi:10.1086/432438.

Hopkins, P.F., Hernquist, L., Hayward, C.C., and Narayanan, D. (2012) Why Are AGN and host galaxies misaligned? *Mon. Not. R. Astron. Soc.*, in press, arXiv:1111.1236.

Hopkins, P.F. and Quataert, E. (2010) How do massive black holes get their gas? *Mon. Not. R. Astron. Soc.*, **407**, 1529–1564, doi:10.1111/j.1365-2966.2010.17064.x.

Hovatta, T., Lister, M.L., Aller, M.F. *et al.* (2012) Faraday rotation in the MOJAVE blazars: 3C 273 a case study, in *Beamed and Unbeamed Gamma-rays from Galaxies, J. Phys.: Conf. Ser.*, **355**(1), 012008, doi:10.1088/1742-6596/355/1/012008.

Howell, K.B. (2001) *Principles of Fourier Analysis*, Studies in Advanced Mathematics, Chapman and Hall/CRC, Boca Raton, London, New York, Washington DC.

Hoyle, F. and Fowler, W.A. (1963) On the nature of strong radio sources. *Mon. Not. R. Astron. Soc.*, **125**, 169.

Hu, E.M. and Cowie, L.L. (2006) High-redshift galaxy populations. *Nature*, **440**, 1145–1150, doi:10.1038/nature04806.

Hubble, E.P. (1926) Extragalactic nebulae. *Astrophys. J.*, **64**, 321–369, doi:10.1086/143018.

Hubble, E.P. (1929) A relation between distance and radial velocity among extragalactic nebulae. *Proc. Natl. Acad. Sci. USA*, **15**, 168–173, doi:10.1073/pnas.15.3.168.

Hughes, P.A., Aller, H.D., and Aller, M.F. (1992) The University of Michigan radio astronomy data base. I – Structure function analysis and the relation between BL Lacertae objects and quasi-stellar objects. *Astrophys. J.*, **396**, 469–486, doi:10.1086/171734.

Icke, V. (1983) A clamshell for Blandford–Rees jets. *Astrophys. J.*, **265**, 648–663, doi:10.1086/160711.

Ikhsanov, N.R. (1989) Generation of infrared synchrotron radiation in the corona of the accretion disk in an X-ray binary source. *Sov. Astron. Lett.*, **15**, 220.

Immler, S. and Lewin, W.H.G. (2003) X-ray supernovae, in *Supernovae and Gamma-Ray*

Bursters, Lecture Notes in Physics, Vol. 598 (ed. K. Weiler) Springer Verlag, Berlin, pp. 91–111.

Ishibashi, W. and Courvoisier, T. (2011) Synchrotron radio emission in radio-quiet AGNs. *Astron. Astrophys.*, **525**, A118, doi:10.1051/0004-6361/201014987.

Ishihara, Y., Nakai, N., Iyomoto, N. *et al.* (2001) Water-vapor maser emission from the Seyfert 2 galaxy IC 2560: Evidence for a super-massive black hole. *Publ. Astron. Soc. Japan*, **53**, 215–225.

Iwasawa, K., Mainieri, V., Brusa, M. *et al.* (2012) Fe*K* emission from active galaxies in the COSMOS field. *Astron. Astrophys. A*, **537**, 86, doi:10.1051/0004-6361/201118203.

Iwasawa, K. and Taniguchi, Y. (1993) The X-ray Baldwin effect. *Astrophys. J. Lett.*, **413**, L15–L18, doi:10.1086/186948.

Jackson, N. and Browne, I.W.A. (1990) Spectral differences between radio galaxies and quasars. *Nature*, **343**, 43–45, doi:10.1038/343043a0.

Jackson, N. and Browne, I.W.A. (1991) Optical properties of quasars. I – Observations. II – Emission-line geometry and radio properties. *Mon. Not. R. Astron. Soc.*, **250**, 414–431.

Jahnke, K., Elbaz, D., Pantin, E. *et al.* (2009) The QSO HE 0450-2958: Scantily dressed or heavily robed? A normal quasar as part of an unusual ULIRG. *Astrophys. J.*, **700**, 1820–1830, doi:10.1088/0004-637X/700/2/1820.

Jahnke, K., Kuhlbrodt, B., and Wisotzki, L. (2004a) Quasar host galaxy star formation activity from multicolour data. *Mon. Not. R. Astron. Soc.*, **352**, 399–415, doi:10.1111/j.1365-2966.2004.07933.x.

Jahnke, K. and Macciò, A.V. (2011) The non-causal origin of the black-hole-galaxy scaling relations. *Astrophys. J.*, **734**, 92, doi:10.1088/0004-637X/734/2/92.

Jahnke, K., Sánchez, S.F., Wisotzki, L. *et al.* (2004b) Ultraviolet light from young stars in GEMS quasar host galaxies at $1.8 < z < 2.75$. *Astrophys. J.*, **614**, 568–585, doi:10.1086/423233.

Jiang, Y.F., Greene, J.E., Ho, L.C. *et al.* (2011) The host galaxies of low-mass black holes. *Astrophys. J.*, **742**, 68, doi:10.1088/0004-637X/742/2/68.

Kaastra, J.S., Petrucci, P.O., Cappi, M. *et al.* (2011) Multiwavelength campaign on Mrk 509. I. Variability and spectral energy distribution. *Astron. Astrophys.*, **534**, A36, doi:10.1051/0004-6361/201116869.

Kallman, T. and Bautista, M. (2001) Photoionization and high-density gas. *Astrophys. J. Suppl.*, **133**, 221–253, doi:10.1086/319184.

Kallman, T.R. (2010) Modeling of photoionized plasmas. *Space Sci. Rev.*, **157**, 177–191, doi:10.1007/s11214-010-9711-6.

Kant, I. (1755) *Allgemeine Naturgeschichte und Theorie des Himmels*, Bei: W. Webel, Zeitz, 1798. Neue Aufl.

Kaspi, S., Brandt, W.N., George, I.M. *et al.* (2002) The ionized gas and nuclear environment in NGC 3783. I. Time-averaged 900 kilosecond Chandra grating spectroscopy. *Astrophys. J.*, **574**, 643–662, doi:10.1086/341113.

Katarzyński, K., Ghisellini, G., Tavecchio, F. *et al.* (2006) Hard TeV spectra of blazars and the constraints to the infrared intergalactic background. *Mon. Not. R. Astron. Soc.*, **368**, L52–L56, doi:10.1111/j.1745-3933.2006.00156.x.

Katz, J.I. (1976) Nonrelativistic Compton scattering and models of quasars. *Astrophys. J.*, **206**, 910–916, doi:10.1086/154455.

Kauffmann, G., Heckman, T.M., Tremonti, C. *et al.* (2003a) The host galaxies of active galactic nuclei. *Mon. Not. R. Astron. Soc.*, **346**, 1055–1077, doi:10.1111/j.1365-2966.2003.07154.x.

Kauffmann, G., Heckman, T.M., White, S.D.M. *et al.* (2003b) Stellar masses and star formation histories for 10^5 galaxies from the sloan digital sky survey. *Mon. Not. R. Astron. Soc.*, **341**, 33–53, doi:10.1046/j.1365-8711.2003.06291.x.

Kawakatu, N. and Wada, K. (2009) Formation of high-redshift ($z > 6$) quasars driven by nuclear starbursts. *Astrophys. J.*, **706**, 676–686, doi:10.1088/0004-637X/706/1/676.

Keel, W.C. (1980) Inclination effects on the recognition of Seyfert galaxies. *Astron. J.*, **85**, 198–203, doi:10.1086/112662.

Kellermann, K.I., Lister, M.L., Homan, D.C. *et al.* (2004) Sub-milliarcsecond imaging of quasars and active galactic nuclei. III. Kinematics of parsec-scale radio jets. *Astrophys. J.*, **609**, 539–563, doi:10.1086/421289.

Kellermann, K.I. and Pauliny-Toth, I.I.K. (1969) The spectra of Opaque radio sources. *Astrophys. J. Lett.*, **155**, L71, doi:10.1086/180305.

Kembhavi, A.K. and Narlikar, J.V. (1999) *Quasars and Active Galactic Nuclei: An Introduction*, Cambridge University Press, Cambridge, ISBN 0521474779.

Kennicutt, Jr., R.C. (1998) Star formation in galaxies along the Hubble sequence. *Annu. Rev. Astron. Astrophys.*, **36**, 189–232, doi:10.1146/annurev.astro.36.1.189.

Khachikian, E.Y. and Weedman, D.W. (1974) An atlas of Seyfert galaxies. *Astrophys. J.*, **192**, 581–589, doi:10.1086/153093.

Khan, F.M., Just, A., and Merritt, D. (2011) Efficient merger of binary supermassive black holes in merging galaxies. *Astrophys. J.*, **732**, 89, doi:10.1088/0004-637X/732/2/89.

Kim, J.H., Wise, J.H., Alvarez, M.A., and Abel, T. (2011) Galaxy formation with self-consistently modeled stars and massive black holes. I. Feedback-regulated star formation and black hole growth. *Astrophys. J.*, **738**, 54, doi:10.1088/0004-637X/738/1/54.

King, A. (2008) Disc accretion in active galactic nuclei. *New Astron. Rev.*, **52**, 253–256, doi:10.1016/j.newar.2008.06.006.

King, A.R., Davies, M.B., Ward, M.J. *et al.* (2001) Ultraluminous X-ray sources in external galaxies. *Astrophys. J. Lett.*, **552**, L109–L112, doi:10.1086/320343.

Kinney, A.L., Bohlin, R.C., Blades, J.C., and York, D.G. (1991) An ultraviolet atlas of quasar and blazar spectra. *Astrophys. J. Suppl.*, **75**, 645–717, doi:10.1086/191546.

Kocevski, D.D., Faber, S.M., Mozena, M. *et al.* (2012) CANDELS: Constraining the AGN-merger connection with host morphologies at $z \sim 2$. Astrophys. J., **744**, 148, doi:10.1088/0004-637X/744/2/148.

Komossa, S. (1998) The complex X-ray spectra of active galaxies with warm absorbers, in *Science with XMM*. published online, ESA Publications Division, Noordwijk.

Komossa, S. and Hasinger, G. (2003) The X-ray evolving universe: (ionized) absorption and dust, from n earby Seyfert galaxies to high-redshift quasars, in *XEUS – Studying the Evolution of the Hot Universe* (eds G. Hasinger, T. Boller, and A.N. Parmer), MPE Report 281, MPE, Garching/Munich, p. 285.

Komossa, S., Voges, W., Xu, D. *et al.* (2006) Radio-loud narrow-line type 1 quasars. *Astron. J.*, **132**, 531–545, doi:10.1086/505043.

Kondratko, P.T., Greenhill, L.J., and Moran, J.M. (2005) Evidence for a geometrically thick self-gravitating accretion disk in NGC 3079. *Astrophys. J.*, **618**, 618–634, doi:10.1086/426101.

Kormendy, J. (1977) Brightness distributions in compact and normal galaxies. II – Structure parameters of the spheroidal component. *Astrophys. J.*, **218**, 333–346, doi:10.1086/155687.

Kormendy, J. and Bender, R. (2011) Supermassive black holes do not correlate with dark matter haloes of galaxies. *Nature*, **469**, 377–380, doi:10.1038/nature09695.

Kormendy, J., Bender, R., and Cornell, M.E. (2011) Supermassive black holes do not correlate with galaxy disks or pseudobulges. *Nature*, **469**, 374–376, doi:10.1038/nature09694.

Koss, M., Mushotzky, R., Veilleux, S., and Winter, L. (2010) Merging and Clustering of the Swift BAT AGN Sample. *Astrophys. J. Lett.*, **716**, L125–L130, doi:10.1088/2041-8205/716/2/L125.

Kotera, K. and Olinto, A.V. (2011) The astrophysics of ultrahigh-energy cosmic rays. *Annu. Rev. Astron. Astrophys.*, **49**, 119–153, doi:10.1146/annurev-astro-081710-102620.

Kovalev, Y.Y., Lister, M.L., Homan, D.C., and Kellermann, K.I. (2007) The inner jet of the radio galaxy M87. *Astrophys. J. Lett.*, **668**, L27–L30, doi:10.1086/522603.

Kozieł-Wierzbowska, D. and Stasińska, G. (2011) FR II radio galaxies in the Sloan Digital Sky Survey: observational facts. *Mon. Not. R. Astron. Soc.*, **415**, 1013–1026, doi:10.1111/j.1365-2966.2011.18346.x.

Kraemer, S.B. and Crenshaw, D.M. (2000) Resolved spectroscopy of the narrow-line region in NGC 1068. III. Physical conditions in the emission-line gas. *Astrophys. J.*, **544**, 763–779, doi:10.1086/317246.

Kraemer, S.B., Schmitt, H.R., and Crenshaw, D.M. (2008) Probing the ionization structure of the narrow-line region in the Seyfert 1 galaxy NGC 4151. *Astrophys. J.*, **679**, 1128–1143, doi:10.1086/587802.

Kraemer, S.B., Schmitt, H.R., Crenshaw, D.M. *et al.* (2011) Multi-wavelength probes of obscuration toward the narrow-line region in Seyfert galaxies. *Astrophys. J.*, **727**, 130, doi:10.1088/0004-637X/727/2/130.

Kriss, G.A., Shull, J.M., Oegerle, W. *et al.* (2001) Resolving the structure of ionized helium in the intergalactic medium with the far ultraviolet spectroscopic explorer. *Science*, **293**, 1112–1116, doi:10.1126/science.1062693.

Kristian, J. (1973) Quasars as events in the nuclei of galaxies: the evidence from direct photographs. *Astrophys. J. Lett.*, **179**, L61, doi:10.1086/181117.

Krolik, J.H. (1999) *Active Galactic Nuclei: from the Central Black Hole to the Galactic Environment*, Princeton University Press, Princeton.

Krolik, J.H. and Begelman, M.C. (1988) Molecular tori in Seyfert galaxies – Feeding the monster and hiding it. *Astrophys. J.*, **329**, 702–711, doi:10.1086/166414.

Krolik, J.H., Hawley, J.F., and Hirose, S. (2005) Magnetically driven accretion flows in the Kerr metric. IV. Dynamical properties of the inner disk. *Astrophys. J.*, **622**, 1008–1023, doi:10.1086/427932.

Krolik, J.H. and Kriss, G.A. (2001) Warm absorbers in active galactic nuclei: A multitemperature wind. *Astrophys. J.*, **561**, 684–690, doi:10.1086/323442.

Krolik, J.H. and Piran, T. (2011) Swift J1644+57: A white dwarf tidally disrupted by a $10^4\ M_\odot$ black hole?, *Astrophys. J.*, **743**, 134, doi:10.1088/0004-637X/743/2/134.

Kukula, M.J., Dunlop, J.S., Hughes, D.H., and Rawlings, S. (1998) The radio properties of radio-quiet quasars. *Mon. Not. R. Astron. Soc.*, **297**, 366–382, doi:10.1046/j.1365-8711.1998.01481.x.

Kundic, T., Turner, E.L., Colley, W.N. *et al.* (1997) A robust determination of the time delay in 0957+561A, B and a measurement of the global value of Hubble's constant. *Astrophys. J.*, **482**, 75, doi:10.1086/304147.

Kwan, J. and Krolik, J.H. (1981) The formation of emission lines in quasars and Seyfert nuclei. *Astrophys. J.*, **250**, 478–507, doi:10.1086/159395.

La Franca, F., Fiore, F., Comastri, A. *et al.* (2005) The HELLAS2XMM survey. VII. The hard X-ray luminosity function of AGNs up to $z = 4$: More absorbed AGNs at low luminosities and high redshifts. *Astrophys. J.*, **635**, 864–879, doi:10.1086/497586.

Laing, R.A. (1988) The sidedness of jets and depolarization in powerful extragalactic radio sources. *Nature*, **331**, 149–151, doi:10.1038/331149a0.

Laing, R.A. and Bridle, A.H. (1987) Rotation measure variation across M84. *Mon. Not. R. Astron. Soc.*, **228**, 557–571.

Lal, D.V. and Ho, L.C. (2010) The radio properties of type 2 quasars. *Astron. J.*, **139**, 1089–1105, doi:10.1088/0004-6256/139/3/1089.

Lal, D.V., Shastri, P., and Gabuzda, D.C. (2011) Seyfert galaxies: Nuclear radio structure and unification. *Astrophys. J.*, **731**, 68, doi:10.1088/0004-637X/731/1/68.

Laor, A. (1991) Line profiles from a disk around a rotating black hole. *Astrophys. J.*, **376**, 90–94, doi:10.1086/170257.

Laor, A. (2000) On black hole masses and radio loudness in active galactic nuclei. *Astrophys. J. Lett.*, **543**, L111–L114, doi:10.1086/317280.

Laor, A. and Behar, E. (2008) On the origin of radio emission in radio-quiet quasars. *Mon. Not. R. Astron. Soc.*, **390**, 847–862, doi:10.1111/j.1365-2966.2008.13806.x.

Laplace, P.S. (1796) *Exposition du système du Monde*, 2em éd., rev. et augm. par l'auteur, Paris.

Lasota, J., Abramowicz, M.A., Chen, X. *et al.* (1996) Is the accretion flow in NGC 4258 advection dominated?, *Astrophys. J.*, **462**, 142, doi:10.1086/177137.

Lasota, J.P., Alexander, T., Dubus, G. *et al.* (2011) The origin of variability of the intermediate-mass black-hole ULX System HLX-1 in ESO 243-49. *Astrophys. J.*, **735**, 89, doi:10.1088/0004-637X/735/2/89.

Lauer, T.R., Tremaine, S., Richstone, D., and Faber, S.M. (2007) Selection bias in observing the cosmological evolution of the $M_{\rm BH}-\sigma$ and $M_{\rm BH}-L$ relationships. *Astrophys. J.*, **670**, 249–260, doi:10.1086/522083.

Lawrence, A. (1991) The relative frequency of broad-lined and narrow-lined active galactic nuclei – Implications for unified schemes. *Mon. Not. R. Astron. Soc.*, **252**, 586–592.

Lawrence, A. (2012) The UV peak in active galactic nuclei: a false continuum from blurred reflection?, *Mon. Not. R. Astron. Soc.*, in press, arXiv:1110.0854.

Lawrence, A. and Elvis, M. (1982) Obscuration and the various kinds of Seyfert galaxies. *Astrophys. J.*, **256**, 410–426, doi:10.1086/159918.

Lawrence, A. and Papadakis, I. (1993) X-ray variability of active galactic nuclei – A universal power spectrum with luminosity-dependent amplitude. *Astrophys. J. Lett.*, **414**, L85–L88, doi:10.1086/187002.

Lefa, E., Rieger, F.M., and Aharonian, F. (2011) Formation of very hard gamma-ray spectra of blazars in leptonic models. *Astrophys. J.*, **740**, 64, doi:10.1088/0004-637X/740/2/64.

Leighly, K.M. (1999) A comprehensive spectral and variability study of narrow-line Seyfert 1 galaxies observed by ASCA. I. Observations and time series analysis. *Astrophys. J. Suppl.*, **125**, 297–316, doi:10.1086/313277.

Leighly, K.M. and Casebeer, D. (2007) Photoionization models of the broad-line region, in *The Central Engine of Active Galactic Nuclei*, Astronomical Society of the Pacific Conference Series, Vol. 373 (eds L.C. Ho and J.-W. Wang), Astronomical Society of the Pacific, San Francisco, p. 365.

Lemoine-Goumard, M., Ferrara, E., Grondin, M. *et al.* (2011) Fermi-LAT detection of gamma-ray emission in the vicinity of the star forming regions W43 and Westerlund 2. *Mem. Soc. Astron. Italiana*, **82**, 739.

León-Tavares, J., Valtaoja, E., Tornikoski, M. *et al.* (2011) The connection between gamma-ray emission and millimeter flares in Fermi/LAT blazars. *Astron. Astrophys.*, **532**, A146, doi:10.1051/0004-6361/201116664.

Levenson, N.A., Packham, C.C., Alonso-Herrero, A. *et al.* (2008) Science of active galactic nuclei with the GTC and CanariCam, in *Society of Photo-Optical Instrumentation Engineers (SPIE) Conference Series*, Vol. 7014, SPIE, Bellingham, Washington, doi:10.1117/12.790123.

Lewis, K.T., Eracleous, M., and Sambruna, R.M. (2003) Emission-line diagnostics of the central engines of weak-line radio galaxies. *Astrophys. J.*, **593**, 115–126, doi:10.1086/376445.

Liang, E.P.T. (1979) On the hard X-ray emission mechanism of active galactic nuclei sources. *Astrophys. J. Lett.*, **231**, L111–L114, doi:10.1086/183015.

Liang, E.P.T. and Price, R.H. (1977) Accretion disk coronae and Cygnus X-1. *Astrophys. J.*, **218**, 247–252, doi:10.1086/155677.

Lintott, C.J., Schawinski, K., Keel, W. *et al.* (2009) Galaxy Zoo: "Hanny's Voorwerp", a quasar light echo?, *Mon. Not. R. Astron. Soc.*, **399**, 129–140, doi:10.1111/j.1365-2966.2009.15299.x.

Lira, P., Arévalo, P., Uttley, P. *et al.* (2010) X-ray, optical, and near-IR long-term monitoring of AGN, in *IAU Symposium*, Vol. 267, Cambridge Univ. Press, Cambridge, pp. 90–95, doi:10.1017/S1743921310005624.

Lira, P., Arévalo, P., Uttley, P. *et al.* (2011) Optical and near-IR long-term monitoring of NGC 3783 and MR 2251-178: evidence for variable near-IR emission from thin accretion discs. *Mon. Not. R. Astron. Soc.*, **415**, 1290–1303, doi:10.1111/j.1365-2966.2011.18774.x.

Lister, M.L. and Homan, D.C. (2005) MOJAVE: Monitoring of jets in active galactic nuclei with VLBA experiments. I. First-epoch 15 GHz linear polarization images. *Astron. J.*, **130**, 1389–1417, doi:10.1086/432969.

Lockman, F.J., Jahoda, K., and McCammon, D. (1986) The structure of galactic HI in directions of low total column density. *Astrophys. J.*, **302**, 432–449, doi:10.1086/164002.

Long, K.S. (1982) The X-ray properties of normal galaxies. *Adv. Space Res.*, **2**, 177–188, doi:10.1016/0273-1177(82)90268-X.

Lotz, J.M., Davis, M., Faber, S.M. *et al.* (2008a) The evolution of galaxy mergers and morphology at $z < 1.2$ in the extended groth strip. *Astrophys. J.*, **672**, 177–197, doi:10.1086/523659.

Lotz, J.M., Jonsson, P., Cox, T.J., and Primack, J.R. (2008b) Galaxy merger morphologies and time-scales from simulations of equal-mass gas-rich disc mergers. *Mon. Not. R. Astron. Soc.*, **391**, 1137–1162, doi:10.1111/j.1365-2966.2008.14004.x.

Lovell, J.E.J., Rickett, B.J., Macquart, J.P. *et al.* (2008) The micro-arcsecond scintillation-induced variability (masiv) survey. II. The first four epochs. *Astrophys. J.*, **689**, 108–126, doi:10.1086/592485.

Lubiński, P., Zdziarski, A.A., Walter, R. *et al.* (2010) Extreme flux states of NGC 4151 observed with INTEGRAL.

Mon. Not. R. Astron. Soc., **408**, 1851–1865, doi:10.1111/j.1365-2966.2010.17251.x.

Luo, B., Brandt, W.N., Xue, Y.Q. *et al.* (2011) Revealing a population of heavily obscured active galactic nuclei at $z \simeq 0.5-1$ in the Chandra deep field-south. *Astrophys. J.*, **740**, 37, doi:10.1088/0004-637X/740/1/37.

Lynden-Bell, D. (1969) Galactic nuclei as collapsed old quasars. *Nature*, **223**, 690–694, doi:10.1038/223690a0.

Lynden-Bell, D. and Rees, M.J. (1971) On quasars, dust and the galactic centre. *Mon. Not. R. Astron. Soc.*, **152**, 461.

Magain, P., Letawe, G., Courbin, F. *et al.* (2005) Discovery of a bright quasar without a massive host galaxy. *Nature*, **437**, 381–384, doi:10.1038/nature04013.

MAGIC Collaboration, Albert, J., Aliu, E., Anderhub, H. *et al.* (2008) Probing quantum gravity using photons from a flare of the active galactic nucleus Markarian 501 observed by the MAGIC telescope. *Phys. Lett. B*, **668**, 253–257, doi:10.1016/j.physletb.2008.08.053.

Magorrian, J., Tremaine, S., Richstone, D. *et al.* (1998) The demography of massive dark objects in galaxy centers. *Astron. J.*, **115**, 2285–2305, doi:10.1086/300353.

Maiolino, R., Marconi, A., Salvati, M. *et al.* (2001) Dust in active nuclei. I. Evidence for "anomalous" properties. *Astron. Astrophys.*, **365**, 28–36, doi:10.1051/0004-6361:20000177.

Malkan, M. (1984) UV and X-ray observations of AGN: Two comparisons with other wave lengths, in *X-ray and UV Emission from Active Galactic Nuclei* (eds W. Brinkmann and J. Truemper), MPE, Garching/Munich, pp. 121–128.

Malkan, M.A. and Sargent, W.L.W. (1982a) The ultraviolet excess of Seyfert 1 galaxies and quasars. *Astrophys. J.*, **254**, 22–37, doi:10.1086/159701.

Malkan, M.A. and Sargent, W.L.W. (1982b) The ultraviolet excess of Seyfert 1 galaxies and quasars. *Astrophys. J.*, **254**, 22–37, doi:10.1086/159701.

Malzac, J. (2001) A quasi-spherical inner accretion flow in Seyfert galaxies?, *Mon. Not. R. Astron. Soc.*, **325**, 1625–1636, doi:10.1046/j.1365-8711.2001.04567.x.

Malzac, J., Beloborodov, A.M., and Poutanen, J. (2001) X-ray spectra of accretion discs with dynamic coronae. *Mon. Not. R. Astron. Soc.*, **326**, 417–427, doi:10.1046/j.1365-8711.2001.04450.x.

Malzac, J. and Petrucci, P. (2002) Reflection at large distance from the central engine in Seyferts. *Mon. Not. R. Astron. Soc.*, **336**, 1209–1216, doi:10.1046/j.1365-8711.2002.05851.x.

Mantovani, F., Bondi, M., and Mack, K.H. (2011) Flux density measurements of a complete sample of faint blazars. *Astron. Astrophys.*, **533**, A79, doi:10.1051/0004-6361/201117328.

Mapelli, M., Ripamonti, E., Zampieri, L., and Colpi, M. (2011) Dynamics of massive stellar black holes in young star clusters and the displacement of ultra-luminous X-ray sources. *Mon. Not. R. Astron. Soc.*, **416**, 1756–1763, doi:10.1111/j.1365-2966.2011.18991.x.

Maraschi, L., Ghisellini, G., and Celotti, A. (1992) A jet model for the gamma-ray emitting blazar 3C 279. *Astrophys. J. Lett.*, **397**, L5–L9, doi:10.1086/186531.

Maraschi, L. and Haardt, F. (1997) Disk-corona models and X-ray emission from seyfert galaxies, in *IAU Colloq. 163: Accretion Phenomena and Related Outflows*, Astronomical Society of the Pacific Conference Series, Vol. 121 (eds D.T. Wickramasinghe, G.V. Bicknell, and L. Ferrario), Astronomical Society of the Pacific, San Francisco, p. 101.

Marchã, M.J.M., Browne, I.W.A., Impey, C.D., and Smith, P.S. (1996) Optical spectroscopy and polarization of a new sample of optically bright flat radio spectrum sources. *Mon. Not. R. Astron. Soc.*, **281**, 425–448.

Marconi, A., Axon, D.J., Maiolino, R. *et al.* (2008) The effect of radiation pressure on virial black hole mass estimates and the case of narrow-line Seyfert 1 galaxies. *Astrophys. J.*, **678**, 693–700, doi:10.1086/529360.

Marconi, A., Risaliti, G., Gilli, R. *et al.* (2004) Local supermassive black holes, relics of active galactic nuclei and the X-ray background. *Mon. Not. R. Astron. Soc.*, **351**, 169–185, doi:10.1111/j.1365-2966.2004.07765.x.

Markevitch, M., Gonzalez, A.H., David, L. *et al.* (2002) A textbook example of a bow shock in the merging galaxy cluster 1E 0657-56. *Astrophys. J. Lett.*, **567**, L27–L31, doi:10.1086/339619.

Marscher, A., Jorstad, S.G., Larionov, V.M. *et al.* (2011) Multi-waveband emission maps of blazars. *J. Astrophys. Astron.*, **32**, 233–237, doi:10.1007/s12036-011-9013-8.

Marscher, A.P. (1987) Synchro-Compton emission from superluminal sources, in *Superluminal Radio Sources* (eds J.A. Zensus and T.J. Pearson), Cambridge Univ. Press, Cambridge, pp. 280–300.

Marscher, A.P. (2005) Multiband impressions of active galactic nuclei. *Mem. Soc. Astron. Italiana*, **76**, 168.

Marscher, A.P. and Gear, W.K. (1985) Models for high-frequency radio outbursts in extragalactic sources, with application to the early 1983 millimeter-to-infrared flare of 3C 273. *Astrophys. J.*, **298**, 114–127, doi:10.1086/163592.

Marshall, F.E., Boldt, E.A., Holt, S.S. *et al.* (1980) The diffuse X-ray background spectrum from 3 to 50 keV. *Astrophys. J.*, **235**, 4–10, doi:10.1086/157601.

Marshall, H.L. (1985) The evolution of optically selected quasars with z less than 2.2 and B less than 20. *Astrophys. J.*, **299**, 109–121, doi:10.1086/163685.

Marshall, H.L. (1987) The optical luminosity function of quasars and low-luminosity active galactic nuclei. *Astron. J.*, **94**, 628–632, doi:10.1086/114496.

Marshall, K., Ryle, W.T., Miller, H.R. *et al.* (2009) Multiwavelength variability of the broad line radio galaxy 3C 120. *Astrophys. J.*, **696**, 601–607, doi:10.1088/0004-637X/696/1/601.

Marshall, N., Warwick, R.S., and Pounds, K.A. (1981) The variability of X-ray emission from active galaxies. *Mon. Not. R. Astron. Soc.*, **194**, 987–1002.

Martocchia, A., Karas, V., and Matt, G. (2000) Effects of Kerr space-time on spectral features from X-ray illuminated accretion discs. *Mon. Not. R. Astron. Soc.*, **312**, 817–826, doi:10.1046/j.1365-8711.2000.03205.x.

Massaro, F., Harris, D.E., and Cheung, C.C. (2011) Large-scale extragalactic jets in the Chandra era. I. Data reduction and analysis. *Astrophys. J. Suppl.*, **197**, 24, doi:10.1088/0067-0049/197/2/24.

Mastichiadis, A. and Kirk, J.G. (1997) Variability in the synchrotron self-Compton model of blazar emission. *Astron. Astrophys.*, **320**, 19–25.

Mathur, S. (2000) Narrow-line Seyfert 1 galaxies and the evolution of galaxies and active galaxies. *Mon. Not. R. Astron. Soc.*, **314**, L17–L20, doi:10.1046/j.1365-8711.2000.03530.x.

Mathur, S., Fields, D., Peterson, B.M., and Grupe, D. (2012) Supermassive black holes, pseudobulges, and the narrow-line Seyfert 1 galaxies. *Astrophys. J.*, submitted, arXiv:1102.0537.

Matt, G., Perola, G.C., and Piro, L. (1991) The iron line and high energy bump as X-ray signatures of cold matter in Seyfert 1 galaxies. *Astron. Astrophys.*, **247**, 25–34.

Matt, G., Porquet, D., Bianchi, S. *et al.* (2005) A changing inner radius in the accretion disc of Q0056-363? *Astron. Astrophys.*, **435**, 857–861, doi:10.1051/0004-6361:20042581.

McClintock, J.E., Narayan, R., Davis, S.W. *et al.* (2011) Measuring the spins of accreting black holes. *Class. Quantum Grav.*, **28**(11), 114 009, doi:10.1088/0264-9381/28/11/114009.

McConnell, N.J., Ma, C.P., Gebhardt, K. *et al.* (2011) Two ten-billion-solar-mass black holes at the centres of giant elliptical galaxies. *Nature*, **480**, 215–218.

McHardy, I. (2010) X-Ray Variability of AGN and Relationship to Galactic Black Hole Binary Systems, in *Lecture Notes in Physics* (ed. T. Belloni), Vol. 794, Springer Verlag, Berlin, p. 203, doi:10.1007/978-3-540-76937-8_8.

McHardy, I.M., Koerding, E., Knigge, C. *et al.* (2006) Active galactic nuclei as scaled-up galactic black holes. *Nature*, **444**, 730–732, doi:10.1038/nature05389.

McHardy, I.M., Papadakis, I.E., Uttley, P. *et al.* (2004) Combined long and short time-scale X-ray variability of NGC 4051 with RXTE and XMM-Newton. *Mon. Not. R. Astron. Soc.*, **348**, 783–801, doi:10.1111/j.1365-2966.2004.07376.x.

McKinney, J.C. (2006) General relativistic magnetohydrodynamic simulations of the jet formation and large-scale propagation from black hole accretion systems. *Mon. Not. R. Astron. Soc.*, **368**, 1561–1582, doi:10.1111/j.1365-2966.2006.10256.x.

McLure, R.J., Dunlop, J.S., de Ravel, L. *et al.* (2011) A robust sample of galaxies at redshifts $6.0 < z < 8.7$: stellar populations, star formation rates and stellar masses.

Mon. Not. R. Astron. Soc., **418**, 2074–2105, doi:10.1111/j.1365-2966.2011.19626.x.

McNamara, B.R. and Nulsen, P.E.J. (2007) Heating hot atmospheres with active galactic nuclei. *Annu. Rev. Astron. Astrophys.*, **45**, 117–175, doi:10.1146/annurev.astro.45.051806.110625.

Meier, D.L. (2003) The theory and simulation of relativistic jet formation: towards a unified model for micro- and macroquasars. *New Astron. Rev.*, **47**, 667–672, doi:10.1016/S1387-6473(03)00120-9.

Merloni, A., Heinz, S., and di Matteo, T. (2003) A fundamental plane of black hole activity. *Mon. Not. R. Astron. Soc.*, **345**, 1057–1076, doi:10.1046/j.1365-2966.2003.07017.x.

Meyer, E.T., Fossati, G., Georganopoulos, M., and Lister, M.L. (2011) From the blazar sequence to the blazar envelope: Revisiting the relativistic jet dichotomy in radio-loud active galactic nuclei. *Astrophys. J.*, **740**, 98, doi:10.1088/0004-637X/740/2/98.

Middelberg, E., Roy, A.L., Nagar, N.M. *et al.* (2004) Motion and properties of nuclear radio components in Seyfert galaxies seen with VLBI. *Astron. Astrophys.*, **417**, 925–944, doi:10.1051/0004-6361:20040019.

Middleton, M., Done, C., and Schurch, N. (2008) High-energy X-ray spectra of Seyferts and Unification schemes for active galactic nuclei. *Mon. Not. R. Astron. Soc.*, **383**, 1501–1505, doi:10.1111/j.1365-2966.2007.12648.x.

Miller, J.S. and Goodrich, R.W. (1990) Spectropolarimetry of high-polarization Seyfert 2 galaxies and unified Seyfert theories. *Astrophys. J.*, **355**, 456–467, doi:10.1086/168780.

Miller, L. and Turner, T.J. (2011) X-ray reverberation in NLS1, in *Narrow-Line Seyfert 1 Galaxies and their Place in the Universe*, Proceedings of Science, Trieste, Italy, id. 19.

Miller, L., Turner, T.J., and Reeves, J.N. (2008) An absorption origin for the X-ray spectral variability of MCG-6-30-15. *Astron. Astrophys.*, **483**, 437–452, doi:10.1051/0004-6361:200809590.

Miller, L., Turner, T.J., and Reeves, J.N. (2009) The absorption-dominated model for the X-ray spectra of type I active galaxies: MCG-6-30-15. *Mon. Not. R. Astron. Soc.*, **399**, L69–L73, doi:10.1111/j.1745-3933.2009.00726.x.

Miller, L., Turner, T.J., Reeves, J.N., and Braito, V. (2010) X-ray reverberation in 1H 0707-495 revisited. *Mon. Not. R. Astron. Soc.*, **408**, 1928–1935, doi:10.1111/j.1365-2966.2010.17261.x.

Miniutti, G. and Fabian, A.C. (2004) A light bending model for the X-ray temporal and spectral properties of accreting black holes. *Mon. Not. R. Astron. Soc.*, **349**, 1435–1448, doi:10.1111/j.1365-2966.2004.07611.x.

Miniutti, G., Panessa, F., de Rosa, A. *et al.* (2009) An intermediate black hole spin in the NLS1 galaxy SWIFT J2127.4+5654: chaotic accretion or spin energy extraction?, *Mon. Not. R. Astron. Soc.*, **398**, 255–262, doi:10.1111/j.1365-2966.2009.15092.x.

Misner, C.W., Thorne, K.S., and Wheeler, J.A. (1973) *Gravitation*, W.H. Freeman and Co., San Francisco.

Mitchell, J. (1784) On the means of discovering the distance, magnitude, &c. of the fixed stars, in consequence of the diminution of the velocity of their light, in case such a dinimution should be found to take place in any of them, and such other data should be procured from observations, as would be farther necessary for that purpose. *Philos. Trans. R. Soc.*, **74**, 35–57.

Miyakawa, T., Ebisawa, K., Terashima, Y. *et al.* (2009) Spectral variation of the Seyfert 1 galaxy MCG-6-30-15 observed with Suzaku. *Publ. Astron. Soc. Japan*, **61**, 1355.

Mohr, P.J., Taylor, B.N., and Newell, D.B. (2008) CODATA recommended values of the fundamental physical constants: 2006. *Rev. Mod. Phys.*, **80**, 633–730, doi:10.1103/RevModPhys.80.633.

Moorwood, A.F.M., Lutz, D., Oliva, E. *et al.* (1996) 2.5–45 μm SWS spectroscopy of the circinus galaxy. *Astron. Astrophys.*, **315**, L109–L112.

Moran, J.M., Greenhill, L.J., and Herrnstein, J.R. (1999) Observational evidence for massive black holes in the centers of active galaxies. *J. Astrophys. Astron.*, **20**, 165, doi:10.1007/BF02702350.

Moretti, A., Campana, S., Lazzati, D., and Tagliaferri, G. (2003) The resolved fraction of the cosmic X-ray background. *Astrophys. J.*, **588**, 696–703, doi:10.1086/374335.

Morgan, C.W., Kochanek, C.S., Morgan, N.D., and Falco, E.E. (2010) The quasar accretion disk size-black hole mass relation. *Astrophys. J.*, **712**, 1129–1136, doi:10.1088/0004-637X/712/2/1129.

Morganti, R., Greenhill, L.J., Peck, A.B. *et al.* (2004) Disks, tori, and cocoons: emission and absorption diagnostics of AGN environments. *New Astron. Rev.*, **48**, 1195–1209, doi:10.1016/j.newar.2004.09.022.

Mortlock, D.J., Warren, S.J., Venemans, B.P. *et al.* (2011) A luminous quasar at a redshift of $z = 7.085$. *Nature*, **474**, 616–619, doi:10.1038/nature10159.

Mücke, A. and Protheroe, R.J. (2001) A proton synchrotron blazar model for flaring in Markarian 501. *Astropart. Phys.*, **15**, 121–136, doi:10.1016/S0927-6505(00)00141-9.

Mücke, A., Protheroe, R.J., Engel, R. *et al.* (2003) BL Lac objects in the synchrotron proton blazar model. *Astropart. Phys.*, **18**, 593–613, doi:10.1016/S0927-6505(02)00185-8.

Mullaney, J.R., Pannella, M., Daddi, E. *et al.* (2011) GOODS-Herschel: the far-infrared view of star formation in active galactic nucleus host galaxies since $z \simeq 3$. *Mon. Not. R. Astron. Soc.*, p. 1756, doi:10.1111/j.1365-2966.2011.19675.x.

Müller, C., Kadler, M., Ojha, R. *et al.* (2011) Dual-frequency VLBI study of Centaurus A on sub-parsec scales. The highest-resolution view of an extragalactic jet. *Astron. Astrophys.*, **530**, L11, doi:10.1051/0004-6361/201116605.

Murphy, K.D. and Yaqoob, T. (2009) An X-ray spectral model for Compton-thick toroidal reprocessors. *Mon. Not. R. Astron. Soc.*, **397**, 1549–1562, doi:10.1111/j.1365-2966.2009.15025.x.

Murray, N., Chiang, J., Grossman, S.A., and Voit, G.M. (1995) Accretion disk winds from active galactic nuclei. *Astrophys. J.*, **451**, 498, doi:10.1086/176238.

Murray, N., Quataert, E., and Thompson, T.A. (2005) On the maximum luminosity of galaxies and their central black holes: Feedback from momentum-driven winds. *Astrophys. J.*, **618**, 569–585, doi:10.1086/426067.

Mushotzky, R. (2004) Ultra-luminous sources in nearby galaxies. *Progr. Theor. Phys. Suppl.*, **155**, 27–44, doi:10.1143/PTPS.155.27.

Nandra, K., George, I.M., Mushotzky, R.F. *et al.* (1997a) ASCA observations of Seyfert 1 galaxies. I. Data analysis, imaging, and timing. *Astrophys. J.*, **476**, 70, doi:10.1086/303600.

Nandra, K., George, I.M., Mushotzky, R.F. *et al.* (1997b) ASCA Observations of Seyfert 1 Galaxies. II. Relativistic iron K_α emission. *Astrophys. J.*, **477**, 602, doi:10.1086/303721.

Nandra, K., Le, T., George, I.M. *et al.* (2000) The origin of the X-ray and ultraviolet emission in NGC 7469. *Astrophys. J.*, **544**, 734–746, doi:10.1086/317237.

Narayan, R. and Yi, I. (1995) Advection-dominated accretion: Underfed black holes and neutron stars. *Astrophys. J.*, **452**, 710, doi:10.1086/176343.

Nelson, C.H. (2000) Black hole mass, velocity dispersion, and the radio source in active galactic nuclei. *Astrophys. J. Lett.*, **544**, L91–L94, doi:10.1086/317314.

Nemmen, R.S., Storchi-Bergmann, T., Eracleous, M., and Yuan, F. (2010) Advection-Dominated Accretion, Jets, and the Spectral Energy Distribution of LINERs, in *IAU Symposium*, vol. 267, Cambridge Univ. Press, Cambridge, pp. 313–318, doi:10.1017/S1743921310006538.

Nenkova, M., Sirocky, M.M., Nikutta, R. *et al.* (2008) AGN dusty tori. II. Observational implications of clumpiness. *Astrophys. J.*, **685**, 160–180, doi:10.1086/590483.

Netzer, H. (1990) AGN emission lines, in *Active Galactic Nuclei* (eds R.D. Blandford, H. Netzer, L. Woltjer, T.J.L. Courvoisier, and M. Mayor), Springer Verlag, Berlin, Heidelberg, New York, pp. 57–160.

Netzer, H., Kaspi, S., Behar, E. *et al.* (2003) The ionized gas and nuclear environment in NGC 3783. IV. Variability and modeling of the 900 kilosecond Chandra spectrum. *Astrophys. J.*, **599**, 933–948, doi:10.1086/379508.

Netzer, H., Lutz, D., Schweitzer, M. *et al.* (2007) Spitzer quasar and ULIRG evolution study (QUEST). II. The spectral energy distributions of Palomar–Green quasars. *Astrophys. J.*, **666**, 806–816, doi:10.1086/520716.

Nicastro, F., Mathur, S., Elvis, M. *et al.* (2005) The mass of the missing baryons in the X-ray forest of the warm-hot intergalactic medium. *Nature*, **433**, 495–498, doi:10.1038/nature03245.

Nilsson, K., Takalo, L.O., Sillanpää, A., and Ciprini, S. (2006) The next outburst of OJ 287, in *Blazar Variability Workshop II: Entering the GLAST Era*, Astronomical Society of the Pacific Conference Series, Vol. 350 (eds H.R. Miller, K. Marshall, J.R. Webb, and M.F. Aller), Astronomical Society of the Pacific, San Francisco, p. 47.

O'Dea, C.P. (1998) The compact steep-spectrum and gigahertz peaked-spectrum radio sources. *Publ. Astron. Soc. Pac.*, **110**, 493–532, doi:10.1086/316162.

O'Dowd, M., Urry, C.M., and Scarpa, R. (2002) The host galaxies of radio-loud active galactic nuclei: The black hole-galaxy connection. *Astrophys. J.*, **580**, 96–103, doi:10.1086/343126.

Oke, J.B. and Sargent, W.L.W. (1968) The nucleus of the Seyfert galaxy NGC 4151. *Astrophys. J.*, **151**, 807, doi:10.1086/149486.

Öpik, E. (1922) An estimate of the distance of the Andromeda Nebula. *Astrophys. J.*, **55**, 406–410, doi:10.1086/142680.

Oppenheimer, J.R. and Volkoff, G.M. (1939) On massive neutron cores. *Phys. Rev.*, **55**, 374–381, doi:10.1103/PhysRev.55.374.

Orienti, M., Dallacasa, D., Tinti, S., and Stanghellini, C. (2006) VLBA images of high frequency peakers. *Astron. Astrophys.*, **450**, 959–970, doi:10.1051/0004-6361:20054656.

Orr, M.J.L. and Browne, I.W.A. (1982) Relativistic beaming and quasar statistics. *Mon. Not. R. Astron. Soc.*, **200**, 1067–1080.

Osterbrock, D.E. (1977) Spectrophotometry of Seyfert 1 galaxies. *Astrophys. J.*, **215**, 733–745, doi:10.1086/155407.

Osterbrock, D.E. (1989) *Astrophysics of Gaseous Nebulae and Active Galactic Nuclei*, University Science Books, Mill Valley.

Osterbrock, D.E. and Pogge, R.W. (1985) The spectra of narrow-line Seyfert 1 galaxies. *Astrophys. J.*, **297**, 166–176, doi:10.1086/163513.

Padovani, P., Giommi, P., Landt, H., and Perlman, E.S. (2007) The deep X-ray radio blazar survey. III. Radio number counts, evolutionary properties, and luminosity function of blazars. *Astrophys. J.*, **662**, 182–198, doi:10.1086/516815.

Paggi, A., Cavaliere, A., Vittorini, V., and Tavani, M. (2009) Power for dry BL Lacertae objects. *Astron. Astrophys.*, **508**, L31–L34, doi:10.1051/0004-6361/200913566.

Pakull, M.W. and Mirioni, L. (2003) Bubble nebulae around ultraluminous X-ray sources, in *Revista Mexicana de Astronomia y Astrofisica Conference Series*, vol. 15 (eds J. Arthur and W.J. Henney), Instituto de Astronomia, Universidad Nacional Autonoma de México, pp. 197–199.

Pakull, M.W., Soria, R., and Motch, C. (2010) A 300-parsec-long jet-inflated bubble around a powerful microquasar in the galaxy NGC 7793. *Nature*, **466**, 209–212, doi:10.1038/nature09168.

Papadakis, I.E. (2004) The scaling of the X-ray variability with black hole mass in active galactic nuclei. *Mon. Not. R. Astron. Soc.*, **348**, 207–213, doi:10.1111/j.1365-2966.2004.07351.x.

Papadakis, I.E., Nandra, K., and Kazanas, D. (2001) Frequency-dependent time lags in the X-ray emission of the Seyfert galaxy NGC 7469. *Astrophys. J. Lett.*, **554**, L133–L137, doi:10.1086/321722.

Pappa, A., Georgantopoulos, I., Stewart, G.C., and Zezas, A.L. (2001) The X-ray spectra of optically selected Seyfert 2 galaxies: are there any Seyfert 2 galaxies with no absorption?, *Mon. Not. R. Astron. Soc.*, **326**, 995–1006, doi:10.1046/j.1365-8711.2001.04609.x.

Paraficz, D. and Hjorth, J. (2010) The Hubble constant inferred from 18 time-delay lenses. *Astrophys. J.*, **712**, 1378–1384, doi:10.1088/0004-637X/712/2/1378.

Pariev, V.I. and Bromley, B.C. (1998) Line emission from an accretion disk around a black hole: Effects of disk structure. *Astrophys. J.*, **508**, 590–600, doi:10.1086/306420.

Patrick, A.R., Reeves, J.N., Lobban, A.P. *et al.* (2011) Assessing black hole spin in deep Suzaku observations of Seyfert 1 AGN. *Mon. Not. R. Astron. Soc.*, **416**, 2725–2747, doi:10.1111/j.1365-2966.2011.19224.x.

Penrose, R. (1969) Gravitational collapse: the role of general relativity. *Nuovo Cim. Riv. Ser.*, **1**, 252.

Penston, M.V., Penston, M.J., Selmes, R.A. *et al.* (1974) Broadband optical and infrared observations of Seyfert galaxies. *Mon. Not. R. Astron. Soc.*, **169**, 357–393.

Perley, D.A., Bloom, J.S., Klein, C.R. *et al.* (2010) Evidence for supernova-synthesized dust from the rising afterglow of GRB 071025 at $z \sim 5$. *Mon. Not. R. Astron. Soc.*, **406**, 2473–2487, doi:10.1111/j.1365-2966.2010.16772.x.

Perley, R.A., Fomalont, E.B., and Johnston, K.J. (1980) Compact radio sources with faint components. *Astron. J.*, **85**, 649–658, doi:10.1086/112723.

Perlman, E.S., Adams, S.C., Cara, M. *et al.* (2011) Optical polarization and spectral variability in the M87 jet. *Astrophys. J.*, **743**, 119, doi:10.1088/0004-637X/743/2/119.

Peterson, B.M. (1997) *An Introduction to Active Galactic Nuclei*, Cambridge University Press, Cambridge, New York, Physical description xvi, 238 p., ISBN 0521473489.

Peterson, B.M. (2007) The masses of black holes in active galactic nuclei, in *The Central Engine of Active Galactic Nuclei*, Astronomical Society of the Pacific Conference Series, Vol. 373 (eds L.C. Ho and J.-W. Wang), Astronomical Society of the Pacific, San Francisco, p. 3.

Peterson, B.M. (2008) The central black hole and relationships with the host galaxy. *New Astron. Rev.*, **52**, 240–252, doi:10.1016/j.newar.2008.06.005.

Peterson, B.M. (2011) Masses of black holes in active galactic nuclei: Implications for narrow-line Seyfert 1 galaxies, in *Proceedings of the conference "Narrow-Line Seyfert 1 Galaxies and their place in the Universe"*. 4–6 April 2011, Milano, Italy, Proceedings of Science, Trieste, Italy, id. 32.

Peterson, B.M. and Horne, K. (2004) Echo mapping of active galactic nuclei. *Astron. Nachr.*, **325**, 248–251, doi:10.1002/asna.200310207.

Peterson, B.M. and Wandel, A. (2000) Evidence for supermassive black holes in active galactic nuclei from emission-line reverberation. *Astrophys. J. Lett.*, **540**, L13–L16, doi:10.1086/312862.

Plotkin, R.M., Markoff, S., Trager, S.C., and Anderson, S.F. (2011) Dynamical black hole masses of BL Lac objects from the sloan digital sky survey. *Mon. Not. R. Astron. Soc.*, **413**, 805–812, doi:10.1111/j.1365-2966.2010.18172.x.

Polko, P., Meier, D.L., and Markoff, S. (2010) Determining the optimal locations for shock acceleration in magnetohydrodynamical jets. *Astrophys. J.*, **723**, 1343–1350, doi:10.1088/0004-637X/723/2/1343.

Polletta, M., Courvoisier, T.J.L., Hooper, E.J., and Wilkes, B.J. (2000) The far-infrared emission of radio loud and radio quiet quasars. *Astron. Astrophys.*, **362**, 75–96.

Polletta, M., Tajer, M., Maraschi, L. *et al.* (2007) Spectral energy distributions of hard X-ray selected active galactic nuclei in the XMM-Newton medium deep survey. *Astrophys. J.*, **663**, 81–102, doi:10.1086/518113.

Polletta, M., Weedman, D., Hönig, S. *et al.* (2008) Obscuration in extremely luminous quasars. *Astrophys. J.*, **675**, 960–984, doi:10.1086/524343.

Polletta, M.d.C., Wilkes, B.J., Siana, B. *et al.* (2006) Chandra and Spitzer unveil heavily obscured quasars in the Chandra/SWIRE survey. *Astrophys. J.*, **642**, 673–693, doi:10.1086/500821.

Pounds, K.A., Turner, T.J., and Warwick, R.S. (1986) Rapid X-ray variability of the Seyfert galaxy MCG-6-30-15. *Mon. Not. R. Astron. Soc.*, **221**, 7P–12P.

Poutanen, J., Krolik, J.H., and Ryde, F. (1997) The nature of spectral transitions in accreting black holes – The case of Cyg X-1. *Mon. Not. R. Astron. Soc.*, **292**, L21–L25.

Poutanen, J. and Stern, B.E. (2011) Fermi Observations of Blazars: Implications for Gamma-ray Production. in *AGN Physics in the CTA Era* (AGN 2011, eds. C. Boisson and H. Sol), Proceedings of Science, Trieste, Italy, id. 15.

Poutanen, J. and Svensson, R. (1996) The two-phase pair corona model for active galactic nuclei and X-ray binaries: How to obtain exact solutions. *Astrophys. J.*, **470**, 249, doi:10.1086/177865.

Predehl, P. and Schmitt, J.H.M.M. (1995) X-raying the interstellar medium: ROSAT observations of dust scattering halos. *Astron. Astrophys.*, **293**, 889–905.

Preto, M., Berentzen, I., Berczik, P., and Spurzem, R. (2011) Fast coalescence of massive black hole binaries from mergers of galactic nuclei: Implications for low-frequency gravitational-wave astrophysics. *Astrophys. J. Lett.*, **732**, L26, doi:10.1088/2041-8205/732/2/L26.

Pringle, J.E. (1981) Accretion discs in astrophysics. *Annu. Rev. Astron. Astrophys.*, **19**, 137–162, doi:10.1146/annurev.aa.19.090181.001033.

Proga, D., Stone, J.M., and Kallman, T.R. (2000) Dynamics of line-driven disk winds in active galactic nuclei. *Astrophys. J.*, **543**, 686–696, doi:10.1086/317154.

Punch, M., Akerlof, C.W., Cawley, M.F. *et al.* (1992) Detection of TeV photons from the active galaxy Markarian 421. *Nature*, **358**, 477, doi:10.1038/358477a0.

Pushkarev, A., Kovalev, Y., and Lobanov, A. (2008) Adiabatic expansion and magnetic fields in AGN jets, in *The Role of VLBI in the Golden Age for Radio Astronomy*, Proceedings of Science, Trieste, Italy, p. 103.

Pushkarev, A.B., Kovalev, Y.Y., and Lister, M.L. (2010) Radio/gamma-ray time delay in the parsec-scale cores of active galactic nuclei. *Astrophys. J. Lett.*, **722**, L7–L11, doi:10.1088/2041-8205/722/1/L7.

Quataert, E., Di Matteo, T., Narayan, R., and Ho, L.C. (1999) Possible evidence for truncated thin disks in the low-luminosity active galactic nuclei M81 and NGC 4579. *Astrophys. J. Lett.*, **525**, L89–L92, doi:10.1086/312353.

Quinn, J., Akerlof, C.W., Biller, S. *et al.* (1996) Detection of gamma rays with $E > 300$ GeV from Markarian 501. *Astrophys. J. Lett.*, **456**, L83, doi:10.1086/309878.

Rafter, S.E., Crenshaw, D.M., and Wiita, P.J. (2011) Radio properties of low-redshift broad-line active galactic nuclei including extended radio sources. *Astron. J.*, **141**, 85, doi:10.1088/0004-6256/141/3/85.

Ramos Almeida, C., Levenson, N.A., Alonso-Herrero, A. *et al.* (2011) Testing the unification model for active galactic nuclei in the infrared: Are the obscuring tori of type 1 and 2 Seyferts different?, *Astrophys. J.*, **731**, 92, doi:10.1088/0004-637X/731/2/92.

Readhead, A.C.S. (1994) Equipartition brightness temperature and the inverse Compton catastrophe. *Astrophys. J.*, **426**, 51–59, doi:10.1086/174038.

Refsdal, S. (1964) The gravitational lens effect. *Mon. Not. R. Astron. Soc.*, **128**, 295.

Refsdal, S. (1966) On the possibility of testing cosmological theories from the gravitational lens effect. *Mon. Not. R. Astron. Soc.*, **132**, 101.

Reid, M.J. and Moran, J.M. (1988) Astronomical masers, in *Galactic and Extragalactic Radio Astronomy*, 2nd edn, Springer-Verlag, Berlin, New York, pp. 255–294.

Reines, A.E., Sivakoff, G.R., Johnson, K.E., and Brogan, C.L. (2011) An actively accreting massive black hole in the dwarf starburst galaxy Henize2-10. *Nature*, **470**, 66–68, doi:10.1038/nature09724.

Remillard, R.A. and McClintock, J.E. (2006) X-ray properties of black-hole binaries. *Annu. Rev. Astron. Astrophys.*, **44**, 49–92, doi:10.1146/annurev.astro.44.051905.092532.

Revnivtsev, M.G., Churazov, E.M., Sazonov, S.Y. *et al.* (2004) Hard X-ray view of the past activity of Sgr A* in a natural Compton mirror. *Astron. Astrophys.*, **425**, L49–L52, doi:10.1051/0004-6361:200400064.

Reyes, R., Zakamska, N.L., Strauss, M.A. *et al.* (2008) Space density of optically selected type 2 quasars. *Astron. J.*, **136**, 2373–2390, doi:10.1088/0004-6256/136/6/2373.

Reynolds, C.S. (1996) *X-ray emission and absorption in active galaxies*, PhD thesis, Cambridge University.

Reynolds, C.S., Fabian, A.C., Brenneman, L.W. *et al.* (2009) Constraints on the absorption-dominated model for the X-ray spectrum of MCG-6-30-15. *Mon. Not. R. Astron. Soc.*, **397**, L21–L25, doi:10.1111/j.1745-3933.2009.00676.x.

Ricci, C., Walter, R., Courvoisier, T.J.L., and Paltani, S. (2011) Reflection in Seyfert galaxies and the unified model of AGN. *Astron. Astrophys.*, **532**, A102, doi:10.1051/0004-6361/201016409.

Richards, G.T., Strauss, M.A., Fan, X. *et al.* (2006) The sloan digital sky survey quasar survey: Quasar luminosity function from data release. *Astron. J.*, **131**, 2766–2787, doi:10.1086/503559.

Richstone, D.O. and Schmidt, M. (1980) The spectral properties of a large sample of quasars. *Astrophys. J.*, **235**, 361–376, doi:10.1086/157640.

Risaliti, G. (2010) The structure of AGNs from X-ray absorption variability, in *IAU Symposium*, Vol. 267, Cambridge

Univ. Press, Cambridge, pp. 299–306, doi:10.1017/S1743921310006514.

Risaliti, G., Nardini, E., Salvati, M. *et al.* (2011) X-ray absorption by broad-line region clouds in Mrk766. *Mon. Not. R. Astron. Soc.*, **410**, 1027–1035.

Risaliti, G., Elvis, M., Fabbiano, G. *et al.* (2005) Rapid Compton-thick/Compton-thin transitions in the Seyfert 2 galaxy NGC 1365. *Astrophys. J. Lett.*, **623**, L93–L96, doi:10.1086/430252.

Risaliti, G., Elvis, M., and Nicastro, F. (2002) Ubiquitous variability of X-ray-absorbing column densities in Seyfert 2 galaxies. *Astrophys. J.*, **571**, 234–246, doi:10.1086/324146.

Robitaille, T.P. and Whitney, B.A. (2010) The present-day star formation rate of the Milky Way determined from Spitzer-detected young stellar objects. *Astrophys. J. Lett.*, **710**, L11–L15, doi:10.1088/2041-8205/710/1/L11.

Rowan-Robinson, M. (1977) On the unity of activity in galaxies. *Astrophys. J.*, **213**, 635–647, doi:10.1086/155195.

Różańska, A. and Madej, J. (2008) Models of the iron K_α fluorescent line and the Compton Shoulder in irradiated accretion disc spectra. *Mon. Not. R. Astron. Soc.*, **386**, 1872–1880, doi:10.1111/j.1365-2966.2008.13173.x.

Rudnick, L. and Edgar, B.K. (1984) Alternating-side ejection in extragalactic radio sources. *Astrophys. J.*, **279**, 74–85, doi:10.1086/161866.

Rumbaugh, N.A., Kocevski, D.D., Gal, R.R. *et al.* (2012) The evolution and environments of X-ray emitting active galactic nuclei in high-redshift large-scale structures. *Astrophys. J.*, **746**, 155, doi:10.1088/0004-637X/746/2/155.

Ruszkowski, M. and Begelman, M.C. (2002) Circular polarization from stochastic synchrotron sources. *Astrophys. J.*, **573**, 485–495, doi:10.1086/340659.

Rybicki, G.B. and Lightman, A.P. (1986) *Radiative Processes in Astrophysics*, Wiley-VCH Verlag GmbH, Weinheim, ISBN 0-471-82759-2.

Ryden, B. (2003) Introduction to Cosmology, Addison Wesley, San Francisco, ISBN 0-8053-8912-1.

Saez, C., Brandt, W.N., Shemmer, O. *et al.* (2011) The X-ray properties of typical high-redshift radio-loud quasars. *Astrophys. J.*, **738**, 53, doi:10.1088/0004-637X/738/1/53.

Salim, S., Rich, R.M., Charlot, S. *et al.* (2007) UV star formation rates in the local universe. *Astrophys. J. Suppl.*, **173**, 267–292, doi:10.1086/519218.

Salpeter, E.E. (1955) The luminosity function and stellar evolution. *Astrophys. J.*, **121**, 161, doi:10.1086/145971.

Salpeter, E.E. (1964) Accretion of interstellar matter by massive objects. *Astrophys. J.*, **140**, 796–800, doi:10.1086/147973.

Samland, M. and Gerhard, O.E. (2003) The formation of a disk galaxy within a growing dark halo. *Astron. Astrophys.*, **399**, 961–982, doi:10.1051/0004-6361:20021842.

Sánchez, S.F., Jahnke, K., Wisotzki, L. *et al.* (2004) Colors of active galactic nucleus host galaxies at 0.5 $< z$1.1 from the GEMS survey. *Astrophys. J.*, **614**, 586–606, doi:10.1086/423234.

Sandage, A. (2005) The classification of galaxies: Early history and ongoing developments. *Annu. Rev. Astron. Astrophys.*, **43**, 581–624, doi:10.1146/annurev.astro.43.112904.104839.

Sanders, D.B. (1999) The “Great Debate”: The case for AGNs. *Astrophys. Space Sci.*, **266**, 331–348.

Sanders, D.B. and Mirabel, I.F. (1996) Luminous infrared galaxies. *Annu. Rev. Astron. Astrophys.*, **34**, 749, doi:10.1146/annurev.astro.34.1.749.

Saxena, S., Summa, A., Elsässer, D. *et al.* (2011) Constraints on dark matter annihilation from M87. Signatures of prompt and inverse-Compton gamma rays. *Eur. Phys. J. C*, **71**, 1815, doi:10.1140/epjc/s10052-011-1815-y.

Saxton, R.D., Turner, M.J.L., Williams, O.R. *et al.* (1993) The soft X-ray excesses of high-luminosity AGN. *Mon. Not. R. Astron. Soc.*, **262**, 63–74.

Sazonov, S., Krivonos, R., Revnivtsev, M. *et al.* (2008) Cumulative hard X-ray spectrum of local AGN: a link to the cosmic X-ray background. *Astron. Astrophys.*, **482**, 517–527, doi:10.1051/0004-6361:20078537.

Sazonov, S., Revnivtsev, M., Krivonos, R. *et al.* (2007) Hard X-ray luminosity function and absorption distribution of nearby AGN: INTEGRAL all-sky survey. *Astron. Astrophys.*, **462**, 57–66, doi:10.1051/0004-6361:20066277.

Sazonov, S.Y. and Revnivtsev, M.G. (2004) Statistical properties of local active galactic nuclei inferred from the RXTE 3–20 keV all-sky survey. *Astron. Astrophys.*, **423**, 469–480, doi:10.1051/0004-6361:20047150.

Sbarrato, T., Ghisellini, G., Maraschi, L., and Colpi, M. (2012) The relation between broad lines and γ-ray luminosities in Fermi blazars. *Mon. Not. R. Astron. Soc.*, 421, 1764–1778, doi:10.1111/j.1365-2966.2012.20442.x.

Sbarufatti, B., Treves, A., and Falomo, R. (2005) Imaging redshifts of BL Lacertae objects. *Astrophys. J.*, **635**, 173–179, doi:10.1086/497022.

Schawinski, K., Evans, D.A., Virani, S. *et al.* (2010) The sudden death of the nearest quasar. *Astrophys. J. Lett.*, **724**, L30–L33, doi:10.1088/2041-8205/724/1/L30.

Schawinski, K., Treister, E., Urry, C.M. *et al.* (2011a) HST WFC3/IR observations of active galactic nucleus host galaxies at $z \sim 2$: Supermassive black holes grow in disk galaxies. *Astrophys. J. Lett.*, **727**, L31, doi:10.1088/2041-8205/727/2/L31.

Schawinski, K., Urry, M., Treister, E. *et al.* (2011b) Evidence for three accreting black holes in a galaxy at $z \sim 1.35$: A snapshot of recently formed black hole seeds?, *Astrophys. J. Lett.*, **743**, L37, doi:10.1088/2041-8205/743/2/L37.

Schawinski, K., Virani, S., Simmons, B. *et al.* (2009) Do moderate-luminosity active galactic nuclei suppress star formation?, *Astrophys. J. Lett.*, **692**, L19–L23, doi:10.1088/0004-637X/692/1/L19.

Schechter, P. (1976) An analytic expression for the luminosity function for galaxies. *Astrophys. J.*, **203**, 297–306, doi:10.1086/154079.

Scheuer, P.A.G. and Readhead, A.C.S. (1979) Superluminally expanding radio sources and the radio-quiet QSOs. *Nature*, **277**, 182–185, doi:10.1038/277182a0.

Schlegel, D.J., Finkbeiner, D.P., and Davis, M. (1998) Maps of dust infrared emission for use in estimation of reddening and cosmic microwave background radiation foregrounds. *Astrophys. J.*, **500**, 525, doi:10.1086/305772.

Schmidt, M. (1963) 3C 273: A star-like object with large red-shift. *Nature*, **197**, 1040, doi:10.1038/1971040a0.

Schmidt, M. (1968) Space distribution and luminosity functions of quasi-stellar radio sources. *Astrophys. J.*, **151**, 393, doi:10.1086/149446.

Schmidt, M. and Green, R.F. (1983) Quasar evolution derived from the Palomar bright quasar survey and other complete quasar surveys. *Astrophys. J.*, **269**, 352–374, doi:10.1086/161048.

Schmidt, M. and Green, R.F. (1986) Counts, evolution, and background contribution of X-ray quasars and other extragalactic X-ray sources. *Astrophys. J.*, **305**, 68–82, doi:10.1086/164229.

Schmitt, H.R., Antonucci, R.R.J., Ulvestad, J.S. *et al.* (2001) Testing the unified model with an infrared-selected sample of Seyfert galaxies. *Astrophys. J.*, **555**, 663–672, doi:10.1086/321505.

Schmoll, S., Miller, J.M., Volonteri, M. *et al.* (2009) Constraining the spin of the black hole in Fairall 9 with Suzaku. *Astrophys. J.*, **703**, 2171–2176, doi:10.1088/0004-637X/703/2/2171.

Schulze, A. and Wisotzki, L. (2011) Selection effects in the black hole-bulge relation and its evolution. *Astron. Astrophys.*, **535**, A87, doi:10.1051/0004-6361/201117564.

Schwarzschild, K. (1916) On the gravitational field of a mass point according to Einstein's theory. *Sitzungsber. Preuss. Akad. Wiss. Berl. (Math. Phys.)*, 189–196.

Schweitzer, M., Lutz, D., Sturm, E. *et al.* (2006) Spitzer Quasar and ULIRG Evolution Study (QUEST). I. The origin of the far-infrared continuum of QSOs. *Astrophys. J.*, **649**, 79–90, doi:10.1086/506510.

Scott, J.E., Kriss, G.A., Brotherton, M. *et al.* (2004) A composite extreme-ultraviolet QSO spectrum from FUSE. *Astrophys. J.*, **615**, 135–149, doi:10.1086/422336.

Seyfert, C.K. (1943) Nuclear emission in spiral nebulae. *Astrophys. J.*, **97**, 28, doi:10.1086/144488.

Shabala, S.S., Kaviraj, S., and Silk, J. (2011) Active galactic nucleus feedback drives the colour evolution of local galaxies.

Mon. Not. R. Astron. Soc., **413**, 2815–2826, doi:10.1111/j.1365-2966.2011.18353.x.

Shakura, N.I. and Sunyaev, R.A. (1973) Black holes in binary systems. Observational appearance. *Astron. Astrophys.*, **24**, 337–355.

Shang, Z., Brotherton, M.S., Green, R.F. *et al.* (2005) Quasars and the big blue bump. *Astrophys. J.*, **619**, 41–59, doi:10.1086/426134.

Shen, Y., Liu, X., Greene, J.E., and Strauss, M.A. (2011) Type 2 active galactic nuclei with double-peaked [O III] lines. II. Single AGNs with complex narrow-line region kinematics are more common than binary AGNs. *Astrophys. J.*, **735**, 48, doi:10.1088/0004-637X/735/1/48.

Shields, G.A. (1978) Thermal continuum from acretion disks in quasars. *Nature*, **272**, 706–708, doi:10.1038/272706a0.

Shields, G.A. (1990) Extragalactic H II regions. *Annu. Rev. Astron. Astrophys.*, **28**, 525–560, doi:10.1146/annurev.aa.28.090190.002521.

Shrader, C.R. and Titarchuk, L. (2003) A method for black hole mass determination in accretion-powered X-ray sources. *Astrophys. J.*, **598**, 168–177, doi:10.1086/378801.

Shu, X.W., Yaqoob, T., and Wang, J.X. (2010) The cores of the FeK_α lines in active galactic nuclei: An extended Chandra high energy grating sample. *Astrophys. J. Suppl.*, **187**, 581–606, doi:10.1088/0067-0049/187/2/581.

Sijacki, D. and Springel, V. (2006) Hydrodynamical simulations of cluster formation with central AGN heating. *Mon. Not. R. Astron. Soc.*, **366**, 397–416, doi:10.1111/j.1365-2966.2005.09860.x.

Sijacki, D., Springel, V., and Haehnelt, M.G. (2009) Growing the first bright quasars in cosmological simulations of structure formation. *Mon. Not. R. Astron. Soc.*, **400**, 100–122, doi:10.1111/j.1365-2966.2009.15452.x.

Sijacki, D., Springel, V., and Haehnelt, M.G. (2011) Gravitational recoils of supermassive black holes in hydrodynamical simulations of gas-rich galaxies. *Mon. Not. R. Astron. Soc.*, **414**, 3656–3670, doi:10.1111/j.1365-2966.2011.18666.x.

Sikora, M., Begelman, M.C., Madejski, G.M., and Lasota, J.P. (2005) Are quasar jets dominated by poynting flux?, *Astrophys. J.*, **625**, 72–77, doi:10.1086/429314.

Sikora, M., Begelman, M.C., and Rees, M.J. (1994) Comptonization of diffuse ambient radiation by a relativistic jet: The source of gamma rays from blazars?, *Astrophys. J.*, **421**, 153–162, doi:10.1086/173633.

Sikora, M., Stawarz, Ł., and Lasota, J.P. (2007) Radio loudness of active galactic nuclei: Observational facts and theoretical implications. *Astrophys. J.*, **658**, 815–828, doi:10.1086/511972.

Sikora, M., Stawarz, Ł., Moderski, R. *et al.* (2009) Constraining emission models of luminous blazar sources. *Astrophys. J.*, **704**, 38–50, doi:10.1088/0004-637X/704/1/38.

Silk, J. and Rees, M.J. (1998) Quasars and galaxy formation. *Astron. Astrophys.*, **331**, L1–L4.

Silverman, J.D., Green, P.J., Barkhouse, W.A. *et al.* (2008) The luminosity function of X-ray-selected active galactic nuclei: Evolution of supermassive black holes at high redshift. *Astrophys. J.*, **679**, 118–139, doi:10.1086/529572.

Silverman, J.D., Kovač, K., Knobel, C. *et al.* (2009) The environments of active galactic nuclei within the zCOSMOS density field. *Astrophys. J.*, **695**, 171–182, doi:10.1088/0004-637X/695/1/171.

Sim, S.A., Long, K.S., Miller, L., and Turner, T.J. (2008) Multidimensional modelling of X-ray spectra for AGN accretion disc outflows. *Mon. Not. R. Astron. Soc.*, **388**, 611–624, doi:10.1111/j.1365-2966.2008.13466.x.

Sim, S.A., Miller, L., Long, K.S. *et al.* (2010a) Multidimensional modelling of X-ray spectra for AGN accretion disc outflows – II. *Mon. Not. R. Astron. Soc.*, **404**, 1369–1384, doi:10.1111/j.1365-2966.2010.16396.x.

Sim, S.A., Proga, D., Miller, L. *et al.* (2010b) Multidimensional modelling of X-ray spectra for AGN accretion disc outflows – III. Application to a hydrodynamical simulation. *Mon. Not. R. Astron. Soc.*, **408**, 1396–1408, doi:10.1111/j.1365-2966.2010.17215.x.

Simonetti, J.H., Cordes, J.M., and Heeschen, D.S. (1985) Flicker of extragalactic radio sources at two frequencies. *Astrophys. J.*, **296**, 46–59, doi:10.1086/163418.

Simpson, C. (2005) The luminosity dependence of the type 1 active galactic nucleus fraction. *Mon. Not. R. Astron. Soc.*, **360**, 565–572, doi:10.1111/j.1365-2966.2005.09043.x.

Slipher, V.M. (1913) The radial velocity of the Andromeda Nebula. *Lowell Obs. Bull.*, **2**, 56–57.

Smith, R.A.N., Page, M.J., and Branduardi-Raymont, G. (2008) Exploring the nuclear environment of the NLS1 galaxy Arakelian 564 with XMM-Newton RGS. *Astron. Astrophys.*, **490**, 103–112, doi:10.1051/0004-6361:200810151.

Smolčić, V., Capak, P., Ilbert, O. *et al.* (2011) The redshift and nature of AzTEC/COSMOS 1: A starburst galaxy at $z = 4.6$. *Astrophys. J. Lett.*, **731**, L27, doi:10.1088/2041-8205/731/2/L27.

Soifer, B.T., Helou, G., and Werner, M. (2008) The Spitzer view of the extragalactic universe. *Annu. Rev. Astron. Astrophys.*, **46**, 201–240, doi:10.1146/annurev.astro.46.060407.145144.

Soldi, S., Ponti, G., Beckmann, V., and Lubiński, P. (2010) AGN variability at hard X-rays, in *Proceedings of "The Extreme sky: Sampling the Universe above 10 keV"*, Otranto, October 2009, Proceedings of Science, Trieste, Italy, id. 31.

Soldi, S., Türler, M., Paltani, S. *et al.* (2008) The multiwavelength variability of 3C 273. *Astron. Astrophys.*, **486**, 411–425, doi:10.1051/0004-6361:200809947.

Soltan, A. (1982) Masses of quasars. *Mon. Not. R. Astron. Soc.*, **200**, 115–122.

Spoon, H.W.W., Marshall, J.A., Houck, J.R. *et al.* (2007) Mid-infrared galaxy classification based on silicate obscuration and PAH equivalent width. *Astrophys. J. Lett.*, **654**, L49–L52, doi:10.1086/511268.

Springel, V., Di Matteo, T., and Hernquist, L. (2005a) Black holes in galaxy mergers: The formation of red elliptical galaxies. *Astrophys. J. Lett.*, **620**, L79–L82, doi:10.1086/428772.

Springel, V., Di Matteo, T., and Hernquist, L. (2005b) Modelling feedback from stars and black holes in galaxy mergers. *Mon. Not. R. Astron. Soc.*, **361**, 776–794, doi:10.1111/j.1365-2966.2005.09238.x.

Springel, V., Frenk, C.S., and White, S.D.M. (2006) The large-scale structure of the Universe. *Nature*, **440**, 1137–1144, doi:10.1038/nature04805.

Sramek, R.A. and Weedman, D.W. (1980) The radio properties of optically discovered quasars. *Astrophys. J.*, **238**, 435–444, doi:10.1086/158000.

Stacy, A., Greif, T.H., and Bromm, V. (2012) The first stars: Mass growth under protostellar feedback. *Mon. Not. R. Astron. Soc.*, **422**, 290–309, arXiv:1109.3147.

Stanghellini, C., O'Dea, C.P., Dallacasa, D. *et al.* (2005) Extended emission around GPS radio sources. *Astron. Astrophys.*, **443**, 891–902, doi:10.1051/0004-6361:20042226.

Steffen, A.T., Barger, A.J., Cowie, L.L. *et al.* (2003) The changing active galactic nucleus population. *Astrophys. J. Lett.*, **596**, L23–L26, doi:10.1086/379142.

Steinle, H., Bennett, K., Bloemen, H. *et al.* (1998) COMPTEL observations of Centaurus A at MeV energies in the years 1991 to 1995. *Astron. Astrophys.*, **330**, 97–107.

Stevens, J.A., Litchfield, S.J., Robson, E.I. *et al.* (1995) The spectral evolution of high-frequency radio outbursts in the blazar PKS 0420-014. *Mon. Not. R. Astron. Soc.*, **275**, 1146.

Stocke, J.T., Danforth, C.W., and Perlman, E.S. (2011) Broad Lyα emission from three nearby BL Lacertae objects. *Astrophys. J.*, **732**, 113, doi:10.1088/0004-637X/732/2/113.

Stocke, J.T., Morris, S.L., Gioia, I.M. *et al.* (1991) The Einstein observatory extended medium-sensitivity survey. II – The optical identifications. *Astrophys. J. Suppl.*, **76**, 813–874, doi:10.1086/191582.

Sturm, E., González-Alfonso, E., Veilleux, S. *et al.* (2011) Massive molecular outflows and negative feedback in ULIRGs observed by herschel-PACS. *Astrophys. J. Lett.*, **733**, L16, doi:10.1088/2041-8205/733/1/L16.

Summons, D.P., Arévalo, P., McHardy, I.M. *et al.* (2007) Timing evidence in determining the accretion state of the Seyfert galaxy NGC 3783. *Mon. Not. R. Astron. Soc.*, **378**, 649–656, doi:10.1111/j.1365-2966.2006.11797.x.

Sun, W. and Malkan, M.A. (1989) Fitting improved accretion disk models to the multiwavelength continua of quasars and active galactic nuclei. *Astrophys. J.*, **346**, 68–100, doi:10.1086/167986.

Sunyaev, R.A. and Titarchuk, L.G. (1980) Comptonization of X-rays in plasma clouds – Typical radiation spectra. *Astron. Astrophys.*, **86**, 121–138.

Swanenburg, B.N., Hermsen, W., Bennett, K. *et al.* (1978) COS B observation of high-energy gamma radiation from 3C 273. *Nature*, **275**, 298, doi:10.1038/275298a0.

Swank, J., Kallman, T., Jahoda, K. *et al.* (2010) Gravity and Extreme Magnetism SMEX (GEMS), in *X-ray Polarimetry: A New Window in Astrophysics*, (eds R. Bellazzini, E. Costa, G. Matt and G. Tagliaferri), Cambridge University Press, Cambridge, p. 251, ISBN: 9780521191845.

Swartz, D.A., Ghosh, K.K., Tennant, A.F., and Wu, K. (2004) The ultraluminous X-ray source population from the Chandra archive of galaxies. *Astrophys. J. Suppl.*, **154**, 519–539, doi:10.1086/422842.

Tadhunter, C. (2008) An introduction to active galactic nuclei: Classification and unification. *New Astron. Rev.*, **52**, 227–239, doi:10.1016/j.newar.2008.06.004.

Tanaka, Y., Nandra, K., Fabian, A.C. *et al.* (1995) Gravitationally redshifted emission implying an accretion disk and massive black hole in the active galaxy MCG-6-30-15. *Nature*, **375**, 659–661, doi:10.1038/375659a0.

Taniguchi, Y. (1999) The minor-merger-driven nuclear activity in Seyfert galaxies: A step toward the simple unified formation mechanism of active galactic nuclei in the local universe. *Astrophys. J.*, **524**, 65–70, doi:10.1086/307814.

Tavecchio, F., Ghisellini, G., Ghirlanda, G. *et al.* (2009) The hard TeV spectrum of 1ES 0229+200: new clues from Swift. *Mon. Not. R. Astron. Soc.*, **399**, L59–L63, doi:10.1111/j.1745-3933.2009.00724.x.

Tavecchio, F., Maraschi, L., and Ghisellini, G. (1998) Constraints on the physical parameters of TeV blazars. *Astrophys. J.*, **509**, 608–619, doi:10.1086/306526.

Tchekhovskoy, A., Narayan, R., and McKinney, J.C. (2011) Efficient generation of jets from magnetically arrested accretion on a rapidly spinning black hole. *Mon. Not. R. Astron. Soc.*, **418**, L79–L83, doi:10.1111/j.1745-3933.2011.01147.x.

Telfer, R.C., Zheng, W., Kriss, G.A., and Davidsen, A.F. (2002) The rest-frame extreme-ultraviolet spectral properties of quasi-stellar objects. *Astrophys. J.*, **565**, 773–785, doi:10.1086/324689.

Teng, S.H., Mushotzky, R.F., Sambruna, R.M. *et al.* (2011) Fermi/LAT observations of Swift/BAT Seyfert galaxies: on the contribution of radio-quiet active galactic nuclei to the extragalactic γ-ray background. Astrophys. J., **742**, 66, doi:10.1088/0004-637X/742/2/66..

Terrell, J. (1977) The luminosity distance equation in Friedmann cosmology. *Am. J. Phys.*, **45**, 869–870, doi:10.1119/1.11065.

Titarchuk, L., Laurent, P., and Shaposhnikov, N. (2009) On the nonrelativistic origin of red-skewed iron lines in cataclysmic variable, neutron star, and black hole sources. *Astrophys. J.*, **700**, 1831–1846, doi:10.1088/0004-637X/700/2/1831.

Trakhtenbrot, B., Netzer, H., Lira, P., and Shemmer, O. (2011) Black-hole mass and growth rate at $z \sim 4.8$: A short episode of fast growth followed by short duty cycle activity. *Astrophys. J.*, **730**, 7, doi:10.1088/0004-637X/730/1/7.

Tran, H.D., Miller, J.S., and Kay, L.E. (1992) Detection of obscured broad-line regions in four Seyfert 2 galaxies. *Astrophys. J.*, **397**, 452–456, doi:10.1086/171801.

Treister, E., Natarajan, P., Sanders, D.B. *et al.* (2010) Major galaxy mergers and the growth of supermassive black holes in quasars. *Science*, **328**, 600, doi:10.1126/science.1184246.

Treister, E., Schawinski, K., Volonteri, M. *et al.* (2011) Black hole growth in the early Universe is self-regulated and largely hidden from view. *Nature*, **474**, 356–358, doi:10.1038/nature10103.

Tripp, T.M., Lu, L., and Savage, B.D. (1998) High signal-to-noise GHRS observations of H 1821+643: O IV associa ted with a group of galaxies at $z = 0.226$ and complex Lyman alpha absorption pro files at low redshift, in *The Scientific Impact of the Goddard High Resolution Spectrograph*, Astronomical Society of the Pacific Conference Series, Vol. 143 (eds J.C. Brandt, T.B. Ake, and C.C. Petersen), Astronomical Society of the Pacific, San Francisco, p. 261.

Trippe, M.L., Crenshaw, D.M., Deo, R.P. *et al.* (2010) A multi-wavelength study of the nature of type 1.8/1.9 Seyfert galaxies. *Astrophys. J.*, **725**, 1749–1767, doi:10.1088/0004-637X/725/2/1749.

Trump, J.R., Impey, C.D., Kelly, B.C. *et al.* (2011) Accretion rate and the physical nature of unobscured active galaxies. *Astrophys. J.*, **733**, 60, doi:10.1088/0004-637X/733/1/60.

Tsang, O. and Kirk, J.G. (2007) The inverse Compton catastrophe and high brightness temperature radio sources. *Astron. Astrophys.*, **463**, 145–152, doi:10.1051/0004-6361:20066502.

Tucker, W. and Giacconi, R. (1985) *The X-ray Universe*, Harvard University Press, Cambridge.

Tueller, J., Mushotzky, R.F., Barthelmy, S. *et al.* (2008) Swift BAT survey of AGNs. *Astrophys. J.*, **681**, 113–127, doi:10.1086/588458.

Türler, M., Chernyakova, M., Courvoisier, T.J.L. *et al.* (2006) A historic jet-emission minimum reveals hidden spectral features in 3C 273. *Astron. Astrophys.*, **451**, L1–L4, doi:10.1051/0004-6361:200600023.

Türler, M., Chernyakova, M., Courvoisier, T. *et al.* (2010) INTEGRAL hard X-ray spectra of the cosmic X-ray background and Galactic ridge emission. *Astron. Astrophys.*, **512**, A49, doi:10.1051/0004-6361/200913072.

Turner, T.J., George, I.M., Nandra, K., and Turcan, D. (1999) On X-ray variability in Seyfert galaxies. *Astrophys. J.*, **524**, 667–673, doi:10.1086/307834.

Turner, T.J. and Miller, L. (2009) X-ray absorption and reflection in active galactic nuclei. *Astron. Astrophys. Rev.*, **17**, 47–104, doi:10.1007/s00159-009-0017-1.

Turner, T.J., Nandra, K., George, I.M. *et al.* (1993) X-ray observations of the warm absorber in NGC 3783. *Astrophys. J.*, **419**, 127, doi:10.1086/173466.

Turner, T.J. and Pounds, K.A. (1989) The EXOSAT spectral survey of AGN. *Mon. Not. R. Astron. Soc.*, **240**, 833–880.

Ueda, Y., Akiyama, M., Ohta, K., and Miyaji, T. (2003) Cosmological evolution of the hard X-ray active galactic nucleus luminosity function and the origin of the hard X-ray background. *Astrophys. J.*, **598**, 886–908, doi:10.1086/378940.

Ulvestad, J.S., Antonucci, R.R.J., and Barvainis, R. (2005a) VLBA imaging of central engines in radio-quiet quasars. *Astrophys. J.*, **621**, 123–129, doi:10.1086/427426.

Ulvestad, J.S., Wong, D.S., Taylor, G.B. *et al.* (2005b) VLBA identification of the milliarcsecond active nucleus in the Seyfert galaxy NGC 4151. *Astron. J.*, **130**, 936–944, doi:10.1086/432034.

Urry, C.M. and Padovani, P. (1995) Unified schemes for radio-loud active galactic nuclei. *Publ. Astron. Soc. Pac.*, **107**, 803, doi:10.1086/133630.

Uttley, P. and McHardy, I.M. (2004) A brief review of long-term X-ray and optical variability in radio-quiet AGN. *Progr. Theor. Phys. Suppl.*, **155**, 170–177, doi:10.1143/PTPS.155.170.

Vagnetti, F., Cavaliere, A., and Giallongo, E. (1991) BL Lacertae objects and radio-loud quasars within an evolutionary unified scheme. *Astrophys. J.*, **368**, 366–372, doi:10.1086/169700.

Valtaoja, E., Teräsranta, H., Tornikoski, M. *et al.* (2000) Radio monitoring of OJ 287 and binary black hole models for periodic outbursts. *Astrophys. J.*, **531**, 744–755, doi:10.1086/308494.

Vanden Berk, D.E., Richards, G.T., Bauer, A. *et al.* (2001) Composite quasar spectra from the sloan digital sky survey. *Astron. J.*, **122**, 549–564, doi:10.1086/321167.

Vaughan, S., Reeves, J., Warwick, R., and Edelson, R. (1999) X-ray spectral complexity in narrow-line Seyfert 1 galaxies. *Mon. Not. R. Astron. Soc.*, **309**, 113–124, doi:10.1046/j.1365-8711.1999.02811.x.

Veilleux, S. (2008) AGN host galaxies. *New Astron. Rev.*, **52**, 289–306, doi:10.1016/j.newar.2008.06.011.

Veilleux, S. and Osterbrock, D.E. (1987) Spectral classification of emission-line galaxies. *Astrophys. J. Suppl.*, **63**, 295–310, doi:10.1086/191166.

Véron-Cetty, M.P. and Véron, P. (2000) The emission line spectrum of active galactic nuclei and the unifying scheme. *Astron. Astrophys. Rev.*, **10**, 81–133, doi:10.1007/s001590000006.

Véron-Cetty, M. and Véron, P. (2010) A catalogue of quasars and active nuclei: 13th edition. *Astron. Astrophys.*, **518**, A10, doi:10.1051/0004-6361/201014188.

Volonteri, M. (2010a) Astrophysics: Making black holes from scratch. *Nature*, **466**, 1049–1050, doi:10.1038/4661049a.

Volonteri, M. (2010b) Formation of supermassive black holes. *Astron. Astrophys. Rev.*, **18**, 279–315, doi:10.1007/s00159-010-0029-x.

Volonteri, M. and Begelman, M.C. (2010) Quasi-stars and the cosmic evolution of massive black holes. *Mon. Not. R. Astron. Soc.*, **409**, 1022–1032, doi:10.1111/j.1365-2966.2010.17359.x.

Wagner, S.J. and Witzel, A. (1995) Intraday variability in quasars and BL Lac objects. *Annu. Rev. Astron. Astrophys.*, **33**, 163–198, doi:10.1146/annurev.aa.33.090195.001115.

Walsh, D., Carswell, R.F., and Weymann, R.J. (1979) 0957+561 A, B – Twin quasistellar objects or gravitational lens. *Nature*, **279**, 381–384, doi:10.1038/279381a0.

Wang, Q.D. (1999) Detection of X-ray-emitting hypernova remnants in M101. *Astrophys. J. Lett.*, **517**, L27–L30, doi:10.1086/312020.

Ward, M., Elvis, M., Fabbiano, G. *et al.* (1987) The continuum of type 1 Seyfert galaxies. I – A single form modified by the effects of dust. *Astrophys. J.*, **315**, 74–91, doi:10.1086/165115.

Wardle, J.F.C., Homan, D.C., Ojha, R., and Roberts, D.H. (1998) Electron–positron jets associated with the quasar 3C 279. *Nature*, **395**, 457–461, doi:10.1038/26675.

Watson, D., Denney, K.D., Vestergaard, M., and Davis, T.M. (2011) A new cosmological distance measure using active galactic nuclei. *Astrophys. J. Lett.*, **740**, L49, doi:10.1088/2041-8205/740/2/L49.

Weaver, H., Williams, D.R.W., Dieter, N.H., and Lum, W.T. (1965) Observations of a Strong Unidentified Microwave Line and of Emission from the OH Molecule. *Nature*, **208**, 29–31, doi:10.1038/208029a0.

Weedman, D.W. (1973) A photometric study of Markarian galaxies. *Astrophys. J.*, **183**, 29–40, doi:10.1086/152205.

Weekes, T.C., Cawley, M.F., Fegan, D.J. *et al.* (1989) Observation of TeV gamma rays from the Crab nebula using the atmospheric Cerenkov imaging technique. *Astrophys. J.*, **342**, 379–395, doi:10.1086/167599.

Wehrle, A.E., Piner, B.G., Unwin, S.C. *et al.* (2001) Kinematics of the parsec-scale relativistic jet in quasar 3C 279: 1991–1997. *Astrophys. J. Suppl.*, **133**, 297–320, doi:10.1086/320353.

Weinberg, S. (1993) *The First Three Minutes: a Modern View of the Origin of the Universe*, Basic Books, New York.

Weymann, R.J., Carswell, R.F., and Smith, M.G. (1981) Absorption lines in the spectra of quasistellar objects. *Annu. Rev. Astron. Astrophys.*, **19**, 41–76, doi:10.1146/annurev.aa.19.090181.000353.

White, R.L., Becker, R.H., Gregg, M.D. *et al.* (2000) The FIRST bright quasar survey. II. 60 nights and 1200 spectra later. *Astrophys. J. Suppl.*, **126**, 133–207, doi:10.1086/313300.

Willott, C.J., Delorme, P., Reylé, C. *et al.* (2010) The Canada–France high-z quasar survey: Nine new quasars and the luminosity function at redshift 6. *Astron. J.*, **139**, 906–918, doi:10.1088/0004-6256/139/3/906.

Winter, L.M., Mushotzky, R.F., and Reynolds, C.S. (2006) XMM-Newton archival study of the ultraluminous X-ray population in nearby galaxies. *Astrophys. J.*, **649**, 730–752, doi:10.1086/506579.

Wisotzki, L., Christlieb, N., Bade, N. *et al.* (2000) The Hamburg/ESO survey for bright QSOs. III. A large flux-limited sample of QSOs. *Astron. Astrophys.*, **358**, 77–87.

Wisotzki, L., Köhler, T., Groote, D., and Reimers, D. (1996) The Hamburg/ESO survey for bright QSOs. I. Survey design and candidate selection procedure. *Astron. Astrophys. Suppl.*, **115**, 227.

Wolf, C., Hildebrandt, H., Taylor, E.N., and Meisenheimer, K. (2008) Calibration update of the COMBO-17 CDFS catalogue. *Astron. Astrophys.*, **492**, 933–936, doi:10.1051/0004-6361:200810954.

Wolf, C., Meisenheimer, K., Rix, H.W. *et al.* (2003) The COMBO-17 survey: Evolution of the galaxy luminosity function from 25 000 galaxies with $0.2 < z < 1.2$. *Astron. Astrophys.*, **401**, 73–98, doi:10.1051/0004-6361:20021513.

Wolf, C., Meisenheimer, K., Röser, H.J. *et al.* (2001) Multi-color classification in the calar alto deep imaging survey. *Astron. Astrophys.*, **365**, 681–698, doi:10.1051/0004-6361:20000064.

Wolter, H. (1952) Spiegelsysteme streifenden Einfalls als abbildende Optiken für Röntgenstrahlen. *Ann. Phys.*, **445**, 94–114, doi:10.1002/andp. 19524450108.

Woltjer, L. (1959) Emission nuclei in galaxies. *Astrophys. J.*, **130**, 38, doi:10.1086/146694.

Wong, K.W., Irwin, J.A., Yukita, M. *et al.* (2011) Resolving the Bondi accretion flow toward the supermassive black hole of NGC 3115 with Chandra. *Astrophys. J. Lett.*, **736**, L23, doi:10.1088/2041-8205/736/1/L23.

Wood, K.S., Meekins, J.F., Yentis, D.J. *et al.* (1984) The HEAO A-1 X-ray source catalog. *Astrophys. J. Suppl.*, **56**, 507–649, doi:10.1086/190992.

Worrall, D.M. (2009) The X-ray jets of active galaxies. *Astron. Astrophys. Rev.*, **17**, 1–46, doi:10.1007/s00159-009-0016-7.

Wright, E.L. (2006) A cosmology calculator for the world wide web. *Publ. Astron. Soc. Pac.*, **118**, 1711–1715, doi:10.1086/510102.

Wright, T. (1750) *An Original Theory or New Hypothesis of the Universe*, H. Chapelle, London.

Wu, C.C. (1977) Ultraviolet observations of 3C 273 by the ANS. *Astrophys. J. Lett.*, **217**, L117–L120, doi:10.1086/182552.

Wu, J., Vanden Berk, D.E., Brandt, W.N. *et al.* (2009) Probing the origins of the C IV and Fe K_α Baldwin effects. *Astrophys. J.*, **702**, 767–778, doi:10.1088/0004-637X/702/1/767.

Wu, Q., Cao, X., and Wang, D.X. (2011a) Evidence for rapidly rotating black holes in Fanaroff–Riley I radio galaxies. *Astrophys. J.*, **735**, 50, doi:10.1088/0004-637X/735/1/50.

Wu, Y.Z., Zhang, E.P., Liang, Y.C. *et al.* (2011b) The different nature of Seyfert 2 galaxies with and without hidden broad-line regions. *Astrophys. J.*, **730**, 121, doi:10.1088/0004-637X/730/2/121.

Yaqoob, T., Murphy, K.D., Miller, L., and Turner, T.J. (2010) On the efficiency of production of the Fe K_α emission line in neutral matter. *Mon. Not. R. Astron. Soc.*, **401**, 411–417, doi:10.1111/j.1365-2966.2009.15657.x.

York, D.G., Adelman, J., Anderson, Jr., J.E. *et al.* (2000) The Sloan digital sky survey: Technical summary. *Astron. J.*, **120**, 1579–1587, doi:10.1086/301513.

Yu, Q. and Tremaine, S. (2002) Observational constraints on growth of massive black holes. *Mon. Not. R. Astron. Soc.*, **335**, 965–976, doi:10.1046/j.1365-8711.2002.05532.x.

Yuan, F. and Narayan, R. (2004) On the nature of X-ray-bright, optically normal galaxies. *Astrophys. J.*, **612**, 724–728, doi:10.1086/422802.

Zacek, V. (2008) Dark matter, in *Fundamental Interactions: Proceedings of the 22nd Lake Louise Winter Institute* (eds. A. Astbury, F. Khanna, and R. Moore), World Scientific Publishing Co. Pte. Ltd., Singapore, pp. 170–206, doi:10.1142/9789812776105_0007.

Zaritsky, D., Kennicutt, Jr., R.C., and Huchra, J.P. (1994) H II regions and the abundance properties of spiral galaxies. *Astrophys. J.*, **420**, 87–109, doi:10.1086/173544.

Zdziarski, A.A., Fabian, A.C., Nandra, K. *et al.* (1994) Physical processes in the X-ray/gamma-ray source of IC 4329A. *Mon. Not. R. Astron. Soc.*, **269**, L55.

Zdziarski, A.A., Johnson, W.N., Done, C. *et al.* (1995) The average X-ray/gamma-ray spectra of Seyfert galaxies from GINGA and OSSE and the origin of the cosmic X-ray background. *Astrophys. J. Lett.*, **438**, L63–L66, doi:10.1086/187716.

Zdziarski, A.A., Lubiński, P., and Smith, D.A. (1999) Correlation between Compton reflection and X-ray slope in Seyferts and X-ray binaries. *Mon. Not. R. Astron. Soc.*, **303**, L11–L15, doi:10.1046/j.1365-8711.1999.02343.x.

Zel'Dovich, Y.B. (1964) The fate of a star and the evolution of gravitational energy upon accretion. *Sov. Phys. Dokl.*, **9**, 195.

Zel'Dovich, Y.B. and Novikov, I.D. (1964) The radiation of gravity waves by bodies moving in the field of a collapsing star. *Sov. Phys. Dokl.*, **9**, 246.

Zhou, H., Wang, T., Yuan, W. *et al.* (2006) A comprehensive study of 2000 narrow line Seyfert 1 galaxies from the sloan digital sky survey. I. The sample. *Astrophys. J. Suppl.*, **166**, 128–153, doi:10.1086/504869.

Zhou, X.L., Zhang, S.N., Wang, D.X., and Zhu, L. (2010) Calibrating the correlation between black hole mass and X-ray variability amplitude: X-ray only black hole mass estimates for active galactic nuclei and ultra-luminous X-ray sources. *Astrophys. J.*, **710**, 16–23, doi:10.1088/0004-637X/710/1/16.

Zoghbi, A., Fabian, A.C., Uttley, P. *et al.* (2010) Broad iron *L* line and X-ray reverberation in 1H 0707-495. *Mon. Not. R. Astron. Soc.*, **401**, 2419–2432, doi:10.1111/j.1365-2966.2009.15816.x.

Zwicky, F. (1937) On the masses of nebulae and of clusters of nebulae. *Astrophys. J.*, **86**, 217, doi:10.1086/143864.

Index